Methods in Molecular Biology

Series Editor
John M. Walker
School of Life and Medical Sciences
University of Hertfordshire
Hatfield, Hertfordshire, UK

For further volumes:
http://www.springer.com/series/7651

For over 35 years, biological scientists have come to rely on the research protocols and methodologies in the critically acclaimed *Methods in Molecular Biology* series. The series was the first to introduce the step-by-step protocols approach that has become the standard in all biomedical protocol publishing. Each protocol is provided in readily-reproducible step-by-step fashion, opening with an introductory overview, a list of the materials and reagents needed to complete the experiment, and followed by a detailed procedure that is supported with a helpful notes section offering tips and tricks of the trade as well as troubleshooting advice. These hallmark features were introduced by series editor Dr. John Walker and constitute the key ingredient in each and every volume of the *Methods in Molecular Biology* series. Tested and trusted, comprehensive and reliable, all protocols from the series are indexed in PubMed.

Lineage Tracing

Methods and Protocols

Edited by

Jorge García-Marqués

CSIC, Cajal Institute, Madrid, Spain

Tzumin Lee

Life Sciences Institute, University of Michigan—Ann Arbor, Ann Arbor, MI, USA

Humana Press

Editors
Jorge García-Marqués
CSIC
Cajal Institute
Madrid, Spain

Tzumin Lee
Life Sciences Institute
University of Michigan–Ann Arbor
Ann Arbor, MI, USA

ISSN 1064-3745 ISSN 1940-6029 (electronic)
Methods in Molecular Biology
ISBN 978-1-0716-4309-9 ISBN 978-1-0716-4310-5 (eBook)
https://doi.org/10.1007/978-1-0716-4310-5

This Humana imprint is published by the registered company Springer Science+Business Media, LLC, part of Springer Nature.
The registered company address is: 1 New York Plaza, New York, NY 10004, U.S.A.

Preface

Understanding the relationship between cell lineage and cell fate has become a cornerstone of modern research, particularly in developmental biology and regenerative medicine. Mapping cell lineages carries profound implications, from unraveling the complexities of embryonic development to devising innovative therapeutic strategies. As we delve deeper into the study of cellular differentiation and lineage commitment, it becomes clear that there is no one-size-fits-all method for all models. Lineage tracing is an art, and methods are as diverse as they are indispensable.

This book aims to be a comprehensive guide and reference for researchers and students, both within and outside the field. It explores a broad spectrum of techniques that have been developed and refined over the years, offering a detailed exposition of their principles, applications, and limitations. From more classic methods, such as genetic lineage tracing and clonal analysis, to cutting-edge technologies like single-cell RNA sequencing and live-cell imaging, this volume encapsulates the state-of-the-art in the field.

Our goal is not only to provide a thorough understanding of each method but also to inspire innovation and interdisciplinary collaboration. By presenting both well-established and emerging techniques, we aim to bridge the gap between theoretical knowledge and practical application, enabling readers to choose the most appropriate tools for their specific research questions.

The preparation of this book has been a collaborative effort, drawing on the expertise of leading scientists from various disciplines. We are grateful for their contributions and insights, which have enriched the content and scope of this work. We also acknowledge the relentless efforts of the researchers whose pioneering studies have laid the foundation for the methods discussed herein.

As we present this book, we envision it as more than just a collection of techniques. It is a testament to the ingenuity and perseverance of the scientific community in its quest to unravel the mysteries of life at the cellular level. We hope that it will serve as a valuable resource, fostering new discoveries and advancements in the exciting field of cell lineage tracing.

Madrid, Spain — *Jorge García-Marqués*
Ann Arbor, MI, USA — *Tzumin Lee*

Contents

Contributors

ANNA ALEMANY • *Department of Anatomy & Embryology, Leiden University Medical Center, Leiden, The Netherlands; The Novo Nordisk Foundation Center for Stem Cell Medicine (reNEW), Leiden, The Netherlands*

DONOVAN J. ANDERSON • *Allen Discovery Center for Lineage Tracing and Department of Laboratory Medicine & Pathology, University of Washington, Seattle, WA, USA*

BLANCA LÓPEZ ARRABAL • *Instituto Cajal, Consejo Superior de Investigaciones Científicas, Madrid, Spain*

INÉS GONZÁLEZ-ASPE • *Instituto Cajal, Consejo Superior de Investigaciones Científicas, Madrid, Spain*

CHLOÉ S. BARON • *Stem Cell Program and Division of Hematology/Oncology, Boston Children's Hospital and Dana-Farber Cancer Institute, Boston, MA, USA; Stem Cell and Regenerative Biology Department, Harvard University, Cambridge, MA, USA*

SARA BIZZOTTO • *Sorbonne Université, Institut du Cerveau (Paris Brain Institute) ICM, Inserm, CNRS, Hôpital de la Pitié Salpêtrière, Paris, France; Université Paris Cité, Imagine Institute, Paris, France*

BROOKE BOSONAC • *Instituto Cajal, Consejo Superior de Investigaciones Científicas, Madrid, Spain*

SARAH BOWLING • *Stem Cell Program, Boston Children's Hospital, Boston, MA, USA; Department of Stem Cell and Regenerative Biology, Harvard University, Cambridge, MA, USA; Department of Developmental Biology, Stanford University School of Medicine, Stanford, United States*

FERNANDO D. CAMARGO • *Stem Cell Program, Boston Children's Hospital, Boston, MA, USA; Department of Stem Cell and Regenerative Biology, Harvard University, Cambridge, MA, USA*

GISELLE CHEUNG • *Institute of Science and Technology Austria (ISTA), Klosterneuburg, Austria*

LAURA DUMAS • *Institute for Neuroscience of Montpellier (INM), University of Montpellier, INSERM, Montpellier, France; Department of Medical Neuroscience, Brain Repair Centre, Faculty of Medicine, Dalhousie University, Halifax, NS, Canada*

JASON DURAND • *Institute for Neuroscience of Montpellier (INM), University of Montpellier, INSERM, Montpellier, France*

A. S. EISELE • *Ecole Polytechnique Fédérale de Lausanne, School of Life Sciences, Institute of Bioengineering, Lausanne, Switzerland*

ISABEL ESPINOSA-MEDINA • *Janelia Research Campus, Howard Hughes Medical Institute, Ashburn, VA, USA*

WEIXIANG FANG • *Department of Biomedical Engineering, Johns Hopkins University School of Medicine, Baltimore, MD, USA; Center for Epigenetics, Johns Hopkins University School of Medicine, Baltimore, MD, USA*

M. FIGUERES-OÑATE • *Instituto Cajal, Consejo Superior de Investigaciones Científicas, Madrid, Spain*

JORGE GARCÍA-MARQUÉS • *Instituto Cajal, Consejo Superior de Investigaciones Científicas, Madrid, Spain*

EUGENIO GENTILE • *Department of Biosystems Science and Engineering, ETH Zürich, Basel, Switzerland*

ZHISONG HE • *Department of Biosystems Science and Engineering, ETH Zürich, Basel, Switzerland*

SIMON HIPPENMEYER • *Institute of Science and Technology Austria (ISTA), Klosterneuburg, Austria*

MARSHALL S. HORWITZ • *Allen Discovery Center for Lineage Tracing and Department of Laboratory Medicine & Pathology, University of Washington, Seattle, WA, USA*

HONGKAI JI • *Department of Biostatistics, Johns Hopkins University Bloomberg School of Public Health, Baltimore, MD, USA*

JAN PHILIPP JUNKER • *Charité Universitätsmedizin Berlin, Berlin, Germany; Max Delbrück Center for Molecular Medicine, Berlin Institute for Medical Systems Biology, Berlin, Germany*

REZA KALHOR • *Department of Biomedical Engineering, Johns Hopkins University School of Medicine, Baltimore, MD, USA; Center for Epigenetics, Johns Hopkins University School of Medicine, Baltimore, MD, USA; Departments of Molecular Biology & Genetics, Medicine, Neuroscience, and Genetic Medicine, Johns Hopkins University School of Medicine, Baltimore, MD, USA*

MARIA KALOGERIDI • *Institute of Molecular Biology and Biotechnology, Foundation for Research and Technology Hellas, Heraklion, Crete, Greece; Department of Biology, University of Crete, Heraklion, Crete, Greece*

TZUMIN LEE • *Life Science Institute, University of Michigan, Ann Arbor, MI, USA; Department of Molecular, Cellular, and Developmental Biology, University of Michigan, Ann Arbor, MI, USA; Howard Hughes Medical Institute, Ann Arbor, MI, USA*

IOANNIS LIASKAS • *Institute of Molecular Biology and Biotechnology, Foundation for Research and Technology Hellas, Heraklion, Crete, Greece*

LAURA LÓPEZ-MASCARAQUE • *Instituto Cajal, Consejo Superior de Investigaciones Científicas, Madrid, Spain*

KARINE LOULIER • *Institute for Neuroscience of Montpellier (INM), University of Montpellier, INSERM, Montpellier, France*

JIAN MA • *IDG/McGovern Institute of Brain Research, Tsinghua University, Beijing, People's Republic of China; New Cornerstone Science Laboratory, Tsinghua University, Beijing, People's Republic of China*

ZIPORA MARX • *Department of Computer Science and Applied Mathematics, Weizmann Institute of Science, Rehovot, Israel*

ASHLEY MAYNARD • *Department of Biosystems Science and Engineering, ETH Zürich, Basel, Switzerland*

YAARA NEUMEIER • *Department of Computer Science and Applied Mathematics, Weizmann Institute of Science, Rehovot, Israel*

A. C. OJALVO-SANZ • *Instituto Cajal, Consejo Superior de Investigaciones Científicas, Madrid, Spain*

FLORIAN M. PAULER • *Institute of Science and Technology Austria (ISTA), Klosterneuburg, Austria*

ANASTASIOS PAVLOPOULOS • *Institute of Molecular Biology and Biotechnology, Foundation for Research and Technology Hellas, Heraklion, Crete, Greece; Janelia Research Campus, Howard Hughes Medical Institute, Ashburn, VA, USA*

YUWEI PENG • *IDG/McGovern Institute of Brain Research, Tsinghua University, Beijing, People's Republic of China; Tsinghua-Peking Center for Life Sciences, Tsinghua University, Beijing, People's Republic of China; School of Life Sciences, Tsinghua University, Beijing, People's Republic of China*

ANGELICA PREVIERO • *Life Science Institute, University of Michigan, Ann Arbor, MI, USA; Department of Molecular, Cellular, and Developmental Biology, University of Michigan, Ann Arbor, MI, USA*

BEATRIZ MADARIAGA PUCHOL • *Instituto Cajal, Consejo 2Superior de Investigaciones Científicas, Madrid, Spain*

BUSHRA RAJ • *Department of Cell and Developmental Biology, Perelman School of Medicine, University of Pennsylvania, Philadelphia, PA, USA; Institute for Regenerative Medicine, University of Pennsylvania Perelman School of Medicine, Philadelphia, PA, USA*

JOHN RALLIS • *Institute of Molecular Biology and Biotechnology, Foundation for Research and Technology Hellas, Heraklion, Crete, Greece; Department of Biology, University of Crete, Heraklion, Crete, Greece; Janelia Research Campus, Howard Hughes Medical Institute, Ashburn, VA, USA*

MICHAEL RATZ • *Department of Cell and Molecular Biology, Karolinska Institute, Stockholm, Sweden*

OFIR RAZ • *Department of Computer Science and Applied Mathematics, Weizmann Institute of Science, Rehovot, Israel*

ALEJO E. RODRIGUEZ FRATICELLI • *Institute for Research in Biomedicine (IRB Barcelona), The Barcelona Institute of Science and Technology, Barcelona, Catalonia, Spain; ICREA, Catalan Institution for Research and Advanced Studies, Barcelona, Catalonia, Spain*

IREPAN SALVADOR-MARTÍNEZ • *Centro Nacional de Análisis Genómico, Barcelona, Spain*

PEDRO SÁNCHEZ SÁNCHEZ • *Institute for Research in Biomedicine (IRB Barcelona), The Barcelona Institute of Science and Technology, Barcelona, Catalonia, Spain*

EHUD SHAPIRO • *Department of Computer Science and Applied Mathematics, Weizmann Institute of Science, Rehovot, Israel*

ZHONGFU SHEN • *IDG/McGovern Institute of Brain Research, Tsinghua University, Beijing, People's Republic of China; Tsinghua-Peking Center for Life Sciences, Tsinghua University, Beijing, People's Republic of China; School of Life Sciences, Tsinghua University, Beijing, People's Republic of China*

SONG-HAI SHI • *IDG/McGovern Institute of Brain Research, Tsinghua University, Beijing, People's Republic of China; Tsinghua-Peking Center for Life Sciences, Tsinghua University, Beijing, People's Republic of China; School of Life Sciences, Tsinghua University, Beijing, People's Republic of China; Beijing Frontier Research Center for Biological Structure, Tsinghua University, Beijing, People's Republic of China; New Cornerstone Science Laboratory, Tsinghua University, Beijing, People's Republic of China*

BASTIAAN SPANJAARD • *Charité Universitätsmedizin Berlin, Berlin, Germany*

D. M. SUTER • *Ecole Polytechnique Fédérale de Lausanne, School of Life Sciences, Institute of Bioengineering, Lausanne, Switzerland*

LIMING TAO • *Department of Computer Science and Applied Mathematics, Weizmann Institute of Science, Rehovot, Israel*

BARBARA TREUTLEIN • *Department of Biosystems Science and Engineering, ETH Zürich, Basel, Switzerland*

LEONIE VON BERLIN • *Department of Cell and Molecular Biology, Karolinska Institute, Stockholm, Sweden*

EFAN WANG • *IDG/McGovern Institute of Brain Research, Tsinghua University, Beijing, People's Republic of China; School of Life Sciences, Tsinghua University, Beijing, People's Republic of China*

CHING-PO YANG • *Life Science Institute, University of Michigan, Ann Arbor, MI, USA; Howard Hughes Medical Institute, Ann Arbor, MI, USA*

YI YANG • *Department of Biomedical Engineering, Johns Hopkins University School of Medicine, Baltimore, MD, USA; Center for Epigenetics, Johns Hopkins University School of Medicine, Baltimore, MD, USA*

Chapter 1

Bayesian Phylogenetic Lineage Reconstruction with Loss of Heterozygosity Mutations Derived from Single-Cell RNA Sequencing

Donovan J. Anderson and Marshall S. Horwitz

Abstract

Mutations are acquired frequently, such t'hat each cell's genome inscribes its history of cell divisions. Loss of heterozygosity (LOH) accumulates throughout the genome, offering large encoding capacity for phylogenetic inference of cell lineage.

In this chapter, we demonstrate a method, using single-cell RNA sequencing, for reconstructing cell lineages from inferred LOH events in a Bayesian manner, annotating the lineage with cell phenotypes, and marking developmental time points based on X-chromosome inactivation. This type of retrospective analysis could be incorporated into scRNA-seq pipelines and was initially developed to investigate Emx1+ cortical projection neuron and glia lineages from C57Bl/6J (B6) and CAST/EiJ (CA) interstrain F1 mice, describing progenitor cells giving rise to multiple cortical cell types through stereotyped expansion and distinct waves of neurogenesis.

Key words Cell lineage, Phylogeny, Loss of heterozygosity, LOH, Bayesian, Development, Single cell, X-inactivation, Method

1 Introduction

An organism's complete cell lineage describes the relationship of all cells produced during development, beginning at the zygote. Cell lineage analysis using dyes or transgenic markers to label a particular cell's, or group of cells', descendants during development is an established approach for inferring these relationships. However, this method is limited in resolution to clonal groups of daughter cells. For example, $>10^{42}$ different possible lineages can describe a clone of just 32 cells [1]. The complete embryonic cell lineage of *C. elegans* [2] represents a milestone in lineage resolution and consequent understanding of developmental biology but relies on the organisms' invariant lineage between individuals and real-time observation of translucent tissue. Lineage reconstruction in

Jorge García-Marqués and Tzumin Lee (eds.), *Lineage Tracing: Methods and Protocols*, Methods in Molecular Biology, vol. 2886, https://doi.org/10.1007/978-1-0716-4310-5_1,

organisms with more cells, more cell types, longer life spans, and opaque tissues has proven more challenging. Specifically, individual cell isolation, cell type classification, detection of distinguishing markers, and reconstruction of lineage all pose obstacles.

For organisms where genetic manipulation is possible, the introduction of engineered recording systems [3, 4] has shown great promise. When combined with single-cell RNA sequencing (scRNA-seq), these methods provide heritable variation and transcriptomic information for high-throughput lineage reconstruction using phylogenetics and cell type identification. In systems where genetic manipulation is not possible (e.g., human samples), phylogenetic lineage reconstruction approaches have relied on naturally occurring, acquired genetic variation arising within development of the individual organism. One approach to map lineage in these systems is to retrospectively infer the sequence of DNA mutations acquired in single cells [5–12]. In fact, normal mutational rates are sufficiently high, such that brute force sequencing of single-cell genomes enables lineage tracing [13].

Single-cell DNA sequencing (scDNA-seq) based lineage determination has limitations, however. Interrogation of multiple loci in a single cell requires sufficient quantities of DNA that can only be achieved through whole genome amplification, which is error prone [6, 14], or clonal expansion in vitro [12, 13, 15], which is difficult to scale, not possible for post-mitotic cell types, and may select for mutations not found in vivo. Moreover, scDNA-seq provides no transcriptomic information for cell type identification. In contrast, thousands of copies of an RNA molecule may be present in a cell. Single nucleotide variant (SNV) detection using scRNA-seq, nevertheless, has its own limitations. Cell type specific transcriptomes comprise only a fraction of the genome, decreasing the number of potentially variable SNV loci detectable and subsequently limiting information available for phylogenetic reconstruction. Polymerase fidelity during transcription and next-generation library construction generate variation as well [16–18], confounding phylogenetic analysis based on individual SNV loci.

Detection of loss of heterozygosity (LOH) from scRNA-seq data offers a workaround. In tumors, LOH is detectable in scRNA-seq expression datasets and has been used to reconstruct their lineage at single-cell resolution [19, 20]. However, such an approach generally requires access to constitutional, non-tumor tissue, in order to identify regions of LOH.

Yet, it is also possible to infer LOH in somatic tissues, even when a constitutional source of DNA is unavailable. LOH can be assayed in single cells independently at sites known to be germline heterozygous [21]. Although point mutations are not uncommon, LOH, arising from recombination or other mechanisms, occurs frequently. Measurements at marker genes indicate that LOH occurs $\sim 10^{-4}$–10^{-5}/locus/cell division [22–24]. In the transcriptome, LOH is recognizable as tracts of apparent monoallelic expression stretched across adjacent genes. LOH regions often occur at megabase scale and consist of >20 SNV loci observed [21, 25–27], which is large enough to counteract the reduction of loci detectable for a specific transcriptome, allow for detection across different cell type transcriptome footprints, and mitigate any variation due to polymerase errors. These germline SNV loci are predictable and more economical to survey than other mutations, which otherwise arise unpredictably anywhere throughout the genome.

Methods for phylogenetic reconstruction of cell lineages vary depending on how genomic variation is treated [28, 29]. Distance methods catalog variation between individual cells that is used to construct a matrix, relying on a measure of difference. Cells are then grouped together in subsequently larger clades based on similarity or some form of hierarchical clustering, such as neighbor joining. Differences between cells are reflected in the branch length connecting the two cells. A tree (lineage) that minimizes the overall branch length is commonly chosen as the representative lineage [30]. Frequentist methods, such as maximum parsimony or maximum likelihood, calculate the most likely lineage, given the observed data, from a sampling of all possible lineage trees (tree space). Parsimony endeavors to minimize the number of changes, while likelihood methods evaluate a model of mutagenesis. Most models are fairly straightforward and are based on the type of variation measured (e.g., point mutations, insertions/deletions, run length of polynucleotide repeats, or mitochondrial DNA variation, just to name a few). These models are well developed in evolutionary theory [29]. In some cases, methods for newly developed recording systems may require their own models of target evolution, such as for the CRISPR/Cas9 based targets. Finally, Bayesian methods of reconstruction are available, as well [31, 32]. In addition to calculating the most likely lineage given the data and model, these methods offer the benefit of calculating the probability of that lineage or sub-lineage (clade) with respect to all lineages simulated in the analysis. The occurrence of LOH events during development of an organism most closely resembles Camin-Sokal parsimony. Character states evolve in a directional manner from absence in the zygote to presence in any daughter cell and are irreversible. We implement this model in a Bayesian framework that allows us to evaluate the most probable lineages and clades given the variation observed.

Here we describe a method for detecting regions of LOH directly from a single-cell RNA sequencing library using a hidden Markov model (HMM), encoding LOH events as appropriate character states for phylogenetic lineage analysis, inferring and evaluating the probability of cell lineages from a single organism, and annotating lineages with cell-specific phenotype information as well as developmental time points. The method was developed to investigate Emx1+ cortical projection neuron and glia lineages from C57Bl/6J (B6) and CAST/EiJ (CA) interstrain F1 mice [21], but can be applied to any samples where heterozygous SNVs are completely phased across the chromosome (e.g., F1 crosses of any inbred strains). Subheadings 3.4 and 3.5 are applicable to any intra-organismal variation, either somatic mutation or genetic recording system, whose alleles behave similarly.

2 Materials

2.1 Programs and Packages

1. STAR 2.5.0c [33]. Single-cell RNA sequencing read alignment and transcript quantification.
2. GATK 4 [34]. Manipulation of scRNA-seq alignment files (BAM, VCF), SNV detection, and filtration of various SNV calls based on sequence quality.
3. R Language Packages; Tidyverse [35], general data manipulation and visualization; HMM [36], generation of LOH hidden Markov models and LOH detection; Seurat [37], cell type identification and visualization of scRNAseq data.
4. MrBayes [31]. Bayesian inference of single-cell lineages.
5. Unix BASH [38]. Text file and program manipulation.
6. iTOL [39]. Online tool for the display, annotation, and management of phylogenetic trees. https://itol.embl.de/.

2.2 Genome Reference Files

1. Genome reference file (.fa) appropriate for organism of interest. The file provided needs to be the same for both STAR and GATK pipelines to ensure compatibility.
2. Transcription model reference (GTF) file appropriate for organism of interest. This file describes the transcriptome and is used in conjunction with the genome reference file by STAR to quantify transcription of genes.

2.3 Single-Cell RNA BAM Files

1. To call variants using RNA-based BAM files (*see* **Note 1**) with GATK, there needs to be some editing and manipulation of the individual read alignments. An individual scRNA-seq BAM file will need read group information added, the file header re-ordered, the file sorted, the RNA alignment cigars split, and read MAPQ values adjusted from the RNA defaults. The final BAMs will be suitable for calling variants for each cell. A GATK command template is given below.

```
#Add read groups to each single-cell BAM.
./gatk AddOrReplaceReadGroups \
   -I SingleCellBarCode.bam \
```

```
    -O SingleCellBarCode_RG.bam \
    --RGID SampleDescription \
    --RGLB LibraryID \
    --RGPL Platform \
    --RGPU PlatformUnit \
    --RGSM SingleCellBarCode
#Reordering single-cell BAM header.
./gatk ReorderSam \
    -I SingleCellBarCode_RG.bam \
    -O SingleCellBarCode_RGReOr.bam \
    -R ReferenceGenome.fa \
    --ALLOW_CONTIG_LENGTH_DISCORDANCE TRUE
#Sorting SingleCellBarCode_RGReOr.bam.
./gatk SortSam \
    -I SingleCellBarCode_RGReOr.bam \
    -O SingleCellBarCode_RGReOrSort.bam \
    --SORT_ORDER coordinate \
    --CREATE_INDEX TRUE
#Splitting SingleCellBarCode_RGReOrSort.bam cigar reads.
./gatk SplitNCigarReads \
    -R ReferenceGenome.fa \
    -I SingleCellBarCode_RGReOrSort.bam \
    -O SingleCellBarCode_RGReOrSortSplit.bam
```

2.4 Sample Specific Heterozygous SNV Loci ROD File

1. One of the advantages of using F1 mice from inbred strains, whose genomes have been sequenced, is that a list of predicted heterozygous SNV loci can be constructed. This list can be passed to GATK as a ROD or guide .vcf file for variant detection. Genotyping only these specific loci substantially reduces computation time (*see* **Note 2**).

2.5 Transcription-Based Cell Type Assignments

1. Determination of lineage relationships between single cells does not require the knowledge of cell type, but interpretation of the results greatly benefits from the information. There are multiple software packages that can group and identify sample cells by transcription profiles provided by an RNAseq aligner such as STAR or 10X Cellranger. In general, we utilize both Seurat [37] and Monocle 3 [40] for this purpose (*see* **Note 8**). If multiple individual organism lineages of the same tissue or population of cells is to be compared, all cells should be clustered and identified collectively. Something to be considered is the role of cell division in the population of cells being studied. In general, clustering should be performed on samples where variation due to cell cycle is accounted for during differential expression analysis and compared to results where cell cycle is not accounted for to assess whether the phase of the cell cycle is driving the clustering result for certain groups of cells.

3 Methods

3.1 SNV Detection in Single Cells with Individual scRNAseq Alignments

Here we use GATK and basic Unix functions to determine the allele (s) transcribed at pre-selected SNV loci across all autosomes of a cell. Each individual cell list of variants (VCF) is gathered with all other cells and genotyped as a group. Variants are then hard filtered for quality. Cell variants passing quality controls are subsequently transformed into a format suitable for analysis in an R environment.

1. Use the GATK HaplotypeCaller tool to determine the expressed allele(s) at heterozygous loci, provided by the ROD guide VCF file, for an individual cell BAM file. The --emit-reference-confidence GVCF option is used to account for loci that are covered in only a subset of cells. Duplicate reads are not included in the analysis.

```
./gatk HaplotypeCaller \
  -R ReferenceGenome.fa \
  -I SingleCellBarCode_RGReOrSortSplit.bam \
  -O SingleCellBarCode.raw.snps.indels.BP.g.vcf.gz \
  -L RodeFile.vcf \
  --emit-ref-confidence GVCF \
  -DF NotDuplicateReadFilter \
  --dont-use-soft-clipped-bases true
```

2. Combine all the individual GVCF files created above into one GVCF representing the entire sample. We can then genotype all cells together, and all cells will have the same number of loci listed whether or not there is coverage in a particular cell. A required input is a list file comprised of the relevant GVCF file path/names, from above, in a single column.

```
./gatk CombineGVCFs \
  -R ReferenceGenome.fa \
  --variant Sample_GVCF.list \
  -O Sample_combGVCF.vcf.gz
```

3. Call genotypes for all cells across all locations using the GATK tool GenotypeGVCFs. We set the minimum confidence threshold for calling to 20 and include non-variant sites, as homozygous reference sites are just as important as variant sites.

```
./gatk GenotypeGVCFs \
  -R ReferenceGenome.fa \
  --variant Sample_combGVCF.vcf.gz \
```

```
--standard-min-confidence-threshold-for-calling 20 \
--include-non-variant-sites true \
-O Sample_combGVCFSites.vcf.gz
```

4. Hard filter SNVs using recommendations from The Broad Institute for RNA sequence-based variants. Specifically, variants with a strand bias score (FS) of greater than 30 and quality-by-depth (QD) value of less than 2 are removed. SNVs that cluster in small windows are also removed to guard against variants that occur due to reads overlapping splice sites.

```
./gatk VariantFiltration \
  -R ReferenceGenome.fa \
  -V Sample_combGVCFSites.vcf.gz \
  -O Sample_combGVCFSites_HardFilt.vcf.gz \
  -window 35 -cluster 3 \
  --filter-name FS --filter-expression "FS > 30.0" \
  --filter-name QD --filter-expression "QD < 2.0"
```

5. Select genotypes that pass filters (*see* **Note 3**):

```
./gatk SelectVariants \
  -R ReferenceGenome.fa \
  -V Sample_combGVCFSites_HardFilt.vcf.gz \
  -O Sample_combGVCFSites_PassHardFilt.vcf.gz \
  --exclude-filtered true
```

6. Extract RNA-based genotypes for analysis in an R environment. For HMM analysis in the R environment, we only need the RNA-based allele calls and not the accompanying information provided in the VCF format. Therefore, we extract the GATK derived RNA-based genotypes 0/0, 0/1, 1/1, and ./. (no coverage) and convert them to integers (3,4,5, and 1, respectively). We separate by chromosome, as the HMM model applies to individual chromosomes for the unit of observation. The objective is to have a tab delimited file with chromosome and position information in the first column and translated allelic expression information for each cell in their own column. A Unix pipe is given below as an example for chromosome 6 (*see* **Note 3**).

```
$ zgrep -E "^chr6|^#C" Sample_combGVCFSites_PassHardFilt.
vcf.gz | \
  awk '{ print $1":"$2 "\t" $0 }' | cut -f 1,11- | \
  awk '{ \
  gsub(/\.\/\.[^\t]*/,"1"); \
```

```
gsub(/0\/0[^\t]*|0\|0[^\t]*/,"3"); \
gsub(/0\/1[^\t]*|0\|1[^\t]*|1\|0[^\t]*/,"4"); \
gsub(/1\/1[^\t]*|1\|1[^\t]*/,"5"); \
print $0 }' > Sample_chr6_GVCF_SNV_PassHardFilt_ForR.txt
```

3.2 Import and Manipulation of Variant Data in R

With respect to LOH determination, we use R for two purposes: hidden state inference and visualization of data. As above, we work with data from a single chromosome at a time. Here we use functions provided in the R Tidyverse package (Fig. 1).

1. Prepare data for analysis and graphing (*see* Fig. 1a). We import the tab delimited chromosome variant files created in the last section with the first column renamed "X" for position.

```
> Sample_chr6_cells <- read.csv("Sample_chr6_GVCF_SNV_Pass
HardFilt_ForR.txt", sep="")
#Remove chromosome information and colon from the first column
```

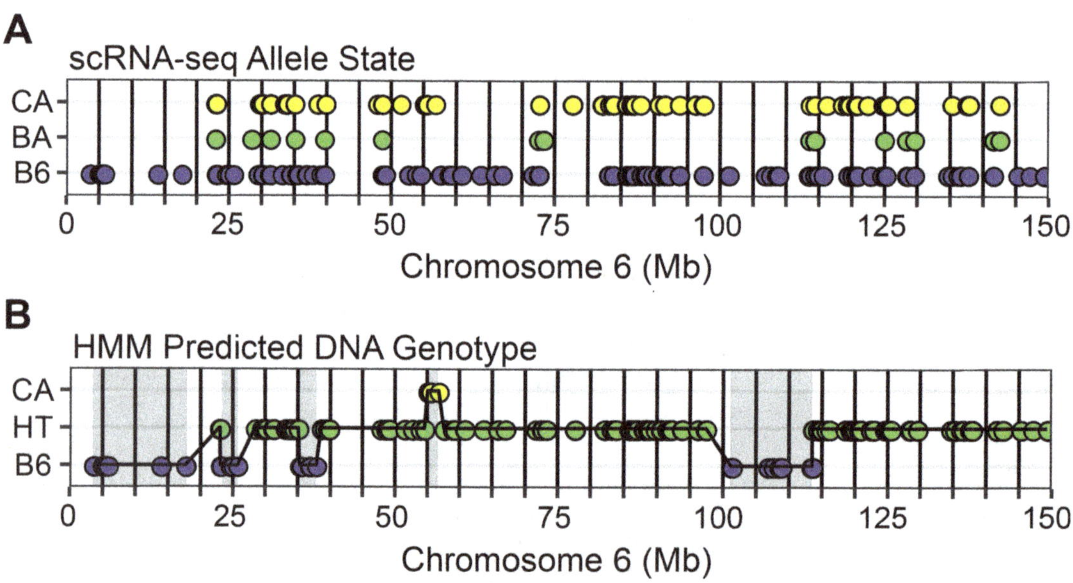

Fig. 1 Inference of LOH events across a chromosome from scRNA-seq derived SNVs. (**a**) Plot of observed scRNA-seq SNV loci and their allele state for chromosome 6 of a 42-day-old mouse. SNV position along chromosome 6 is shown on the *X*-axis. The GATK called allele state is plotted on the *Y*-axis. B6 = monoallelic B6, CA = monoallelic CA. BA = biallelic B6/CA. Each dot is a unique SNV locus and has been color coded for clarity. B6 = blue, CA = yellow, BA = green. (**b**) Plot of LOH HMM predicted DNA genotype (hidden) states for each SNV loci observed in (**a**). Again, the SNV position is shown along the *X*-axis. The HMM genotype is now plotted on the *Y*-axis. B6 (blue) = homozygous B6, CA (yellow) = homozygous CA, HT (green) = heterozygous B6/CA. Here, a switch away from the HT level along the *Y*-axis indicates a hidden state change and therefore an LOH event. The length of the event is indicated by consecutive SNVs plotted on either the CA or B6 levels. A black line has been added to aid in visualization of the state transitions, and gray regions mark individual LOH events

```
and convert the value to integer.
> Sample_chr6_cells$X <- as.integer(sub('......', '', (Sam
ple_chr6_cells$X)))
#Create a long format version of the data frame for plotting.
> tdyCells <- gather(Sample_chr6_cells[1:ncol(Sample_chr6_
cells)], "Cell", "state", -X)
#Sample plot of allele state across chromosome faceted by cell
> ggplot(filter(tdyCells, state >= 3), aes(x=X, y=state,
fill=factor(state))) +
   geom_point(size=3,shape=21) + facet_grid(Cell ~ .) +
   scale_fill_manual(values= setNames(c(3,4,5), c("blue",
   "green","yellow"), name="Allele") +
   scale_y_continuous(limits=c(2,6), breaks=c(3,4,5),mino
   r_breaks = NULL) +
   scale_x_continuous(limits=c(0,#ChrLength),
   breaks=c(seq(0,#ChrLength,by=5000000)),
   minor_breaks=NULL,expand=c(0,0)) +
   theme_bw() + theme(panel.spacing.y = unit(2, "lines"),
   legend.position="none")
```

3.3 HMM Inferred Loss of Heterozygosity

1. Hidden Markov model description and matrix construction. A cell's genotype at a particular locus can be inferred from scRNAseq-based variant transcription calls by modeling their observation as an HMM process. The underlying genotype is considered the hidden state while the observed RNA-based transcription state is the emission state. We base our model on observed frequencies of interhomolog chromosomal exchanges generating LOH [23] and observed frequencies of monoallelic or biallelic transcription [21, 41]. The transition matrix between hidden genotype states is defined and shown in Table 1.

   The probability of observing a particular SNV in the heterozygous (biallelic) state or either homozygous (monoallelic)

Table 1
Loss of Heterozygosity HMM hidden state transition matrix

	Homozygous B6	Heterozygous	Homozygous CA
Homozygous B6	0.99999	0.00001	0
Heterozygous	0.00010	0.99980	0.00010
Homozygous CA	0	0.00001	0.99999

Table 2
Loss of heterozygosity HMM emission state matrix

		Emission state		
		Monoallelic B6	Biallelic	Monoallelic CA
Hidden state	Homozygous B6	1	0	0
	Heterozygous	0.45	0.10	0.45
	Homozygous CA	0	0	1

state given a particular hidden genotype state is described by the emission state matrix shown in Table 2.

The Markov chain's starting state probability is given as 0.05 for each of the homozygous states and as 0.90 for the heterozygous state.

2. Create the LOH HMM model with the HMM R package.

```
> library("HMM")
> library('tidyverse')
#Define hidden states and emission symbols.
> HiddenState <- c("B6","HT","CA")
> EmissionSym <- c("3","4","5")
#Define model probabilities.
> StartProb <- c(0.05,0.9,0.05)
> TransProb <- matrix(c(0.99999, 0.00001, 0.0, 0.0001,
0.9999, 0.0001, 0.0, 0.00001, 0.99999), nrow=3, byrow=
TRUE)
> TransProb
[,1] [,2] [,3]
[1,] 0.99999 0.00001 0.00000
[2,] 0.00010 0.99990 0.00010
[3,] 0.00000 0.00001 0.99999
> EmissionProb <- matrix(c(1.0, 0.0, 0.0, 0.45, 0.1, 0.45,
0.0, 0.0, 1.0), nrow=3, byrow=TRUE)
> EmissionProb
[,1] [,2] [,3]
[1,] 1.00 0.0 0.00
[2,] 0.45 0.1 0.45
[3,] 0.00 0.0 1.00
#Initialize model.
> LOH_HMM <- initHMM(HiddenState, EmissionSym, StartProb,
TransProb, EmissionProb)
```

3. Infer the hidden genotype state (*see* Fig. 1b). The most likely sequence of hidden states, given a set of observations and model, is called the Viterbi path, and is implemented in the R HMM package. We provide the sequence of observed scRNAseq-based transcription calls for any one chromosome of any given cell as emission states, and the function viterbi() returns the most probable genotype state at each observed locus. The Viterbi path describes the amalgamation of both homologous chromosomes and is assumed to have a copy number of 2. Hemizygosity will appear as LOH (B6 or CA) while duplications will appear as heterozygous (HT). An R function that calculates the Viterbi path for multiple cells in a data set is shown in **Note 4**.

```
#Predict the most likely hidden state for each cell across the chromosome.
> Sample_chr6_cells_GT <- cnloh.gt(Sample_chr6_cells, LOH_HMM)
#Create long form version of _GT data frame for visualization.
> tdyCells_GT <- gather(Sample_chr6_cells_GT[1:ncol(Sample_chr6_
cells_GT)], "Cell", "state", -POS)
#Sample plot of most likely hidden state across the chromosome faceted by
#cell.
> ggplot(tdyCells_GT, aes(x=POS, y=state, fill=factor(state))) +
  geom_point(size=3,shape=21) + facet_grid(Cell ~ .) +
  scale_fill_manual(values=setNames(c("B6","HT","CA"),c("yellow",
  "green","blue"),
  name= "Allele State",breaks=c("CA","HT","B6")) +
  scale_y_discrete(limits=c("B6","HT","CA")) +
  scale_x_continuous(limits=c(0,#ChrLength),
  breaks=c(seq(0,#ChrLength,by=5000000)),
  minor_breaks=NULL,expand=c(0,0)) +
  theme_bw() + theme(panel.spacing.y = unit(2, "lines"),legend.positi
  on="none")
```

4. Collect LOH run information for each cell across the chromosome. The function cnloh.gt returns a data frame where the most likely hidden state (heterozygous (HT) or homozygous B6 (B6) or CA (CA)) replaces the observed RNA allele state for each locus observed in the particular cell (column). Consecutive occurrences of either B6 or CA indicate a run of LOH while HT indicates regions of heterozygosity. A function (get.gt.runs()) that detects the presence or absence of any LOH runs, describes their start and end locations, strain of origin, and number of SNV loci observed in the run is provided in **Note 5**. This function produces a data frame that can be exported as a table and used to determine LOH mutations

for all cells along the chromosome. These mutations are then used for phylogenetic lineage determination.

```
#Summarize LOH runs across the chromosome for each cell.
>Cells_GT_PosRunsTable<-get.gt.runs(Sample_chr6_cells_GT)
> write.table(Cells_GT_PosRunsTable, file="Sample_chr6_
GTRunTable.txt", quote=FALSE, sep="\t", row.names=FALSE)
```

3.4 Determining LOH Alleles and Converting to Two-State Characters

For our phylogenetic analysis of lineage we model the evolution of individual chromosomes within the organism. We treat each chromosome as independently evolving, lacking LOH events as a zygote and acquiring unique heritable events through development. Therefore, at the time of sampling, each chromosome in a particular cell will have some array of LOH alleles. We can simplify the encoding of these different LOH alleles for phylogenetic analysis by defining a number of evolving states corresponding to each unique LOH event and recording their presence or absence in binary form [42].

1. Determine and compile LOH events. Transcriptional bursting kinetics or epigenetic silencing on one allele can account for short runs of either B6 or CA alleles, thereby confounding our results. To control for such events, we remove homozygous runs that span a region <1 Mb. Due to non-uniform gene expression and scRNA-seq coverage across cells, we group the borders of LOH events in 2 Mb bins, starting at the beginning of each chromosome. Events are characterized by their beginning and ending bins along with their parental allele identity. For example, a homozygous run on chromosome 6, starting at position 3,000,000 and ending at position 7,500,000, consisting exclusively of B6 variants, would be described as chr6_2B64. A chromosome with no LOH events is coded as HT, designating the chromosome as apparently heterozygous (e.g., for chromosome 6, the designation would be chr6_HT). A chromosome for any one cell can contain multiple LOH events. We compile a tab delimited file listing all filtered LOH events across all autosomes for all cells. The first column of the file indicates the chromosome number that events in that row describe. The number of rows for any particular chromosome is determined by the maximum number of LOH events for any cell on that chromosome. The remaining columns list the LOH events for each cell. Because not all cells will share the same number of events, empty rows of an individual column are allowed. This ensures an equal number of rows, which is required in R data frames. Our basic nomenclature for these files is Sample_AllelesForBinaryTrans.txt.

Table 3
Binary encoding of LOH allele characters

Cell\state	chr6_1CA3	chr19_15CA26	chr19_17CA23
Cell01	0	0	0
Cell02	1	0	1
Cell03	1	1	0

2. Encode LOH alleles into a binary state format. To facilitate phylogenetic lineage inference, unique LOH events are encoded as binary allele states with 1 indicating the presence of the allele and 0 absence. A fictional example comprised of three cells across chromosome six is given below.

   Cell01: chr6_HT

   Cell02: chr6_1CA3, chr19_17CA23

   Cell03: chr6_1CA3, chr6_15CA26

   There are three unique LOH events with one cell lacking any event. These cells would be factored and encoded as shown in Table 3.

   Using the file created in the previous step we can determine all the unique alleles for any sample, create their state, and code the presence or absence of those alleles for every cell in R. We then export the final data frame in a format compatible with downstream phylogenetic analysis.

```
> library("tidyverse")
> Sample_2Mb_Alleles <- read.delim("Sample_AllelesForBin
aryTrans.txt", stringsAsFactors=FALSE)
#Retrieve all unique alleles across all cells and sort
#ascending.
> uniqAlleles <- unique(unlist(Sample_2Mb_Alleles[2:ncol
(Sample_2Mb_Alleles)]))
> uniqAllSort <- sort(uniqAlleles)
#Remove chr##_HT alleles as they are coded as absent
#(0) for any allele state.
> uniqAllSortHT <- uniqAllSort[!grepl("HT", uniqAllSort)]
#Remove "allele" that is created from empty positions in
#data frame and create vector of
#all unique alleles for encoding.
> MasterAlleles <- uniqAllSortHT[uniqAllSortHT != ""]
#Create new data frame for all cells (columns) with rows
#(alleles) where presence of
#allele state is indicated with 1 and absence 0.
> celltoget <- as.list(names(Sample_2Mb_Alleles[2:ncol
(Sample_2Mb_Alleles)]))
> Sample_2Mb_PhyBi <- data.frame("Alleles" = MasterAlleles)
```

```
> for (i in celltoget) { +
   Sample_2Mb_PhyBi[,i] <- as.integer(MasterAlleles %in%
   Sample_2Mb_Alleles[,i]) +
   }
#Transpose data frame for NEXUS format and write table.
> tFinFrame <- as.data.frame(t(as.matrix(Sample_2Mb_PhyBi)))
> write.table(tFinFrame[1:nrow(tFinFrame),],
   file="Sample_2Mb_BinaryTransWAlleles.txt", quote=FALSE,
   sep="\t", row.names=TRUE, col.names=FALSE)
```

3.5 Phylogenetic Lineage Determination

As indicated above, autosomal LOH events can be coded as a series of two-state characters. The occurrence of these events are discrete and irreversible for any one chromosome, with the absence of any event in the zygote. This process can be modeled by Camin-Sokal parsimony [43], with each chromosome evolving independently. We adapt the restriction site model in MrBayes to capture this dynamic for our lineage reconstructions. A Bayesian analysis allows us to not only calculate the most likely lineage, given the data, but simulate multiple lineages and infer the posterior probability of any given bipartition observed.

1. Prepare a sample NEXUS data file for use by MrBayes. The character state data created in Subheading 3.4 **step 2**, with some modification, is imported into MyBayes using the NEXUS data file format. An initial block of text is inserted at the head of the file describing the input data matrix. The "ntax" variable is set to the number of cells in the sample plus one for the ancestral zygote added to the data matrix (*see* **Note 6**). The "nchar" variable is set to the number of character states (unique LOH allele states) in the data matrix. The "datatype" variable is set to restriction to indicate the restriction site model and the data matrix is not interleaved. The data matrix consists of all of the Sample_2Mb_BinaryTransWAlleles.txt file minus the first row. The theoretical zygote (*see* **Note 6**) can be inserted as the first cell (taxa) in the data matrix. A single semicolon is appended to the next line after the last line of the data matrix with one tab indentation. The last line consists of "end;" with no indentation. Cell names (taxa) are limited to ten characters. The file name requires a .nex extension (e.g., Sample_2Mb_RestrictionBin.nex). An example is given below with a truncated data matrix for clarity.

```
#NEXUS
begin data;
   dimensions ntax=## nchar=##;
   format datatype=restriction interleave=no;
   matrix
```

```
Zygote____ 0 0 0 ...
A64326____ 0 0 0 ...
...
LastCell__ 0 0 0 ...
   ;
end;
```

2. Create a Camin-Sokal parsimony-inspired Bayesian evolution model in MrBayes. The input .nex file is imported into MrBayes first and sets the data type model as restriction site.

```
MrBayes > execute Sample_2Mb_RestrictionBin.nex
#Set the character coding to "all."
MrBayes > lset coding=all
#Fix state frequencies to reflect that mutations are
irreversible.
MrBayes > prset statefreqpr=fixed(0.01,0.99)
#The rest of the parameters are left at the default
#settings. Review the model.
MrBayes > showmodel
```

3. Set Markov chain Monte Carlo parameters and run simulation. In general, we attempt to run simulations until the standard deviation of split frequencies between independent runs is below 0.01. The number of generations required to accomplish this is dependent on the number of cells in the analysis and the informative variation in the data set. Two million generations is a good starting point. The simulation can be extended if the split frequency standard deviation is not at the desired level after the selected number of generations. We leave the other parameters in their default setting to begin with as well.

```
#Setting Markov chain Monte Carlo simulation parameters.
MrBayes > mcmcp ngen=2000000 diagnfreq=10000 sample
freq=1000 printfreq=1000
#Run the MCMC simulation. By default two independent runs
#are initialized.
MrBayes > mcmc
```

The standard deviation of split frequencies will be reported every 10,000 generations. At the end of the requested number of generations the option of continuing the simulation will appear. The simulation can be extended if the split frequency value is not acceptable. If the value is acceptable stop the simulation.

```
#Summarize inferred model parameters.
MrBayes > sump
```

```
#Summarize tree statistics and set the consensus tree to
#include all compatible #bifurcations.
MrBayes > sumt contype=allcompat
```

4. Interpretation of phylogenetic results (Fig. 2). The sump command will output several files that allow for examination of statistics describing both MCMC simulations. We point the reader to the MrBayes manual (*see* **Note 7**) for further information. As we have fixed the state transition probabilities in our model, the only model parameter sampled is the overall tree length. The sumt command provides us information about the sampled lineage trees themselves and, more importantly, a consensus tree (Sample.con.tre) file that contains a compatible majority rule tree along with branch lengths, clade posterior probabilities, and associated descriptive statistics in the NEXUS file format. A list of trees searched and their posterior probabilities is also provided in the Sample.trprobs file. The consensus tree gives an overall estimate of the lineage and will converge with the most probable tree as its probability increases. One prediction that should hold is that the representative zygote added to the sample should always be the outgroup and have a very short branch length. Branch lengths are also informative because they are estimates of the amount of change along the particular edge. Mutations that are correlated with cell division can estimate the mitotic history of particular cell types or clades of cells because longer branch lengths would indicate more cell divisions. Determining meaningful posterior probability cutoffs for individual clades in a cell lineage is an active area of research and will be affected by the type of variation observed. One useful comparison would be inferring lineages of a sample comprised of cells from two different individuals [21]. Comparing the posterior probabilities of clades consisting of only one individual versus mixed clades and the overall separation of individuals across the tree topology should be informative.

3.6 Annotating Phylogenetic Lineages

1. Add cell type information to lineages (*see* Fig. 2). Cell types determined by various clustering analysis of single-cell transcriptomes (*see* Subheading 2.5) can be used to "decorate" the phylogenetic lineage tree. Depending on the system studied, patterns of cell types across clades, whether monophyletic or paraphyletic, can inform models of development, suggest new hypotheses, or elucidate the existence of early disease. The Interactive Tree of Life (iTOL, https://itol.embl.de) provides a powerful tool for manipulating phylogenies generated by MrBayes. NEXUS files, with accompanying statistical data, can be directly uploaded and manipulated into different styles of tree. Trees can further be annotated with phenotypic data

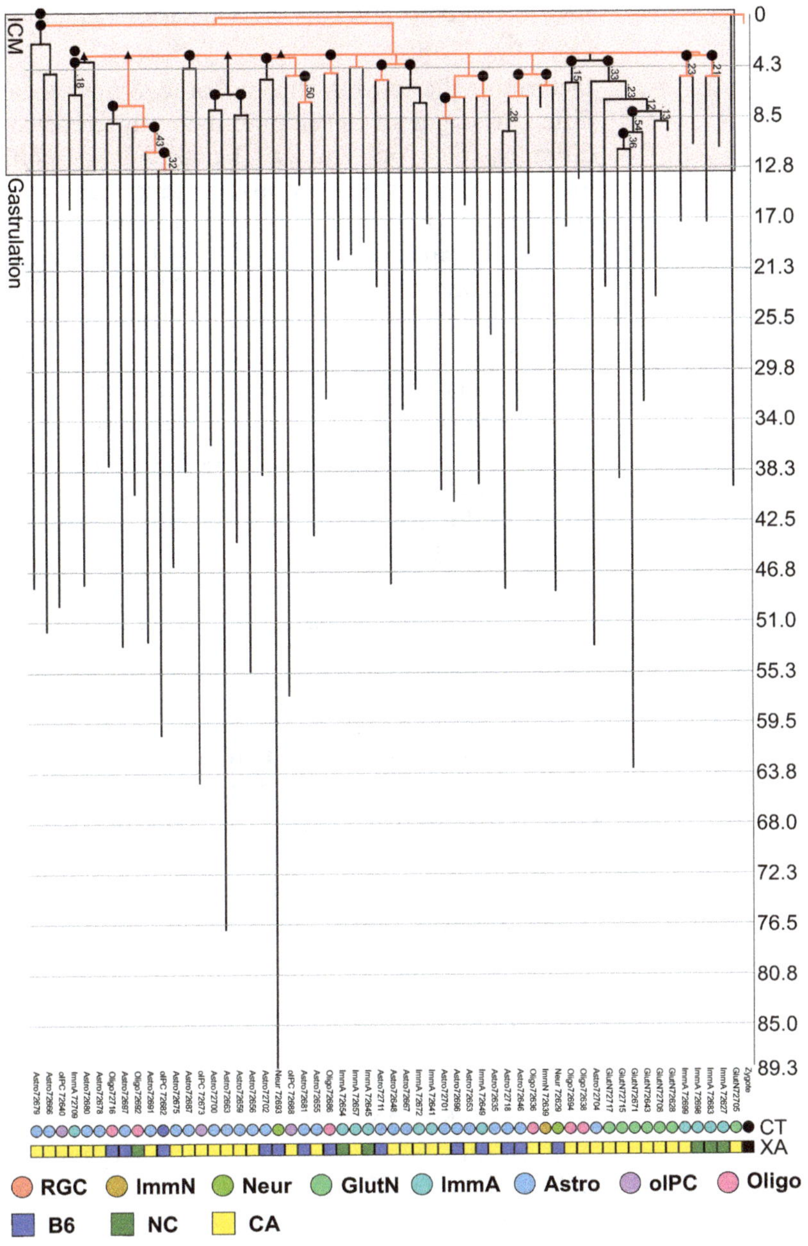

Fig. 2 Annotated phylogenetic lineage of Emx1+ cortical cells plus reference zygote. A Bayesian inferred consensus phylogeny of 56 cortical neurons and glial cells plus the control zygote is shown, with the base of the lineage at the top. The *Y*-axis represents the amount of genetic change as measured by LOH events. Leaf tips of the tree are annotated with scRNA-seq library ID (text), cell type (colored circles), and active X chromosome (colored squares). Clonal LOH events are marked with black circles on internal nodes. Black triangles mark LOH events, which occur in $n - 1$ members of the indicated clade. Bifurcations with Bayesian posterior probability >0.10 are resolved; probabilities >0.15 are shown along their branch lengths. For mixed clades, composed of cells with either active X chromosome, the internal nodes (most recent common cell) are inferred to occur about or before the time of gastrulation (based on its temporal relationship with X-chromosome inactivation), and their respective preceding branches are colored red. The pink box represents branches of the lineage tree predicted to occur during expansion of the epiblast from the inner cell mass

such as cell type, and exported in various graphical formats (*see* **Note 8**).

2. Determine the active X chromosome. X inactivation happens at different stages of development for different species and for different cell types and can be used to mark the timing of internal nodes as well as provide secondary conformation of clonality of clades (*see* Fig. 2). In our F1 crosses of inbred strains, we can easily determine which X chromosome is active by calculating the percentage of RNA monoallelic loci, correlating to either strain, compared to all loci observed. In general, we consider 75% or greater monoallelic expression of one strain or the other as indicative of that chromosome being active [21]. In mice, X-chromosome inactivation occurs around the time of gastrulation for most tissues. Clade nodes with mixed active X chromosomes, either B6 or CA, would set a lower boundary for the timing of gastrulation on an individual lineage (*see* Fig. 2).

4 Notes

1. The 10X Genomics platform provides one BAM file for all cells in the sample. One can create individual BAM files for each cell using the filter function in BamTools [44], the filtered list of GEM barcodes provided by 10X, and the Unix cat function.

```
$ cat barcodes.tsv | bamtools filter -in possorted_geno
me_bam.bam -tag "CB:$1" \
  -out $1.bam
```

2. The Mouse Genomes Project hosts an extensive collection of inbred strain genomes (https://www.mousegenomes.org/) aligned to C57Bl/6J. The Broad Institute's recommended hard filters for DNA SNVs are QD < 2.0, QUAL < 30.0, SOR > 3.0, FS > 60.0, MQ < 40, MQRankSum < −12.5, and ReadPosRankSum < −8.0. The GATK VariantFiltration and VariantSelection tools can be used to create cross-specific ROD SNV files from these alignments using the --select-type-to-include SNP option.
3. We use hard filtered SNV loci in our ROD files to increase the efficiency of genotyping as well as the quality. Sometimes

Fig. 2 (continued) (ICM) to gastrulation. CT = cell type, XA = active X chromosome. RGC = radial glia cell, ImmN = immature neuron, Neur = neuron, GlutN = glutamatergic neuron, ImmA = immature astrocyte, Astro = astrocyte, oIPC = oligodendrocyte intermediate progenitor cell, Oligo = oligodendrocyte. B6 = B6 active X chromosome, NC = biparental active X chromosome/not called, CA = CA active X chromosome

multiple variants (GATK called "multiallelic") are detected at a particular loci. Predominantly these loci are near splice junctions and are read alignment errors. They will show up in VCF genotype fields as a value of 2 or higher. We remove these loci from further analysis, but they can be later inspected further as possible somatic mutations.

4. R function cnloh.gt(). Requires libraries Tidyverse and HMM.

```
#Genotyping cells from scRNAseq data. chrdata.GT <- cnloh.gt()
#Designed to take input data frame and calculate the most
#probable genotype state for
#any observed position, given an HMM model from the HMM
#package.
#chrdata = input data frame (e.g. Sample_chr6_cells)
#model = defined HMM model (e.g. LOH_HMM)
#returns a data frame with inferred genotype states for each
#cell.
cnloh.gt <- function(chrdata, model) {
  gt.df <- data.frame("POS" = chrdata$X)
  celltoget <- as.list(names(chrdata[2:ncol(chrdata)]))
  for (i in celltoget) {
  cell.df <- as.data.frame(chrdata[chrdata[,i] >= 3,c("X", i)])
  cell.df$GT <- viterbi(model,as.character(cell.df[,i]))
  gt.df[,i] <- cell.df[match(gt.df$POS, cell.df$X),3]
}
gt.df
}
```

5. R function get.gt.runs(). Requires libraries Tidyverse and HMM.

```
#Retrieving consecutive HMM inferred genotype run
#information chrdata_PosRunsTable
#<- get.gt.runs()
#Takes in data frame with inferred genotypes and returns
#any consecutive run of
#genotypes with corresponding positional information for
#each cell.
  get.gt.runs <- function(chrdata) {
  celltoget <- as.list(names(chrdata[2:ncol(chrdata)]))
  Cell <- c()
  Starts <- c()
  Ends <- c()
  State <- c()
  Length <- c()
  for (i in celltoget) {
    RunFrame <- as.data.frame(chrdata[chrdata[,i] %in% c
    ("B6","HT","CA"), c("POS", i)])
```

```
        runs <- rle(RunFrame[,i])
        myruns <- which(runs$values %in% c("B6","CA","HT"))
        #need for starts
        RunsSumIndex <- cumsum(runs$lengths)
        Ends <- c(Ends, RunFrame$POS[RunsSumIndex])
        startIndex <- ifelse(myruns > 1, myruns - 1, 0)
        c_starts <- RunFrame$POS[RunsSumIndex[startIndex] + 1]
        if (0 %in% startIndex) c_starts = c(RunFrame$POS[1], c_starts)
        Starts <- c(Starts, c_starts)
        Cell <- c(Cell, rep(i, length(runs$values)))
        State <- c(State, runs$values)
        Length <- c(Length, runs$lengths)
    }
    cellRuns <- data.frame("Cell"=Cell,"rStart"=Starts,"rEnd"
    =Ends,"rState"=State,"rLength"=Length)
    cellRuns
}
```

6. The theoretical zygote is simply a constructed cell (taxa) whose state is 0 (ancestral) for every character observed in the data set. It is added to the NEXUS file as a control. Under Camin-Sokal assumptions, it should appear as an outgroup with a very short branch length and root the tree. In phylogenetic terms, it is the most recent common ancestor of the organism from which the cells are sampled.
7. The MrBayes manual can be found at: https://github.com/NBISweden/MrBayes/blob/develop/doc/manual/Manual_MrBayes_v3.2.pdf
8. Both the Satija (https://satijalab.org/seurat/) and Trapnell (https://cole-trapnell-lab.github.io/monocle3/) labs have well-documented methods and recommendations for cell identification using scRNA-seq libraries, and we point the reader to these resources. Other R packages may prove useful. The phangorn package [45] function densiTree [46] allows visualization of multiple phylogenies sampled from a single MrBayes simulation. This is very useful for assessing multiple phylogenies with similar posterior probabilities.

Acknowledgments

Supported by the Paul G. Allen Frontiers Group (University of Washington). DJA wishes to dedicate this chapter to the memory of Donald R. Anderson.

References

1. Salipante SJ, Horwitz MS (2007) A phylogenetic approach to mapping cell fate. Curr Top Dev Biol 79:157–184
2. Sulston JE, Schierenberg E, White JG, Thomson JN (1983) The embryonic cell lineage of the nematode Caenorhabditis elegans. Dev Biol 100:64–119
3. McKenna A, Findlay GM, Gagnon JA, Horwitz MS, Schier AF, Shendure J (2016) Whole-organism lineage tracing by combinatorial and cumulative genome editing. Science 353:aaf7907
4. Frieda KL, Linton JM, Hormoz S, Choi J, Chow KK, Singer ZS, Budde MW, Elowitz MB, Cai L (2017) Synthetic recording and in situ readout of lineage information in single cells. Nature 541(7635):107–111. https://doi.org/10.1038/nature20777. Epub 2016 Nov 21. PMID: 27869821; PMCID: PMC6487260.
5. Carlson CA, Kas A, Kirkwood R, Hays LE, Preston BD, Salipante SJ, Horwitz MS (2011) Decoding cell lineage from acquired mutations using arbitrary deep sequencing. Nat Methods 9:78–80
6. Frumkin D, Wasserstrom A, Itzkovitz S, Harmelin A, Rechavi G, Shapiro E (2008) Amplification of multiple genomic loci from single cells isolated by laser micro-dissection of tissues. BMC Biotechnol 8:17
7. Ju YS, Martincorena I, Gerstung M, Petljak M, Alexandrov LB, Rahbari R, Wedge DC, Davies HR, Ramakrishna M, Fullam A et al (2017) Somatic mutations reveal asymmetric cellular dynamics in the early human embryo. Nature 543:714–718
8. Lee-Six H, Øbro NF, Shepherd MS, Grossmann S, Dawson K, Belmonte M, Osborne RJ, Huntly BJP, Martincorena I, Anderson E et al (2018) Population dynamics of normal human blood inferred from somatic mutations. Nature 561:473–478
9. Lodato MA, Woodworth MB, Lee S, Evrony GD, Mehta BK, Karger A, Lee S, Chittenden TW, D'Gama AM, Cai X et al (2015) Somatic mutation in single human neurons tracks developmental and transcriptional history. Science 350:94–98
10. Ludwig LS, Lareau CA, Ulirsch JC, Christian E, Muus C, Li LH, Pelka K, Ge W, Oren Y, Brack A et al (2019) Lineage tracing in humans enabled by mitochondrial mutations and single-cell genomics. Cell 176:1325–1339.e22
11. Salipante SJ, Horwitz MS (2006) Phylogenetic fate mapping. Proc Natl Acad Sci USA 103: 5448–5453
12. Spencer Chapman M, Ranzoni AM, Myers B, Williams N, Coorens THH, Mitchell E, Butler T, Dawson KJ, Hooks Y, Moore L et al (2021) Lineage tracing of human development through somatic mutations. Nature 595:85–90
13. Behjati S, Huch M, van Boxtel R, Karthaus W, Wedge DC, Tamuri AU, Martincorena I, Petljak M, Alexandrov LB, Gundem G et al (2014) Genome sequencing of normal cells reveals developmental lineages and mutational processes. Nature 513:422–425
14. Sabina J, Leamon JH (2015) Bias in whole genome amplification: causes and considerations. Methods Mol Biol 1347:15–41
15. Fasching L, Jang Y, Tomasi S, Schreiner J, Tomasini L, Brady MV, Bae T, Sarangi V, Vasmatzis N, Wang Y et al (2021) Early developmental asymmetries in cell lineage trees in living individuals. Science 371:1245–1248
16. Ding J, Lin C, Bar-Joseph Z (2019) Cell lineage inference from SNP and scRNA-Seq data. Nucleic Acids Res 47:e56
17. Ji JP, Loeb LA (1992) Fidelity of HIV-1 reverse transcriptase copying RNA in vitro. Biochemistry 31:954–958
18. Minoche AE, Dohm JC, Himmelbauer H (2011) Evaluation of genomic high-throughput sequencing data generated on Illumina HiSeq and genome analyzer systems. Genome Biol 12:R112
19. Fan J, Lee H-O, Lee S, Ryu D-E, Lee S, Xue C, Kim SJ, Kim K, Barkas N, Park PJ et al (2018) Linking transcriptional and genetic tumor heterogeneity through allele analysis of single-cell RNA-seq data. Genome Res 28:1217–1227
20. Serin Harmanci A, Harmanci AO, Zhou X (2020) CaSpER identifies and visualizes CNV events by integrative analysis of single-cell or bulk RNA-sequencing data. Nat Commun 11: 89. https://doi.org/10.1038/s41467-019-13779-x
21. Anderson DJ, Pauler FM, McKenna A, Shendure J, Hippenmeyer S, Horwitz MS (2022) Simultaneous brain cell type and lineage determined by scRNA-seq reveals stereotyped cortical development. Cell Syst 13(6): 438–453
22. LaFave MC, Sekelsky J (2009) Mitotic recombination: why? when? how? where? PLoS Genet 5:e1000411

23. Larson JS, Yin M, Fischer JM, Stringer SL, Stringer JR (2006) Expression and loss of alleles in cultured mouse embryonic fibroblasts and stem cells carrying allelic fluorescent protein genes. BMC Mol Biol 7:36
24. Moynahan ME, Jasin M (2010) Mitotic homologous recombination maintains genomic stability and suppresses tumorigenesis. Nat Rev Mol Cell Biol 11:196–207
25. Melcher R, Hartmann E, Zopf W, Herterich S, Wilke P, Müller L, Rosler E, Kudlich T, Al-Taie O, Rosenwald A et al (2011) LOH and copy neutral LOH (cnLOH) act as alternative mechanism in sporadic colorectal cancers with chromosomal and microsatellite instability. Carcinogenesis 32:636–642
26. Mohamedali A, Gäken J, Twine NA, Ingram W, Westwood N, Lea NC, Hayden J, Donaldson N, Aul C, Gattermann N et al (2007) Prevalence and prognostic significance of allelic imbalance by single-nucleotide polymorphism analysis in low-risk myelodysplastic syndromes. Blood 110:3365–3373
27. O'Keefe C, McDevitt MA, Maciejewski JP (2010) Copy neutral loss of heterozygosity: a novel chromosomal lesion in myeloid malignancies. Blood 115:2731–2739
28. Gong W, Granados AA, Hu J, Jones MG, Raz O, Salvador-Martínez I, Zhang H, Chow K-HK, Kwak I-Y, Retkute R et al (2021) Benchmarked approaches for reconstruction of in vitro cell lineages and in silico models of C. elegans and M. musculus developmental trees. Cell Syst 12:810–826.e4
29. Felsenstein J (2004) Inferring phylogenies. Sinauer Associates, Inc., Sunderland
30. Day WHE (1983) Computationally difficult parsimony problems in phylogenetic systematics. J Theor Biol 103:429–438
31. Ronquist F, Teslenko M, van der Mark P, Ayres DL, Darling A, Höhna S, Larget B, Liu L, Suchard MA, Huelsenbeck JP (2012) MrBayes 3.2: efficient Bayesian phylogenetic inference and model choice across a large model space. Syst Biol 61:539–542
32. Bouckaert R, Vaughan TG et al (2019) BEAST 2.5: an advanced software platform for Bayesian evolutionary analysis. PLoS Comput Biol 15(4):e1006650
33. Dobin A, Davis CA, Schlesinger F, Drenkow J, Zaleski C, Jha S, Batut P, Chaisson M, Gingeras TR (2013) STAR: ultrafast universal RNA-seq aligner. Bioinformatics 29:15–21
34. Keane TM, Goodstadt L, Danecek P, White MA, Wong K, Yalcin B, Heger A, Agam A, Slater G, Goodson M et al (2011) Mouse genomic variation and its effect on phenotypes and gene regulation. Nature 477:289–294
35. Wickham H, Averick M, Bryan J, Chang W, McGowan LD et al (2019) Welcome to the tidyverse. J Open Source Softw 4(43):1686. https://doi.org/10.21105/joss.01686
36. Himmelmann LM (2022) University of Applied Sciences Rapperswill, Switzerland. https://cran.r-project.org/web/packages/HMM/index.html
37. Stuart T, Butler A, Hoffman P, Hafemeister C, Papalexi E, Mauck WM 3rd, Hao Y, Stoeckius M, Smibert P, Satija R (2019) Comprehensive integration of single-cell data. Cell 177:1888–1902.e21
38. GNU, Pl (2007) Free Software Foundation. https://www.fsf.org
39. Ivica L, Peer B (2021) Interactive Tree Of Life (iTOL) v5: an online tool for phylogenetic tree display and annotation. Nucleic Acids Res 49 (W1):W293–W296
40. Cao J, Packer JS, Ramani V, Cusanovich DA, Huynh C, Daza R, Qiu X, Lee C, Furlan SN, Steemers FJ et al (2017) Comprehensive single-cell transcriptional profiling of a multicellular organism. Science 357:661–667
41. Borel C, Ferreira PG, Santoni F, Delaneau O, Fort A, Popadin KY, Garieri M, Falconnet E, Ribaux P, Guipponi M et al (2015) Biased allelic expression in human primary fibroblast single cells. Am J Hum Genet 96:70–80
42. Kluge AG, Faris JS (1969) Quantitative phyletics and the evolution of anurans. Syst Zool 18:1–32
43. Camin JH, Sokal RR (1965) A method for deducing branching sequences in phylogeny. Evolution 19:311–326
44. Barnett DW, Garrison EK, Quinlan AR, Strömberg MP, Marth GT (2011) BamTools: a C++ API and toolkit for analyzing and managing BAM files. Bioinformatics 27(12): 1691–1692
45. Schliep KP (2011) phangorn: phylogenetic analysis in R. Bioinformatics 27(4):592–593
46. Bouckaert RR (2010) DensiTree: making sense of sets of phylogenetic trees. Bioinformatics 26: 1372–1373

Chapter 2

Tracking Somatic Mutations for Lineage Reconstruction

Yaara Neumeier, Ofir Raz, Liming Tao, Zipora Marx, and Ehud Shapiro

Abstract

The human genome is composed of distinct genomic regions that are susceptible to various types of somatic mutations. Among these, Short Tandem Repeats (STRs) stand out as the most mutable genetic elements. STRs are short repetitive polymorphic sequences, predominantly situated within noncoding sectors of the genome. The intrinsic repetition characterizing these sequences makes them highly mutable in vivo. Consequently, this characteristic provides the chance to unravel the natural developmental history of human viable cells retrospectively. However, STRs also introduce stutter noise in vitro amplification, which makes their analysis challenging. Here we describe our integrated biochemical-computational platform for single-cell lineage analysis. It consists of a pipeline whose inputs are single cells and whose output is a lineage tree of input cells.

Key words Short tandem repeats, Lineage tree, Single cell, Human, Cancer, Retrospective

1 Introduction

A cell lineage tree is a binary tree with a single root, where nodes represent cells and edges' progeny relations. Additional annotations such as physical location, surface markers, or cell type can be added. Knowing what the human cell lineage tree holds would answer many open questions in biology and medicine. The structure, dynamics, and variability of the human cell lineage tree along development or during disease could deepen our understanding. For example, what is the origin of cancer metastasis? Is recurrence after chemotherapy derived from cancer cells that randomly evaded the treatment or perhaps from cancer "stem cells" with a slow enough proliferation rate to avoid therapy [1, 2]? Are beta cells/oocytes/neurons/heat muscle cells regenerate during adulthood [3]? These and other types of questions can be answered with knowledge of the human cell lineage tree.

Cutting-edge extant methodologies for cell lineage tracing are deemed impractical in the context of human applications due to their reliance on interventions, whether prospective or

Jorge García-Marqués and Tzumin Lee (eds.), *Lineage Tracing: Methods and Protocols*, Methods in Molecular Biology, vol. 2886, https://doi.org/10.1007/978-1-0716-4310-5_2,

retrospective [4, 5]. In contrast, exploration of human cell lineages can only rely on retrospective analysis of naturally occurring somatic mutations. To be effective, such analysis should be applied to highly mutable regions of the genome and may include single nucleotide variations (SNVs), copy number variations (CNVs), mitochondrial heteroplasmy, or slippage mutations in short tandem repeats (STRs).

STRs, also known historically as microsatellites, encompass short genomic sequences of a repeated unit consisting of 1–6 base pairs (bp). These repetitive units are arranged in tandem fashion and collectively account for approximately 3% of the entire human genome. Due to their repetitive composition, they are susceptible to polymerase slippage occurrences during DNA replication. This phenomenon subsequently engenders a proclivity for recurrent contractions or expansions within the repeated sequence. Consequently, the mutational rate of STRs surpasses other genomic variations by several orders of magnitude [6]. Given the heightened mutagenic activity observed in vivo, STRs have garnered notable attention within the expanding domain of lineage phylogenetic analysis, which seeks to delineate evolutionary relationships within an organism. In particular, a major interest in this field is the construction of lineage trees of cancer tumors [7].

Several aspects of the analysis of STR in general and specifically for lineage reconstruction make the computational process of STR analysis particularly challenging [8]. First, similar to the high mutation rate in vivo, STR undergoes noisy amplification in vitro, which forms an artifact often described as stutter noise (*see* Fig. 1). When depicting the STR length of the multiple reads per locus for each specific cell, a smeared and skewed signal is often observed.

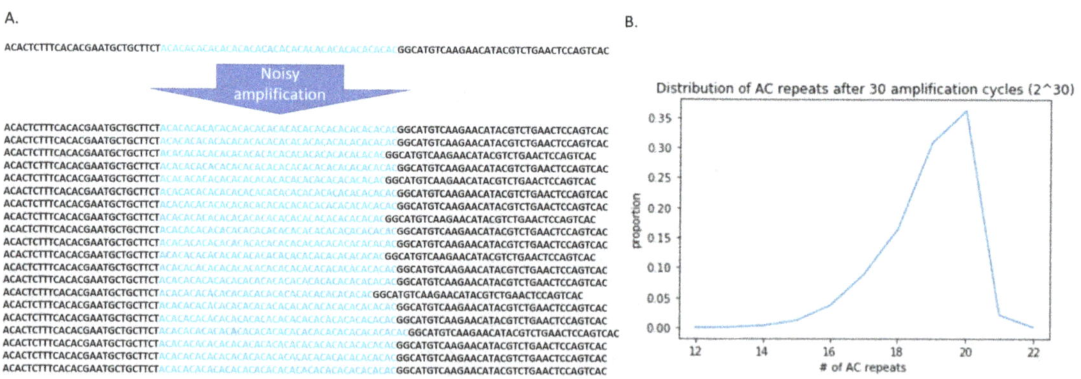

Fig. 1 STR stutter mutations during acquisition. Unlike point mutations, which affect only a single nucleotide, microsatellite mutations lead to the gain or loss of an entire repeat unit and sometimes two or more repeats simultaneously. (**a**) An STR of 20 AC originally accumulates mutations during PCR amplification. (**b**) A typical distribution of repeat numbers for a 20 AC STR following 30 PCR amplification cycles. Note that the *X* axis indicates length in units of "AC" whereas using an axis of single base units. The characteristic "stutter" would be observed as the case where mutations of less than a whole repeat unit are rare

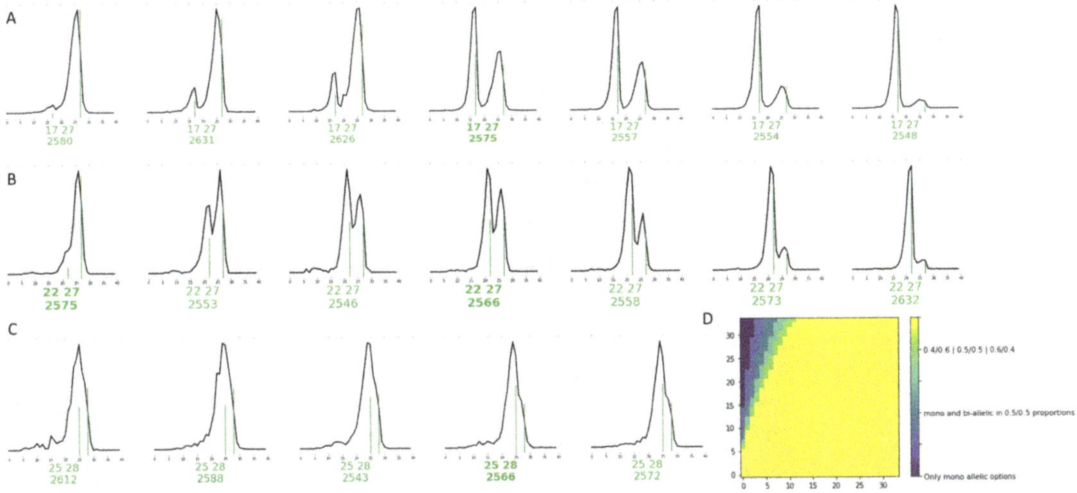

Fig. 2 Histogram of overlaid biallelic STRs. (**a**) Recognizing overlapping alleles spanning 17 and 27 AC repeats at various proportions. The green vertical bars indicate the genotyped alleles and their proportions attributed the signal prior to PCR amplification. (**b**) Examples of 22,27 AC histograms. (**c**) Examples of 25,28 AC histograms. (**d**) Genotyping grows more difficult and less accurate the closer the maternal and paternal alleles get and the longer the STR becomes. We therefore exclude combinations of close and long STR allelic pairs from consideration as depicted in the heatmap where the *x* axis indicates the distance between the alleles and *y* axis indicates the length of the smaller allele

Moreover, when genotyping non-homozygous loci, the number of reads of each allele varies significantly due to amplification imbalance. This is the case for autosomal chromosomes, X chromosomes in females, and copy number variation (CNV) cases (which are common in cancer patients that are of great interest in the context of cell lineage construction).

The combination of these two aspects poses a significant challenge. For instance, in the case where the histogram shows two peaks of similar height and adjacent values (*see* Fig. 2), several hypotheses should be considered:

1. The two peaks correctly represent the biallelic signal.
2. The true biallelic signal is of a single value (i.e., the length of the STR in both strands is identical) which is represented by one of the peaks, where the other peak is of stutter noise.
3. Due to imbalance amplification, one of the biallelic signals is lost and is not represented by either of the main peaks. The other signal is represented by one peak, and the other peak is of stutter noise.

Similarly, in the case where the histogram depicts one high peak surrounded by relatively small peaks (*see* Fig. 2c), several hypotheses should be considered:

1. The main peak correctly represents the biallelic signal representing two identical lengths.
2. The main peak correctly represents the biallelic signal representing two identical lengths. One of the biallelic signals is completely lost due to imbalance amplification, and the peak represents only one of the signals.
3. Due to imbalance amplification, one of the biallelic signals is relatively smaller than the higher peak.

To address this issue, the different aspects of Whole-genome amplification (WGA) should be considered. Several studies have addressed different aspects of amplification factors, which can be utilized to calibrate amplification factors and conditions [9–12]. Recently, in [13], different WGA kits were compared in the context of STR-targeted sequencing, concluding that the best overall yield is achieved using RepliG-SC. However, Ampli1 demonstrates the highest degree of allelic uniformity and cross-cell reproducibility, making it the most suitable choice for conducting comparative heterozygous analysis on single-cell genomes. As a result, when addressing the genotyping problem of the case in Fig. 2, the used WGA kit should be considered, and in particular, if Ampli1 is used, it is more probable that two identical signals (of STR length) were detected, while in the case of other kits, this option is less probable.

It was also demonstrated that the rate of STR mutations, both within living organisms (in vivo) and within laboratory settings (in vitro), is directly influenced by two primary factors: (a) the type of repeat unit, where shorter unit STRs (i.e., mono- and di-repeats) exhibit a higher propensity for mutability compared to longer unit types, and (b) the length of the STR, wherein STRs with an increased number of repeats display a greater degree of mutability when contrasted with their shorter counterparts [14].

Some contemporary tools primarily focus on mapping reads to the reference genome [9, 15, 16]. However, their algorithms for stutter error correction predominantly rely on calibrating statistical models. These models are constructed based on indirect measurements, encompassing STR distributions within progeny populations and/or user-defined datasets.

Other single-cell lineage reconstruction approaches are based on DNA mutations, including single-nucleotide polymorphism (SNP) [17, 18] and mitochondrial mutations [19], and single-cell RNA-seq [20] (*see* **Note 1**).

Other reconstruction methods are based on chromatin morphology [21], methylomes [22], proteome [23], as well as multimodal measurements of the same cells, for instance, SNP and RNA-Seq [24, 25]. However, the number of mutations that are represented using these methods is relatively limited compared to STR. Some overviews of different single-cell lineage approaches can be found in [25, 26].

In [8], we introduced a methodology for conducting controlled assessments of stutter noise phenomena throughout the amplification process, encompassing a range of STR types and lengths. Using these measurements, a mathematical model that accurately captures and predicts the stutter pattern of in vivo amplification was devised and calibrated. Based on the genotyping process, where the biallelic length of each STR loci is calculated, a phylogenetic phase can be utilized, where the biallelic data of STR length for a set of loci is used to reconstruct a cell lineage tree, uncovering the relations between the cells [27, 28].

As a validation of our proposed methodology, we replicated a documented lineage history of cells derived from a previously established benchmark tree of DU145 ex vivo [27]. In summary, a single cell (SC) originating from the DU145 human male prostate cancer cell line underwent clonal cultivation. Subsequently, SCs were systematically selected from preceding cultures and subjected to clonal cultivation, thereby generating an in vitro lineage tree spanning nine generations, each cultivated for 12–15 cell divisions. During each iteration, SCs were meticulously chosen using the CellCelector, annotated with their respective plate and well coordinates in the lineage tree, and whole-genome amplification (WGA) products were prepared for downstream analyses.

Figure 3a depicts the in vitro reference tree, colors assigned by shared ancestors labeled by sampling wells prefix, *y* axis indicated sampling generations. Figure 3b depicts the reconstructed lineage tree that is the output of our pipeline, with the same coloring scheme as Fig. 3a.

In this chapter, we describe our end-to-end lineage reconstruction framework, comprising both a biological pipeline and a computational pipeline. Using this framework, a cell lineage tree can be reconstructed from a set of single cells.

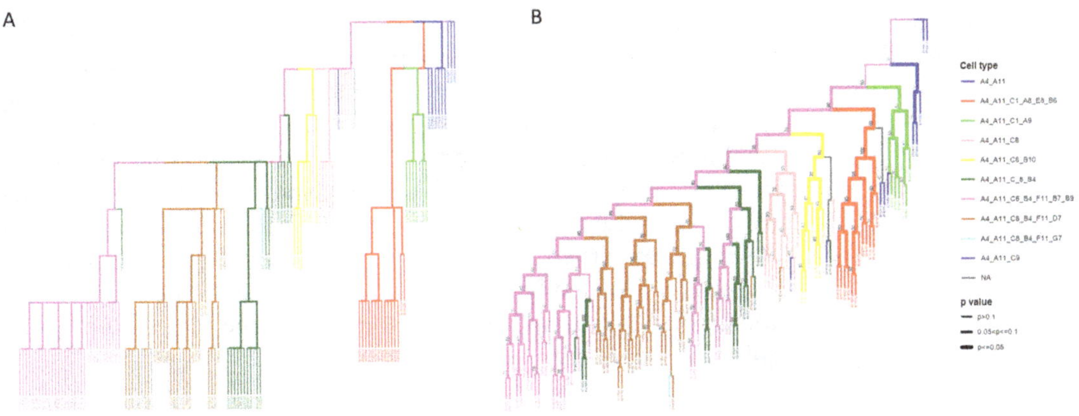

Fig. 3 Ex vivo reference tree and a corresponding Lineage reconstructed tree. (**a**) The in vitro reference tree, colors assigned by shared ancestors labeled by sampling wells prefix, *y* axis indicated sampling generations. (**b**) The reconstructed lineage tree that is the output of our pipeline, with the same coloring scheme as (**a**)

2 Materials

2.1 Reagents for Probe Preparation

1. Customized Oligo Pool; KOD DNA Polymerase; 10 mM dNTP; SYBR 100X.

2.2 Reagents for Library Construction

1. 5 M Betaine solution.
2. Phusion High-Fidelity DNA Polymerase.
3. Ampligase 10X Reaction Buffer.
4. Ampligase DNA Ligase.
5. Exonuclease I (*E. coli*).
6. Exonuclease III (*E. coli*).
7. RecJf Exonuclease; Exonuclease T.
8. T7 Exonuclease; Lambda Exonuclease.
9. NEBNext Ultra II Q5 MasterMix.
10. MinElute PCR Purification Kit.
11. Qubit® dsDNA HS Assay Kit.
12. Agencourt Ampure XP Beads; 2.
13. 2% Agarose, dye-free, BluePippin, 100–600.
14. TapeStation Screen Tape.
15. TapeStation Reagents.

2.3 Reagents for Sequencing

1. MiSeq Reagent Kits (300 cycle).
2. NextSeq 500/550 High Output Kit v2.5 (300 cycles).

3 Methods

3.1 An End-to-End Overview of Our Cell Lineage Pipeline

An overview of the biological pipeline is given in Fig. 4 [29]. Briefly, our STR targeting platform was developed using duplex Molecular inversion probes (MIPs, alternatively known as padlock probes) derived from precursors that were synthesized on a high-resolution microarray. Tens of thousands of duplex MIPs targeting highly mutable STRs are hybridized to single-cell whole-genome-amplified genomic DNA, gap-filled, ligated. Standard Illumina libraries are constructed by PCR with unique barcoded primers, then sequenced by either NextSeq or NovaSeq (*see* **Note 2**). The raw reads were then fed into our one-stop clineage pipeline [30] for bioinformatic analysis and phylogenetic tree evaluation, fine-tuning, and plotting.

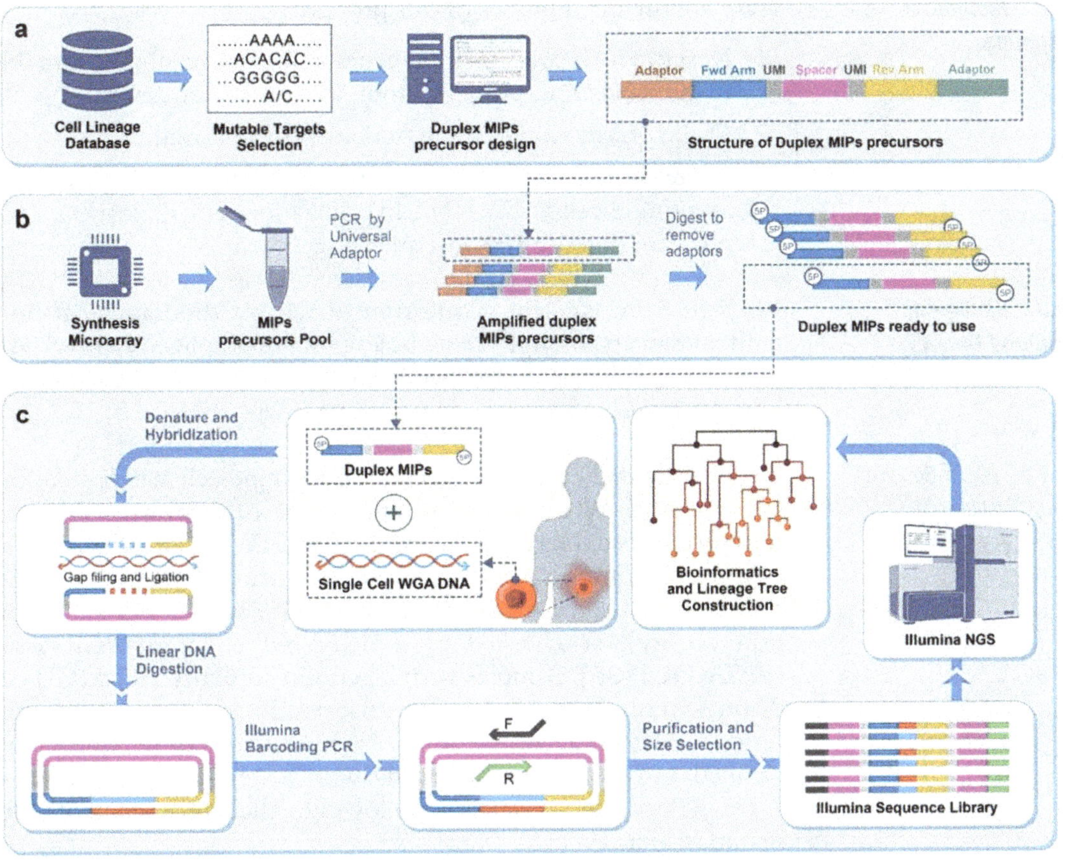

Fig. 4 (**a**) Duplex MIPs-based cell lineage workflow. Design of duplex MIPs precursor: MIPs have been employed as a high-throughput and cost-effective approach for cell lineage tracing. These MIPs are constituted by single-stranded DNA molecules, featuring two distant targeting arms tethered via an internal linker region. This design offers significant advantages, including the potential for enhanced multiplexing capabilities and superior specificity in contrast to conventional multiplex Polymerase Chain Reaction (PCR) methods. (**b**) Duplex MIPs preparation: The synthesis of duplex MIP precursors is initiated on a microarray platform, followed by their consolidation and subsequent amplification through Polymerase Chain Reaction (PCR). The resulting PCR product is subjected to digestion by MlyI enzyme to eliminate the universal adaptors. Following this enzymatic treatment, purification is undertaken, and the resultant material is appropriately diluted to yield active duplex MIPs, ready for subsequent utilization. (**c**) Gap filling and Illumina sequencing: MIPS and template DNA are mixed together to promote the annealing of the targeting arms (colored blue and yellow) to regions flanking the specific target sequence within the template DNA. Subsequently, the resulting products undergo circularization, a process facilitated by the action of DNA polymerase and ligase enzymes. To eliminate linear DNA, which includes surplus MIPs and template DNA, exonuclease digestion is performed. Subsequent to these preparations, an individual Illumina sequencing library is generated for each sample through a PCR step, which includes the incorporation of adaptors and barcodes. These libraries are subsequently pooled and subjected to sequencing using the Illumina next-generation sequencing platform. The acquired raw sequencing data are then subjected to detailed analysis to detect mutations, which are subsequently employed to infer the cell lineage tree

3.2 Biological Pipeline

Here are the main phases of this process:

1. Design of duplex MIPs precursor including highly mutable targets selection, primer design, and precursor design.
2. Probe production by PCR and adaptor removal.
3. Target enrichment library preparation through duplex MIPs and single-cell WGA DNA hybridization, circularization, exonuclease digestion, and sequencing.

3.3 Biological Method Details

Adapted from efficient acquisition of tens of thousands of short tandem repeats in single-cell whole-genome-amplified DNA STAR protocol [31] and Retrospective cell lineage reconstruction in humans by using short tandem repeats [32] (*see* **Note 3**).

3.3.1 Whole-Genome Amplification

We have evaluated several commercial single-cell whole-genome amplification (WGA) kits [13], and several of them work decently for our purpose. Here we discuss the Ampl1WGA kit, which works best for our lineage platform in our evaluation. WGA from single cells was performed following the manual [33, 34]. The amplification quality was checked by a dedicated endpoint PCR assay (WGA-QC) [35]. Samples with a genome integrity index (GII) of 3 and 4 were selected for lineage analyses. In addition, to eliminate residual oligo-nucleotides and reagents, sc WGA samples were purified with AMPure XP beads with a bead-to-DNA ratio of 1.8X (Beckman Coulter) following the protocol of the manufacturer.

3.3.2 Target-Specific Duplex MIPs Design and Preparation

STRs Loci Selection

Hypermutable STRs were chosen based on the reference human genome annotation. The selection criteria were the AC-and AG-type STRs should be longer than 10 repeats (*see* **Note 4**).

The A- and G-type STRs should be longer than six repeats; amplicons containing TTAA sequences were ruled out to comply with the design of the Ampli1 WGA kit; amplicons size were set as 100~200 bp.

Duplex MIPs Design

Primer3-based python script was used to design the targeting primers in parallel. Only top-scored primers were chosen as candidates for duplex MIPs precursor design. Additional filters were applied to the selected candidates: Precursors harboring MlyI restriction sites were ruled out; the total length of the precursors should not be longer than 150 bp, which was the limit of the oligonucleotide synthesis provider. Each duplex MIP comprises multiple functional elements in the following structure: 5′--[Mly1_F] + [FW_PRIMER_SEQ] + [UMI] + [Backbone] + [UMI] + [(RV_PRIMER_SEQ) in reverse-complement orientation] + [(Mly1_R) in reverse-complement orientation]-3′. The mentioned elements were designed as: Mly1_F: 5′GTCTAT GAGTGTGGAGTCGTTGC3′; Mly1_R: 5′CTAGCTTCCTGAT GAGTCCGATG 3′; FW_PRIMER_SEQ, RV_PRIMER_SEQ: variable primers designed to chosen targets; UMI: 5′-NNN-3′ (N stands for the randomized base over A, C, G, T); Backbone 5′- AGATCGGAAGAGCACACGTCTGAACTCTTTCCCTA CACGACGCTCTTCCGATCT-3′.

Duplex MIPs Precursor Synthesis by Microarray

The 10,000~100,000 duplex MIPs precursors are synthesized as a oligo pool. Upon receiving, the pool of oligos was stored at −80 °C or diluted to a final concentration of 1 ng/μL as templates for probe preparation.

Probe PreAmp PCR

PCR primers (OM4_Mly_F:5′ GTCTATGAGTGTG GAGTCGTTGC 3′; OM4_Mly_R:5′ CTAGCTTCCTGAT GAGTCCGATG 3′). Amplification of the precursor oligonucleotide pool was done in a 96 well-plate by PCR with the following reagents: 0.2 ng/μL template, 0.3 pmol/μL of each OM4_Mly_F and OM4_Mly_R primers, 1X KOD Polymerase MIX prepared according to manufacturers' instructions, with 0.5X SYBR green I. The reaction volume was 45 μL, typically 4–16 reactions per panel. The PCR was executed in the following program: 95 °C for 2 min (denaturation), 18 cycles of 95 °C for 20 s, 60 °C for 10 s, 70 °C for 5 s; final elongation at 70 °C for 5 min. Every four wells were pooled and purified by one MinElute PCR purification column, eluting in 45 μL DDW. Sample concentration was measured by Qubit dsDNA HS Assay Kit and further diluted to 1 ng/μL as a template for the next step.

Probe Production PCR

Production PCR reactions (final volume 45 μL) were done the same as in the PreAmp PCR, except reducing total amplification to 12 cycles, typically 32–48 reactions were done per panel. The same purification procedure was executed as the PreAmp step. The purified products were pooled in one tube, and the concentration was measured using the NanoDrop spectrophotometer. The pool of purified PCR products was further diluted to ~30 ng/μL.

Universal Adaptors Removal

MlyI digestion was carried out by mixing 84 μL of the diluted DNA (30 ng/μL), 10 μL of 10× CutSmart Buffer, and 6 U MlyI (final concentration of 0.6 U/μL MlyI) in a 100 μL reaction. Several reactions were performed, 20 μL of the undigested sample was saved for quality control. The reactions were incubated in a thermal cycler at 37 °C for 12 h, inactivated at 80 °C for 20 min, and finally kept at 4 °C. Subsequently, each reaction was purified using the MinElute PCR Purification Kit and eluted in 25 μL DDW.

Quality Control of the Duplex MIPs

All digested and purified products were pooled, and its concentration was measured using the Qubit dsDNA HS assay kit. The size of the digested product (~105 bp) and the undigested sample (~150 bp) was verified by TapeStation. If unwanted by-products (represented by minor peaks in the TapeStation) were detected in samples after digestion, BluePippin size selection method can be used for further refinement of the MIPs for desired size (~105 bp).

Preparation of Duplex MIPs Working Solution

Based on the average size of the duplex MIPs (105 bp) and measured concentration, the purified digestion product was diluted to a final concentration of 80 nM (80 fmol/μL) (equivalent to 5.8 ng/μL for ~10 K panels). This stock solution can be saved at −80 °C for the long term storage. The stock solution was further diluted 10 times as a working solution (8 nM), stored at −20 °C.

3.3.3 Duplex MIPs-based Targeted Enrichment Pipeline

Hybridization

Single-cell WGA product concentration is generally 100–200 ng/μ L. 2 μL of single-cell WGA DNA was mixed with 8 μL of hybridization in total volume of 10 μL. The plate was incubated in a thermal cycler with the following program: 100 °C lid temperature, 98 °C for 3 min, followed by a gradual decrease in temperature of 0.01 ° C/s to 56 °C and incubated at 56 °C for 17 h. Alternatively, PCR strategy can be applied: 100 °C lid temperature; denature at 98 °C for 3 min; decreased by 0.1 °C in a step-wise way; hold for 15 s for each step, until 56 °C then incubate at 56 °C for 17 h.

Gap Filling

Prepare the gap filling mix before the end of the hybridization, with a final concentration of: 0.3 mM each dNTP, 2 mM NAD (freshly thawed from −80 °C), 1.1 M betaine, 1X Ampligase buffer, 0.5 U/ μL Ampligase and 0.8 U/μL Phusion® High-Fidelity DNA Polymerase,10 μL mix per reaction, kept at 56 °C in a heat block. After the hybridization, the plate was moved from the PCR machine to a 56 °C heat block. 10 μL gap filling mix was added to each well, mixed by pipette, sealed tightly, and quickly returned to the PCR machine. Incubate at 56 °C for 4 h, followed by 68 °C for 20 min, and hold at 4 °C until the next step. Optionally, the gap-filled plate can be stored at 4 °C for up to 2 days.

Digestion of Linear Template DNA and Probes

Prepare 2 μL/reaction digestion mix with final concentration of: 3.5 U/μL Exonuclease I, 18 U/μL Exonuclease III, 4 U/μL T7 Exonuclease, 0.4 U/μL Exonuclease T, 3 U/μL RecJf, and 0.2 U/μL Lambda Exonuclease. Add 2 μL per well and mix well. The reaction plate was sealed, spun down, and incubated at 37 °C for 60 min, 80 °C for 10 min, and 95 °C for 5 min. The reactions can be stored at −20 °C for months.

3.3.4 Sequencing Library Preparation

Sample-Specific Barcoding PCR

Sample-specific barcoding PCR was performed using dual-index barcoding primers and NEBNext Ultra II Q5 DNA Polymerase. The structure of the dual-index Illumina barcoding primers was designed based onTrueSeq HT primers.I5-index-primer: AATGATACGGCGACCACCGAGATCTACAC [i5-8bp-index] ACACTCTTTCCCTACACGACGCTCTTCCG ; I7-index-primer: CAAGCAGAAGACGGCATACGAGAT [i7-8bp-index] GTGACTGGAGTTCAGACGTGTGCTCTTCCG ; 2 μL of the target enriched product was used as template, 0.5 pmol/mL (final concentration) of each unique barcoding pair of dual-index Illumina primers for each sample, 1X NEBNext Ultra II Q5 Master Mix; 0.5X SYBER Green in a 20 mL PCR reaction. The PCR was performed in a Roche 480 LightCycler using the following program: 98 °C, 30 s (denaturation), 5 cycles of 98 °C for 10 s, 56 °C for 30 s, and 65 °C for 45 s; 17 cycles of 98 °C for 10 s, 65 °C for 75 s; followed by 65 °C for 5 min (elongation); cool down (*see* **Note 5**).

Purification and Pooling of Barcoding PCR Samples

The barcoding PCR product was purified by adding 0.8X (by volume) AMPure XP SPRI magnetic beads (Beckman Coulter) according to the manufacturer's manual and eluted in 38 mL DDW. The cleaned PCR products were transferred by a robot (Bravo, Agilent) to a 384-well plate, and an equal volume of each of them pooled by Echo (Echo550, Labcyte). The pool was concentrated by the MinElute column (Qiagen) to 35 μL.

Size Selection for Diagnostic Sequencing

The concentrated pool (30 mL) was loaded on a 2% V1 cassette for BluePippin (Sage Science) size selection with a setting range of 240–340 bp according to the manufacturer's protocol, while 3 mL of the concentrated pool was kept for later quality control. The size-selected elution was collected, cleaned by MinElute, and eluted with 15 mL DDW. The concentration was measured by the Qubit dsDNA HS assay kit. The size distribution of the concentrated pool before and after BluePippin was checked by TapeStation dsDNA screen tape. The size-selected pool with a single peak around 300 bp was used to prepare 12 mL of 4 nM (4 fmol/mL) ready-to-run library for Illumina NGS.

Diagnostic Sequencing

The ready-to-run library was sequenced on MiSeq Nano (MS-103-1001, Illumina) or Micro flow cell (MS-103-1002, Illumina) with 151×2 paired-end run parameters according to the manufacturer's manual, the default sequencing primers were used.

Production Sequencing

Based on the diagnostic sequencing results, a new pool was produced by Echo550 with a volume adjustment for each sample in order to equalize the reads. The pool was concentrated by MiniElute to 35 μL and used to prepare a sequencing library as in the step "size selection for diagnostic sequencing." The normalized library was sequenced on NextSeq 500 (FC–404-2004 Illumina) flow cell with 151×2 paired-end reads, and 8×2 indexes with default sequencing primers according to the manufacturer's manual.

3.4 Computational Pipeline Details

The computational pipeline uses the measurements of STR lengths for different cells of the same individual to reconstruct the cell lineage tree, based on somatic mutations of these STR loci.

3.4.1 Determining STR Repeat Number for Each Sample and Loci

Reads are first processed using PEAR [34] to facilitate merging of R1 and R2 and obtain consensus sequence and combined quality score (*see* Fig. 5). As the NGS progresses along each read, the quality tends to diminish. This effect is especially pronounced when reading through the STR part of the read.

The successfully merged reads are mapped against an index of STR variations, consisting of all the targeted panel's amplicons according to their sequence in the reference genome and padded by N characters on both sides. Each amplicon's reference sequence is repeated in the index with variations in the number of repeat units that make up the STR parts of the amplicon (*see* Fig. 6). Each row can be thought of as a "chromosome," labeled by an amplicon ID and the repeat numbers attributed to the STRs. The mapping itself is done using bwa-mem [35], favored for its performance but can be swapped to bowtie2 or other mappers. The result of the mapping process attributes STR lengths to the reads via the chromosome to which they were mapped.

An all-encompassing implementation of the mapping and genotyping methodologies detailed herein, from the initial stages to the culmination with the R&B genotyper, is accessible at [36].

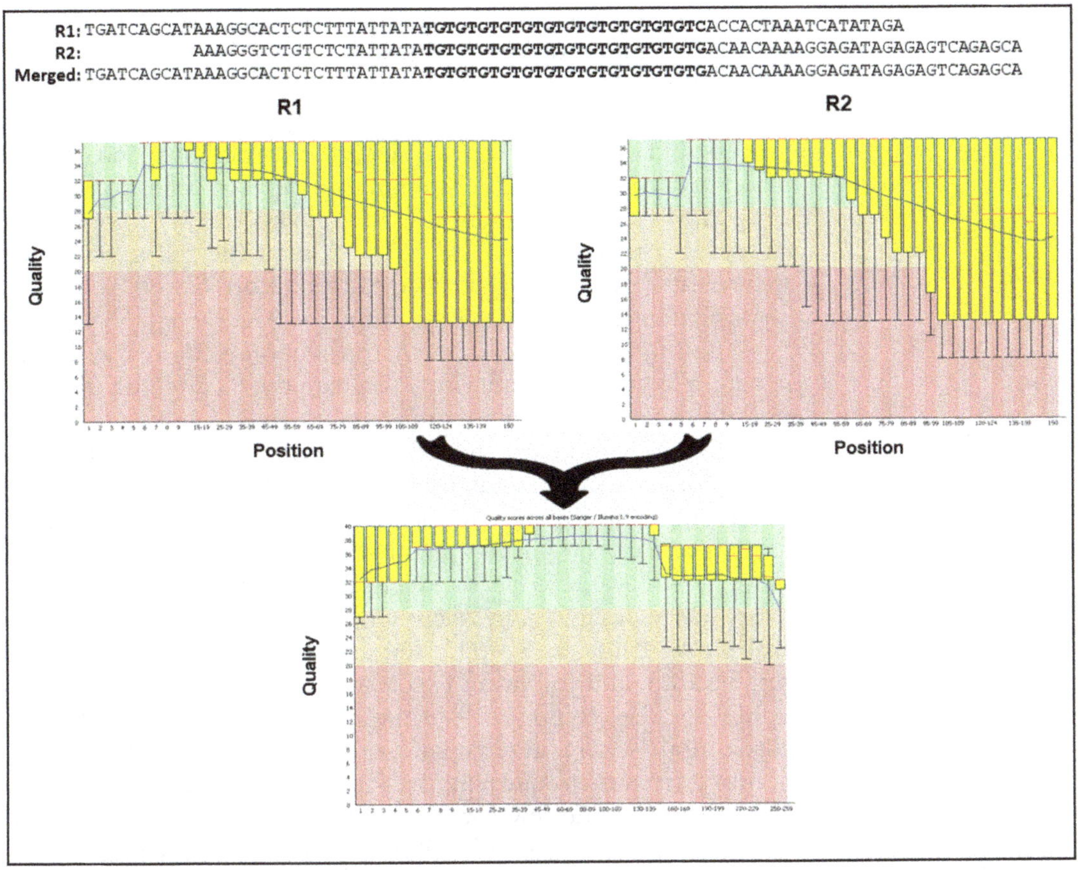

Fig. 5 Paired-end quality score enhancement. Example of the quality score of each of the paired-end reads and the resulting merge (using FastQC)

ATTTACTATGCCATGCTG CTG TCTGGCATGTGCAAACAAGAACAGATGCC
ATTTACTATGCCATGCTG CTG CTG TCTGGCATGTGCAAACAAGAACAGATGCC
ATTTACTATGCCATGCTG CTG CTG CTG TCTGGCATGTGCAAACAAGAACAGATGCC
ATTTACTATGCCATGCTG CTG CTG CTG CTG TCTGGCATGTGCAAACAAGAACAGATGCC
ATTTACTATGCCATGCTG CTG CTG CTG CTG CTG TCTGGCATGTGCAAACAAGAACAGATGCC
ATTTACTATGCCATGCTG CTG CTG CTG CTG CTG CTG TCTGGCATGTGCAAACAAGAACAGATGCC
ATTTACTATGCCATGCTG CTG CTG CTG CTG CTG CTG CTG TCTGGCATGTGCAAACAAGAACAGATGCC
ATTTACTATGCCATGCTG CTG CTG CTG CTG CTG CTG CTG CTG TCTGGCATGTGCAAACAAGAACAGATGCC
ATTTACTATGCCATGCTG CTG CTG CTG CTG CTG CTG CTG CTG CTG TCTGGCATGTGCAAACAAGAACAGATGCC
ATTTACTATGCCATGCTG CTG CTG CTG CTG CTG CTG CTG CTG CTG CTG TCTGGCATGTGCAAACAAGAACAGATGCC
ATTTACTATGCCATGCTG CTG CTG CTG CTG CTG CTG CTG CTG CTG CTG CTG TCTGGCATGTGCAAACAAGAACAGATGCC
ATTTACTATGCCATGCTG CTG CTG CTG CTG CTG CTG CTG CTG CTG CTG CTG CTG TCTGGCATGTGCAAACAAGAACAGATGCC
ATTTACTATGCCATGCTG CTG CTG CTG CTG CTG CTG CTG CTG CTG CTG CTG CTG CTG TCTGGCATGTGCAAACAAGAACAGATGCC
ATTTACTATGCCATGCTG CTG CTG CTG CTG CTG CTG CTG CTG CTG CTG CTG CTG CTG CTG TCTGGCATGTGCAAACAAGAACAGATGCC

Fig. 6 Customized STR variations mapping index. A custom "reference genome" is used, containing only the amplicons targeted by the panel, each expanded with all possible variants of the STR sequence within it (in red)

3.4.2 Signal Processing

This phase is where each histogram of the samples is compared to the Markov model of typical STR stutter noise [8]. The comparison is performed using KDtree algorithm which is efficient in terms of runtime (*see* **Note 6**).

To obtain the model histograms, we have designed and ordered a library of plasmids, each containing a unique combination of STR type and length, spanning all naturally occurring mono- and di-repeats (namely: A, C, AC, AG, AT) in the full spectrum of their natural genomic occurrence (*see* also **Note 6**).

The implementation can be found in [37].

The single step transition matrix, denoted M, is a stochastic matrix, allowing us to efficiently simulate the STR length distribution resulting from n amplification cycles as M_n.

The chosen mutation model and its parameters are the result of an exploratory experiment charting the behavior of all naturally occurring mono and di-repeat STRs under common in vitro amplification scenarios [8]. Briefly, synthetic amplicons bearing representative STR repeat unit types and length were synthesized in a ready-to-sequence design. Those were sequenced following three amplification scenarios, as-is, following a single round of typical PCR amplification and following two rounds of typical PCR amplification. Candidate step models including existing literature and simplistic models were all parameter-tuned to fit the stutter patterns resulting from the sequencing scenarios of the library and a model with a good fit and few parameters was chosen, 1up3dw linear as shown in Fig. 7. This step is followed by naive allelotyping strategy where it was assumed that the retrived signal indeed consists of both alleles (*see* **Notes 7– 9**).

3.4.3 Lineage Tree Reconstruction

The tree reconstruction is performed using Fasttree2 algorithm [38].

Originally, FastTree2 is designed to infer approximately-maximum-likelihood phylogenetic trees from aligned sequences of nucleotides or amino acids. We employ fasttree2 for its scalable and accurate reconstruction, by translating STR repeat numbers to arbitrary amino acids and concatenating the panel's targeted STRs to a sequence-like input. Numerous alternative reconstruction methods are available for this step (*see* **Note 10**).

3.4.4 Bootstrapping

Given a lineage tree, it is of interest to evaluate this tree, and in particular, its stability. That is, it is of interest to assess that minor changes in the data do not result in major changes in the resulting tree. For this purpose, the bootstrap methodology can be used [36].

Bootstrap is a statistical test that uses random sampling to mimic the sampling process. Sampling techniques can be with or without replacement. Bootstrap is done with replacement, that is, in our case, each loci can be sampled multiple times.

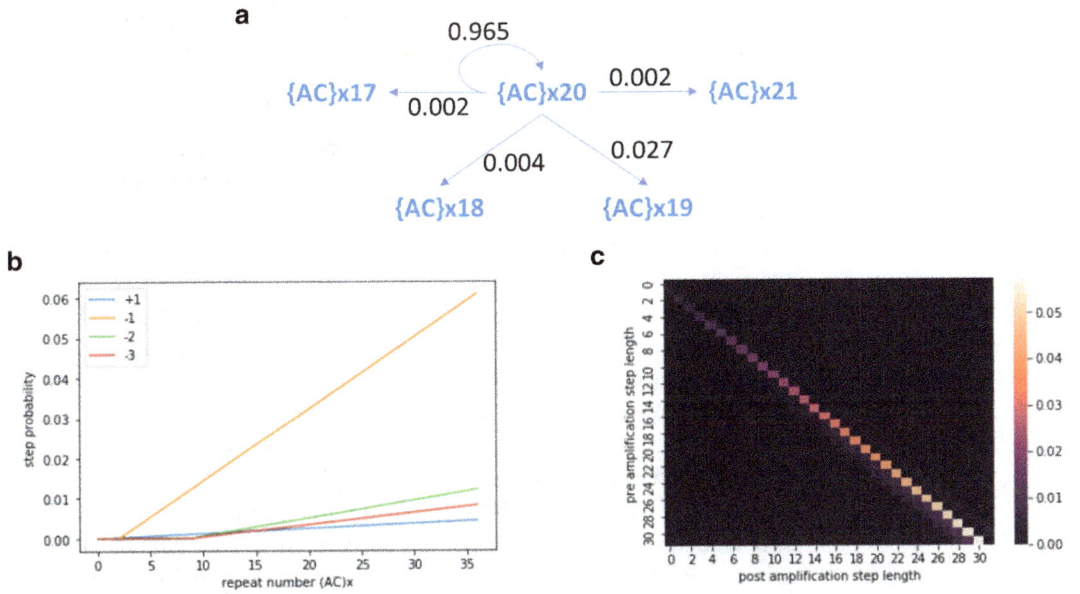

Fig. 7 "AC" STR mutation probabilities during amplification (in vitro). (**a**) The slippage mutation probabilities of a typical AC STR, 20 repeats long. The mutations allowed for an individual amplification cycle are +1, −1, −2, and −3. (**b**) The mutation probability tradeoff with the number of STR repeats, shown for each of the allowed mutation steps. (**c**) Heatmap depiction of (**b**)

Split reproducibility is a critical facet of the bootstrap methodology, emphasizing the capacity to replicate the results through the division of the original dataset into distinct subsets. This technique involves partitioning the data multiple times, followed by the application of the bootstrap resampling procedure to each subset independently. The aim is to assess the robustness and reliability of the derived statistical estimates across various partitions. The concept of split reproducibility ensures that the findings obtained from bootstrap resampling remain consistent and replicable, even when the dataset is subjected to different grouping or segmentation strategies.

In the realm of cell lineage analysis, bootstrap resampling provides a means to estimate the stability of inferred lineage relationships. By resampling subsets of observed data with replacement, the bootstrap generates multiple replicate datasets, enabling the assessment of the stability of lineage tree structures.

Specifically, for lineage trees, the sampling is done such that n loci are sampled out of the n loci in the full data, but with replacement. Then, for each such partition, the whole computational pipeline is recalculated. Finally, the Transfer Bootstrap Expectation (TBE) score is calculated per node in the tree. It is common to assume that TBE >70% is stable [39]. The TBE can be calculated using the dedicated tool from [40].

3.4.5 Evaluating Clustering Quality

Another aspect of evaluating the resulting tree is quantifying the quality of the cluster separation. This kind of evaluation is based on the availability of some biological classifications of the samples. For in vivo trees, it could be the type of cells, whereas for ex vivo trees (for instance, *see* Fig. 3), it can be based on the hierarchical structure of the real structure of the tree. We used the Hypergeometric score (also termed p-value of Fisher exact test) to test for statistically significant clustering of a predefined group of cells on the reconstructed lineage tree.

We compute a hypergeometric score for each internal node within the tree through a defined procedure [41]. In our experimental context, we classify N cells, wherein s cells pertain to group A, and the remaining $(n - s)$ cells are allocated to the complementary group B. For each internal node established within the reconstructed lineage tree, we apply a hypergeometric test to evaluate the null hypothesis positing no association between the sub-tree and the aforementioned classification. Specifically, considering a sub-tree of cells, the associated branch's p-value signifies the probability of observing an equal or greater number of cells of a particular type, assuming a random sampling process from the pool of N cells.

$$p = f(n, N, s, x) = \frac{\binom{s}{N}\binom{N-s}{x-s}}{\binom{N}{x}}$$

3.4.6 Tree Plotting

The reconstructed lineage tree (with or without bootstrap values), which is stored as a standard NEWICK file, can be depicted using any dedicated tool. Specifically, we developed the LineagePlotter platform in R. It allows us to color the branches of the tree according to predefined classifications and to display the hypergeometric p-value score of the branches as the width of each branch. Bootstrap values can be displayed over the trees as well as additional information that can be displayed over the trees in various ways. For instance, the lineage tree in Fig. 3 was created using LineagePlotter. LineagePlotter is available at [42].

4 Notes

1. The potential to do multiomics with single-cell lineage tracing.
   (a) Several methods have been published to get both mRNA and single-cell genomic DNA from the same cell (G&T: [43], scTrio-seq: [44], etc.).
   (b) Several commercial single-cell multiomics kits are also available (Ampli1WGA plus Ampli1WTA).

(c) For the single-cell mRNA profiling, we can use Smart-Seq2 for full-length coverage, or other cheaper methods like MARS-seq2.0: [45] and SCRB-seq: [46].

(d) Our lineage analysis pipeline is python based, it has a jupyter notebook interface to customize the analysis and take advantage of single-cell RNA analysis packages like scanpy [47] in the same computational environment. We can annotate single-cell lineage trees and cell state labels, SNV mutations, patients' clinical information, etc.

2. The timeline of a typical duplex MIPs experiment

(a) The duplex MIPs are usually prepared on a large scale, prepared once, and can last for thousands of reactions.

(b) Whole-genome amplification of single cells is done in advance, and the amplified product can be easily stored, and transferred safely.

(c) The wet lab part from hybridization to sequencing can take days. However, the hands-on time is not that long. We start the hybridization reaction in the late afternoon, and let it run overnight. The next morning, we prepare the gap-filling mix, and add the mix to each well. It takes 4 h to fill the gap. We set the temperature as 4 °C to hold after gap-filling in the PCR program. We can continue the digestion step in the afternoon or save the plate in the freezer. It is a safe pause point. The digestion step takes 1 h, then the barcoding library PCR takes 2 h, pool and size selection takes 2 h. We can safely stop at any step after digestion. Miseq sequencing can be done on site or sent for service. Normalization of different samples after diagnosis Miseq is critical to get uniform sequencing coverage across samples. We can do it manually for one or two plates, for large-scale experiments, liquid handlers are highly recommended.

3. Panel size and sequencing coverage

(a) The panel size of STR loci depends on the project. For example, Melanoma is known to be highly mutable, 10K STR loci can do a decent job for lineage reconstruction. But for Leukemia or healthy samples, we will need 100K STR loci, or even more to get sufficient power.

(b) We can also include probes that target patient-specific SNV, INDEL in the sample duplex MIPs reactions.

(c) Two or more different suplex MIPs panels can be mixed by ratio to cover more targets, it is pretty flexible.

(d) We need at least five reads per locus to get accurate genotyping. If we need more reads for certain samples, we do another sequencing to gain more reads. Sequencing reads from two runs can be aggregated for the lineage analysis.

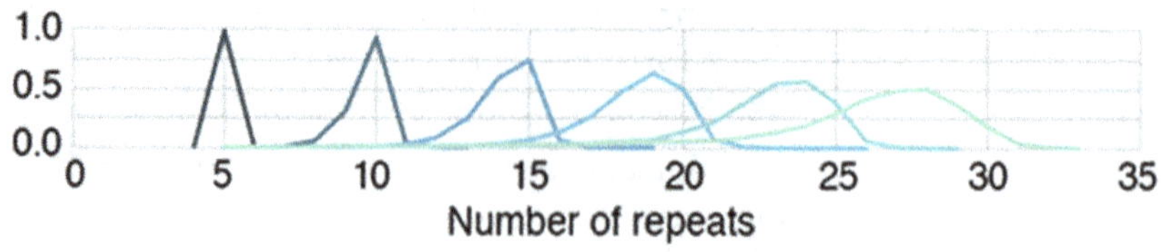

Fig. 8 Stutter patterns resulting from a library of synthetic amplicons bearing a range of AC repeats following extensive PCR amplification

4. Types of hypermutable STR targets and their adverse effects

   To broadly chart one's considerations when planning which STR loci to target, maximizing sensitivity via somatic mutation rates is usually the primary goal. However, three issues have to be considered:

   (a) Stutter noise tends to increase with in vivo mutability, requiring more reads to accurately genotype (*see* Fig. 8).

   (b) Highly mutable STRs tend to be selected against PCR amplification, maintaining good panel-wide coverage requires balanced planning.

   (c) Allelotyping of biallelic signals is affected both by the germline variance of STR lengths (maintaining distinct maternal/paternal alleles) and by the stutter distortions that can create overlaps and occlusions between the two allelic signals.

5. Throughput and cost

   (a) For target enrichment throughput, panel size can be increased to 100K or even larger, thanks to the high multiplexity property of MIPs reactions.

   (b) For cell throughput, with liquid handlers, hundreds to thousands of cells can be handled. Millions of cells are out of reach even with the assistance of liquid handlers, where split-pool approaches like [48].

   (c) The wet lab cost is already pretty low since we use all in-house protocols, and the reaction volume can be further reduced. Now we do 5 or 10 μL reactions. With a nano dispenser, we can do 1 μL or less.

   (d) The sequencing costs are project dependent. A lower mutation rate means more loci, which translates to more sequencing. The good news is more and more sequencers are available other than Illumina sequencers. The sequencing costs are getting lower with these competitions. We have tested our libraries on BGI sequencer and Avidity sequencer, and all working decently.

6. About the signal processing phase:

This phase in our computational pipeline addresses the key issue of stutter noise. STR undergoes noisy amplification in vitro as well as in vivo replication slippage. The resulting stutter noise can be observed as a histogram of lengths (for instance, *see* Fig. 1). A key challenge is thus to dig the cases where replication slippage occurred (i.e., the length of the STR truly changed) out of all the noisy signals.

In tackling the stutter issue, conventional noise models, exemplified by the highest peak analysis, are frequently utilized for genotyping PCR-free NGS libraries or for slowly evolving STR loci characterized by repeat units of three bases or more. However, these straightforward models may not be applicable to highly polymorphic STRs, notably mono- and di-repeats, particularly in instances where substantial amplification occurs. Employing such models in such scenarios is susceptible to yielding inaccuracies in genotyping, thereby leading to erroneous results.

In particular, the shape of the histogram that represents the different read lengths does not take the form of a normal distribution, which can be seen in the example in Fig. 1. First, it tends to be asymmetrical, where erases are more likely to happen. Moreover, the particular histogram depends on the repeat unit, as well as the repeat number.

To address this issue, our signal processing is uniquely based on controlled amplification noise measurement of a synthetic STR library spanning all naturally occurring mono- and di-repeats (namely: A, C, AC, AG, AT) in the full spectrum of their natural genomic occurrence and [8].

The set of all the one-allelic model histograms can be described as

$$S_{\mathrm{H}} = \{H(u, l, c)\}_{u \in \{\mathrm{A,C,AC,AG,AT}\},\ l \in \{3,\ldots,50\},\ c \in \{1,\ldots,50\}}$$

where u represents the repeat unit, l stands for the repeat length, and c represents the amplification cycles number.

Given the resulting set of model histograms and a histogram that was achieved from a biological experiment, we need to match the detected histogram to a model histogram which is the best match. In [8], many metrics were tested, and it was shown that the norm "Linear1up3dw" is the most suitable. An advantage in terms of computational complexity in runtime is that this search can be computed using Kd-tree.

This model addresses the case of a one-allelic signal. To address the bi-allelic signal, we build the set of model histograms such that a subset of bi-allelic models is formed for each possible pair of one-allelic models. The bi-allelic histogram model follows the following definition:

$$S'_{H} = \{H_1, H_2, p\}_{H_1,\ H_2 \in\ S_H,\ p\ \in\ \{0,0.1,0.2,\ldots,1\}} = \{pH_1 + (1-p)\,H_2\}$$

Therefore, for the biallelic case, we have a much bigger set of histogram models. This case can still be handled in a similar form to the one-allelic case, by searching for the model histogram that is the closest to the detected model. However, future work may address this case differently, considering that on the one hand, several model histograms can be very similar to each other, making it hard to distinguish. However, on the other hand, multiple histogram models are related to the same (H_1, H_2) pair.

7. Allelotyping

   A common scenario—interpreting genotyped STR lengths for a single locus, across a population of single cells derived from a single individual, where there is no additional information that can assist in discerning the maternal and paternal identities. WGA typically adds additional noise in the form of imbalance amplification of the two alleles. Ambiguity in allelic identities should therefore be kept in mind during downstream analysis. Adaptive distance metrics, iterative phylogenetic reconstruction, allelotyping and parameterized confidence intervals are some possible ways to approach the issue. However, it is often the case that random allelotyping "noise," e.g., attributing the difference between the maternal and paternal alleles to mutations between cells, is handled well by phylogenetic reconstruction algorithms. The resulting tree presents longer-than-expected edges between the leaves and their immediate ancestors, but the remainder of the tree's edges and its overall topology are approximately correct. One might then only note the allelotyping issues when recalculating the edge length of the tree, a much easier computational problem [49].

8. Sex differences in allelotyping

   One trivial special case of allelotyping is the single X chromosome in males, similar to the Y chromosome but with a negligible fraction of the genome. One can therefore focus on male patients and design target only X chromosome STRs to altogether avoid allelotyping issues. It could also be employed for self-checks and calibrations when compared to females or other chromosomes within the same experimental system.

9. Genetic diversity (distance between parents)

   Allelotyping is almost trivial when for loci where the maternal and paternal alleles differ in their STR length by more than the margins that can form by somatic mutations and stutter noise. This often correlates to parents from more distant genetic backgrounds. Such a background can also improve the chances of identifying a neighboring SNV on the STR-bearing amplicon. Phased sequencing of SNV and STR loci within the same amplicon easily resolves the allelotyping issue.

10. Phylogenetic algorithms for tree reconstruction: The problem of constructing a lineage tree is close to the problem of constructing a phylogenetic tree. The main difference is that phylogenetic trees describe the evolution over the course of generations of organisms, while lineage trees describe the ancestral relationships within a particular organism. As a result, phylogenetic trees can be constructed based on somatic mutations that are relatively slow such as nucleotide or protein sequences, whereas lineage requires much more mutable data, such as STR. However, any known phylogenetic algorithm can be adapted for lineage by mapping the STR repeat numbers to arbitrary amino acids. Here, we used Fastree2 for the reconstruction, which is considered to yield good results in a reasonable rum time. However, other algorithms can be considered, as well.

    The phylogenetic reconstruction of evolutionary history from molecular data commonly employs four primary phylogenetic methods, each based on three optimization criteria. These methods include Neighbor-Joining (NJ), Minimum Evolution (ME), Maximum Parsimony (MP), and Maximum Likelihood (ML). These techniques serve as analytical frameworks for inferring the evolutionary relationships among biological entities, such as species or genes, by utilizing genetic sequence information. The Neighbor-Joining method constructs a tree by iteratively joining the most similar pairs of taxa. Minimum Evolution seeks the tree that requires the least total evolutionary change. Maximum Parsimony aims to find the tree that requires the fewest evolutionary events. Maximum Likelihood estimates the tree that maximizes the probability of the observed molecular data given a particular model of evolution. Each of these methods provides distinct advantages and considerations, and their application depends on the characteristics of the data and the specific goals of the evolutionary analysis [50]. Specifically, Fastree2 is an approximate maximum likelihood phylogenetic algorithm (AML), which has the capacity to approximate the Maximum Likelihood with better time complexity.

References

1. Blagosklonny MV (2006) Target for cancer therapy: proliferating cells or stem cells. Leukemia 20(3):385–391
2. Shlush L, Chapal-Ilani N, Adar R et al (2012) Cell lineage analysis of acute leukemia relapse uncovers the role of replication-rate heterogeneity and microsatellite instability. Blood 120(3):603–612
3. Reizel Y, Itzkovitz S, Adar R et al (2012) Cell lineage analysis of the mammalian female germline. PLoS Genet 8(2):e1002477

4. Michlits G, Hubmann M, Szu-Hsien W et al (2017) CRISPR-UMI: single-cell lineage tracing of pooled CRISPR–Cas9 screens. Nat Methods 14(12):1191–1197
5. Salvador-Martínez I, Grillo M, Michalis A et al (2019) Is it possible to reconstruct an accurate cell lineage using CRISPR recorders? elife 8: e40292
6. Gymrek M (2017) A genomic view of short tandem repeats. Curr Opin Genet Dev 44:9–16
7. Quintanal-Villalonga Á, Joseph MC, Helena A et al (2020) Lineage plasticity in cancer: a shared pathway of therapeutic resistance. Nat Rev Clin Oncol 17(6):360–371
8. Raz O, Biezuner T, Spiro A et al (2019) Short tandem repeat stutter model inferred from direct measurement of in vitro stutter noise. Nucleic Acids Res 47(5):2436–2445
9. Fungtammasan A, Guruprasad A, Suzanne EH (2015) Accurate typing of short tandem repeats from genome-wide sequencing data and its applications. Genome Res 25(5): 736–749
10. Byrd C, Ohtsuka MW, Khorana HG (1965) Synthetic deoxyribo-oligonucleotides as templates for the DNA polymerase of Escherichia coli: new DNA-like polymers containing repeating nucleotide sequences. Proc Natl Acad Sci USA 53(1):79–86
11. Shinde D, Yinglei L, Fengzhu S et al (2003) Taq DNA polymerase slippage mutation rates measured by PCR and quasi-likelihood analysis: (CA/GT)n and (A/T)n microsatellites. Nucleic Acids Res 31(3):974–980
12. Hite JM, Eckert KA, Cheng KC (1996) Factors affecting fidelity of DNA synthesis during PCR amplification of d(CA)n·d(GT)n microsatellite repeats. Nucleic Acids Res 24(12):2429–2434
13. Raz O, Liming T, Biezuner T et al (2022) Whole-genome amplification—surveying yield, reproducibility, and heterozygous balance, reported by STR-targeting MIPs. Int J Mol Sci 23(11):6161
14. Ellegren H (2004) Microsatellites: simple sequences with complex evolution. Nat Rev Genet 5(6):435–445
15. Gymrek M, Golan D, Rosset S et al (2012) lobSTR: a short tandem repeat profiler for personal genomes. Genome Res 22(6): 1154–1162
16. Highnam G, Franck C, Martin A et al (2013) Accurate human microsatellite genotypes from high-throughput resequencing data using informed error profiles. Nucleic Acids Res 41(1):e32–e32
17. Budowle B, Van Daal A (2008) Forensically relevant SNP classes. Biotechniques 44(5): 603–610
18. Monteford J, Bowers TW, Iacob P et al (2021) What's a SNP between friends: the lineage of Clostridioides difficile R20291 can affect research outcomes. Anaerobe 71:102422
19. Evans SM, Yelon D, Conlon FL et al (2010) Myocardial lineage development. Circ Res 107(12):1428–1444
20. Klattenhoff CA, Scheuermann JC, Surface LE et al (2013) Braveheart, a long noncoding RNA required for cardiovascular lineage commitment. Cell 152(3):570–583
21. Ulicna K, Vignal G, Courtin C et al (2021) Automated deep lineage tree analysis using a Bayesian single-cell tracking approach. Front Comput Sci 3:734559
22. Karemaker ID, Vermeulen M (2018) Single-cell DNA methylation profiling: technologies and biological applications. Trends Biotechnol 36(9):952–965
23. Plum T, Wang X, Ronnsch M et al (2020) Human mast cell proteome reveals unique lineage, putative functions, and structural basis for cell ablation. Immunity 52(2):404–416
24. Ding J, Li C, Bar-Joseph Z (2019) Cell lineage inference from SNP and scRNA-Seq data. Nucleic Acids Res 47(10):e56
25. Stuart T, Satija R (2019) Integrative single-cell analysis. Nat Rev Genet 20(5):257–272
26. Wagner DE, Klein AM (2020) Lineage tracing meets single-cell omics: opportunities and challenges. Nat Rev Genet 21(7):410–427
27. Biezuner T, Spiro A, Raz O et al (2016) A generic, cost-effective, and scalable cell lineage analysis platform. Genome Res 26(11): 1588–1599
28. Liming T, Raz O, Marx Z et al (2017) A duplex MIPs-based biological-computational cell lineage discovery platform. bioRxiv:191–296. https://doi.org/10.1101/191296
29. Liming T, Raz O, Marx Z et al (2021) Retrospective cell lineage reconstruction in humans by using short tandem repeats. Cell Rep Methods 1(3):100054
30. https://github.com/shapirolab/clineage
31. Liming T, Marx Z, Raz O et al (2021) Efficient acquisition of tens of thousands of short tandem repeats in single-cell whole-genome-amplified DNA. STAR Protoc 2(4):100828
32. Klein CA, Schmidt-Kittler O, Schardt JA et al (1999) Comparative genomic hybridization, loss of heterozygosity, and DNA sequence analysis of single cells. Proc Natl Acad Sci USA 96(8):4494–4499

33. Polzer B, Medoro G, Pasch S (2014) Molecular profiling of single circulating tumor cells with diagnostic intention. EMBO Mol Med 6(11):1371–1386. https://doi.org/10.15252/emmm.201404033
34. Zhang J, Kobert K, Flouri T et al (2014) PEAR: a fast and accurate Illumina Paired-End reAd mergeR. Bioinformatics 30(5):614–620
35. Li H (2013) Aligning sequence reads, clone sequences and assembly contigs with BWA-MEM. arXiv preprint arXiv:1303.3997
36. Berry V, Gascuel O (1996) On the interpretation of bootstrap trees: appropriate threshold of clade selection and induced gain. Mol Biol Evol 13(7):999–1011
37. https://github.com/ofirr/clineage/tree/standalone
38. Price MN, Dehal PS, Arkin AP (2010) FastTree 2–approximately maximum-likelihood trees for large alignments. PLoS One 5(3): e9490
39. Lemoine F, Domelevo Entfellner JB, Wilkinson E (2018) Renewing Felsenstein's phylogenetic bootstrap in the era of big data. Nature 556(7702):452–456
40. https://booster.pasteur.fr/
41. Spiro A, Shapiro E (2016) Accuracy of answers to cell lineage questions depends on single-cell genomics data quality and quantity. PLoS Comput Biol 12(6):e1004983
42. https://github.com/yaaraneumeier/LineagePlotter/release
43. Macaulay IC, Haerty W, Kumar P et al (2015) G&T-seq: parallel sequencing of single-cell genomes and transcriptomes. Nat Methods 12(6):519–522
44. Hou Y, Guo H, Cao C et al (2016) Single-cell triple omics sequencing reveals genetic, epigenetic, and transcriptomic heterogeneity in hepatocellular carcinomas. Cell Res 26(3): 304–319
45. Keren-Shaul H, Kenigsberg E, Jaitin DA et al (2019) MARS-seq2.0: an experimental and analytical pipeline for indexed sorting combined with single-cell RNA sequencing. Nat Protoc 14(6):1841–1862
46. Bagnoli JW, Ziegenhain C, Janjic A et al (2018) Sensitive and powerful single-cell RNA sequencing using mcSCRB-seq. Nat Commun 9(1):2937
47. Wolf FA, Angerer P, Theis FJ (2018) SCANPY: large-scale single-cell gene expression data analysis. Genome Biol 19(1):15
48. Mulqueen RM, Pokholok D, O'Laughlin BJ et al (2021) High-content single-cell combinatorial indexing. Nat Biotechnol 39(12): 1574–1580
49. Sankoff D (1975) Minimal mutation trees of sequences. SIAM J Appl Math 28(1):35–42
50. Rosenberg MS, Kumar S (2001) Traditional phylogenetic reconstruction methods reconstruct shallow and deep evolutionary relationships equally well. Mol Biol Evol 18(9):1823–1827

Chapter 3

Multicolor Cell Lineage Tracing Using MAGIC Markers Strategies

Laura Dumas, Jason Durand, and Karine Loulier

Abstract

Multicolor MAGIC Markers strategies are useful lineage tracing tools to study brain development at a multicellular scale. In this chapter, we describe an in utero electroporation method to simultaneously label multiple neighboring progenitors and their respective progeny using these multicolor reporters. In utero electroporation enables the introduction of any gene of interest into embryonic neural progenitors lining the brain ventricles through a simple pipeline consisting of a micro-injection followed by the application of electrical pulses. Successful in utero electroporation requires a concise yet complete understanding of each step of the surgical protocol, spanning from the preoperative preparation to the postoperative care, as well as the MAGIC Markers tool outlined in this study. Besides a detailed protocol, we present non-integrative and integrative approaches to demonstrate the range of cell and lineage tracking possibilities of multicolored progenitors and their descent over time.

Key words In utero electroporation, Neural progenitor, Lineage, Episome, Transposase, Multicolor, Development

1 Introduction

In utero electroporation (IUE) is a powerful technique for introducing genetic material into progenitor cells in living organisms, without the tedious crossing of multiple transgenic mice. In this chapter, we describe a step-by-step methodology for tracing multiple lineages during cerebral cortex development using color combinatorial MAGIC Markers. Our IUE protocol allows the expression of MAGIC Markers transgenes in pallial progenitors that are lining the dorsal side of the embryonic lateral ventricles and are known to give rise to both excitatory neurons and glia. In particular, we describe non-integrative and integrative approaches to label embryonic progenitors and their progeny either restricted to pyramidal neurons (non-integrative strategy) or including both neuronal and glial descents (integrative strategy), respectively. The non-integrative approach relies on the transient expression of

Jorge García-Marqués and Tzumin Lee (eds.), *Lineage Tracing: Methods and Protocols*, Methods in Molecular Biology, vol. 2886, https://doi.org/10.1007/978-1-0716-4310-5_3,

episomal DNA carried out by plasmid vectors that will be diluted after successive division of the progenitor. It can be used to either mark or transiently perturb a targeted step of cortical neurogenesis but will only recapitulate a short temporal window of the lineage of a neural progenitor. On the other hand, the integrative approach is based on the stable integration of transgenes into the genome of the electroporated progenitor using integrase enzymes such as PiggyBac or Tol2 transposases. As the integrative approach allows to recapitulate the entire descent of progenitors over multiple rounds of cell divisions, it can be used to assess the potentiality of these progenitors to give rise to neurons and glia from embryonic to adult stages. We illustrate the principle and a representative result for each approach.

2 Materials

All laboratory supplies, solutions, and surgical instruments used for IUE must be sterile. All experiments involving animals must be conducted in accordance with institutional and governmental ethics. Animal protocols have been approved by the Languedoc Roussillon animal experimentation ethical board (CEEACD/N° 36). Successful completion of courses in rodent handling and restraint, intraperitoneal and subcutaneous injections, as well as rodent anesthesia and experimental surgery training is required prior to any manipulation and survival surgery on mice.

2.1 Laboratory Supplies

1. 1mm diameter glass capillary purchased, or micro-glass needle pulled using a micropipette puller.
2. Absorbable suture (suture size: 4/0, suture length: 45 cm, needle length: 19 mm, needle shape: 3/8, needle type: circle reverse cutting).
3. Plastic-coated bench coat.
4. Laboratory tape.
5. Sterile 35mm diameter Petri dish.
6. Electrode gel (Free-Med 15–60).
7. Sterile compress.
8. Alcohol pad.
9. Individual sterile transfer pipette.
10. Surgical coat, gloves, mask.
11. 1-mL insulin syringes.
12. Preoperative clean cage.
13. Postoperative clean cage with accessible food and a simple lying tissue to assess the nesting ability of mice after surgery, as a

read-out of the well-being of the pregnant mice, with extra nesting materials once the mouse has fully recovered from surgery.

14. Pen and paper to keep track of electroporated embryos during the experiment.
15. Anesthesia and analgesia monitoring booklet.

2.2 Solutions

1. Pre-warmed sterilized saline solution at 0.9 g/L (w/v): 9 g NaCl in 1 L distilled water.
2. Filtered fast green solution at 0.01% (w/v): 0.1 g fast green in 10 mL phosphate buffered saline (without Ca^{2+}/Mg^{2+}).
3. Endotoxin-free DNA plasmid mix with 0.01% fast green solution (*see* Subheading 3.1 for details).
   (a) PB-CAG-Cytbow plasmid (Addgene ID 158995).
   (b) Tol2-CAG-Nucbow plasmid (Addgene ID 158992).
   (c) CAG-SeCre plasmid.
   (d) CAG-PBase plasmid.
   (e) CAG-Tol2ase plasmid.
4. Freshly prepared anesthetic cocktail: 5 mg/mL Ketamine and 1 mg/mL Xylazine in sterile saline solution.
5. Freshly prepared analgesic solution: 0.03 mg/mL of Buprenorphine in sterile saline solution.
6. Iodine solution (10% solution).
7. Ocular gel.

2.3 Dissecting Instruments

1. Arched tip forceps, 10 cm.
2. Straight Graefe forceps.
3. Curved Graefe forceps.
4. Adson forceps.
5. Straight iris scissors.
6. Olsen-Hegar needle holder.

2.4 Equipment

1. Electroporator (Sonidel, NEPA 21).
2. Tweezers electrode (3-mm, Sonidel, CUY650P3).
3. Micro-injector controlled by foot pedal (INJECT+MATIC).
4. Fiber optic light source.
5. Bead glass sterilizer.
6. Warming pads.
7. 10 μL micropipette and corresponding tip.

8. Hair clipper.
9. Water bath at +40 ± 2 °C to keep the saline solution warm.

2.5 Animal Model

IUE is usually performed during mouse development between the embryonic day (E)12.5 and E15.5. The vaginal plug of the timed mated pregnant mouse is detected the morning following a breeding night and defined as E0.5. This experimental procedure can be carried out on all laboratory mice regardless of their inbred or outbred genetic background.

3 Methods

3.1 Preparation of DNA Plasmids Mix (Total DNA Quantity: Between 8 and 12 μg for the Entire IUE Procedure per Pregnant Female Depending on the Number of Embryos per Litter)

The various DNA plasmids (MAGIC Markers constructs, SeCre and PBase/Tol2ase plasmid vectors) have been prepared using an endotoxin-free protocol to ensure high electroporation efficiency and absence of toxicity. Up to 1 mg of endotoxin-free plasmid can be purified from a maxi preparation using commercially available kit. Each DNA concentration is measured three times on a Nanodrop spectrophotometer, averaged and kept at −20 °C until being used.

The multi-addressable genome-integrative color (MAGIC) Markers strategy is used to simultaneously label multiple neural progenitors with distinct combinatorial markers using IUE (Fig. 1, [1]). The color code generated by the combinatorial expression of fluorescent proteins in both the cytoplasm and nucleus of electroporated progenitors and their descents is used as a reliable marker for the tracking of their incomplete (non-integrative approach) or complete (integrative approach) progeny from the day of the IUE (Figs. 1 and 2).

In the non-integrative approach (Fig. 2A), three plasmids are simultaneously electroporated into neural progenitors. These plasmid vectors include cytoplasmic (PB-CAG-Cytbow) and nuclear (Tol2-CAG-Nucbow) multicolor reporter constructs and a self-excising Cre recombinase construct [1]. MAGIC Markers multicolor reporters consist of four genes encoding fluorescent proteins with distinct spectral features (EBFP, tdTomato/mCherry, EYFP, mCerulean/mTurquoise2), flanked by three pairs of incompatible lox sites (loxP, lox2272, and loxN). The self-excising Cre recombinase plasmid (SeCre) subjects itself to the excision of its coding sequence therefore minimizing both the intensity and duration of its activity. Using SeCre plasmids and the short-lived action of this recombinase decreases the probability of Cre recombinase binding identical pairs of lox sites located several copies apart. Minimizing Cre recombinase action is critical to prevent the excision of entire MAGIC Markers transgenes, that might reduce the numbers of MAGIC Markers copies and subsequently impoverish the color

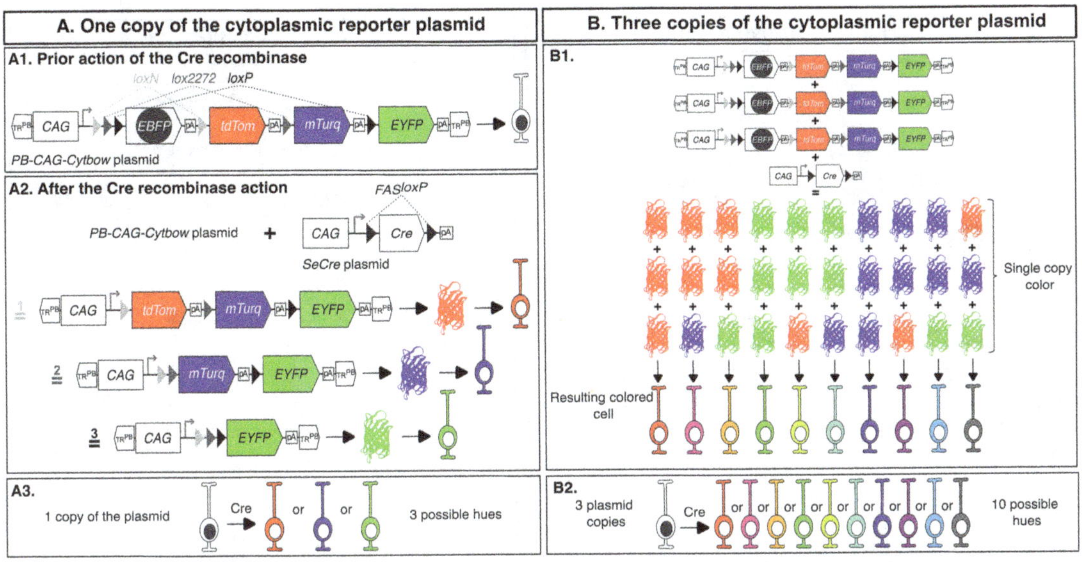

Fig. 1 Principle of the MAGIC Markers labeling. Comparison of the potential color diversity expressed by neural progenitors after electroporation of one (**A**) or three (**B**) MAGIC Markers cytoplasmic constructs along with the SeCre plasmid. Prior to Cre recombinase action, the first gene after the *CAG* promoter is expressed: electroporated progenitors exhibit EBFP-positive nuclei by default (**A1**). The recombinase randomly recognizes compatible pairs of *lox* sites (either two *loxN* (light gray) or *lox2272* (middle gray) or *loxP* (black) sites) and excises the DNA fragment flanked by these sites (**A2**). When electroporated cells express only one copy of the cytoplasmic reporter plasmid, they exhibit either tdTomato, mTurquoise2, or EYFP-positive cytoplasm (**A2**–**A3**). When multiple copies of the cytoplasmic MAGIC Markers constructs have been electroporated into cells, the Cre recombinase rearranges each copy independently, resulting in the expression of various combinations of fluorescent proteins and a diverse palette of colors (**B1**). Three copies of cytoplasmic MAGIC Markers constructs result in ten possible color hues (**B2**). The large variety in color shades helps differentiate between adjacent neural progenitors and their progeny, for example, by producing different nuances of blue (blue, light blue, and cyan) or red (red, magenta, and orange). Of note, fluorescent proteins are coded with false colors in the scheme (EBFP in gray, tdTomato in red, mTurquoise 2 in blue, EYFP in green) to help with the interpretation of combinatorial colors. *Abbreviations used*: CAG: composite promoter composed of the fusion of CMV (cytomegalovirus promoter) enhancer, chicken ß actin promoter, and rabbit ß globin splice acceptor site; Cre: recombinase; EBFP: EBFP2; mTurq: mTurquoise2; pA: polyadenylation stop signal; SeCre: self-excising Cre recombinase; tdTom: tdTomato; TR^{PB}: terminal repeats recognized by piggyBac transposase

palette at our disposal. Prior to Cre recombinase action, the first gene downstream from the ubiquitous CAG promoter is expressed and therefore successfully electroporated progenitors exhibit EBFP-positive nuclei. When a cell expresses only one copy of the plasmid: the Cre recombinase has three random possibilities of recombination resulting in the exclusive expression of either tdTomato/mCherry, or EYFP, or mTurquoise2/mCerulean within cytoplasm/nucleus. However, the electroporation typically introduces several copies of the MAGIC Markers plasmid into targeted cells [1]. Each copy of the plasmid is independently rearranged by the Cre recombinase, leading to the combinatorial expression of

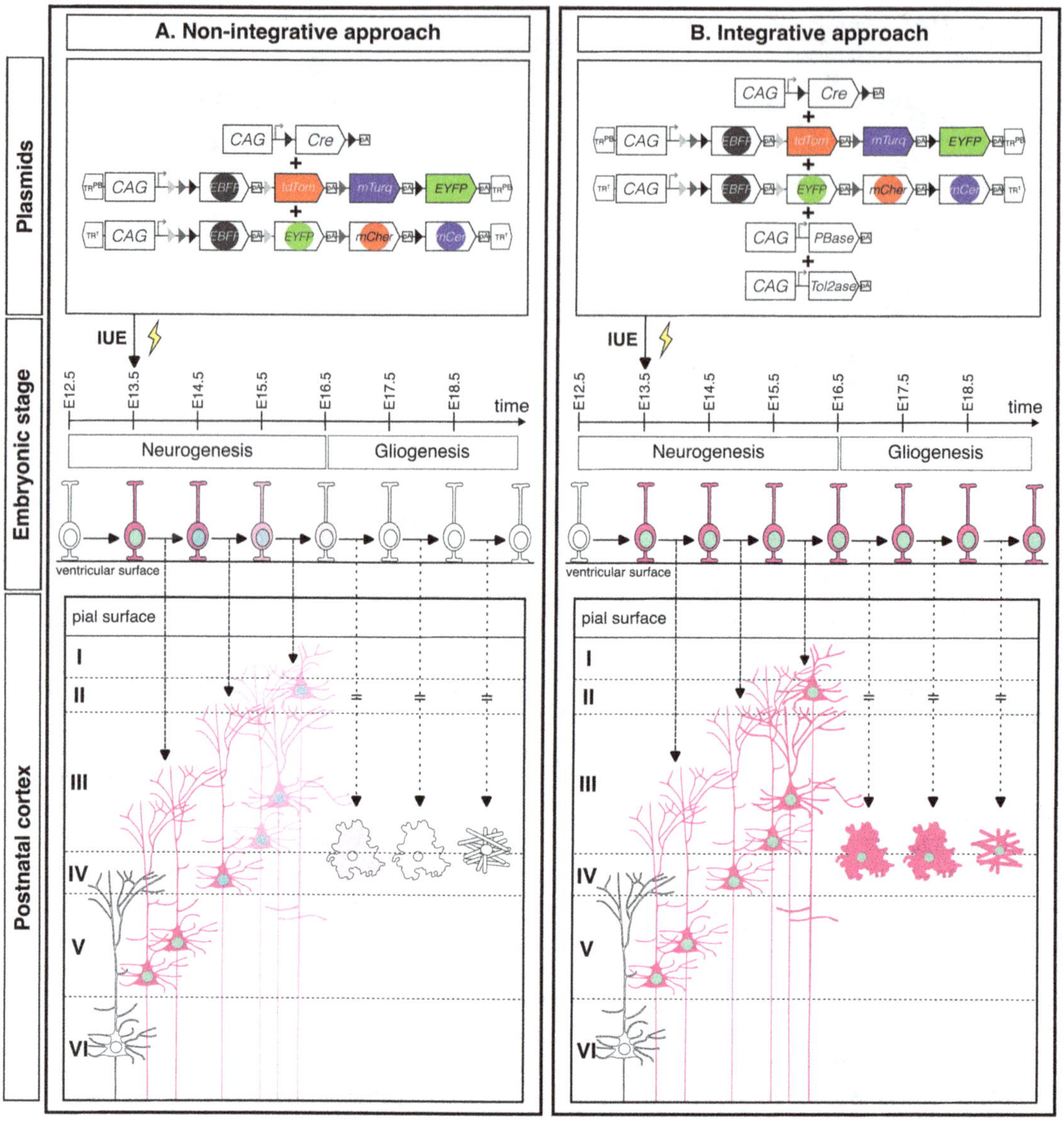

Fig. 2 Non-integrative (**A**) and integrative (**B**) approaches for tracing the lineage of pallial progenitors electroporated with MAGIC Markers at E13.5. The diagram illustrates the progressive dilution of the multicolor reporter over the successive divisions of the labeled embryonic pallial progenitor in the non-integrative approach ((**A**) *middle panel*), compared to a maintained multicolor reporter observed with the integrative approach ((**B**) *middle panel*). In the non-integrative approach ((**A**) *bottom panel*), the multicolor reporter produced by co-injecting the SeCre, PB-CAG-Cytbow and Tol2-CAG-Nucbow plasmids is gradually lost over time: late-born pyramidal neurons of the upper cortical layers (II, III) show a faded color compared to the deep pyramidal neurons (cortical layers V and IV) born at the time of the electroporation. After co-electroporation of the SeCre, PB-CAG-Cytbow, Tol2-CAG-Nucbow, CAG-PBase, and CAG-Tol2ase plasmids (integrative approach, (**B**) *bottom panel*), upper pyramidal neurons, astrocytes, oligodendrocytes are still expressing the MAGIC Markers. *Abbreviations used*: CAG: composite promoter composed of the fusion of CMV (cytomegalovirus promoter) enhancer, chicken ß actin promoter and rabbit ß globin splice acceptor site; Cre: recombinase; E: embryonic day; EBFP: EBFP2; mCer: mCerulean; IUE: in utero electroporation; mCher: mCherry; mTurq: mTurquoise2; pA: polyadenylation stop signal; SeCre: self-excising Cre recombinase; tdTom: tdTomato; TR^{PB}: terminal repeats recognized by piggyBac transposase, TR^{T}: terminal repeats recognized by Tol2 transposase

Table 1
Final concentration and ratio used to prepare the DNA mix solution for the IUE

	Non-integrative approach	Integrative approach
DNA mix solution	Final concentration (μg/μL)	Final concentration (μg/μL)
PB-CAG-Cytbow	0.8	0.8
Tol2-CAG-Nucbow	0.8	0.8
CAG-SeCre	0.16	0.16
CAG-PBase	0	0.4
CAG-Tol2ase	0	0.4

multiple fluorescent proteins in both cellular compartments, also known as a color code.

In the integrative approach (Fig. 2B), two additional plasmids encoding PiggyBac (PB) and Tol2 transposases are added to the mixture described above. These two transposases insert the cytoplasmic and nuclear MAGIC Markers sequences located between specific transposition feet, called transposons, directly into the genome of the targeted cells. It ensures the labeling of the entire progeny of electroporated neural progenitors through identical replication of the targeted-cell genome, including integrated MAGIC Markers sequences, by the cellular machinery during successive cell divisions. Without transposase and stable integration of the foreign DNA into the cell genome, the MAGIC Markers-expressing plasmid vectors, remaining in the cytoplasm as episomes, are progressively diluted after each division until being undetectable (non-integrative approach).

The principle of MAGIC Markers strategies for the non-integrative and integrative approaches is explained in Figs. 1 and 2.

A total volume of 20 μL of DNA mix is typically prepared for the IUE of a single pregnant female.

For the electroporation, we use a 2:1 MAGIC Markers plasmid and corresponding transposase ratio. As previously mentioned, a restricted Cre recombinase activity is essential to obtain an extensive color palette. It also implies a reduced concentration of SeCre plasmid with a ratio of 1:9 to the total MAGIC Markers plasmid concentration (*see* Table 1).

Dilute the DNA mix solution at a 9:1 ratio with 0.01% fast green solution for injection.

3.2 Preparation of the Pre-operating and Operating Areas for the Surgery

1. Isolate the time mated pregnant mouse before the surgery from the grouped female cage. If imported from another animal facility, the pregnant females should be acclimatized for at least 1 week before surgery.

Table 2
Transfer pulse parameters to electroporate C57BL/6 or SWISS embryos

Mouse strain	Voltage (V)	Duration (ms)	Interval (ms)	Pulse number	Pulse decay (%)	Polarity
C57BL/6	30	50	950	3	10	+
SWISS	35	50	950	4	10	+

2. Keep the areas as uncluttered as possible.
3. The preoperative materials include a hair clipper, 1-mL insulin syringes, freshly prepared anesthetic and analgesic solutions, a weighting scale to measure the body weight of the timed-pregnant mouse, sterile gauze, an eye moist ointment, ethanol swabs, and iodine solution.
4. The operative room (from skin opening to wound closure) needs to be equipped with an electroporator with tweezers electrodes, a micro-injector with a glass capillary, a lamp, a bead glass sterilizer, a heating pad covered with a plastic-coated examination sheet to absorb the heated saline solution that will keep the embryos hydrated throughout the operation, a 10 μL micropipette with tips, and a water bath to maintain saline solution warm. Also, make sure you have the necessary miscellaneous and surgical items: a petri dish, electrode gel, autoclaved dissecting tools, and a clean cage with a tissue placed at the bottom of the clean cage to easily observe any bleeding signs of the mouse during the hours following the surgery. Switch on the glass bead sterilizer, warming pad, and water bath to respectively 300 and 40 °C.
5. To electroporate C57BL/6 or SWISS mouse embryos from the embryonic day (E)12.5 to E15.5, set up the following parameters on the electroporator (NEPA21 device) using the transfer pulse option (*see* Table 2). The voltage and pulse number delivered to the embryo must be optimized for each mouse strain.
6. Add one drop of electrode gel to a petri dish; this gel will facilitate current conduction during the pulse delivery.
7. Add 1 μL reference drop of the initial 0.01% fast green solution (DNA free) in a petri dish; this 1 μL drop will help to set the parameters of the micro-injector to deliver the appropriate volume of DNA mix into the embryonic lateral ventricles.
8. Place the glass capillary into the micro-injector holder, aspirate the DNA mix solution, and gently break the tip of the capillary with forceps until the micro-injector's pressure parameter generates a drop equivalent in size to the reference drop.

9. In case several pregnant mice have to be successively electroporated, re-sterilize each surgical tool tip in the hot glass bead sterilizer for 30 s after cleaning them with water and ethanol between each surgical procedure.

3.3 Surgical Preparation of the Mouse

The surgical preparation of the mouse takes place in the pre-operating room. The surgeon must wash his/her/their hands and wear the required personal protective equipment before entering the animal facility.

1. Weigh the pregnant mouse and record the preoperative body weight on the anesthesia log.
2. Inject intraperitoneally the anesthetic solution (12.5 μL/g of body weight) and carefully replace the pregnant mouse in its original cage.
3. Monitor the decrease in the respiratory rate and check for the absence of withdrawal response after toe pinch 5 min after the injection.
4. Once anesthetized, inject subcutaneously the mouse with a prophylactic analgesic solution (1.6 μL/g of body weight).
5. Generously add ocular lubricant on both eyes to prevent dry eye during surgery.
6. Shave the lower abdomen skin of the timed-pregnant mouse using a hair clipper. Remove the hairs from the surgical area.
7. Perform a gentle surgical scrub of the shaved area by wiping it alternately with first iodine and then ethanol, following a circular inner-outside motion in order to push away any residual hair from the surgical site and therefore prevent infection.
8. Place the mouse on its back on the warming pad covered with an absorbent plastic-coated tissue in the operating area under the light source.
9. Place sterile gauzes around the shaved, cleaned, and sanitized around its lower abdominal midline where the incision will be done.

Throughout the whole surgery, regularly check the breathing pattern and rate of the anesthetized animal, as well as its surgical plan (toe pinch response).

3.4 Laparotomy

1. Using scissors, make a 2-cm vertical incision through the skin and muscle layers along the midline of the lower abdomen, taking care to lift each tissue layer (skin then muscle) with forceps as you make the incision to avoid damaging the underlying organs.
2. Pull the uterine horns out of the abdominal cavity by gently manipulating the embryonic bags.

3. Identify the cervix, connecting the lower part of the two uterine horns to the vagina, and count the number of embryonic bags existing on both sides. At these embryonic stages, the uterine horns are transparent. Draw a diagram illustrating the embryonic sacs organization on either side of the cervix. Note any pertinent observations and information on the embryo's diagram, such as deceased embryo or the (un)successful micro-injection and/or electroporation at an embryonic stage. This embryonic map is particularly useful for tracking electroporated embryos over time and for monitoring any embryo resorption after the surgery.

3.5 Intraventricular Micro-injection

1. If necessary, gently re-orient the embryo to observe the bregma reference point (triple vascular intersection connecting the large blood vessels running along the cerebral scissures) and the eye on the chosen side (*see* Fig. 3). Imagine a virtual line connecting those points and introduce the loaded capillary between them. Inject 1 μL of the DNA mix solution in one hemisphere by pressing the foot pedal of the calibrated micro-injector.
2. If the injection is above or below the intended lateral ventricle, the dye will spread and quickly diffuse before disappearing (Fig. 3A). If the DNA solution has been successfully injected, the ventricles will be filled with a blue-green color after the micro-injection (Fig. 3B).

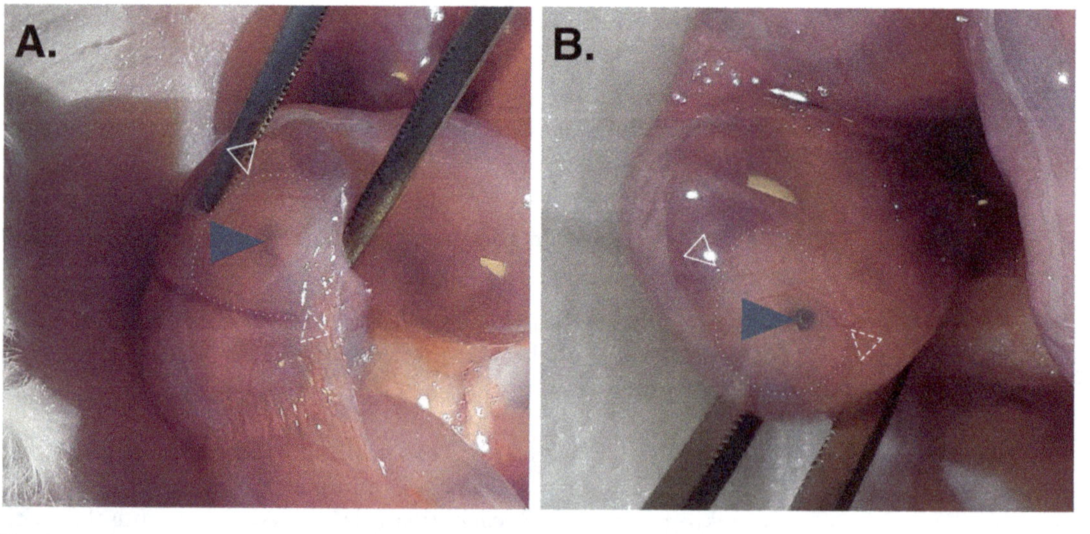

Fig. 3 Unsuccessful (**A**) versus successful (**B**) micro-injection of the DNA mix into the lateral ventricle of E13.5 SWISS embryos, prior to electroporation. The injected area is indicated by a blue-green arrowhead, while the reference points for the bregma and ipsilateral unpigmented eye are respectively indicated by a white dashed arrowhead and a full line white arrowhead. The injected hemisphere is also outlined

Of note, the cerebrospinal fluid volume is limited at this age (up to 15 μL is collected from a mouse litter at E14.5 [2]), thus the maximal injected volume cannot exceed 1 μL to avoid edema.

3.6 Electroporation

1. Orient the electrodes, previously dipped in electrode gel, on both sides of the selected embryo: The anode should cover the pial surface of the targeted cortical hemisphere, while the cathode should cover the caudal part of the embryo.
2. Press the pedal of the electroporator to deliver the preset electroporation parameter (*see* Table 2). The blue-green ventricle observed after the micro-injection will exhibit a darker shade after the delivery of electroporation pulses. During this step, the electroporation mix is transferred from the ventricle toward the neural progenitors lining up the dorsal wall of the lateral ventricle through membrane pores made by electric pulses applied by the tweezers. Indeed, the negatively charged DNA is simultaneously directed toward the anode located at the brain surface and away from the tail end, where the negative cathode is located.
3. Throughout the entire procedure, keep the embryos hydrated by frequently adding warmed saline solution. Keeping the uterine moisture is essential for both efficient electroporation and high embryos survival rates.
4. Repeat the cycle of micro-injection and electroporation steps for each targeted embryo.
5. Gently use forceps to replace the uterine horns within the abdomen. Then, fill up the abdominal cavity with a warmed saline solution. This bolus of physiological fluid is a preventive postoperative measure helping the operated mouse to recover.

3.7 Wound Suture

1. By gently opposing the muscle layer edge with a slight eversion, promptly realize a continuous suture of the muscular layer. A continuous suture is sufficient because the muscle layer is not directly accessible to the mice, preventing them from chewing up the sutures.
2. Make successive surgical nodes to close up the skin layer using absorbable sutures. This interrupted pattern is a safer knot against potential scratching or biting from the operated mouse. To ensure a fast and safe suturing, follow the rule of halves: place the initial interrupted knot at the center of the skin incision, place the next ones in the middle of the skin incision delimited by the previous knot, and so on.
3. Place the animal inside a clean cage, resting on its side atop a clean tissue. To help the anesthetized mouse to keep their body temperature, maintain the clean cage half on the warming pad until a movement is observed.

3.8 Postoperative Care

1. The day after the surgery, observe and record any signs or behaviors that may indicate discomfort or pain, without manipulating the operated mouse at first. These include abnormal posture, reduced mobility, decreased grooming, and blood spots on the tissue. The creation of a (complex) nest is a good recovery sign.
2. Observe and record any signs of clinical infection of the sutured wound (swelling, redness, discharge, dehiscence) by gentle and rapid manipulation of the operated mouse.
3. Regular postoperative observation for at least 3 consecutive days is recommended.

4 Remarks

1. Judging a putative pregnancy by the presence of a vaginal plug alone may lead to pregnancy misinterpretation. To accurately distinguish pregnant from non-pregnant mice prior to the surgery, record the weight of the mouse once the vaginal plug is detected (E0.5) and 10 days later (E10.5). Even if the pregnancy is not visible macroscopically, a pregnant mouse typically gains an average 3.62 g of body weight between E0.5 and E10.5 (C57BL/6 strain, Fig. 4A), whereas a non-pregnant one maintains approximately the same body weight (Fig. 4B).
2. Once the pregnancy is established (*see* **Note 1**), the number of embryos to be electroporated can be estimated as each embryo is associated with an average maternal weight gain of 0.47 g at E10.5 (Fig. 4C).
3. Be thoughtful about grouping females prior to mating. Females housed together have disrupted estrus rhythms [3], which can impact mating outcomes. However, this negative effect can be overcome by exposing them to bedding soaked with male urine for 3 days before mating [4]. This practice will help synchronize their estrus cycles and increase the likelihood of obtaining a positive mating plug.
4. Postnatal maternal care may differ between nulliparous (no living births) and primi- to multiparous (one or many living births) mice. It is therefore recommended to use at least a primiparous mouse if the analysis is performed postnatally to increase the survival chance of the litter.
5. In utero electroporation can be performed with any inbred and outbred mouse lines (C57BL/6, SWISS, OF1). However, the resulting embryo and litter size will vary between mouse strains. For example, the average litter size of the inbred

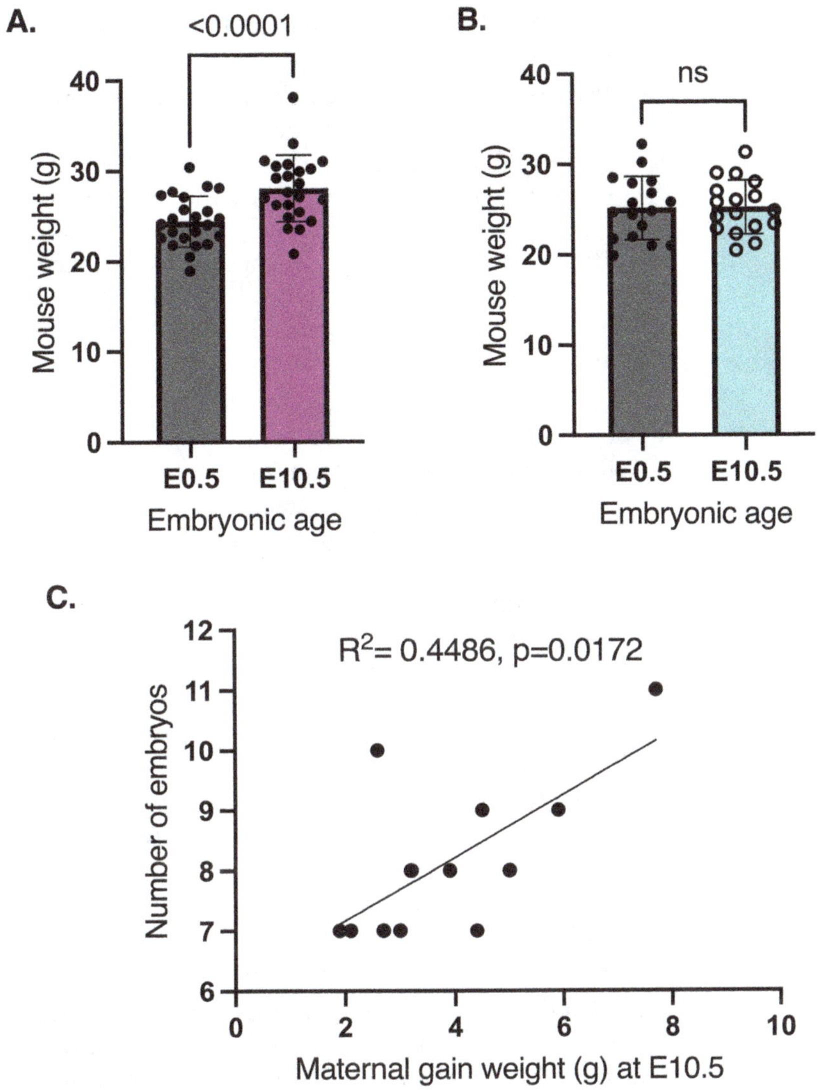

Fig. 4 Evolution of the body weight 10 days after detecting a positive vaginal plug in C57BL/6 female to determine the pregnancy status. (**A**) Pregnant mice exhibited a significant average weight gain of 3.62 ± 0.29 g between mating (E0.5) and 10 days later (mean ± SEM, 23 mice, paired *t* test). (**B**) Non-pregnant mice do not significantly change weight (0.08 ± 0.34 g of body weight) within the same timeframe (mean ± SEM, 17 mice, paired *t* test). (**C**) Significant positive correlation between the number of embryos and the maternal gain weight at E10.5 (12 mice, simple linear regression)

C57BL/6 strain is about 6 pups, whereas the average litter size of outbred SWISS strain is 11 pups. Although, the number of embryos is likely to be superior to the postnatal pup number. However, the use of an injectable anesthesia limits the surgical procedure duration to 30–35 min, hence electroporation of each embryo is often not possible, especially when learning the procedure. Just select the embryos that are in the right orientation to be electroporated easily, with minimal manipulation.

6. It is crucial to select a female that is below 6 months of age, as the reproductive capacity declines with age [5]. Furthermore, for inbred mouse strains, a better survival rate of electroporated embryos will be observed when the pregnant female is multipara, as she will be more likely to feed and nurture her litter after delivery. Finding an appropriate balance between the expected litter size, influenced by the maternal age and the likelihood of postnatal survival, related to the primi- or multiparous status, is an important point to consider.
7. Given that the suggested integrative approach leads to the expression from both transient/episomal and stable/integrated reporter plasmids, it is important to exercise caution when assessing the neurogenic and gliogenic clonal properties of electroporated neural progenitors as neurons produced at the time of the IUE will express combinatorial markers issued from both episomal and integrative vectors. Alternatively, the integration-coupled On genetic switch (iOn) strategy, which allows the expression of three monochromatic vectors only upon their integration into the host cell genome, can be used to faithfully label the entire neuronal and glial progeny of electroporated progenitors [6]. In the presently described MAGIC Markers strategy, cells born at the time of electroporation express a broad range of colors (e.g., deep layer pyramidal neurons generated at E13.5 in Fig. 5) due to the expression of both episomal and integrative plasmids, which will likely differ from the color of their glial sister cells expressing combinatorial colors only from the genome-integrated plasmids.
8. MAGIC Markers expressing tissue can be imaged using epifluorescence, confocal, or multiphoton microscopes, without amplification of the endogenous fluorescent signal. Epifluorescence microscopy is useful to quickly verify that the area of interest has indeed been electroporated prior to further histological processing. Confocal microscope is the tool of choice for imaging MAGIC Markers samples, as it enables a tight control of emission and excitation wavelengths and can be used to optically section any brain tissue up to 100 μm thick [7]. Following a sequential excitation of mTurquoise2/mCerulean, EYFP and tdTomato/mCherry with 440, 514, and 561 nm lasers respectively, the emitted signal from each fluorescent protein is then collected between 455–500 nm for mTurquoise2/mCerulean, 520–550 nm for EYFP, and 575–620 nm for tdTomato/mCherry. As the tissue volume imaged by a confocal microscope is limited, this approach may be unsuitable for studying progenitors whose descent may disperse far away from their site of origin. Above 100 μm

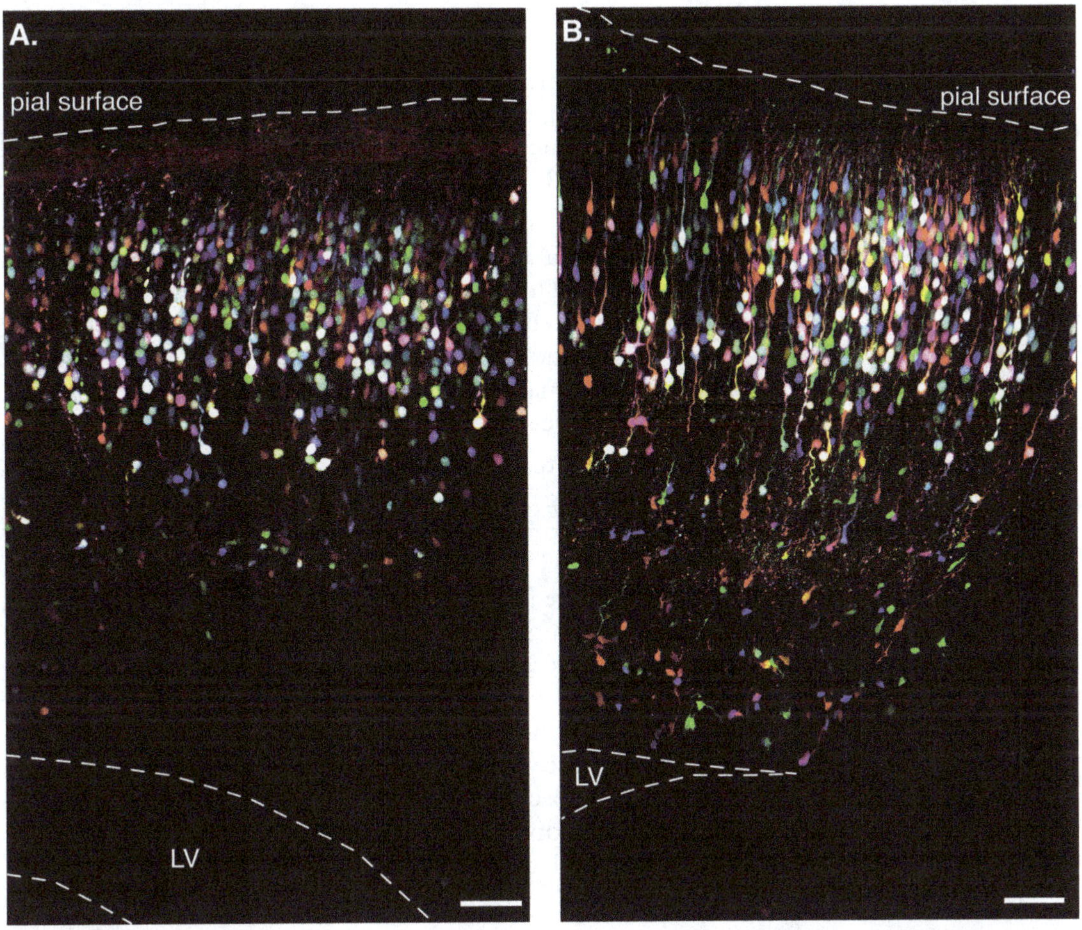

Fig. 5 Sagittal section of the cerebral cortex of an E18.5 SWISS embryo electroporated at E13.5 with MAGIC Markers following a non-integrative (**A**) or integrative (**B**) approach and imaged on a confocal microscope. In both approaches, it should be noted that most of the cells born at the time of the electroporation exhibit a white color. This hue results from a mixture of red, green, and blue corresponding to the initial episomal burst of multiple fluorescent proteins expression. In contrast, the majority of later-generated cells lose the episomal expression and do not display this complex hue. Multicolored pallial progenitors are still observed five days after the electroporation only when the integrative approach is used. Scale bar: 50 μm. *Abbreviations used*: LV: lateral ventricle

and up to half a millimeter in thickness, a multiphoton microscope equipped with lasers set to synchronized femtosecond pulses at the wavelengths of mTurquoise2/mCerulean and tdTomato/mCherry, allows simultaneous excitation of EYFP and thus collection of the full color palette offered by MAGIC Markers strategies [8]. In combination with automated serial tissue sectioning, it is now possible to image a continuous three-dimensional multicolor volume of several mm^3 of mouse brain labeled with MAGIC markers [9], thus offering the possibility to track related cells over long distances.

9. For volumetric imaging using a confocal microscope equipped with a long working-distance objective or a multiphoton microscope, fluorescent proteins expressed from MAGIC Markers construct have been documented to be compatible with hydrophilic tissue clearing reagents, such as Scale [10], See DB2 [11], or CUBIC [12].
10. Besides lineage tracing, the MAGIC Markers approach also offers the advantage of distinguishing between multicolored adjacent cells. This allows the study of their individual morphology [13], cellular interactions and dynamics [1], or their clonal organization [14]. In addition, MAGIC Markers can be used for functional analysis by associating a specific color with a gene activity level [1].
11. Besides the acute transgenesis model created by electroporation, pallial progenitors can also be labeled using a transgenic strategy. This strategy relies on the crossing of the MAGIC Markers mice [1] with cell-specific Cre recombinase lines, such as the Emx1-Cre [15] or Emx1-Cre^{ERT2} lines [16]. The pallial progenitor expresses the MAGIC Markers either constitutively (Cre line) or after tamoxifen administration at a specific developmental stage for timely labeling of the progeny (Cre^{ERT2} line). It should be noted that constitutive recombinase lines generally produce fewer colors than inducible lines. For inducible lines, a dose-response to tamoxifen must be made to achieve an appropriate balance between the diversity of colors and number of labeled cells.

Acknowledgments

This work is supported by the INSERM ATIP-Avenir program, FRM ENV202109013951, Agence Nationale de la Recherche (ANR) under the project "ANR-22-CE16-0012-01." We thank the imaging (Montpellier Ressources Imagerie) and animal (RAM-Neuro) core facilities and the support services of the Institute for Neuroscience of Montpellier. We also thank Dorine Thobois for her technical assistance.

References

1. Loulier K, Barry R, Mahou P et al (2014) Multiplex cell and lineage tracking with combinatorial labels. Neuron 81(3):505–520
2. Zappaterra MW, LaMantia AS, Walsh CA et al (2013) Isolation of cerebrospinal fluid from rodent embryos for use with dissected cerebral cortical explants. J Vis Exp (73):50333
3. Van Der Lee S, Boot LM (1955) Spontaneous pseudopregnancy in mice. Acta Physiol Pharmacol Neerl 4(3):442–444
4. Whitten WK (1956) Modification of the oestrous cycle of the mouse by external stimuli associated with the male. J Endocrinol 13(4): 399–404

5. Finn CA (1963) Reproductive capacity and litter size in mice: effect of age and environment. Reproduction 6(2):205–214
6. Kumamoto T, Maurinot F, Barry-Martinet R et al (2020) Direct readout of neural stem cell transgenesis with an integration-coupled gene expression switch. Neuron 107(4):617–630.e6
7. Wimmer VC, Möller A (2010) High-resolution confocal imaging in tissue. In: Hewitson TD, Darby IA (eds) Histology protocols, Methods in molecular biology, vol 611. Humana Press, Totowa, pp 183–191
8. Mahou P, Zimmerley M, Loulier K et al (2012) Multicolor two-photon tissue imaging by wavelength mixing. Nat Methods 9(8): 815–818
9. Abdeladim L, Matho KS, Clavreul S et al (2019) Multicolor multiscale brain imaging with chromatic multiphoton serial microscopy. Nat Commun 10(1):1662
10. Hama H, Kurokawa H, Kawano H et al (2011) Scale: a chemical approach for fluorescence imaging and reconstruction of transparent mouse brain. Nat Neurosci 14(11):1481–1488
11. Ke MT, Nakai Y, Fujimoto S et al (2016) Super-resolution mapping of neuronal circuitry with an index-optimized clearing agent. Cell Rep 14(11):2718–2732
12. Susaki EA, Tainaka K, Perrin D et al (2014) Whole-brain imaging with single-cell resolution using chemical cocktails and computational analysis. Cell 157(3):726–739
13. Dumas L, Clavreul S, Durand J et al (2020) In utero electroporation of Multiaddressable Genome-Integrating Color (MAGIC) Markers to individualize cortical mouse astrocytes. J Vis Exp (159):61110
14. Clavreul S, Abdeladim L, Hernández-Garzón E et al (2019) Cortical astrocytes develop in a plastic manner at both clonal and cellular levels. Nat Commun 10(1):4884
15. Gorski JA, Talley T, Qiu M et al (2002) Cortical excitatory neurons and glia, but not GABAergic neurons, are produced in the Emx1-expressing lineage. J Neurosci 22(15): 6309–6314
16. Kessaris N, Fogarty M, Iannarelli P et al (2006) Competing waves of oligodendrocytes in the forebrain and postnatal elimination of an embryonic lineage. Nat Neurosci 9(2): 173–179

Chapter 4

Single-Cell Lineage Tracing and Clonal State-Fate Analysis

Alejo E. Rodriguez Fraticelli and Pedro Sánchez Sánchez

Abstract

Lineage tracing has significantly advanced our comprehension in many areas of biology, such as development or immunity, by precisely measuring cellular processes like migration, division, or differentiation across labeled cells and their progeny. Traditional recombinase-based prospective lineage tracing is limited by the need for a priori cell type information and is constrained in the numbers of clones it can simultaneously track. In this sense, clonal lineage tracing with integrated random barcodes offers a robust alternative, enabling researchers to label and track a vast array of cells and their progeny over time. Moreover, clonal lineage tracing can be combined with single-cell omics technologies to study cell states and their maintenance over time. Key steps in these protocols include stable barcode integration, cell division to expand clones, and simultaneous capture of cellular properties with barcode information. Here, we comment on those steps and summarize important parameters to take into account during the design of single-cell lineage tracing experiments. Also, we present the main features for various available lentiviral libraries of expressed barcodes than can be captured alongside the transcriptome of individual cells. We cover other crucial aspects of experimental design, such as the optimization of cellular sampling, library diversity, and the minimization of clonal dropouts. Regarding sequencing data analysis, we provide some tips based on our experience, as well as available computational tools for the assignment of clonal identities and the identification of fate determinants. We finally discuss limitations of current methodologies and use an example step-by-step protocol to illustrate key points during the process. In sum, we provide a roadmap for considering and implementing single-cell lineage tracing studies to comprehensively explore fate determinants and their mechanisms.

Key words Clones, Lineage tracing, Barcode, Single cell, Transcriptome, Sequencing, State-fate, Clonal memories

1 General Principle of Clonal Lineage Tracing with Single-Cell Sequencing Readout

Prospective *lineage tracing* has revolutionized our understanding of development, immunity, and cancer, thanks to its capacity to accurately measure a diversity of critical processes such as cell division, differentiation, and movements [1]. But, prospective lineage tracing has a limitation: One must either choose and know a priori the type of cells that will be traced (e.g., Lgr5+ cells to trace intestinal stem cells, and therefore design an Lgr5-CreER cassette),

Jorge García-Marqués and Tzumin Lee (eds.), *Lineage Tracing: Methods and Protocols*, Methods in Molecular Biology, vol. 2886, https://doi.org/10.1007/978-1-0716-4310-5_4,

or one has to randomly label cells without capturing the identity of the cell of origin that gives rise to a particular lineage. In addition, although doable, it is impractical (and error-prone) to trace many *clones* (i.e., thousands) using conventional prospective lineage tracing, due to the imprecisions of label dilution or the lack of sufficient allele diversity potential (e.g., the Confetti mouse model). Clonal lineage tracing with expressed integrated *barcodes* provides a possible solution to many of these issues and has quickly become a powerful tool for developmental and cancer biologists to tackle important questions, such as lineage bias and origins of therapy resistance [2–15].

The general principle for clonal lineage tracing is to label a cell pool (from hundreds but up to millions of cells) with individual, unique DNA barcodes, allow them to divide (for a few cell generations) to generate clones, and then randomly split the cell pool into two or more parts. In one part of the cell pool (which we call the *initial state*), some property of the cells is measured (e.g., the *transcriptome*) simultaneously with their barcode information, thus connecting barcodes (clones) with a specific early state at the first sampling point [16, 17]. Then, the rest of the cells are tested at a second time point (the *late state*, or *fate*) for whatever cellular properties the user deems of interest (differentiation into various lineages, cell death propensity, migration, or cell state). The fate properties are again measured together with the barcode information, allowing the researcher to connect initial states with late fate properties.

The most critical parts for this general principle to work is to be able to: (a) label cells with a stable, heritable, genome-integrated barcode of sufficient diversity for the system of study, (b) allow the cells to divide to generate clones of sufficient size, and (c) simultaneously capture the cellular properties together with the barcode information. If the barcode is part of an expressed construct, the analysis of the initial state can be performed at massive scale with *single-cell* RNA *sequencing* kits, measuring any cellular contents that can be barcoded in droplet-based (e.g., 10× Chromium) or slide-spot-based (e.g., Slide-seq) reactions, in thousands of cells at a time. It can be less fashionable, but still very useful (and many-fold less expensive), to profile any property of the initial or late states through labeling (e.g., fluorescently-labeled antibody) and then separating (e.g., by FACS, fluorescence-activated cell sorting) the cell pool into high versus low-labeled fractions, for which the barcode contents (or rather their distributions) can then be quantified by isolating the DNA, a simple barcode PCR and NGS sequencing (Fig. 1a).

Importantly, this principle of analysis is based on the assumption that daughter cells retain some amount of information (some property) through cell differentiation that then influences (or at least correlates with) the fates that we intend to study. While this

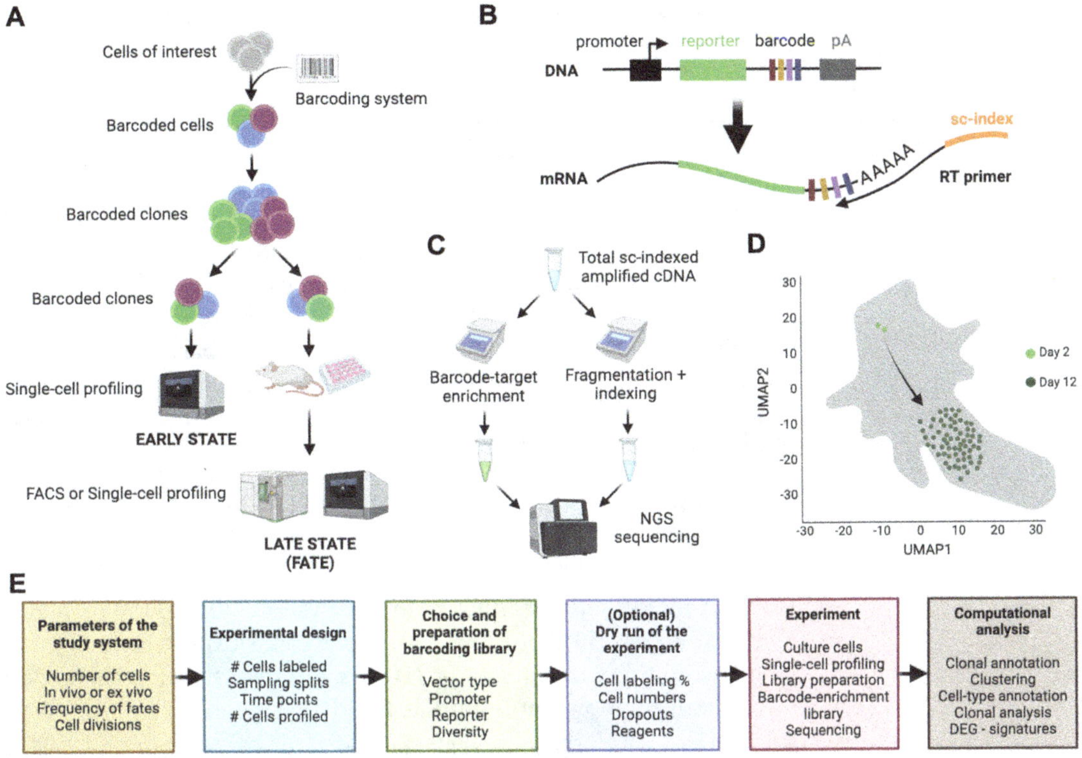

Fig. 1 (**a**) General pipeline for state-fate experiment using clone splitting and single-cell lineage tracing. Cells of interest are labeled with a barcoding system and then allowed to expand in vivo or ex vivo. Cells are then split into at least two pools, one that is profiled at the initial *early state*, and one that is profiled at a *late state*, after assessing some cellular function (differentiation, migration, metastasis, cancer therapy resistance, etc.). (**b**) Elements of an expressed barcode system, allowing detection in 3'polyA capture based single-cell transcriptomics platforms. (**c**) Common library preparation steps. After the single-cell-indexed amplified cDNA library is generated, part of it is used for building the whole transcriptome library (fragmentation, end repair, and indexing), while a second part (10–25% of total volume) is used to perform a targeted PCR to enrich the barcode-containing cDNA sequences and align the reads to increase coverage of the full barcode sequence. (**d**) An example low-dimensional representation of the different single-cell transcriptomes, highlighting an HSC clone with cell states profiled at day 2 and profiled at day 12 during myeloid differentiation. (**e**) Dominating state-fate experiments requires domain knowledge, technical awareness, a careful experimental design and execution, and a detailed computational analysis to answer the biological question of interest

may be true for many fate properties of interest to the researcher, some processes (e.g., random asymmetric partitioning of cell contents during cell division) can limit the detection capabilities of these clonal assays. The method has so far excelled in measuring processes where at least some amount of short-to-long-term memory is present as the clones are generated. This is obvious in a unidirectional rapid differentiation hierarchy, such as during embryonic development [18, 19]. More surprisingly, various researchers have identified clonal memory in a variety of systems, including hematopoietic stem and progenitor cells, T cells, and cancer cells, with clones retaining a large amount of transcriptional identity even after many cell divisions [16, 20–34].

2 Parameters of the Dynamic Process Under Study

As with any assays that we aim to apply to our study subject, it is critical to know as many parameters from the system of study as possible. This includes the number of cells that we aim to track, their location, their cell cycle dynamics, their movement, and so on, but also parameters about the efficiency of the barcoding process itself. In our lab, we typically ask ourselves the following questions and measure at least the following parameters before beginning a new project:

- Timing of the process (how long does it take to generate the outcome)—How many days does it take until metastatic cancer cells reach the liver?
- Number of cells of interest in the system (and their growth/death across time)—How many cells can I expect to have after growing them in culture for 10 days and then treating them with a chemotherapy compound at LD90?
- Transgenesis efficiency—How easy is it to transduce relevant cells in our system? One can test this in a "dry run" of the experiment, testing different amounts of the integration barcode library directly in the cells of interest.
- Number of cell divisions—Do all cells divide or only some? How many cells divide per day? Are division cycles changing during the course of the experiment?
- Tracking of the cells—Where do the cells of interest move? This can be done using Luciferase or fluorescent protein reporter systems.
- Cellular states—How many cell types are there in my sample, and what are the cell types of interest for the system of study? These can be verified sometimes with some simple flow cytometry or histology/immunofluorescence at different time points.

The more parameters we know about the system of study, the better the experimental design will be, and the higher the chances of success.

3 Types of Available Expressed Barcode Libraries

One of the most important decisions before beginning any single-cell lineage tracing experiment is the barcoding system that will be used. Barcoding systems can either be endogenous, if the system is integrated already as a transgene inside the genome of the organism of study, or exogenous, if the barcodes are delivered through an integration-competent vector (transposon, lentivirus, or similar). Endogenous barcoding systems usually generate the allelic diversity through mutagenesis, transposon-mobilization, recombination, or gene-editing, and these processes are activated through temperature, compounds, or light. In contrast, in exogenous barcoding systems, the allelic diversity is created during the molecular cloning process, and cells are modified with a library of barcodes. In both systems, ideally, every cell acquires a unique DNA sequence (or combination of DNA sequences) that will allow the user to identify each cell and its progeny (the clone) unequivocally. Barcoding of multiple independent cells with the same barcode can lead to drastic misinterpretation of data due to false positives. Therefore, a system with sufficient diversity is necessary. In addition to diversity, an important consideration is the expression cassette. Each barcoding system can also contain a different promoter, reporter, and polyadenylation (polyA) signal, and all of these are going to impact the stability of the barcode-containing transcript (Fig. 1b).

Here, we briefly discuss a series of readily available expressible lentiviral or transposon-based barcoding systems, which are exogenous barcoding systems that most researchers can rapidly incorporate into their subjects of study with little effort. Many of these are available at repositories such as Addgene.

LARRY—This lentiviral barcoding library contains the barcode within an EGFP expression cassette with a strong polyA signal (bovine growth hormone, bGH) [16, 22]. The expression cassette is in the negative strand of the viral genome, to allow efficient viral production. Expression is driven by the full EF1-alpha promoter, which is a very strong and ubiquitous promoter in most mammalian systems. Due to the proximity of the barcode to the polyA cassette, the detection of LARRY barcodes in single-cell RNA sequencing platforms is excellent, with over 90% of capture in profiled GFP+ cells. In addition, it displays little to no silencing during differentiation (with a 10% dropout in detection efficiency only in some lineages of low RNA content). The original library, expanded and available at Addgene, contains a diversity of ~200,000 unique barcodes. A recently published SISBAR library is similar in design and contains a larger total diversity [28]. The TraCe-seq library also uses the same design and incorporates the puromycin-resistance gene in the expression cassette [25].

CellTag—This lentiviral barcoding library contains an 8-nt barcode within an EGFP expression cassette with a strong polyA signal (SV40 polyA) [21, 35, 36]. The expression cassette is placed in the negative strand of the viral genome, allowing efficient viral production despite having a polyA element. Expression of EGFP-barcode is driven by the CMV promoter, which is a medium strength promoter but with ubiquitous expression in mammalian systems. Similar to LARRY, due to the placement of the barcode right before the polyA signal, the detection of CellTag barcodes is excellent (over 90% of profiled EGFP+ cells when using the most recent 10× Chromium 3′ capture beads). CellTag researchers originally deposited three libraries (V1, V2, V3) with different indexes (six fixed bases at the 5′ of the barcode sequence) allowing researchers to multiplex experiments or perform sequential re-barcoding of cells during a differentiation process (increasing phylogenetic resolution). A recent CellTagMulti library contains an additional Nextera s5 sequence in the 5′ end of the barcode, allowing amplification of the barcode in single-cell chromatin accessibility or Multiome reactions [31]. Using CellTagMulti, authors report efficient mapping of both chromatin accessibility and the CellTag barcode in the same single-cell. The CellTagMulti libraries were constructed with a longer and more diverse barcode than the previous CellTag barcodes, with 18–25 random nucleotide positions. The deposited libraries, available at Addgene, contain ~80,000 unique barcodes. There is an interesting web-app that helps visualize results and run simulations (celltag.org).

STICR—This lentiviral barcoding contains the barcode within an EGFP expression cassette, with a WPRE element and a strong polyA signal (bGH), reaching very high expression levels [37]. The expression cassette is in the negative strand of the viral genome, to allow efficient viral production. Expression is driven by the chimeric CAG promoter, which is very strong and ubiquitous in most mammalian systems. Although this system contains, in principle, the strongest combination of expression elements (CAG, WPRE, bGHpA), the reported barcode recovery rate is ~65–75%, which is below LARRY or CellTag. However, it is possible that this is specific to differentiating human neurons, which is the only model where the barcodes were employed so far. The STICR library contains 42 random nucleotide positions (separated in 3 stretches of 14 bases), and the deposited libraries available at Addgene contain ~40–50 million unique barcodes. The STICR team further created three libraries, with different indexes (fixed sequences in the 5′ of the barcode that allows users to differentiate each library and even multiplex experiments or perform sequential barcoding).

SPLINTR—Three versions of these libraries are available, with EGFP, mCherry, or EBFP as the reporters [24]. The expression cassette is driven by an EF1 alpha promoter and is oriented in the positive strand of the viral genome. The barcode is placed after a WPRE element and before the 3'LTR, and uses the viral 3'LTR sequence (post integration) as the only polyadenylation signal. This results in the barcode being at a longer distance from the start of the polyA tract in the transcript, and the minimal barcode-containing transcript size is about 500 bases long. This longer length, compared to other libraries, could affect clustering in

NextSeq or HiSeq Illumina platforms, and thus, sequencing on NovaSeq is recommended for barcode sequencing. Researchers report >70% of mappability of barcodes after single-cell quality control. TREX libraries use a similar design architecture [38]. Commercial vectors that follow a similar architecture are also available from Cellecta (CloneTracker).

Watermelon—This lentiviral barcoding library contains a 30-nt semirandom barcode within an EGFP expression cassette and a strong polyA signal (bGH), reaching high expression levels [39]. The expression cassette is in the negative strand of the viral genome, to allow efficient viral production. Expression is driven by the human PGK promoter, which is very strong in human cells (note: low activity in mouse cells). In addition to the barcode, the system also contains a Tet-regulated H2B-mCherry cassette, which allows fluorescent histone pulse-chasing to track cellular divisions by label dilution. This allows to separate clones based on high versus low cycling rates, for instance, to compare quiescent-dormant versus cycling therapy-resistant cancer clones.

ClonMapper—This lentiviral barcoding library is based on the popular CROP-seq architecture, and it uses sgRNA sequences instead of semirandom nucleotide barcodes [40]. In the absence of Cas9, the barcodes do not pose any risk for editing the genome, yet their sequence can still be obtained thanks to the placement of the RNApol III sgRNA cassette within the 3'LTR sequence. The CROP-seq vector is further modified to express BFP as a reporter, and it uses the EF1 alpha promoter to drive the expression of both reporter and sgRNA-barcode. The major advantage of ClonMapper is its possibility to combine it with COLBERT, a system that allows the user to retrieve clones of interest through transduction of the clonal pool with CRISPRa (dCas9-VPR) and a specific sgRNA-guided reporter (the Recall plasmid) [41]. The Recall plasmid contains the sgRNA sequence of the clone of interest, plus a minimal CMV promoter, driving EGFP expression. COLBERT lights up the clone of interest with EGFP, allowing FACS-based sorting to isolate it and characterize it further. The main limitation of ClonMapper lies in the necessity to expand the clones sufficiently in order to be able to transduce enough cells to sort. Thus, the method is still mostly restricted to isolating large enough clones.

FateMap/Rewind—This lentiviral barcoding library contains a large 100-nt semirandom barcode within an EGFP expression cassette driven by the EFS Promoter [23, 27, 42]. The cassette is in the plus strand of the viral genome, and expression uses the endogenous 3'LTR polyA signal, similar to SPLINTR and TREX libraries. In the FateMap library, the barcode is placed before the WPRE element of the lentiviral scaffold, at least 1000 bases upstream of the polyA signal, which results in a more limited capture when using 3′-biased reverse transcription single-cell methods (e.g., 10×). Target-enrichment of the barcode by PCR results in a very large amplicon and sequencing on NovaSeq or long-read platforms is recommended. The major advantage of the FateMap library is its longer barcode, which allows users to design a pool of RNA FISH probes to hybridize and detect clones of interest at their initial state, a method that the authors call Rewind [23]. This enables users to sort even extremely rare cell states/fates (1:1000 or even 1:10000) and measure pretty much any content (genome, transcriptome, immunofluorescence) in the clones of interest, which would require sequencing many millions of single-cells in a conventional single-cell lineage tracing design.

iTracer, macsGESTALT, and PCT48—These libraries contain both integration barcodes, as well as a scratchpad of sgRNA-guided editable sequences [43–45]. This strategy is sometimes termed an "evolvable" barcode, which is quite appropriate considering the need to perform alignment and phylogenetic reconstruction based on similarity of edits to reconstruct the most likely phylogenetic ordering of resulting cells. These systems are far more complex to construct, due to the need to modify cells with multiple vectors: an inducible Cas9 vector (typically doxycycline-inducible), an inductor vector (typically expressing rtTA), a guide-RNA expression vector (driving expression of multiple different sgRNAs), and a scratchpad vector (containing both the integration barcode as well as the sgRNA-editable evolvable barcode).

4 Sampling Parameters, Library Diversity, and Clonal Dropout

Once we establish the basic parameters of the dynamic process under study, we must then proceed to the experimental design. The experiment must be designed to maximize the sampling events that carry lineage information (i.e., enough profiled cells with common barcodes at early and late time points to have adequate statistical power for analyses). Three parameters need to be optimized: the number of cells that will be labeled, the number of cells that will be sampled, and the time points at which they will be sampled. The choice of these parameters must take into consideration two factors that will affect our detection capacity: (1) the rarity of fates that we aim to study, (2) the number of fate replicates that are required, and (3) the dropouts that are expected through the sample processing.

It is useful to think of an example. Imagine that a researcher desires to analyze the initial clonal states of the cancer cells that will survive a chemotherapy that will kill 99% of cells (which means the rarity of the fate is 1%). If they start with a labeled population of 100,000 cells, allow them all to divide once (100,000 clones of 2 cells each), and then perfectly split them randomly into 2 pools (1 initial state, and 1 fate). Then, the chance to detect any clone in both the initial state and the fate becomes $0.5 \times 0.01 = 0.005$.

Thus, if we profile every single one of the 100,000 cells in the initial state, and every single cell after treatment at the late time point (fates), we can expect on average to find only 500 cells at the initial state for which we can also detect their final fate (in this case, therapy resistance).

In a more realistic scenario, clone proliferation is heterogeneous, cells get stuck to pipette tips or tube walls, splitting is imperfect, and single-cell profiling (or even FACS sorting) results in loss of at least 10–50% of the sample. And this is without taking into account transplantation bottlenecks if in vivo readout is necessary. All of these processes, which we refer to as clone/sample dropout, severely impact our ability to detect *informative* clones that will help us identify fate determinants. It is useful for researchers to create a mathematical model where sampling and random processing dropout of cells is accounted for, in order to estimate the number of clones (and cells) that will be informative for the single-cell lineage tracing analysis according to their specific experimental design (see supplementary methods in [22]).

One solution to clone dropout problems is to generate a larger clonal library, through longer culture/expansion of the initially labeled sample. While this can be practical for cancer cell lines, this is rarely possible with more precious primary samples. Moreover, cellular states (and thus any fate-regulatory information) can vary dramatically with more cell divisions, introducing noise in the analysis, which prevents precise identification of fate determinants. Despite this, we and others have witnessed that certain cell types (e.g., hematopoietic stem cells) have extremely long-lived transcriptional/epigenetic clonal memory and this can be measured even after many cell divisions. The expansion of labeled cells can and should always be adjusted to maximize the statistical power for the type of question (and thus clonal memory) that one seeks to address (short, intermediate, or long-term). In addition to time, specific treatments or cell culture conditions may aid in maintaining or losing this memory and this must also be taken into consideration.

Another way to deal with clonal dropouts is to use the most adequate technical profiling method for the problem in question. For instance, when dealing with extremely rare cell samples, it is preferential to process through plate-based approaches (e.g., SmartSeq-3), where processing dropout can be smaller than with droplet-based methods. We also find it useful to coat tips and tubes with PBS 0.1% BSA (or use low-binding plastics) before using them for processing samples. Centrifugation speeds, times, and brake-speeds should be adjusted precisely for the sample of interest. All of these steps can be carefully accounted for in a "dry run" of the final experiment, where users can measure the cell numbers at each step, to accurately estimate sample dropouts.

In addition to clonal dropouts, barcode detection dropouts are another frequent source of data-loss in single-cell lineage tracing. It is critical to properly identify the barcodes in the sequencing library, not only to avoid dropouts, but also to avoid false positives. While the whole transcriptome library will usually contain them, this library is randomly fragmented, and the localization of the barcode sequence may be close or far from the point of sequencing adapter ligation. When performing short-read next-generation sequencing (e.g., Illumina), this means that reads may contain partial sequences of the barcode, or might miss the random part of the barcode. In addition, in our experience, the reads mapping to the barcode are rarely more than 2% of the total whole-transcriptome reads. To increase the number of detected barcode reads per cell and optimize the barcode alignments, it is advisable to perform a target-enrichment PCR on a fraction of the post-amplification cDNA (Fig. 1c). This takes advantage of UMIs to correct PCR duplicates and efficiently quantify and assign the correct barcode to each cell. We typically use 25% of the volume of post-amplification cDNA to perform the barcode target-enrichment reaction, but less can be used if barcode expression levels are very high.

In sum, researchers should focus their experimental design to choose the optimal number of cells, labeling frequencies, and sampling time points, and to use a protocol that minimizes clone dropouts and barcode detection dropouts. Considering the costly nature of single-cell sequencing experiments, this will maximize the resources of the researchers.

5 Computational Analysis

Once sequencing files are obtained at the end of the experiment, then computational analysis begins. Here we will focus on single-cell lineage tracing, but FACS-based barcode analysis has been covered elsewhere [46]. It must be noted that computational analyses for single-cell lineage tracing are moderately intensive and, depending on resources, can easily take more than 1 week to complete and require access to a high-performance computing device. Computational analysis first involves creating a barcode whitelist (as a reference for alignment), preprocessing the sequenced files, mapping reads (to transcriptome, barcodes, etc.), and then quantification. Finally, clonal identities can be assigned to cells in a process that takes into account their gene expression distribution (across all cells), and minimizes false-positives (false assignment of clonal identity) and removes cells (droplets) with barcode collisions (which can either be doublets, barcode swaps, or contamination). There are several recent pipelines to do this available on code repositories (i.e., GitHub), including *BARtab* & *bartools* [47], *CellBarcode* [48], *CellTagR* [21, 31, 35, 36], *kite*

[49] (from the *kallisto* alignment suite), *NextClone* & *CloneDetective* [50], *pychasier* [40], *timemachine* [23], and *cloneRanger* [51]. Due to the high expression level of certain barcodes, it is possible that samples with a lot of cell death will result in single-cell libraries where the background threshold should be set at a higher count number. We have found it useful in these cases to use normalization-based approaches or to count with an internal control that can allow determining the true background (e.g., by mixing two samples carrying different barcode libraries, or by mixing male and female cells, within the same single-cell encapsulation reaction).

After assigning clonal identities, the user needs to also assign cellular state identities. This can be done with a number of single-cell sequencing suites, including *Seurat* [52–56], *Scanpy* [57–59], and others [60, 61]. After cell clustering, it is useful to then annotate cell identities either manually or automatically based on prior knowledge of the field. There are certain recent packages that perform this automatically (using cell atlases as reference) and most work well for cell types for which there is abundant ground truth information [62, 63].

After assigning clonal identities to single-cells, the next recommended step is to visualize the data obtained (Fig. 1d). When many fates are assayed at once, it is advisable to use multivariable analysis mathematical tools, such as data normalization, transformation, scaling, and principal component analysis, and then choose a distance metric to build a graph/network, which can then be visualized using 2D/3D embeddings such as UMAP or tSNE. This can be performed using clones as vectors, and the results can then be clustered to identify the most common fate patterns. A recently developed tool, ClonoCluster allows users to visualize and integrate both clonal-level and single-cell-level information into embeddings [64]. Another way to annotate clones with multiple fates is to use differential proportion/abundance analysis. These analyses are particularly useful when users lack knowledge that the fates of every clone have been sufficiently captured (i.e., there is undersampling or significant dropout). Manual annotation of fates is also possible, especially when there are few fate categories (e.g., drug-resistant vs. non-resistant clones after therapy).

To find fate determinants, it can be useful to transfer the fate annotations for each clone back to the single-cell matrix and then explore the initial state single-cell data using 2D/3D embeddings. One typical way to find fate-associated markers is to compare initial state cells based on their fate categories (using differential expression/enrichment analysis). As with all biological experiments, it is crucial to count with sample replicates to calculate p-values and identify significant markers. One possibility is to pool all cells falling into a certain fate category (irrespective of clone) and rely on fully independent biological specimens to calculate p-values. Another possibility is to use each clone as a co-dependent biological variable to increase statistical power. Based on our experience, considering the sparsity of scRNA-seq data, and the biases that can be introduced by differences in total RNA counts or mitochondrial gene counts detected, we still recommend comparing completely independent biological and experimental replicates. Also, a phylogenetic analytical framework has been recently defined to quantitate key aspects regarding transcriptional heritability [65].

One major contributor to error in identification of fate markers is the sparsity of fate capture due to aforementioned dropouts, especially when clones are small. CoSpar is a recently developed package that specifically deals with this fate-detection sparsity and increases the power of fate marker identification [66]. CoSpar accurately detected fate markers in a recent mouse study using clonal barcoding to study hematopoietic stem cell heterogeneity in vivo [67].

6 Limitations

While single-cell lineage tracing can be a powerful tool, it has important limitations that need to be carefully considered before choosing it as the solution to a particular question.

First, most single-cell sequencing is destructive and therefore state-fate experiments do not contain information about the real fates of single cells. Instead, it relies on clonal analysis, and therefore can only fate-map clonally heritable states. Circumventing this pitfall, researchers have recently developed non-destructive single-cell RNA sequencing technologies (Live-seq), but they are still in early development and cannot process large numbers of cells [68].

Second, not every model system can be manipulated with integration barcodes or transgenes, and even so, this requires optimization of promoters and capture devices. Moreover, the true power of clonal lineage tracing is connecting states and fates, which requires sampling clones at least for two timepoints. The necessary labeling and isolation of cells thus introduces perturbations, which can further affect interpretation of results.

Third, barcode integrations or editing can affect cellular behavior, and this can alter the interpretation of results. Especially when considering very rare fates, a sufficient number of experimental and biological replicates is necessary to avoid integration-based biases. We also recommend sequencing of integration sites across independent experiments.

Fourth, due to the high costs of single-cell library preparation, the number of cells traced (initial state) is still typically low (much lower than with traditional recombinase-tracing systems) and therefore many errors can be introduced because of undersampling and dropouts (of early states and of fates).

Finally, the methods are limited by their capacity to detect fate determinants. This is especially concerning when using highly sparse capture methods (single-cell RNA or ATAC sequencing). Increasing the cell numbers per clone solves this issue, but this results in a higher cost of sequencing of the initial state and becomes quickly prohibitive for very rare fates. Some elegant solutions have been implemented for this, including using barcodes to design fluorescent hybridization probe panels that bind to the desired clone cells (Rewind) [23], or capturing the clones of interest through transducing the early state cells with clone-specific fluorescent reporters (ClonMapper, CloneSifter, and CaTCH) [40, 69, 70]. Both methods allow users to sort and sequence only the clones of interest, greatly reducing costs and increasing statistical power.

7 Example Protocol: Tracing of Hematopoietic Progenitor Cells in Myeloid Differentiation Conditions

7.1 Step 1—Isolation and LARRY Barcoding of Mouse Long-Term (LT) HSCs

Isolate mononuclear cells from the femurs and tibias of C57BL6J mice and stain with lineage, Sca-1, cKit, CD48, CD150, and EPCR markers. Sort 5,000 LT-HSCs (LSK48-150+EPCR+) using a fluorescence-activated cell sorting (FACS) device directly into a single-well of a Fibronectin-coated 96-well culture plate containing F12-medium (GIBCO) supplemented with 0.1% PVA (Sigma) and recombinant cytokines (10 ng/mL Stem Cell Factor (SCF), 100 ng/mL Thrombopoietin (TPO), and 10 ng/mL Flt3 Ligand). BSA-free reagents and recombinant cytokines are crucial for maintaining stemness and preventing premature differentiation. Add 20 μL of LARRY lentiviral prep (titer 10e6/mL, as performed in 293T cells) to the culture well containing the cells. This is approximately 20 infective particles per HSC, which should represent an MOI of 0.2. Incubate at room temperature for 30 min. Add additional media up to 200–220 μL. Mix the well through pipetting up and down twice. Centrifuge the plate at room-temp for 1 h at 750 g, with low brake speed. Add PBS to at least 30 other wells of the plate, to help maintain humidity. Transfer immediately to a 32 °C 5% CO2 incubator and keep overnight.

7.2 Step 2—Expansion of Long-Term HSCs

Then transfer the cells to a low-oxygen (hypoxia) incubator (2% O2, 5% CO2, 37 °C), change media after 5 days, and keep culturing the cells for 10 days, to allow clones to expand. Hypoxia allows HSC clones to maintain their stemness for a prolonged time and delays differentiation. After 10 days, you can expect to recover 50,000 cells, of which approximately 20% (10,000) will be EGFP+.

7.3 Step 3—Sample Splitting for the Initial State and Fate Assays

Transfer plates on ice. Add 100 μL of cold PBS to the culture wells and set the pipette at 100 μL volume. Pipet up and down the well to resuspend the HSCs and move to a 1.5 mL tube (low-bind). Use the FACS sorter (yield mode) to sort the EGFP+ cells into a new 1.5 mL tube containing 500 μL of PBS. Keep everything on ice or at 4 °C. Take 250 μL of the sorted cells (approximately half the sample), and transfer to a new 1.5 mL tube on ice. Centrifuge at 500 g 4 °C for 5 min. Remove as much as possible of the supernatant, and resuspend in 50 μL volume with IMDM media (Gibco). Add 1 mL of methylcellulose-based myeloid differentiation media (MethoCult, STEMCELLTech—M3434) and vortex for 30 s. Transfer the mix to a single well of a 12-well plate. Transfer plates to a 37 °C 5% CO2 incubator and culture for 10 days.

7.4 Step 4—Single-Cell Profiling (Initial State)

Take 250 μL of the sorted cells (approximately half the sample), and transfer to a new 1.5 mL tube on ice. Centrifuge at 500 g 4 °C for 5 min. Remove as much as possible of the supernatant, leaving only 30 μL volume. Take 5 μL to count the cell concentration. Load the remainder into a single-cell sequencing reaction (e.g., for 10× Chromium v3 Gene Expression kits, we load 25 μL). Adjust volumes as necessary and follow manufacturer's instructions for library preparation.

7.5 Step 5—Single-Cell Profiling (Fate)

Take the differentiation plate from the incubator and check under the microscope for the appearance of colonies of cells with various shapes (some bound to plastic, some larger, smaller, different opacities), confirming differentiation. Transfer plate on ice. Add 1 mL of cold PBS to the culture wells and set the pipette at 1 mL volume. Slowly pipet up and down the well to resuspend the cells and move to a 15 mL tube. Add 10 mL of PBS. Centrifuge at 700 g 4 °C for 15 min. Remove as much as possible of the supernatant and wash again with 10 mL PBS. Transfer to a 1.5 mL tube and wash again with 1 mL PBS 0.1%BSA. After centrifugation, leave only 50 μL volume. Take 5 μL to count cell numbers. Add PBS 0.1%BSA to adjust concentration of cells (depending on single-cell device manufacturer's instructions). Load the remainder into one or more single-cell sequencing reactions, as necessary (e.g., for 10× Chromium v3 Gene Expression kits).

7.6 Step 6—Library Preparation, Barcode Enrichment, and Indexing

The goal of this step is to generate a specific LARRY barcode sequencing library through PCR-based enrichment. This helps sequence more LARRY sequences per cell, but it also helps localize the LARRY barcode sequence within reach of the sequencing cycles of short-read NGS platforms, significantly improving assignment of barcodes to single-cells. To create this enriched LARRY library, we recommend to take a split of the initial cDNA library, before any tagmentation/fragmentation and indexing/amplification. Usually using 25% of this cDNA as a template is enough for most barcoding systems, but more is better. It is critical to perform very few cycles to avoid amplification biases (which could result in poor-diversity sequencing libraries and, in turn, many cells without assigned barcodes). Optional: perform a test qPCR with only 1% of the cDNA library, and then adjust the number of cycles based on the amplification curve. Ideally, the reaction should be stopped at a number of cycles right after the logarithmic amplification phase begins. LARRY barcode target-enrichment libraries should use different sequencing (e.g., TruSeq) indexes, to easily demultiplex from gene expression libraries. The remaining cDNA library can be stored if this reaction fails, and it can even be further amplified if necessary, but this comes at the expense of sequence biases and reduced representation.

7.7 Step 7—Sequencing

Sequence the single-cell transcriptome and LARRY barcode libraries using a next-generation sequencing platform, such as Illumina NovaSeq. Aim for a depth that ensures adequate coverage of all barcode sequences and gene transcripts for accurate clonal tracking and transcriptomic profiling. We typically perform an initial pass sequencing at 20,000 reads per cell (for the transcriptome) and 2000 reads per cell (for the LARRY), and then we resequence based on the saturation results, calculated using unique molecular identifiers. No PhiX is required when pooling with other samples, but if all LARRY libraries represent over 5% of total molecules to cluster, we recommend spiking in with 10% PhiX (Illumina platforms).

7.8 Step 8—Computational Analysis

After base calling and demultiplexing, load resulting *fastq* files into an alignment software package for constructing single-cell count matrices (e.g., CellRanger for 10× Genomics libraries). We use *cloneRanger* to assign feature barcodes (i.e., CITE-seq, TotalSeq, Multi-seq, or Perturb-seq) and clonal barcodes (e.g., LARRY barcodes) to single-cells alongside their gene expression. By default, the Snakemake workflow creates the barcode reference given to 10× CellRanger by merging the fastq files of the LARRY-enriched runs and finding the unique barcodes by collapsing within a hamming distance of 4. With minor modification, the sequencing file of the whole viral library can also be used. Once the alignments are done, we typically request a minimum 5–10 reads per UMI and 3 UMIs per cell, but we explore many parameters for every analysis. It is important to set a threshold to properly call clones and avoid errors. In that regard, we label each cell under the most expressed barcode, since the majority of the cells only present one integration. We know that clones coming from a cell with more than one integrated barcode could be split, but since they are few, we prefer this scenario rather than creating bigger, spurious clones by taking into account other barcodes.

Once each cell has been assigned a clone/barcode, the workflow creates a report with some basic results after standard transcriptome analysis (quality control, normalization, feature selection, dimensionality reduction, clustering, etc.). This automatic report also includes minimal barcode analysis. This allows the analyst to have a glimpse of how the data look like, before manually calling doublets (for which feature barcodes and clone labels can be used), integrate data from different samples or batches, and annotate cell types.

With the fully annotated counts matrix, we proceed to calculate the cell type proportions for each clone, after normalizing for the clone size. A useful parameter to calculate is the observed versus expected ratio (O/E), which can be calculated by creating an expected distribution of cell type proportions based on random sampling of the clone size across the different cell type/annotation according to its total frequency in the entire dataset. Typically, we plot the observed fractions divided by the median of the expected distribution [16], and p-values can also be calculated using a one-sample t-test. Another analysis that we typically perform is to calculate a fate bias, which is a ratio between the clonal proportion of one cell type, e.g., Megakaryocytes, and all the rest (Megakaryocyte bias). When a clone is stem-cell biased, we refer to it as "low-output," since we reason that this bias is related to a reduced output/differentiation activity per-stem-cell [22].

After calculating different fate biases and O/E ratios, we can proceed to analyze expressed genes that correlate with any of these parameters. Typically, gene expression data and clonal dropouts make it hard to perform quantitative correlations with gene expression at the single-cell level. Thus, we frequently use clonal metacells (i.e., pooling all cells of a clone from a certain cluster, and use this as a pseudo-bulk), or pooling all cells from a certain cluster that share the same fate annotation (e.g., Megakaryocyte-biased HSCs) [22]. For large enough clones, another possibility is to look for clonally variable genes, or "mnemes," which have gene expression memory states that are sufficiently unique to identify and even track a clonal identity.

8 Final Summary

Single-cell lineage tracing can be a very powerful method when used appropriately. Here, we provided the basic considerations and steps that need to be taken to implement these technologies to address a biological question (Fig. 1e). Tracking the fates of hundreds or thousands of individual cells at a time, while simultaneously capturing their whole transcriptome, can be a very efficient way to discover new gene expression states connected with specific differentiation fates. We envision these tools to continue their growing adoption across cellular systems and biology fields and particularly excelling in those where little prior knowledge or tools exist. In addition, we envision that integrating barcoded clonal analysis with more single-cell modalities in addition to the transcriptome (such as protein levels, phosphorylation-states, chromatin states) will completely revolutionize our understanding of fate-determinants and fate mechanisms across a wide diversity of models.

Acknowledgments

This work has been supported by the Cris Foundation Excellence Award (PR_EX_2020-24), the ERC Starting Grant MemOriStem (101042992), the Spanish National Research Agency (PID2020-114638RA-I00), the Agencia de Gestio d'Ajuts Universitaris i de Recerca (AGAUR, 2017 SGR 1322), and the CERCA Program/ Generalitat de Catalunya. A.R.F. acknowledges support from the Institut Catalá de Recerca i Estudis Avançats (ICREA), the American Society of Hematology (ASH) Scholar Award, the Leukemia Lymphoma Society Special Fellow Career Development Program Award (3391–19), the NIH NHLBI K99/R00 transition to independence award (K99 HL146983), the *"la Caixa" Foundation under the project code [LCF/PR/HR23/52430009]*, the Ministry of Science Ramon y Cajal Fellowship, and the LaCaixa Junior Fellows Incoming Fellowship.
Illustrations were created with Biorender.

Conflicts of Interest A.E.R.F. is an advisor for Retro Bio. The authors declare no competing interests.

Contributions A.E.R.F. wrote the manuscript with the assistance of P.S.S.

References

1. Garcia-Marques J, Espinosa-Medina I, Lee T (2021) The art of lineage tracing: from worm to human. Prog Neurobiol 199:101966
2. Naik SH, Schumacher TN, Perié L (2014) Cellular barcoding: A technical appraisal. Exp Hematol 42:598–608
3. Kebschull JM, Zador AM (2018) Cellular barcoding: lineage tracing, screening and beyond. Nat Methods 15:871–879
4. Kester L, Van Oudenaarden A (2018) Single-cell transcriptomics meets lineage tracing. Cell Stem Cell 23:166–179
5. McKenna A, Gagnon JA (2019) Recording development with single cell dynamic lineage tracing. Development 146:dev169730
6. Baron CS, Van Oudenaarden A (2019) Unravelling cellular relationships during development and regeneration using genetic lineage tracing. Nat Rev Mol Cell Biol 20:753–765
7. Wagner DE, Klein AM (2020) Lineage tracing meets single-cell omics: opportunities and challenges. Nat Rev Genet 21:410–427
8. Morgan D, Jost TA, De Santiago C, Brock A (2021) Applications of high-resolution clone tracking technologies in cancer. Curr Opin Biomed Eng 19:100317
9. VanHorn S, Morris SA (2021) Next-generation lineage tracing and fate mapping to interrogate development. Dev Cell 56:7–21

10. Serrano A, Berthelet J, Naik SH, Merino D (2022) Mastering the use of cellular barcoding to explore cancer heterogeneity. Nat Rev Cancer 22:609–624
11. Sankaran VG, Weissman JS, Zon LI (2022) Cellular barcoding to decipher clonal dynamics in disease. Science 378:eabm5874
12. Chen C, Liao Y, Peng G (2022) Connecting past and present: single-cell lineage tracing. Protein Cell 13:790–807
13. Jones MG, Yang D, Weissman JS (2023) New tools for lineage tracing in cancer in vivo. Ann Rev Cancer Biol 7:111–129
14. Howland KK, Brock A (2023) Cellular barcoding tracks heterogeneous clones through selective pressures and phenotypic transitions. Trends Cancer 9:591–601
15. Haghverdi L, Ludwig LS (2023) Single-cell multi-omics and lineage tracing to dissect cell fate decision-making. Stem Cell Rep 18:13–25
16. Weinreb C, Rodriguez-Fraticelli A, Camargo FD, Klein AM (2020) Lineage tracing on transcriptional landscapes links state to fate during differentiation. Science 367:eaaw3381
17. Tian L, Tomei S, Schreuder J et al (2021) Clonal multi-omics reveals Bcor as a negative regulator of emergency dendritic cell development. Immunity 54:1338–1351.e9
18. Alemany A, Florescu M, Baron CS et al (2018) Whole-organism clone tracing using single-cell sequencing. Nature 556:108–112
19. Wagner DE, Weinreb C, Collins ZM et al (2018) Single-cell mapping of gene expression landscapes and lineage in the zebrafish embryo. Science 360:981–987
20. Yao Z, Mich JK, Ku S et al (2017) A single-cell roadmap of lineage bifurcation in human ESC models of embryonic brain development. Cell Stem Cell 20:120–134
21. Biddy BA, Kong W, Kamimoto K et al (2018) Single-cell mapping of lineage and identity in direct reprogramming. Nature 564:219–224
22. Rodriguez-Fraticelli AE, Weinreb C, Wang S-W et al (2020) Single-cell lineage tracing unveils a role for TCF15 in haematopoiesis. Nature 583:585–589
23. Emert BL, Cote CJ, Torre EA et al (2021) Variability within rare cell states enables multiple paths toward drug resistance. Nat Biotechnol 39:865–876
24. Fennell KA, Vassiliadis D, Lam EYN et al (2022) Non-genetic determinants of malignant clonal fitness at single-cell resolution. Nature 601:125–131
25. Chang MT, Shanahan F, Nguyen TTT et al (2022) Identifying transcriptional programs underlying cancer drug response with TraCe-seq. Nat Biotechnol 40:86–93
26. Bandler RC, Vitali I, Delgado RN et al (2022) Single-cell delineation of lineage and genetic identity in the mouse brain. Nature 601:404–409
27. Goyal Y, Busch GT, Pillai M et al (2023) Diverse clonal fates emerge upon drug treatment of homogeneous cancer cells. Nature 620:651–659
28. You Z, Wang L, He H et al (2023) Mapping of clonal lineages across developmental stages in human neural differentiation. Cell Stem Cell 30:473–487.e9
29. Radtke S, Enstrom MR, Pande D et al (2023) Stochastic fate decisions of HSCs after transplantation: early contribution, symmetric expansion, and pool formation. Blood 142: 33–43
30. Ceresa D, Alessandrini F, Lucchini S et al (2023) Early clonal extinction in glioblastoma progression revealed by genetic barcoding. Cancer Cell 41:1466–1479.e9
31. Jindal K, Adil MT, Yamaguchi N et al (2024) Single-cell lineage capture across genomic modalities with CellTag-multi reveals fate-specific gene regulatory changes. Nat Biotechnol 42:946–959
32. Jain N, Goyal Y, Dunagin MC et al (2024) Retrospective identification of cell-intrinsic factors that mark pluripotency potential in rare somatic cells. Cell Syst 15:109–133.e10
33. Mold JE, Weissman MH, Ratz M et al (2024) Clonally heritable gene expression imparts a layer of diversity within cell types. Cell Syst 15:149–165.e10
34. Schaff DL, Fasse AJ, White PE et al (2024) Clonal differences underlie variable responses to sequential and prolonged treatment. Cell Syst 15:213–226.e9
35. Kong W, Biddy BA, Kamimoto K et al (2020) CellTagging: combinatorial indexing to simultaneously map lineage and identity at single-cell resolution. Nat Protoc 15:750–772
36. Guo C, Kong W, Kamimoto K et al (2019) CellTag Indexing: genetic barcode-based sample multiplexing for single-cell genomics. Genome Biol 20:90
37. Delgado RN, Allen DE, Keefe MG et al (2022) Individual human cortical progenitors can produce excitatory and inhibitory neurons. Nature 601:397–403
38. Ratz M, Von Berlin L, Larsson L et al (2022) Clonal relations in the mouse brain revealed by single-cell and spatial transcriptomics. Nat Neurosci 25:285–294

39. Oren Y, Tsabar M, Cuoco MS et al (2021) Cycling cancer persister cells arise from lineages with distinct programs. Nature 596:576–582
40. Gutierrez C, Al'Khafaji AM, Brenner E et al (2021) Multifunctional barcoding with ClonMapper enables high-resolution study of clonal dynamics during tumor evolution and treatment. Nat Cancer 2:758–772
41. Al'Khafaji AM, Deatherage D, Brock A (2018) Control of lineage-specific gene expression by functionalized gRNA barcodes. ACS Synth Biol 7:2468–2474
42. Harmange G, Hueros RAR, Schaff DL et al (2023) Disrupting cellular memory to overcome drug resistance. Nat Commun 14:7130
43. He Z, Maynard A, Jain A et al (2022) Lineage recording in human cerebral organoids. Nat Methods 19:90–99
44. Simeonov KP, Byrns CN, Clark ML et al (2021) Single-cell lineage tracing of metastatic cancer reveals selection of hybrid EMT states. Cancer Cell 39:1150–1162.e9
45. Yang D, Jones MG, Naranjo S et al (2022) Lineage tracing reveals the phylodynamics, plasticity, and paths of tumor evolution. Cell 185:1905–1923.e25
46. Bramlett C, Jiang D, Nogalska A et al (2020) Clonal tracking using embedded viral barcoding and high-throughput sequencing. Nat Protoc 15:1436–1458
47. Holze H, Talarmain L, Fennell KA et al (2024) Analysis of synthetic cellular barcodes in the genome and transcriptome with BARtab and bartools. Cell Rep Methods 4:100763
48. Sun W, Perkins M, Huyghe M et al (2024) Extracting, filtering and simulating cellular barcodes using CellBarcode tools. Nat Comput Sci 4:128–143
49. Booeshaghi AS, Min KH (Joseph), Gehring J, Pachter L (2024) Quantifying orthogonal barcodes for sequence census assays. Bioinformatics. Advances 4:vbad181
50. Putri GH, Pires N, Davidson NM et al (2023) Extraction and quantification of lineage-tracing barcodes with NextClone and Clone-Detective. bioRxiv
51. Singh I, Fernandez-Perez D, Sanchez-Sanchez P, Rodriguez-Fraticelli AE (2024) Pre-existing stem cell heterogeneity dictates clonal responses to acquisition of cancer driver mutations. bioRxiv
52. Satija R, Farrell JA, Gennert D et al (2015) Spatial reconstruction of single-cell gene expression data. Nat Biotechnol 33:495–502
53. Butler A, Hoffman P, Smibert P et al (2018) Integrating single-cell transcriptomic data across different conditions, technologies, and species. Nat Biotechnol 36:411–420
54. Stuart T, Butler A, Hoffman P et al (2019) Comprehensive integration of single-cell data. Cell 177:1888–1902.e21
55. Hao Y, Hao S, Andersen-Nissen E et al (2021) Integrated analysis of multimodal single-cell data. Cell 184:3573–3587.e29
56. Hao Y, Stuart T, Kowalski MH et al (2023) Dictionary learning for integrative, multimodal and scalable single-cell analysis. Nat Biotechnol 42:293
57. Wolf FA, Angerer P, Theis FJ (2018) SCANPY: large-scale single-cell gene expression data analysis. Genome Biol 19:15
58. Virshup I, Bredikhin D, Heumos L et al (2023) The scverse project provides a computational ecosystem for single-cell omics data analysis. Nat Biotechnol 41:604–606
59. Heumos L, Schaar AC, Lance C et al (2023) Best practices for single-cell analysis across modalities. Nat Rev Genet 24:550–572
60. Amezquita RA, Lun ATL, Becht E et al (2020) Orchestrating single-cell analysis with Bioconductor. Nat Methods 17:137–145
61. Lun ATL, Kancherla J (2023) Powering single-cell analyses in the browser with WebAssembly. J Open Source Softw 8:5603
62. Ianevski A, Giri AK, Aittokallio T (2022) Fully-automated and ultra-fast cell-type identification using specific marker combinations from single-cell transcriptomic data. Nat Commun 13:1246
63. Clarke ZA, Andrews TS, Atif J et al (2021) Tutorial: guidelines for annotating single-cell transcriptomic maps using automated and manual methods. Nat Protoc 16:2749–2764
64. Richman LP, Goyal Y, Jiang CL, Raj A (2023) ClonoCluster: a method for using clonal origin to inform transcriptome clustering. Cell Genom 3:100247
65. Schiffman JS, D'Avino AR, Prieto T et al (2022) Defining ancestry, heritability and plasticity of cellular phenotypes in somatic evolution. bioRxiv
66. Wang S-W, Herriges MJ, Hurley K et al (2022) CoSpar identifies early cell fate biases from single-cell transcriptomic and lineage information. Nat Biotechnol 40:1066–1074
67. Li L, Bowling S, McGeary SE et al (2023) A mouse model with high clonal barcode diversity for joint lineage, transcriptomic, and epigenomic profiling in single cells. Cell 186: 5183–5199.e22
68. Chen W, Guillaume-Gentil O, Rainer PY et al (2022) Live-seq enables temporal

transcriptomic recording of single cells. Nature 608:733–740

69. Umkehrer C, Holstein F, Formenti L et al (2021) Isolating live cell clones from barcoded populations using CRISPRa-inducible reporters. Nat Biotechnol 39:174–178

70. Feldman D, Tsai F, Garrity AJ et al (2020) CloneSifter: enrichment of rare clones from heterogeneous cell populations. BMC Biol 18:177

Chapter 5

Lineage Recording in Human Brain Organoids with iTracer

Eugenio Gentile, Ashley Maynard, Zhisong He, and Barbara Treutlein

Abstract

Induced pluripotent stem cell (iPSC)-derived organoids provide models to study human organ development. Single-cell transcriptomics enables highly resolved descriptions of cell states within these systems; however, approaches are needed to directly determine the lineage relationship between cells. Here we provide a detailed protocol (Fig. 1) for the application of iTracer (He Z, Maynard A, Jain A, et al., Nat Methods 19:90–99, 2022), a recently published lineage recorder that combines reporter barcodes with inducible CRISPR–Cas9 scarring and is compatible with single-cell and spatial transcriptomics. iTracer is used to explore clonality and lineage dynamics during brain organoid development. More broadly, iTracer can be adapted to any iPSC-derived culture system to dissect lineage dynamics during normal or perturbed development.

Key words Lineage tracing, Brain organoid, Single-cell RNA sequencing, Inducible scarring

1 Introduction

Experimental lineage tracing methods seek to provide a detailed understanding of cellular interrelatedness within a biological system. This involves creating cell lineage trees to aid the comprehension of biological processes such as cell development, tissue maintenance, and disease pathogenesis. Notably, recent advancements have prioritized the incorporation of distinctive, heritable genetic identifiers, referred to as "barcodes" or "scars," into cell genomes, detectable through techniques like single-cell RNA sequencing (scRNAseq) [1–6]. Barcodes are synthetic DNA sequences with a specific length and random base composition. They are typically integrated into a cell's genome along with DNA sequences encoding fluorescent reporters [2, 5]. As a result, daughter cells inherit the barcode and express the fluorescent protein, simplifying their identification. In contrast, DNA scars are cell-specific genomic barcodes generated through CRISPR-Cas9-induced random insertion-deletions (indels) [3, 4, 6, 7]. Indels are generated at a target site on DNA, making them suitable for lineage

Jorge García-Marqués and Tzumin Lee (eds.), *Lineage Tracing: Methods and Protocols*, Methods in Molecular Biology, vol. 2886,
https://doi.org/10.1007/978-1-0716-4310-5_5,

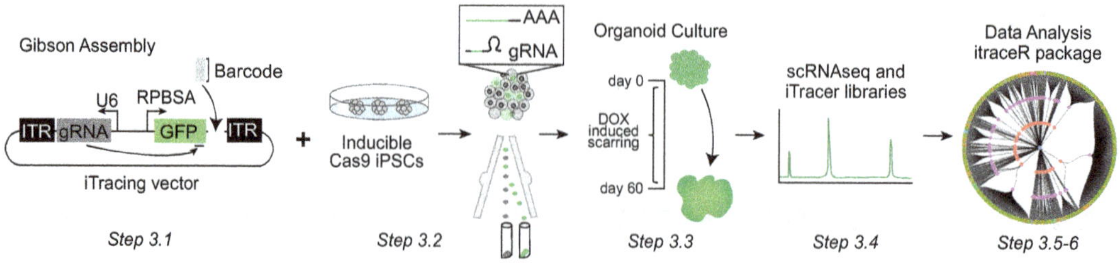

Fig. 1 Overview schematic showing steps of the iTracer protocol from Gibson assembly of the barcodes into the iTracer backbone (**step 3.1**) to data analysis using the itraceR package (**step 3.6**)

tracing, similar to barcodes. Unlike naturally occurring cell-specific single nucleotide polymorphisms (SNPs), which require high-depth sequencing for their detection, barcode and scar sequences are synthetically created and easily incorporated within known constructs and can therefore be more easily detected (Fig. 1).

Lineage tracing techniques can be employed in a static or dynamic manner. Static tracing introduces a barcode or scar at a single time point, offering a straightforward experimental setup. The mark is inherited by all progeny of the labeled cell, resulting in direct parent-progeny trees. While limited in the information they provide, these static trees are valuable, particularly when assessing the potency of progenitor cells or testing differentiation protocols. Dynamic lineage tracing, in contrast, allows for the introduction of genetic marks at chosen time points, enabling the creation of higher-resolution, branching lineage trees. Several studies have enhanced the resolution of Cas9 scar lineage trees by allowing the generation of multiple scars at different time points [7, 8]. This approach helps resolve branching lineage relationships, and when scarring is induced by external factors like drugs or heat shock, it reveals the timing of key differentiation steps in various tissues.

Here we describe iTracer [1], a recently introduced lineage tracing construct that contains both a static and a dynamic tracing component. iTracer allows for an initial barcode-based cell segregation as well as a subsequent lineage recording at user-set time points by an inducible Cas9 scarring system. Below we describe the use of iTracer to trace the lineages generated by induced pluripotent stem cells (iPSCs) when differentiating toward neural lineages in brain organoids [9]. In conjunction with the iTracer plasmid, this system requires iPSCs with a stable integration of a doxycycline-inducible Cas9 [10]. Upon induction, Cas9 forms a complex with the guide RNA on the iTracer plasmid that then scars an area within a reporter gene. This protocol describes the workflow to create barcoded iTracer plasmids, establishment of iTracer cells, and scarring schemes in brain organoids. We also describe how to generate sequencing libraries to capture barcodes and scars from single-cell RNA sequencing data. Finally, we detail how to compile these data to integrate with existing analysis software packages using our itraceR package and how to create lineage trees.

2 Materials

2.1 Prepare Barcoded iTracer Plasmid

1. pSBbi-iTracer-R (RFP, European Plasmid Repository, Plasmid #349).
2. pSBbi-iTracer-G (GFP, European Plasmid Repository, Plasmid #350).
3. Barcode template: GACGAGCTGTACAAGT GATCCGWNNNNNNNNNNWCACCCAGCTTTCTTGTA CAAAGTGGCTGTG.
4. Barcode forward primer: GACGAGCTGTACAAGTGATC.
5. Barcode reverse primer: CACAGCCACTTTGTACAAGA.
6. Phusion High-Fidelity PCR Master Mix with HF Buffer (ThermoFisher, #F531L).
7. Miniprep Kit.
8. NucleoSpin Gel and PCR Clean-up kit.
9. NEB® 5-alpha Competent E. coli (NEB, #C2987H).
10. S.O.C. buffer.
11. EcoRI-HF (NEB, #R3101S).
12. Cutsmart buffer (NEB, #B7204).
13. EcoRI-HF (NEB, #R3101S).
14. NEBuilder® HiFi DNA Assembly Master Mix (NEB, #E2621S).
15. Midiprep kit.
16. NanoDrop or Qubit Instrument.

2.2 Generate iTracer Cells and Cell Culture

1. iCRISPR Cas9 cell line (*see* **Note 1**).
2. Sleeping Beauty Plasmid (addgene, #34879).
3. Cell Culture 6-well plate.
4. ROCK Inhibitor/Y27632.
5. mTeSR™ Plus Basal Medium.
6. mTeSR™ Plus 5× Supplement.
7. PBS.
8. Knockout™ DMEM/F-12.
9. T75 Flask.
10. 4D-Nucleofector (Lonza™).
11. Matrigel.

12. P3 Primary Cell 4D-Nucleofector™ Kit (Lonza™, #V4XP-3024).
13. TrypLE™ Express Enzyme.
14. 30 μM filter.
15. Doxycycline.
16. Bioanalyzer (or similar Fragment Analyzer).

2.3 Single-Cell iTracer Capture and Sequencing

1. Primers:

Primer name	Sequence
Root PCR GFP	GACGACGGCAACTACAAGACC
Nest barcode GFP	GTGACTGGAGTTCAGACGTGTGCTCTTCCGATC TGGATCACTCTCGGCATGGA
Nest scar GFP	GTGACTGGAGTTCAGACGTGTGCTCTTCCGATC TAGGTGAACTTCAAGATCCGCC
Root PCR RFP	CGGCACGCTGATCTACAAGG
Nest barcode RFP	GTGACTGGAGTTCAGACGTGTGCTCTTCCGATC TTGGAACAGTACGAGCGCTC
Nest scar RFP	GTGACTGGAGTTCAGACGTGTGCTCTTCCGATC TGAGTTCAAGACCATCTACATGGCC
Universal primer	CTACACGACGCTCTTCCGATCT

2. Phusion High-Fidelity PCR Master Mix with HF Buffer (ThermoFisher, #F531L).
3. DMSO.
4. SPRIselect beads.
5. Ethanol.
6. Buffer EB.

2.4 iTracer Analysis

1. High-performance computer.
2. Cell Ranger (≥4.0.0).
3. R (≥4.0.0).
4. R package: itraceR (https://github.com/quadbio/itraceR).

3 Methods

3.1 Preparation of iTracer Plasmid-Barcode Integration with Gibson Assembly

1. Resuspend all oligos to a concentration of 100 μM, and store these stocks at −20°C. Dilute stock primers to 10 μM. Dilute barcode template 50 μM.
2. Amplify barcodes by combing the following in a PCR tube: 1.25 μL, 10 μM barcode forward primer, 1.25 μL, 10 μM barcode reverse primer, 0.5 μL 50 μM barcode template, 12.5 μL Phusion Master Mix, 9.5 μL water.
3. Run PCR according to the following program: 98 °C for 2 min, then 25 cycles of (98 °C for 10 s, 58 °C for 20 s, 72 °C for 20 s), 72 °C for 5 min.
4. Run 1% agarose gel with 1 μL of amplified barcode product to check for successful amplification.
5. Clean up barcodes with column cleanup and Nanodrop or Qubit.
6. Prepare transformation reaction to amplify plasmid: combine 50 μL NEB® 5-alpha Competent E. coli + 2 μL iTracer plasmid, place sample on ice for 15 min, heat shock the cells for 45 s at 42 °C, place tube on ice for 1 min, add 350 μL S.O.C. buffer to reaction, incubate by shaking (~250 RPM) for 45 min at 37 °C.
7. Add reaction mix to 3 mL of 2YT buffer with Ampicillin.
8. Incubate overnight (~16 h) at 37 °C.
9. Harvest the cells from the liquid culture and use a QIAGEN® Miniprep column to purify the plasmid and elute in 35 μL. Nanodrop or Qubit the purified plasmid to obtain the concentration.
10. To cut the plasmid for barcode insertion first digest the plasmid with EcoRI-HF by combining in a PCR tube: 5 μg amplified iTracer plasmid, 5 μL Cutsmart buffer, 3 μL EcoRI-HF, fill to 50 μL with water.
11. Incubate 37 °C overnight.
12. Run 0.5% agarose gel with digested and undigested plasmid (control) to check for complete cutting.
13. Perform a gel extraction cleanup of EcoRI fragmented plasmid and Nanodrop or Qubit.
14. Calculate the pmol concentration of 1 μL of the amplified barcodes and the pmol of 1 μL of the plasmid fragment. In this example, we will use 1 μL of barcodes for 1 μL vector in the Gibson assembly (*see* **Note 2**).
15. Prepare a Gibson barcoded sample: 1 μL of the calculated volume of EcoRI fragmented plasmid, 1 μL of the calculated

volume of barcodes, 10 μL NEBuilder® HiFi DNA Assembly Master Mix, fill to 20 μL with water.

16. Prepare a Gibson control sample: 1 μL of the calculated volume of EcoRI fragmented plasmid, 10 μL NEBuilder® HiFi DNA Assembly Master Mix, fill to 20 μL with water.
17. Incubate 15 min at 50 °C, then place on ice.
18. Perform one control and two barcode Gibson assembly transformation reactions. For each reaction: combine 50 μL NEB® 5-alpha Competent E. coli + 10 μL Gibson reaction sample, incubate samples on ice for 15 min, heat shock the cells for 45 min at 42 °C, place tubes on ice for 1 min, add 350 μL S.O.C. buffer to each sample, incubate for 45 min at 37 °C.
19. For each barcode assembled sample, combine the reactions into one tube and equally plate the total reaction volume (~550 μL each) on LB with Ampicillin plates (×3 15 cm square plates). If necessary, add 200–250 μL S.O.C. buffer to help evenly spread the volume on the plates.
20. For the control plasmid only reaction, plate ~273 μL of total reaction volume. If necessary add 200–250 μL S.O.C. buffer to help evenly spread the volume on the plate.
21. Incubate plates overnight (~16 h) at 37 °C.
22. Visually inspect control plate versus barcode plates. There should be a notable difference, where the control plate has few colonies compared to the barcode reaction plates. If this is observed, continue to the next step (*see* **Note 3**).
23. To collect Gibson assembled plasmids, use 10 mL of 2YT buffer and a scraper to collect all the colonies in a 15 mL conical vial.
24. Wash the plate with another 5–7 mL 2YT buffer and collect in the same tube as above.
25. Perform three QIAGEN® Midiprep reactions and elute in a final volume of 200 μL.
26. Pool final midiprep reactions (final volume ~600 μL) and Nanodrop or Qubit for final concentration (*see* **Note 4**).

3.2 Generating iTracer Cells

Generating iTracer cells requires integration of the plasmid randomly into the genome of iCRISPR cells. Here, we describe the integration using nucleofection (*see* **Note 5**), after integration, we select iTracer positive cells with fluorescence-activated cell sorting (FACS).

1. Prepare a 1.5 mL tube with 10 μg iTracer plasmid, 1 μg sleeping beauty plasmid, and 100 μL of nucleofection solution.
2. Precoat a T75 with 4 mL of Matrigel, place in a 37 °C incubator for no less than 30 min. Remove excess Matrigel by

aspirating and add 7.5 mL of prewarmed mTeSR™ Plus with Rock Inhibitor (RI) (1:1000). Place the flask back into the incubator until ready to use.

3. For one nucleofection reaction, remove one 6-well plate well at ~70% cell confluency. Lift the cells and singularize them: remove media, wash with 1 mL PBS for 30 s by tilting the plate slightly, remove PBS, add 1 mL of TrypLE™ solution and incubate in a 37 °C incubator for 3–4 min (*see* **Note 6**), remove TrypLE™ by aspiration, add 1–2 mL Knockout™ DMEM/F-12 media to the well, and singularize cells by pipetting gently.
4. Move cells to a 15 mL conical tube together with 4 mL Knockout™ DMEM/F-12. Spin down the tube at 200 ×*g* for 5 min.
5. Remove supernatant from centrifuged 15 mL conical tube. Do not disturb the cell pellet. Flick the tube to loosen the pellet. Resuspend the pellet in 100 μL nucleofector solution with plasmids added (from **step 1**).
6. Pipette suspension into Lonza™ nucleofector cuvette. Run the "hES cell, H9" nucleofector program (*see* **Note 7**).
7. After the run completes, immediately remove the cuvette from the instrument. With the provided plastic pipette, take 500 μL prewarmed mTeSR™ Plus with RI (1:1000) media and add to the cuvette.
8. Slowly take up all cells with media and add cells to the prepared T75. Place the T75 in an incubator.
9. Let cells recover and grow for 3 days before proceeding to FACS sorting (*see* **Note 8**).
10. To FACS iTracer positive cells, lift the cells as described above. Briefly, remove media, wash with 4 mL PBS for 30 s by tilting the plate slightly, remove PBS, add 4 mL of TrypLE™ solution, and incubate in a 37 °C incubator for 3–4 min, remove TrypLE™ by aspiration, add 4 mL Knockout™ DMEM/F-12 media to the well, and singularize cells by pipetting gently. Move cells to a 15 mL conical tube together with 4 mL Knockout™ DMEM/F-12. Spin down the tube at 200 ×*g* for 5 min.
11. Remove supernatant from centrifuged 15 mL conical tube. Do not disturb the cell pellet. Flick tube to loosen the pellet and resuspend the pellet in 1 mL of cold mTeSR™ Plus with RI (1: 200). Filter the cells through a 30 μM filter and transfer cells to a FACS tube.
12. Sort iTracer-positive cells (GFP- or RFP-positive depending on iTracer plasmid used, Fig. 2) into a collection tube with mTeSR™ Plus with RI (1:200).
13. After sorting, spin cells in the collection tube at 200 ×*g* for 5 min, aspirate media, and resuspend in mTeSR™ Plus with RI (1:200). Plate the cells in the appropriately sized well (*see* **Note 9**).

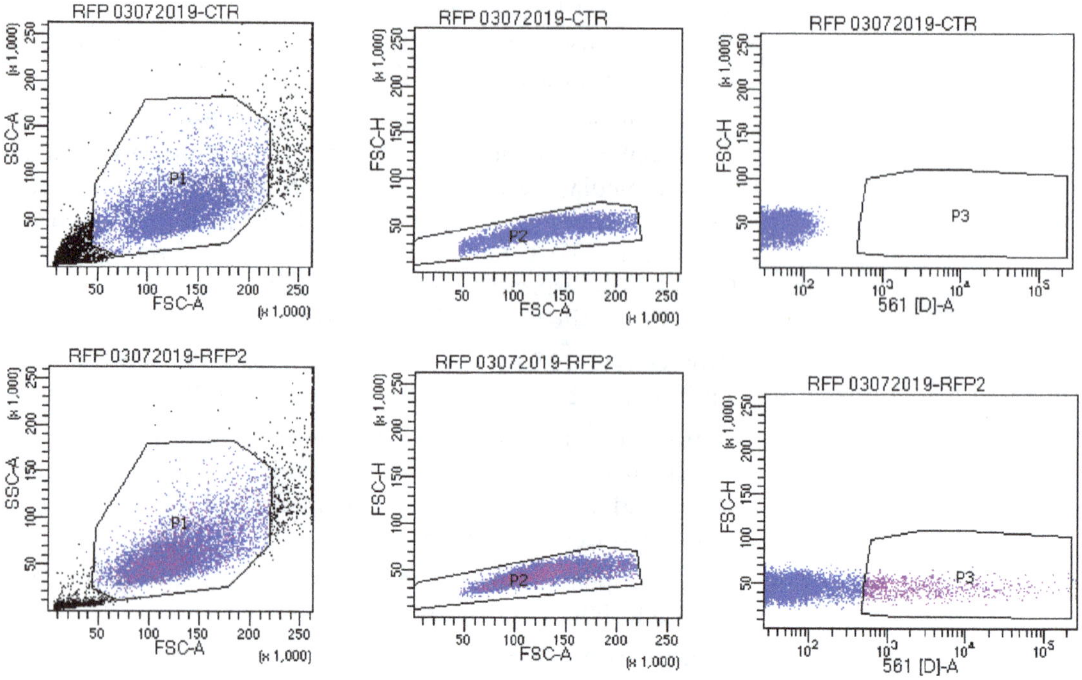

Tube: RFP2

Population	#Events	%Parent	%Total
All Events	10,000	####	100.0
P1	8,239	82.4	82.4
P2	8,239	100.0	82.4
P3	728	8.8	7.3

Fig. 2 Example FACS plots for control cells (top row) and iTracer nucleofected cells (bottom row). In this example, cells in the P3 gate should be selected for collection

3.3 Organoid Culture and Scarring

1. We generated organoids following the unguided brain organoid protocol outlined below [1, 11].
2. Addition of doxycycline (DOX) induces Cas9 expression and consequently scarring of the iTracer fluorescent reporter. Scarring is achieved by adding DOX to the media at any time during organoid generation for 24 h. The media used for scarring is dependent on the time point targeted for scarring. The concentration of DOX must be optimized for the size and type of the organoid being used (*see* **Note 10**).
3. Remove and add fresh media without DOX after 24 h and continue with organoid culture until the desired collection day.

3.4 Collection and Processing of iTracer Organoids

Organoids can be collected at any desired time point. iTracer cells can be measured with standard dissociation and scRNAseq methods (3' mRNA, 10× Genomics) or with sequencing-based spatial transcriptomics (Visium, 10× Genomics). The leftover cDNA yielded from either method is used to amplify the iTracer readouts.

1. Normalize cDNA input to 50 ng in a volume of 17.75 μL (*see* **Note 11**).
2. For each sample prepare 32.25 μL of Root PCR mix: 1.5 μL 100% DMSO, 2.5 μL 10 μM 10× Root PCR Forward Primer (*see* **Note 11**), 2.5 μL 10 μM 10× Universal Primer, 25 μL Phusion Master Mix, 0.75 μL EvaGreen.
3. Add 32.25 μL per well of Root PCR mix to a qPCR plate.
4. To each well add 17.75 μL of normalized cDNA.
5. Run qPCR according to the following program: 98 °C for 30 s, then cycles of (98 °C for 15 s, 66 °C for 15 s, 72 °C for 20 s). The number of cycles used is dependent on when the reaction levels off after the exponential phase, when this happens, stop the reaction.
6. Clean-up reaction using SPRI beads: Thoroughly resuspend SPRI beads by shaking or vortexing. Add 50 μL (1×) of SPRI select reagent to the sample. Mix thoroughly by pipetting 15×. Incubate for 5 min at room temperature. Place the reaction vessel on a PCR tube or plate magnet and wait until the solution clears. Remove and discard the supernatant. While still in the magnet, wash the beads with 180 μL 80% ethanol solution. Wait for 30 s then discard the ethanol without disturbing the beads. Repeat the ethanol wash step once more before centrifuging the reaction briefly. Replace the samples onto the magnet and remove remaining ethanol using a p10 pipette. Remove from the magnet and add 30.5 μL Qiagen EB buffer and mix thoroughly by pipetting. Incubate for 2 min at room temperature. Replace the reaction onto the magnet and wait until the solution clears. Transfer 30 μL sample to a clean tube or plate.
7. For each sample prepare 45 μL of Nested Barcode reaction mix: 25 μL Phusion Master Mix, 2.5 μL 10 μM Nest Barcode PCR Forward Primer (*see* **Note 12**), 2.5 μL 10 μM 10× Universal Primer, 0.75 μL EvaGreen, and 14.25 μL water.
8. Add 45 μL per well of reaction mix to qPCR plates.
9. Add 5 μL of Root PCR product to each well.
10. Run qPCR according to the following program: 98 °C for 30 s, then cycles of (98 °C for 15 s, 65 °C for 15 s, 72 °C for 20 s). The number of cycles is dependent on when the reaction levels off after the exponential phase, when this happens, stop the reaction.
11. Clean up each reaction with 1:1 SPRI beads (50 μL beads added) and elute with 30 μL EB buffer as described above.

12. Obtain the sample concentrations with a Nanodrop or Qubit.
13. For each sample prepare 45 μL of Nested Scar reaction mix: 25 μL Phusion Master Mix, 2.5 μL 10 μM Nest Scar PCR Forward Primer (*see* **Note 12**), 2.5 μL 10 μM 10× Universal Primer, 0.75 μL EvaGreen, and 14.25 μL water.
14. Add 45 μL per well of reaction mix to qPCR plates.
15. Add 5 μL of Root PCR product to each well.
16. Run PCR according to the following program: 98 °C for 30 s, then cycles of (98 °C for 15 s, 66 °C for 15 s, 72 °C for 20 s). The number of cycles is dependent on when the reaction levels off after the exponential phase, when this happens stop the reaction.
17. Clean up each reaction with 1:1 SPRI beads (50 μL beads added) and elute with 30 μL EB buffer as described above.
18. Obtain the sample concentrations with a Nanodrop or Qubit.
19. Prepare samples by diluting nested reaction products to 60 ng of input in a final volume should be 15 μL.
20. To each normalized sample, add 25 μL of Phusion Master Mix.
21. Add 10 μL of 10× dual index primers such that indexes are not repeated (*see* **Note 13**).
22. Run PCR according to the following program: 98 °C for 30 s, then 8 cycles of (98 °C for 15 s, 55 °C for 15 s, 72 °C for 20 s), 72 °C for 2 min.
23. Clean up each reaction with a SPRI double-sided bead clean up: Add 50 μL of EB to each reaction to have a total volume of 100 μL. Vortex thoroughly to resuspend the SPRI beads. Add 60 μL SPRI beads (0.6×) to each sample and pipette mix 15×. Incubate for 5 min at room temperature. Place the tubes on a magnet until the solution clears. Transfer 150 μL supernatant to a new tube. Add 20 μL SPRI beads (0.8×) to each sample and pipette mix 15×. Incubate for 5 min at room temperature. Place the tubes on a magnet until the solution clears. Remove 165 μL of supernatant. With the tube still in the magnet, add 200 μL 80% ethanol to the pellet and wait 30 s. Remove the ethanol and repeat the wash once more. Centrifuge briefly and place samples on a magnet. Remove remaining ethanol using a p10 pipette. Remove from the magnet, add 35.5 μL EB, and mix by pipetting 15×. Incubate 2 min at room temperature, then place the sample on a magnet until the solution clears. Transfer 35 μL to a new tube. Samples can be stored at 4 °C for up to 72 h or at −20 °C for long-term storage.
24. Check the quality of the libraries by diluting 1 μL sample with 9 μL of water and load on a Bioanalyzer (Fig. 3).

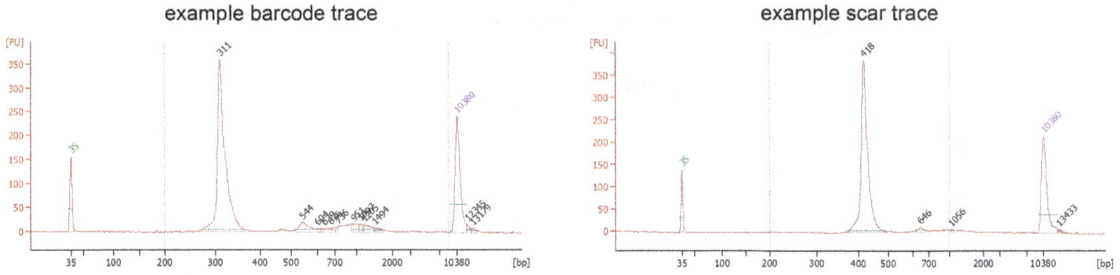

Fig. 3 Example Bioanalyzer traces for barcodes (left) and scars (right). Barcodes are expected to be in the 300 base pair range for barcode libraries and in the 400 base pair range for scar libraries

3.5 Extraction and Filtering of iTracer Readouts from Single-Cell RNA Sequencing Data

1. Use Cell Ranger (10× Genomics) to demultiplex base-call files to FASTQ files and align reads. Default alignment parameters are used to align reads to a modified human reference including the fluorescent reporter (GFP or RFP) from the cell lineage recorder (hg38).
2. Build a custom reference genome with the cell ranger mkref pipeline. Genome FASTA and GTF files, as well as code, are available on our github (https://github.com/quadbio/iTracer_analysis/).
3. Align gene expression and barcode and scar libraries using default parameters, with the exception that the force-cells argument is set to 200,000.
4. Use the R package itraceR to extract and filter the iTracer readouts. Open an R session (R, rstudio, jupyter notebook, etc).
5. If the R package "devtools" is not yet installed, install it with:

```
install.packages('devtools')
```

6. If itraceR is not yet installed, install it using "devtools":

```
devtools::install_github('quadbio/itraceR')
```

7. Load the itraceR package in your R session:

```
library(itraceR)
```

8. Use the following R code to extract and filter the iTracer readout:

```
# Extract barcode UMIs
#Note: replace file_barcode with the path to the BAM file
output by CellRanger mapping of the barcode library (e.g.,
```

```
barcode/outs/possorted_genome_bam.bam)

df_barcode <- extract_barcodes(file_barcode)

# Extract scar UMIs
# Note: replace file_scar with the path to the BAM file output
by CellRanger mapping of the scar library (e.g., scar/outs/
possorted_genome_bam.bam)

df_scar <- extract_scars(file_scar)

# Filter barcode UMIs based on number of reads, barcode
sequences, etc.
# Note: do_plot=T to plot the read number per UMI distribu-
tion; remove it if such plot is not needed

df_barcode <- do_filter_barcodes(df_barcode, do_plot=T)

# Filter scar UMIs based on number of reads, etc.
# Note: do_plot=T to plot the read number per UMI distribu-
tion; remove it if such plot is not needed

df_scar <- do_filter_scars(df_scar , do_plot=T)

# Merge barcode UMIs and scar UMIs data frames into the final
iTracer readout data frame, by taking only barcode UMIs and
scar UMIs sharing the same cell barcode and UMI barcode

df_barcode_scar <- merge_barcode_scars(df_barcode, df_scar)
```

The "do_plot=T" parameters above will provide histograms as the following example to show the estimated threshold of number of reads per UMI to exclude barcode/scar UMIs with too few supporting reads (Fig. 4).

3.6 Merging iTracer Readouts with Single-Cell RNA Sequencing Data and Analysis

1. Add the filtered barcode and scar information (**step 3.5**) to your single-cell RNA sequencing data (*see* **Note 14**), it can be added as metadata columns to the single-cell dataset obtained from the sequencing of the iTracer experiment. This can be done with the "incorporate_to_seurat" function in the "itraceR" package:

```
# Incorporate the iTracer readout data frame obtained above to
the metadata of the corresponding Seurat object
# Note: by default the barcode information is added to the new
column 'GeneBarcodeMerged', and the scar is added to 'Scar-
Merged'
```

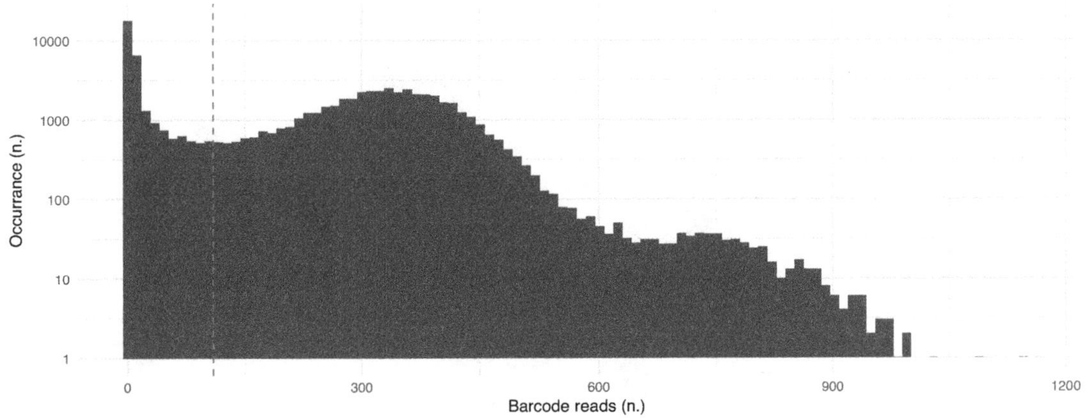

Fig. 4 The do_plot=T parameter setting yields a bar plot showing the frequency of barcodes with a given number of reads. The filtering cutoff for low read number barcodes is represented with a dotted line

```
seurat_obj <- incorporate_to_seurat(seurat_obj, df_barcode_s-
car)
```

2. Plot the lineage tree with the incorporated iTracer readout in the meta.data slot of the seurat object, optionally color the cells by the annotated cell type (*see* **Note 14**, Fig. 5) if col_lab is not given, all cells are colored the same:

```
# Plot the lineage tree
# Note: if col_lab is not specified, the function assumes no
cell type annotation, and all cells will be colored the same

plot_tree(seurat_obj@meta.data, col_lab = 'annot')
```

4 Notes

1. iTracer can be used with any iPSC line that has an inducible Cas9 protein.
2. Example Calculation:

   barcodes = 63 bp with 0.0238 μg/μL = 0.57 pmol DNA (in 1 μL)

   plasmid fragment (iTracer-R) = 5613 bp with 0.2044 μg/μ L = 0.055 pmol (in 1 μL)

   barcodes should be 10-fold the molar excess of an insert, therefore: 10× 0.055 pmol (the plasmid fragment) = 0.55 pmol

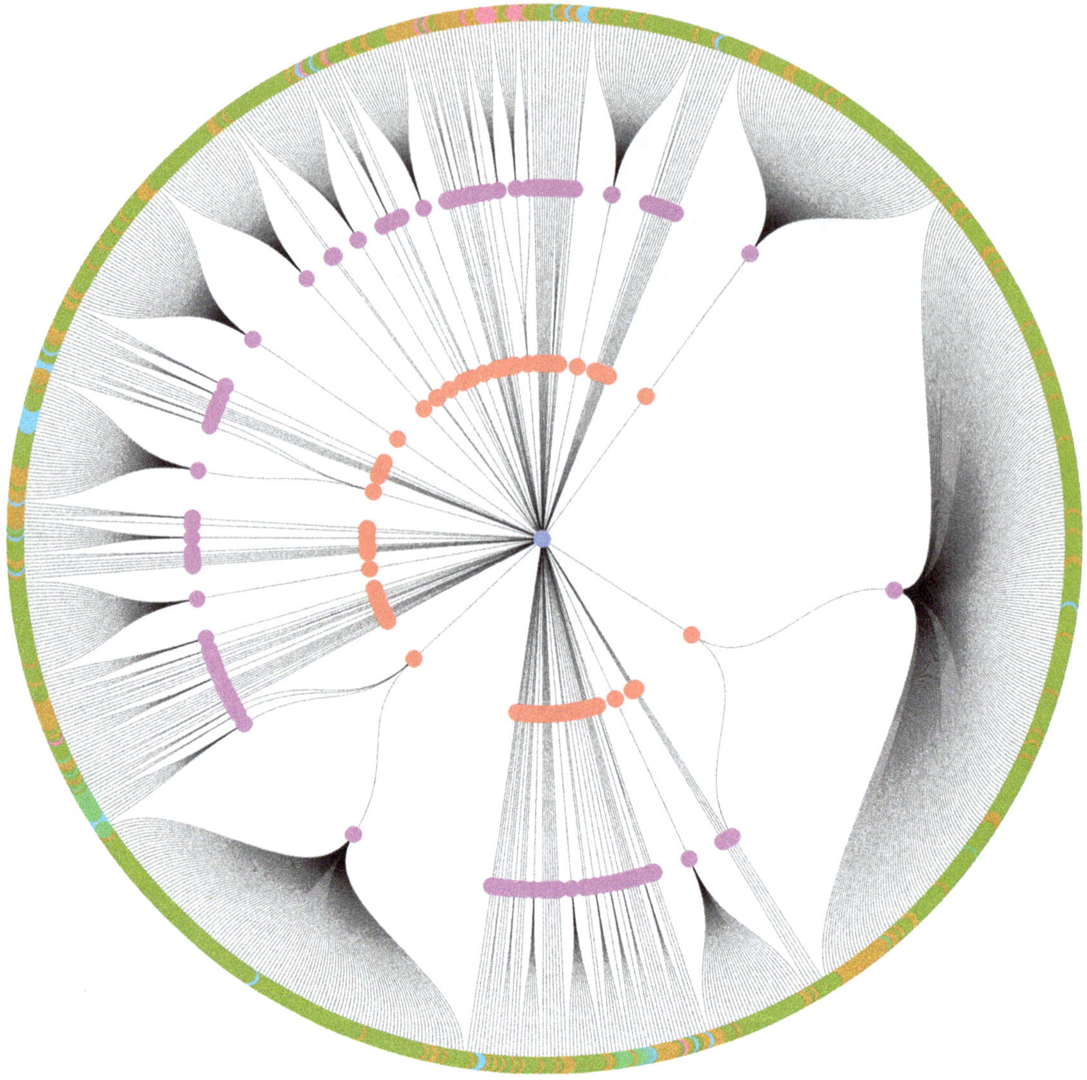

Fig. 5 The plot_tree function yields a graphical representation of the lineage relationships deduced from the single-cell RNA sequencing data. The tree emanates from a pseudonode (blue) that symbolizes the sequenced organoid as a whole. The first layer of nodes (pink) represents barcode sequences, the second layer of nodes (purple) represents scar patterns, and the last layer represents individual cells which are colored by their cell type label

3. If this is not the case, you may have an error and will need to troubleshoot. Control plates with many colonies may have plasmid carryover from the backbone fragment cleanup, do these steps and repeat.
4. Bulk amplicon sequencing can be done from the pool of plasmids to assess the diversity of barcodes available in the pool.

(a) Primers:

GFP Forward Primer: TCGTCGGCAGCGTCA GATGTGTATAAGAGACAGGAGAAGCGCGAT CACATGG

RFP Forward Primer: TCGTCGGCAGCGTCA GATGTGTATAAGAGACAGTCCCACAACGAG GACTACAC

Reverse Primer: GTCTCGTGGGCTCGGAGATGTGTA TAAGAGACAGAGTGCCTTCCAGGGTCAAG GAAGG

(b) Amplify target area using the appropriate primers (amplify barcodes with either GFP or RFP primers). Normalize DNA input to 50ng/μL. For each sample prepare reaction mix: 5 μL Buffer B (KAPA Biosystems, #KK5005), 0.5 μL dNTPs, 1.25 μL 10 μM PCR Forward Primer, 1.25 μL 10 μM PCR Reverse Primer, 1 μL input DNA, and fill to 25 μL with water.

(c) Run PCR according to the following program: 95 °C for 3 min, then 35 cycles of 95 °C for 15 s, 65 °C for 15 s, 72 °C for 15 s, 72 °C for 1 min.

(d) Run a 2% gel to check for proper amplification.

(e) Clean up each reaction with 1:1 SPRI beads (25 μL beads added) and elute with 25 μL EB buffer as described above.

(f) Index the reaction with Nextera XT indexes (Illumina, #FC-131-1096). For each sample prepare reaction mix: 12.5 μL Phusion Master Mix (ThermoFisher, #F531L), 1.25 μL each Nextera index (i5, and i7), 1 μL amplified DNA, and 9 μL water (25 μL total volume).

(g) Run PCR according to the following program: 98 °C for 30 s, then 25 cycles of 98 °C for 10 s, 58 °C for 20 s, 72 °C for 20 s, 72 °C for 5 min.

(h) Run a 2% gel to check for proper amplification.

(i) Clean up each reaction with a SPRI double sided bead clean up. Add 75 μL of EB to each reaction to have a total volume of 100 μL. Vortex thoroughly to resuspend the SPRI beads. Add 60 μL SPRI beads (0.6×) to each sample and pipette mix 15×. Incubate for 5 min at room temperature. Place the tubes on a magnet until the solution clears. Transfer 150 μL supernatant to a new tube. Add 20 μL SPRI beads (0.8×) to each sample and pipette mix 15×. Incubate for 5 min at room temperature. Place the tubes on a magnet until the solution clears. Remove 165 μL of supernatant. With the tube still in the magnet, add 200 μL 80% ethanol to the pellet and wait 30 s.

Remove the ethanol and repeat the wash once more. Centrifuge briefly and place samples on a magnet. Remove remaining ethanol using a p10 pipette. Remove from the magnet, add 35.5 μL EB, and mix by pipetting 15×. Incubate 2 min at room temperature, then place the sample on a magnet until the solution clears. Transfer 35 μL to a new tube.

5. Incorporation of iTracer plasmid along with Sleeping Beauty plasmid can be achieved using other methods such as electroporation or lipofection.
6. Timing for lifting the cells may vary. It is advised to check the cells after 2 min and every minute thereafter to make sure the cells have not lifted off the plate. If cells do lift off the plate due to over incubation, you can directly add media to the well to quench the reaction and proceed to the next step.
7. We optimized this step and found this protocol worked best for our cells. Depending on the cell line used, this step can be optimized for optimum results and survivability.
8. Three days is the minimum time to allow cells to recover. If left for less than 3 days then there is risk that transcriptionally active, non-integrated plasmid is left in the cells and will result in false positives. However, the longer the cells are left to recover and multiply the more clonal amplifications that can occur. This could lead to diversity loss if left too long.
9. It is recommended that 100,000 cells are plated per 12-well or 300,000 cells are plated per 6-well. Multiple wells can be used to plate depending on the number of iTracer cells sorted.
10. The concentration of DOX should be optimized for the time points targeted. To accomplish this, one can use multiple concentrations of DOX and assess the scarring efficiency and survival of cells using bulk amplicon sequencing and/or fluorescent signal.
11. Lower input is possible but not recommended. If necessary, add as much volume as possible without exceeding 17.75 μL.
12. Primers used are dependent on the iTracer plasmid chosen, either use the GFP or RFP depending on which is used/selected.
13. Care should be taken to not have repeated indexes used between iTracer barcode and scar libraries, if sequencing iTracer libraries with 10× 3' single-cell libraries, then it is important to also not have repeated indexes with those samples as well.
14. This step assumes you have constructed a seurat object from the gene expression data called "seurat_obj," with the annotated cell type information stored as a metadata column called "annot." To this object you will merge the dataframe created in **step 3.5** called "df_barcode_scar."

References

1. He Z, Maynard A, Jain A et al (2022) Lineage recording in human cerebral organoids. Nat Methods 19:90–99
2. Weinreb C, Rodriguez-Fraticelli A, Camargo FD, Klein AM (2020) Lineage tracing on transcriptional landscapes links state to fate during differentiation. Science (1979). https://doi.org/10.1126/science.aaw3381
3. Spanjaard B, Hu B, Mitic N, Olivares-Chauvet P et al (2018) Simultaneous lineage tracing and cell-type identification using CRISPR-Cas9-induced genetic scars. Nat Biotechnol 36: 469–473
4. Chan MM, Smith ZD, Grosswendt S et al (2019) Molecular recording of mammalian embryogenesis. Nature 570:77–82
5. Biddy BA, Kong W, Kamimoto K, Guo C, Waye SE, Sun T, Morris SA (2018) Single-cell mapping of lineage and identity in direct reprogramming. Nature 564:219–224
6. McKenna A, Findlay GM, Gagnon JA, Horwitz MS, Schier AF et al (2016) Whole-organism lineage tracing by combinatorial and cumulative genome editing. Science (1979). https://doi.org/10.1126/science.aaf7907
7. Bowling S, Sritharan D, Osorio FG et al (2020) An engineered CRISPR-Cas9 mouse line for simultaneous readout of lineage histories and gene expression profiles in single cells. Cell 181:1410–1422.e27
8. Raj B, Wagner DE, McKenna A, Pandey S et al (2018) Simultaneous single-cell profiling of lineages and cell types in the vertebrate brain. Nat Biotechnol 36:442–450
9. Lancaster MA, Renner M, Martin CA et al (2013) Cerebral organoids model human brain development and microcephaly. Nature 501:373–379
10. Riesenberg S, Maricic T (2018) Targeting repair pathways with small molecules increases precise genome editing in pluripotent stem cells. Nat Commun. https://doi.org/10.1038/s41467-018-04609-7
11. Sutcliffe M, Lancaster MA (2019) A simple method of generating 3D brain organoids using standard laboratory equipment. In: Methods in molecular biology. Humana Press, pp 1–12

Chapter 6

Clonal Tracking in the Mouse Brain with Single-Cell RNA-Seq

Michael Ratz and Leonie von Berlin

Abstract

Lineage tracing methods enable the identification of all progeny generated by a single cell. High-throughput lineage tracing in the mammalian brain involves parallel labeling of thousands of progenitor cells with genetic barcodes in vivo followed by single-cell RNA-seq of lineage relations and cell types. Here we describe the generation of barcoded lentivirus, microinjections into the embryonic day 9.5 mouse forebrain, dissociation of 2-week-old mouse brain tissue, single-cell RNA-seq library preparation, and data analysis using a custom software. Compared to traditional methods based on sparse fluorophore labeling of progenitor cells, lineage tracing with genetic barcodes and single-cell RNA-seq has a >100-fold higher throughput while using >10 times fewer mice.

Key words Genetic barcodes, Lentivirus, Mouse brain, Neural stem cells, Single-cell RNA-seq, Lineage tracing

1 Introduction

Lineage tracing, the identification of all cells generated by an individual cell [1], is an essential tool for understanding the developmental principles of complex tissues such as the mammalian brain where few homogenous progenitor cells generate many diverse cell types [2]. Prospective lineage tracing involves the experimental labeling of a single cell with a heritable marker that is transmitted to all future generations of cells resulting in a group of labeled cells called a clone. Most common approaches rely on the introduction of a genetically encoded fluorescent protein into a progenitor cell using plasmid transfection, retrovirus infection, or site-specific recombination [2]. This is followed by confocal fluorescence microscopy of tissue sections collected at a later timepoint for quantitative clonal analysis and cell phenotyping using anatomical and/or molecular features. However, conventional approaches are laborious and low-throughput, because the number of progenitor

Jorge García-Marqués and Tzumin Lee (eds.), *Lineage Tracing: Methods and Protocols*, Methods in Molecular Biology, vol. 2886, https://doi.org/10.1007/978-1-0716-4310-5_6,

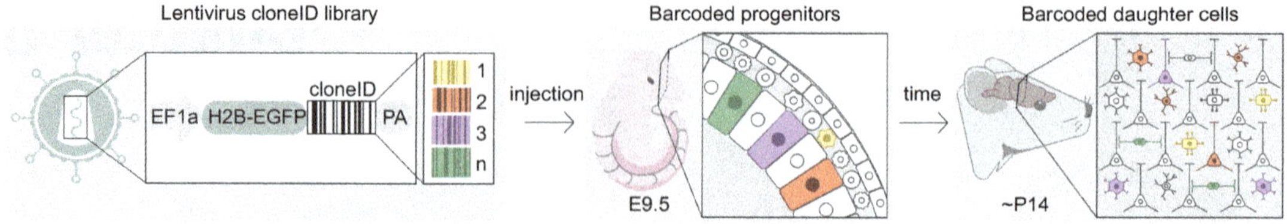

Fig. 1 Overview of clonal tracking and gene expression (TREX) profiling in the mouse brain. Millions of diverse cloneIDs are used to label progenitor cells in the embryonic mouse brain in vivo. Each progenitor cell transmits its cloneID to its progeny and cloneIDs are queried using single-cell RNAseq. (Adapted from [9])

cells that can be traced in parallel is limited by the number of spectrally distinguishable fluorophores, hence requiring sparse labeling of few (<10) progenitor cells per mouse brain [3–7]. Moreover, cataloguing the full diversity of related brain cells is challenging using fluorescence imaging of molecular markers and an unbiased, transcriptome-based taxonomy of mammalian brain cell types would be preferred [8].

Here we describe protocols that enable combined clonal tracking and gene expression profiling (TREX) in the mammalian brain in vivo [9] (Fig. 1). TREX is based on the lentiviral delivery of genetic barcodes (from here on called cloneIDs) into progenitor cells and single-cell RNA-seq for combined cloneID and whole transcriptome readout. TREX utilizes random 30 bp cloneIDs transcribed as part of the 3'UTR of a nuclear-localized enhanced green fluorescent protein (EGFP) driven by a strong, ubiquitous EF1a promoter. We have used TREX for high-throughput tracing of embryonic day 9.5 progenitor fates in the mouse forebrain [9, 10] and embryonic day 8.5 progenitor fates in mouse neural crest [11]. We have achieved combined barcode and gene expression readout using droplet-based [9, 11] and plate-based [10] single-cell RNA-seq as well as spatial transcriptomics [9]. In the following sections, we provide detailed protocols for (1) the generation and characterization of barcoded lentivirus preparations, (2) the performance of microinjections into the mouse forebrain at embryonic day 9.5, (3) single-cell RNA-seq library preparation of barcoded mouse brain tissue, and (4) data analysis using a custom software. Our protocol is intended to be used by developmental (neuro)-biologists that have expertise with animal models and that want to learn more about the generation of cloneID libraries and data analysis.

2 Materials

2.1 Small and Large Agar Plates

Measure 35 g of pre-mixed LB-agar high salt powder per liter of water. Sterilize by autoclaving, let cool down to 60 °C and add ampicillin at a final concentration of 100 μg/mL. Distribute across four large square bioassay dishes (245 mm × 245 mm) by pouring 225 mL of liquid agar into each dish. Distribute the remaining agar equally across four small Petri dishes (diameter 90 mm). One large and one small plate is needed per library transformation. Unused plates can be stored at 4 °C until use.

2.2 DNA Oligos

Cloning and characterization of cloneID library via next-generation sequencing requires oligos listed in Table 1. The enrichment of cloneID sequences from full-length cDNA obtained during single-cell RNA-seq library preparation requires oligos listed in Table 2.

2.3 Dissociation of Mouse Brain Tissue

1. Xite Fluorescence Flashlight, Royal Blue (NIGHTSEA).
2. Stereo microscope.

Table 1
DNA oligos required for cloning and characterization of cloneID library

Name	5′-sequence-3′	Purpose
MRX1642	CGGCATGGACGAGCTGTACAAG	Forward oligo for PCR on MRX1680 ultramer
MRX1643	GTGGCTAAGATCTACAGCTGCC	Reverse oligo for PCR on MRX1680
MRX1680	CGGCATGGACGAGCTGTACAAGTAAATGCA NNNNNNNNNNNNNNNNNNNNNNNNNNNNNN CTTTAAGACCAATGACTTA CAAGGCAGCTGTAGATCTTAGCCAC	Ultramer™ DNA Oligo with random 30 bp barcode
MRX1693	ACACTCTTTCCCTACACGACGCTCTTCCGATCT CGGCATGGAC GAGCTGTACAAG	Forward oligo for PCR1 on viral cDNA
MRX1694	AGACGTGTGCTCTTCCGATCT GTGGCTAAGATCTACAGCTGCC	Reverse oligo for PCR1 on viral cDNA
MRX142	AATGATACGGCGACCACCGAGATCTACAC **TAGATCGC** ACACT CTTTCCCTACACGACG	Forward oligo for indexing PCR (i5 index N501)
MRX614	CAAGCAGAAGACGGCATACGAGAT **AGGAGTCC** GTGACTGGAGTTC AGACGTGTGCTCTTCCGATCT	Reverse oligo for indexing PCR (i7 index N705)

Table 2
DNA oligos required for enrichment of cloneID sequences from cDNA

Name	5′-sequence-3′	Purpose
MRX14	CTGGTCGAGCTGGACGGCGACG	CloneID enrichment from cDNA, PCR1
MRX1587	CTACACGACGCTCTTCCGATCT	CloneID enrichment from cDNA, PCR1/2
MRX1588	GTGACTGGAGTTCAGACGTGTGCTCTTCCGATCT CATGGTCCTGCTGGAGTTCGTGAC	CloneID enrichment from cDNA, PCR2
MRX1589	AATGATACGGCGACCACCGAGATCTACACTCTTTCC CTACACGACGCTCTTCCGATCT	CloneID enrichment from cDNA, PCR3

3. Dissection equipment.
   - Forceps, Curved, Serrated (World Precision Instruments).
   - Dumont #5/45 Forceps (Fine Science Tools).
4. 1 mm acrylic brain matrix for mouse (World Precision Instruments).
5. Pasteur pipettes 150 mm.
6. Black Sylgard dissection dish (Scintica).
7. Artificial cerebral fluid (ACSF): 87 mM NaCl, 2.5 mM KCl, 1.25 mM NaH_2PO_4, 26 mM $NaHCO_3$, 75 mM sucrose, 20 mM glucose, 2 mM $MgSO_4$, 2 mM $CaCl_2$. Saturate solution with carbogen (5% CO2/95% O2) for 15 min on ice and adjust pH to 7.3–7.4 before adding divalent cations. Sterilize with membrane filter before use and always keep on ice.
8. ACSF/BSA: Prepare 50 mL of ACSF containing 1% BSA.
9. Papain: Reconstitute lyophilized papain (Worthington) in vial with 5 mL of aCSF and vortex briefly. Incubate at 37 °C to ensure full solubility and activity.
10. DNase I: Reconstitute DNase I vial (Worthington) content with 0.5 mL of aCSF and vortex briefly. Add 250 μL to 5 mL papain solution.
11. Ovomucoid protease inhibitor—stock: Reconstitute vial content with 32 mL of aCSF. Solubilize completely by stirring on a magnetic stirrer and sterilize with membrane filter (0.22 μm pore size).
12. Ovomucoid protease inhibitor—dilution: Dilute 1 mL of ovomucoid protease inhibitor stock solution in 9 mL of ACSF.

2.4 Barcode Extraction and Clone Reconstruction

1. Software to process raw scRNAseq data such as *Cell Ranger* (10x Genomics Chromium Platform https://support.10xgenomics.com/single-cell-gene-expression/software/

downloads/latest), *Space Ranger* (10x Genomics Visium Platform https://www.10xgenomics.com/support/software/space-ranger/downloads) or *zUMIs* (Smart-Seq2/3 platforms https://github.com/sdparekh/zUMIs).

2. Software to analyze single-cell genomics data such as *Seurat* or *Scanpy* (https://satijalab.org/seurat/ or https://scanpy.readthedocs.io/en/stable/).
3. Python 3.7 or newer (https://www.python.org/downloads/).
4. Version control system Git (https://git-scm.com/).
5. scRNAseq output files.
6. Fasta file containing the sequence of the barcode construct.

3 Methods

3.1 Generation and Characterization of Lentivirus Barcode Library

We utilize a lentivirus vector expressing a random 30 bp barcode within the 3'-UTR of nuclear-localized EGFP transgene under the EF1a promoter (Fig. 2). We perform Gibson assembly [12] to generate a library of lentivirus transfer plasmids that is subjected to lentiviral packaging followed by library characterization using next-generation sequencing. All steps involved in this process are described below.

3.1.1 Restriction Digestion of Backbone

Digest plasmid DNA by combining the following reagents in 1.5 mL DNA LoBind® Tube:

Reagent	Volume [μL]
ddH2O	35
Plasmid DNA (1 μg/μL)	5
rCutSmart® Buffer (10×)	5
NsiI-HF® (20 units/μL)	2.5
KpnI-HF® (20 units/μL)	2.5
Total	**50**

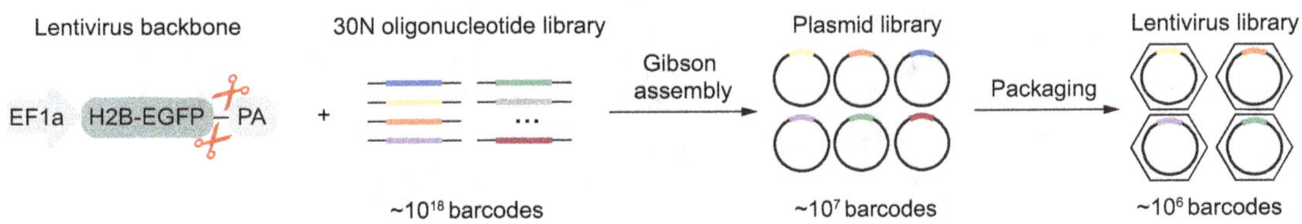

Fig. 2 Barcoded lentivirus generation. Gibson assembly is used to generate an initial lentivirus transfer plasmid library followed by packaging of high titer lentivirus. (Adapted from [9])

Mix by flicking the tube 2–3 times, spin down briefly and incubate at 37 °C for 1 h.

Add 1 μL of rSAP to the digestion mix and incubate at 37 °C for another 1 h.

Heat inactivate rSAP and NsiI at 65 °C for 20 min.

Add 10 μL of 6× Gel Loading Dye (comes with restriction enzymes).

Mix by flicking the tube 2–3 times, spin down briefly.

Load 60 μL and 1 kb DNA ladder on separate lanes of a 1% agarose gel and run gel electrophoresis.

Cut 8.2 kb DNA band and purify DNA using QIAquick Gel Extraction Kit.

Elute in 32 μL ddH2O and use 2 μL to measure DNA concentration at NanoDrop.

Final volume is 30 μL and expected yield is 50–100 ng/μL.

3.1.2 Low-Cycle PCR to Amplify Random Barcodes

Combine the following reagents in a PCR tube:

Reagent	Volume [μL]
ddH2O	21.3
Q5® High-Fidelity 2X Master Mix	25
MRX1642 (100 μM)	0.1
MRX1643 (100 μM)	0.1
MRX1680 ultramer (1 μM)	1
DMSO	2.5
Total	**50**

Mix by flicking the tube 2–3 times, spin down briefly, transfer to thermocycler, and amplify using the following PCR protocol:

Temperature [°C]	Time	# of cycles
98	5 min	
98 65 72	20 s 20 s 10 s	10 (*See* **Note 1**)
72	5 min	
4	hold	

Add 10 μL of 6× Gel Loading Dye, mix by flicking the tube 2–3 times, spin down briefly.

Load 10 μL and 100 bp DNA ladder on separate lanes of a 3% agarose gel and run gel electrophoresis.

The expected length of the amplified DNA fragment is 108 bp.

Purify DNA from the remaining 50 μL using MinElute PCR Purification Kit.

Elute in 12 μL ddH2O and use 2 μL to measure DNA concentration at NanoDrop.

Final volume is 10 μL and expected yield is 50–100 ng/μL (*See* **Note 2**).

Dilute to 18 ng/μL using ddH2O.

3.1.3 Barcode Library Cloning Using Gibson Assembly

Set up two reactions in two separate PCR tubes on ice. One reaction will serve as a "no insert" negative control to assess the background of bacterial colonies that do not contain a barcode due to re-ligation and/or undigested backbone.

Mix the following reagents in tube 1 for Gibson assembly of the barcode library:

Reagent	Volume [μL]
ddH2O	7
pMR506 (100 ng/μL) prepared in Subheading 3.1.1)	2
Barcodes (18 ng/μL)	1 (*See* **Note 3**)
2x Gibson Master Mix	10

Mix the following reagents in tube 2 for the "no insert" negative control:

Reagent	Volume [μL]
ddH2O	8
pMR506 (100 ng/μL) prepared in Subheading 3.1.1)	2
2x Gibson Master Mix	10

Transfer tubes to a preheated thermocycler, incubate at 50 °C for 1 h.

Transfer content of each tube (20 μL) to a new DNA LoBind® Tube on ice.

Add 4 μL of glycogen (5 μg/μL), 2.4 μL of Sodium Acetate (3 M, pH 5.5) solution (0.1× sample volume) and 66 μL of absolute ethanol (2.5× sample volume).

Store both tubes at −80 °C for 30 min.

Centrifuge in a precooled microcentrifuge at 14,000 rpm (16,873 g) at 4 °C for 15 min.

A small pellet will be visible at the bottom of the tube after centrifugation.

Gently aspirate the supernatant and discard it.

Wash pellet by adding 200 μL cold 70% ethanol and briefly vortexing.

Centrifuge in a precooled microcentrifuge at 14,000 rpm (16,873 g) at 4 °C for 5 min.

Gently aspirate the supernatant and discard it.

Dry the pellet at 37 °C for 2–5 min.

Add 10 μL ddH2O to the pellet, dissolve by vortexing followed by incubation at 37 °C for 5 min.

3.1.4 Transformation of E. coli

To clone and maintain lentiviral barcode libraries with high efficiency, we use Endura electrocompetent cells and electroporation for transformation. Follow this protocol to transform both the barcode library and negative control from the previous step:

Thaw Recovery Medium (comes with Endura cells) in a 37 °C water bath.

Prepare two Nunc™ Square Dishes and two Nunc™ Petri Dishes with LB-ampicillin agar.

Pre-chill two 0.1 cm MicroPulser Cuvettes on ice.

Thaw one vial of 50 μL Endura cells, transfer 25 μL cells into each cuvette and keep on ice for 15 min.

Add the entire 10 μL of the Gibson Assembly reaction products (Subheading 3.1.3).

Gently mix by flicking the bottom of the cuvette a few times.

Place the cuvette into MicroPulser Electroporator and use the program "Ec1" ($V = 1.8$ kV) to electroporate.

Add 975 μL of prewarmed Recovery Medium directly to the cuvette, resuspend bacteria, and transfer the total volume of 1010 μL to a 1.5 mL DNA LoBind® Tube.

Place the tube in a TS-100 Thermo-Shaker and mix at 700 rpm at 37 °C for 1 h.

Serially dilute 10 μL of the bacterial suspension with 90 μL recovery medium using a dilution factor of 10 to create a 1:100, 1:1000, and 1:10000 dilution of the main suspension.

Plate entire suspension of the 1:10000 dilution on Nunc™ Petri Dish.

Plate the remaining 1 mL of bacterial suspension on Nunc™ Square Dishes.

Incubate the plates at 37 °C for 12–16 h (*See* **Note 4**).

3.1.5 Library Size Quantification and Plasmid Preparation

The large square dish will contain many colonies that will be difficult and time-consuming to count. To estimate the total number of colonies on the large dish, count the number of colonies on the small Petri dish and multiply with the dilution factor. A typical result is shown in Table 3.

These results provide information about (1) the total number of colonies and therefore potential barcodes and (2) background caused by re-ligated and/or undigested backbone without barcodes.

We usually obtain 1–2 million colonies for a single square dish which is sufficient for most downstream in vivo applications. If needed, the total number of colonies can be increased by setting up multiple Gibson assembly reactions and subsequent transformations for the barcode library.

Ideally, the control plate should contain zero colonies. In practice, this is rarely the case, but it is sufficient if the number of colonies on the control plate is <10% of the number of colonies on the library plate. The example results in Table 3 correspond to a background of about 1.4% and we therefore continue with plasmid preparation using the following protocol:

A. Scrape the cells from the square dish into 100 mL PBS and distribute suspension equally among two 50 mL centrifuge tubes.

B. Spin bacterial suspensions at 3000 g for 15 min at room temperature.

Table 3
Typical results for the number of colonies after library transformation

Plate	# colonies on Petri dish	Dilution factor	# colonies on square dish
Control	2	10,000	20,000
Library	146	10,000	1,460,000

C. Remove supernatants and isolate transfection-grade plasmid DNA using the QIAGEN Plasmid Plus Midi Kit and following the manufacturer's instructions.

D. Total expected yield of plasmid DNA is 0.5–1 mg.

3.1.6 Barcoded Lentivirus Production

We outsourced packaging of VSV-G pseudotyped lentivirus particles to a local facility (VirusTech Core Facility, Karolinska Institute, Stockholm, Sweden) or to a commercial provider (GEG-Tech, Paris, France). In general, it is critical to obtain a lentivirus preparation that is of high titer ($\geq 10^9$ TU/mL) to maximize the number of labeled cells in vivo and of high purity as impurities (e.g., debris from the packaging cells) may cause inflammatory responses in vivo [13]. Once generated, lentivirus aliquots (5–10 μL) can be cryopreserved at −80 °C. We typically use the virus within 1 year.

3.1.7 Next-Generation Sequencing of Barcoded Lentivirus

To determine the number of barcodes and to assess library distribution in a typical lentivirus preparation, we use next-generation sequencing (*See* **Note 5**). This involves (1) isolation and reverse transcription of lentiviral RNA, (2) PCR amplification and indexing of the barcode region, and (3) sequencing and data analysis.

To isolate lentiviral RNA, use the QIAamp Viral RNA Mini Kit and follow manufacturer's instructions. We typically use 1 μL of barcoded lentivirus preparation (titer $\geq 10^9$ TU/mL) as input and elute in 40 μL buffer AVE (RNase-free water with 0.04% NaN3, Qiagen). Next, combine the following components in a tube on ice to set up a reverse transcription reaction:

Reagent	Volume [μL]
ddH2O	2
5X VILO™ Reaction Mix	12
10X SuperScript® Enzyme Mix	6
Viral RNA	40
Total	**60**

Gently mix tube contents, spin down briefly, transfer to thermocycler, and reverse transcribe into cDNA using the following protocol:

Temperature [°C]	Time [min]
25	10
42	60
85	5

To amplify the barcode region from lentiviral cDNA via PCR, combine the following reagents in a tube:

Reagent	Volume [μL]
cDNA	60
ddH2O	10
MRX1693 (100 μM)	2
MRX1694 (100 μM)	2
5X Q5 Reaction Buffer	20
Q5® High-Fidelity DNA Polymerase	1
DMSO	5
Total	**100**

Mix by flicking the tube 2–3 times, spin down briefly, transfer to thermocycler, and amplify using the following PCR protocol:

Temperature [°C]	Time	# of cycles
98	5 min	
98 65 72	20 s 20 s 10 s	12
72	5 min	
4	hold	

The resulting amplicon has a size of 158 bp. Purify the PCR product using double-sided size selection and SPRIselect bead cleanup (*See* **Note 6**):

A. Add 60 μL of SPRIselect beads to the tube (0.6× ratio).

B. Mix by vortexing and incubate at room temperature for 5 min.

C. Transfer tube to DynaMag™ and incubate at room temperature until solution clears.

D. Transfer 160 μL supernatant to a new tube and discard beads (do NOT discard supernatant).

E. Add 40 μL of SPRIselect to the tube (final ratio is 1.0×).

F. Mix by vortexing and incubate at room temperature for 5 min.

G. Transfer tube to DynaMag™ and incubate at room temperature until solution clears.

H. Remove supernatant without discarding any beads.

I. Wash twice with 200 μL of 80% ethanol for 30 s.

J. Centrifuge briefly, place on DynaMag™ and remove remaining ethanol.

K. Air dry at room temperature for 2 min.

L. Remove from DynaMag™ and add 23 μL of water.

M. Vortex for 10 s and incubate for 2 min at room temperature.

N. Place on DynaMag™ until the solution clears and transfer 22 μL to a PCR tube.

O. Determine DNA concentration using Qubit 1x dsDNA HS Assay Kit and 2 μL of eluate.

P. Expected yield is 1–2 ng/μL.

Set up indexing PCR by combining the following reagents in a tube:

Reagent	Volume [μL]
2x Q5 master mix	25
ddH2O	19.5
PCR product	1
MRX142 (10 μM)	1
MRX614 (10 μM)	1
DMSO	2.5
Total	**50**

Mix by flicking the tube 2–3 times, spin down briefly, transfer to thermocycler, and amplify using the following PCR protocol:

Temperature [°C]	Time	# of cycles
98	5 min	
98 65 72	20 s 20 s 10 s	10
72	5 min	
4	hold	

The indexed library will have a size of 240 bp. Purify the indexed PCR product using SPRIselect bead cleanup:

A. Add 50 μL of SPRIselect beads (1.0× ratio).

B. Mix by vortexing and incubate at room temperature for 5 min.

C. Transfer tube to DynaMag™ and incubate at room temperature until solution clears.

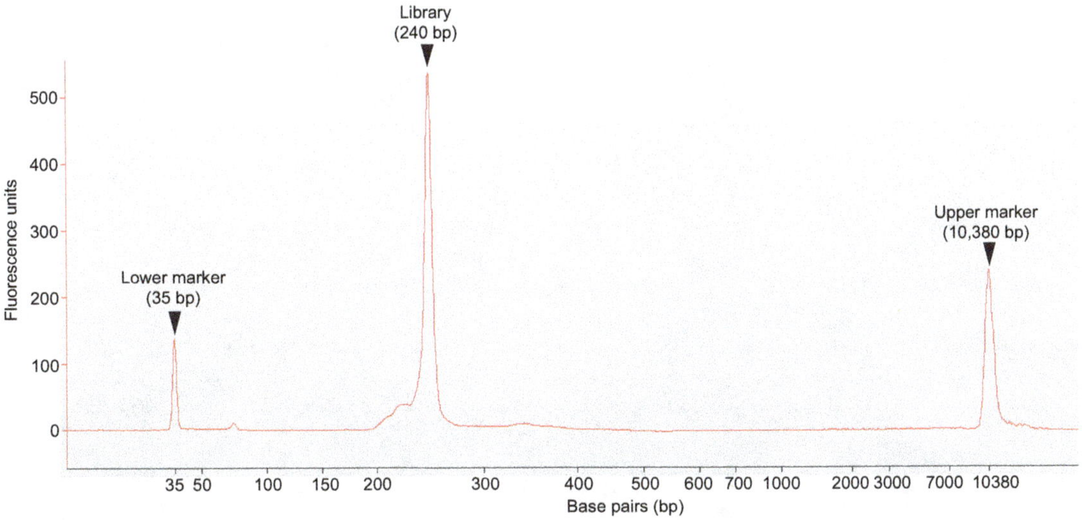

Fig. 3 Representative Bioanalyzer trace for indexed library (240 bp peak) prepared from a typical barcoded lentivirus preparation. A single peak at around 240 bp indicates a high-quality library that is ready for next-generation sequencing

D. Discard supernatant and wash twice with 200 μL of 80% ethanol for 30 s.

E. Centrifuge briefly, place on DynaMag™, and remove remaining ethanol.

F. Air dry at room temperature for 2 min.

G. Remove from DynaMag™ and add 23 μL of water.

H. Vortex for 10 s and incubate for 2 min at room temperature.

I. Place on DynaMag™ until the solution clears and transfer 22 μL to a PCR tube.

J. Determine DNA concentration using Qubit 1x dsDNA HS Assay Kit and 2 μL of eluate.

K. Expected yield is 0.5–2 ng/μL.

L. Load 1 μL of indexed library to check library quality via Bioanalyzer High Sensitivity DNA Analysis (Fig. 3).

The resulting library is sequenced on an Illumina sequencer (e.g., MiSeq, NextSeq550) using 124 cycles (read 2) and 5--10 million reads. Read alignment against a reference containing the 30 bp barcode and flanking regions is done using the BWA-MEM algorithm [14] followed by barcode extraction and counting using SAMtools [15]. The following Bash script can be used to extract unique barcodes and corresponding read counts:

```
#!/bin/bash -l


# indexing of reference "EGFP-BC" stored in file ref.fa in FASTA format

# >EGFP-BC

CAAGCAGAAGACGGCATACGAGATTCGCCTTAGTGACTGGAGTTCAGACGTGTGCTCTTCCGATCTGTGGCTAAG

ATCTACAGCTGCCTTGTAAGTCATTGGTCTTAAAGNNNNNNNNNNNNNNNNNNNNNNNNNNNNNNTGCATTTACT

TGTACAGCTCGTCCATGCCGAGATCGGAAGAGCGTCGTGTAGGGAAAGAGTGTGCGATCTAGTGTAGATCTCGGT

GGTCGCCGTATCATT

bwa index ref.fa


# align reads against reference

bwa mem egfp-bc.fa reads.fastq > aligned.sam


# Converting SAM to BAM

samtools view -S -b aligned.sam > aligned.bam


# sorting

samtools sort aligned.bam -o aligned.sorted.bam


# indexing

samtools index aligned.sorted.bam


# save alignments EGFP-BC region in new BAM file

samtools view -b aligned.sorted.bam EGFP-BC:111-140 > barcode.region.bam

# indexing

samtools index barcode.region.bam
```

```
# convert to FASTQ file
samtools bam2fq barcode.region.bam > barcode.region.fastq

# extract barcode region
grep "TAAAG.*TGCAT" barcode.region.fastq > barcode.region.txt

# remove sequences upstream and downstream of barcodes to reduce size to 40
bp
sed 's/^.*\(TAAAG..............................TGCAT\).*$/\1/'
barcode.region.txt > all.barcodes.txt

# extract barcodes with length of 30 bp
# with 5bp flanking regions on both sites, this means to extract 40 bp long
sequences
awk 'length==40' all.barcodes.txt > 40mer_barcodes.txt

# remove 5 bp flanking regions
cut -c5-35 40mer_barcodes.txt > 30mer_barcodes.txt
# extract unique barcode sequences
sort 30mer_barcodes.txt | uniq -c > 30mer_barcodes.unique.counts.txt
```

The resulting file will contain the number of unique barcodes and the number of reads per barcode. These results can be used to determine (1) library size (= number of unique barcodes), (2) library uniformity (= distribution of read counts among barcodes), and (3) Hamming distance between barcodes (Fig. 4). In our hands, a typical lentivirus library (V = 1 μL, titer = 2×10^9 TU/mL) contained around 1.57 million barcodes/μL with a highly uniform distribution (Gini index = 0.2;

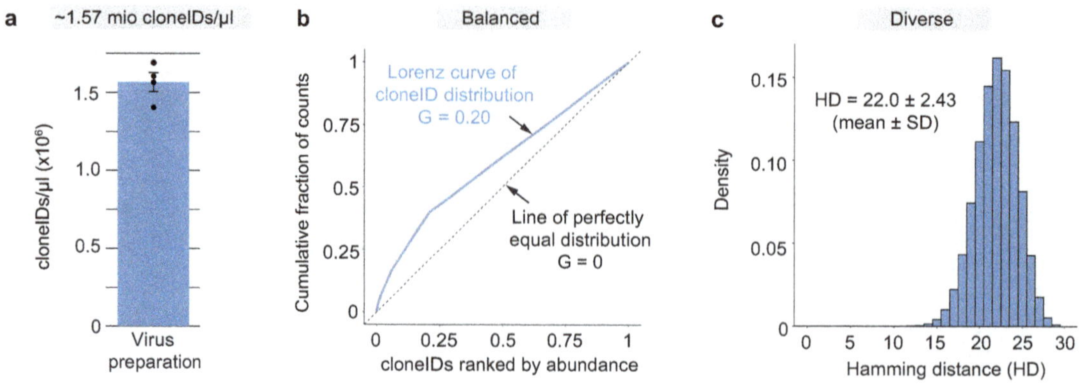

Fig. 4 Characterization of barcoded lentivirus preparations. A typical lentivirus preparation contains on average 1.57×10^6 cloneIDs/µL (**a**) with uniform distribution (**b**) and high diversity (**c**). (Adapted from [9])

Gini index of 0 represents perfect uniformity) and an average pairwise Hamming distance between barcodes of 22 indicating high barcode diversity [9].

3.2 Considerations for Ultrasound-Guided In Utero Injection of Mouse Embryos

We inject barcoded lentivirus into the forebrain ventricle of mouse embryos at embryonic (E) day 9.5. This complex procedure requires skills in mouse surgery, and we refer the reader to excellent methodological descriptions and visualized experiments published elsewhere [16–18].

Therefore, instead of describing ultrasound-guided in utero injection in detail again, we want to briefly discuss efficiency and uniqueness of progenitor cell labeling upon in vivo barcoding at E9.5 (Ref. [9]). Please also note that while intraventricular injections during early developmental stages (≤E10.5) require ultrasound guidance, it is possible to inject viruses (or plasmids, dyes, etc.) into the brains of older mouse embryos (≥E11.5) routinely without ultrasound.

We typically inject 0.6 µL of barcoded lentivirus at E9.5 into the forebrain ventricle which corresponds to about 940,000 barcodes and which leads to labeling of about 1.83% of progenitor cells (Fig. 5). We estimated that this corresponds to about 2600 barcoded progenitor cells at E9.5 in total of which 99.6% of cells are labeled with a unique barcode [9]. Given the large size and high uniformity of the barcode library, many thousands of cells could be barcoded uniquely in parallel. For example, our library could be used for labeling around 6300 cells of which 99% will have a unique barcode [9]. However, it is currently challenging to achieve such a high labeling density using microinjections of high titer lentivirus into the mouse forebrain at E9.5.

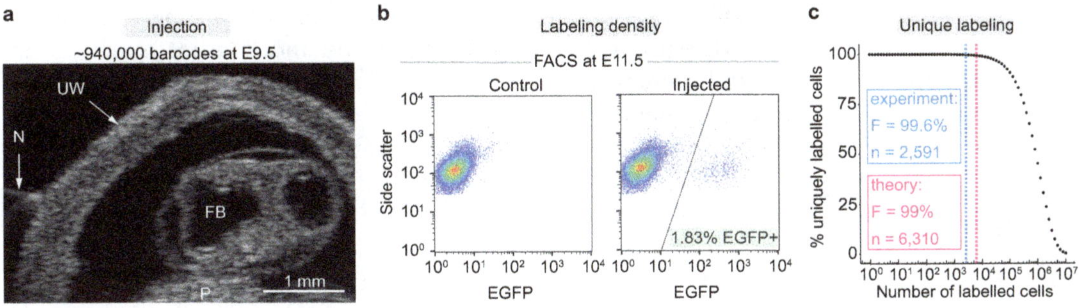

Fig. 5 Ultrasound-guided in utero injection of barcoded lentivirus. (**a**) Ultrasound image of E9.5 mouse embryo showing forebrain (FB), placenta (P), uterine wall (UW), and injection needle (N). A typical experiment involves the injection of ~940,000 barcodes into the FB. (**b**) Flow cytometry analysis of dissociated forebrain tissue shows that ~1.83% of cells are labeled with a barcode 2 days after injection. (**c**) A total of around 2600 progenitor cells is labeled of which 99.6% express a unique barcode (blue). Technically, ~6300 cells can be labeled with a typical lentivirus preparation (magenta). (Adapted from [9])

3.3 Tissue Dissociation and Single-Cell RNA-Seq Library Preparation

3.3.1 Extract and Dissect Mouse Brains

We focus our studies on brains from ~2-week-old mice (*See* **Note 7**). To identify EGFP+ mice that have been injected with a barcoded lentivirus at E9.5, use a fluorescence flashlight and corresponding filter glasses. Sacrifice mice with an overdose of isoflurane followed by transcardial perfusion with ice-cold artificial cerebrospinal fluid (ACSF), decapitation, and brain extraction. Keep brains in ACSF on ice all the time and prepare 1 mm coronal slices using an acrylic mouse brain matrix and single edge razor blades. Transfer slices to a black sylgard dissection dish containing cooled ACSF and transfer the dissection dish to a larger dish containing wet ice. Move to a stereo microscope and identify regions of interest using a reference atlas such as the Allen Reference Atlas (https://mouse.brain-map.org/static/atlas) or the atlas by Paxinos and Franklin [19]. Microdissect the region of interest, for example, neocortex, by holding a brain slice in place with curved blunt forceps and by cutting around the area of interest using sharp Dumont #5/45 forceps (*See* **Note 8**). Transfer the tissue pieces to a new 1.5 mL tube containing 1 mL of cold ACSF and keep on ice.

3.3.2 Dissociate Brain Tissue

Let the tissue settle at the bottom of the tube and carefully remove ACSF completely. Add 1 mL of papain solution to the tube with tissue pieces and mix by slowly inverting 2–3 times. Transfer tube to TS-100 Thermo-Shaker and incubate at 37 °C for 20 min. Invert the tube several times every 5 min to resuspend settled pieces of tissue. In the meantime, prepare fire-polished glass Pasteur pipette through direct exposure to a Bunsen flame and by constantly rotating the pipette to blunt the edges and to narrow down the opening to about half of its original diameter. Use the fire-polished pipette for careful trituration of tissue pieces by pipetting up and down 15 times. Inspect the suspension and if larger tissue clumps are still present, incubate another 5 min 37 °C followed by trituration to generate a single-cell suspension.

Wet a sterile 30 μm pre-separation filter by passing through 1 mL of cold ACSF/BSA. Add 9 mL cold ACSF/BSA to a 15 mL centrifuge tube, place the prewetted filter on the tube and add the single-cell suspension for filtration. Mix the suspension well by closing the tube and inverting 2–3 times. Pellet cells in a precooled centrifuge at 300 g for 5 min at 4 °C. Carefully remove the supernatant and resuspend cell pellet in 1 mL diluted ovomucoid protease inhibitor solution. Prepare a discontinuous density gradient by carefully overlaying 2 mL of undiluted ovomucoid protease inhibitor stock with 1 mL of cell suspension in a 15 mL tube followed by centrifugation at 100 g for 6 min at 4 °C. Carefully remove the supernatant and resuspend the cell pellet in 1 mL cold ACSF/BSA and transfer the cell suspension to a round-bottom tube.

Dilute 20 μL of cell suspension in 20 μL trypan blue solution, mix well by flicking the tube several times, load 20 μL of the mix onto the gridded area of a hemocytometer, and apply a coverslip. Count the total number of cells under a microscope and assess cell viability by determining the number of live cells impermeable for trypan blue. Cell viability should be ≥90% before moving on to the next step (*See* **Note 9**).

3.3.3 Isolate Barcoded Cells

Load the sample into the cooled tube holder of a BD Influx™ cell sorter equipped with a 140 μm nozzle. Fill a 1.5 mL DNA LoBind tube with 500 μL ACSF/BSA and place it into the cooled sample collection holder of the sorter. Set gates such that single cells are included while smaller particles and debris as well as doublets and larger aggregates are excluded. Sort barcoded, EGFP+ cells into the DNA LoBind tube. Depending on how many cells have been sorted, the final volume of ACSF/BSA in the tube can change dramatically. We determined a drop volume of 8.63 nl when using a 140 μm nozzle at the BD Influx™ which means that sorting, for example, 100,000 cells will lead to a volume increase of 863 μL in the collection tube.

Centrifuge the 1.5 mL tube with purified cells in a cooled microcentrifuge at 300 g for 5 min. Carefully remove the supernatant and resuspend the cell pellet in 42 μL of ACSF/BSA. Mix 2 μL of cell suspension with 2 μL of trypan blue and determine cell number as well as cell viability (≥90% ideally, ≥70% acceptable). The final number of cells that are recovered at this step corresponds to 50–60% of the number of cells that have been counted in the flow cytometer, for example, sorting of 100,000 cells should yield about 50,000–60,000 cells in total.

3.3.4 Prepare Single-Cell RNA-Seq Libraries

We have successfully used droplet-based [9, 11] and plate-based single-cell RNA-seq methods [10] for the simultaneous capture of whole transcriptomes and barcodes. The existing protocols for library preparation are detailed and the methodology is robust. We therefore refer the reader to specific protocols for droplet-based (e.g., "Chromium Next GEM Single Cell 3′ Reagent Kits v3.1 User Guide," 10x Genomics) and plate-based single-cell RNA-seq [20].

3.4 PCR Enrichment of Barcodes From Full Length cDNA

When preparing single-cell libraries using a droplet-based method, we recover barcodes from whole transcriptome libraries and from amplicon libraries enriched for barcodes (Fig. 6). Compared to barcode recovery from transcriptome libraries only, barcode enrichment from cDNA increases the proportion of barcode-positive cells by 5–15% (Ref. [9]).

This step involves three PCRs. To set up the first PCR, combine the following components in a tube on ice:

Reagent	Volume [μL]
ddH2O	15.5
2x Q5 master mix	25
DMSO	2.5
MRX1587 (10 μM)	1
MRX14 (10 μM)	1
cDNA	5
Total	**50**

Gently mix tube contents, spin down briefly, transfer to thermocycler, and run PCR1 using the following protocol:

Temperature [°C]	Time	# of cycles
98	5 min	
98 69 72	20 s 20 s 30 s	10
72	5 min	
4	hold	

The resulting PCR product will have a size of 988 bp. Purify the PCR product using SPRIselect bead cleanup:

A. Add 40 μL of SPRIselect beads (0.8× ratio).

B. Mix by vortexing and incubate at room temperature for 5 min.

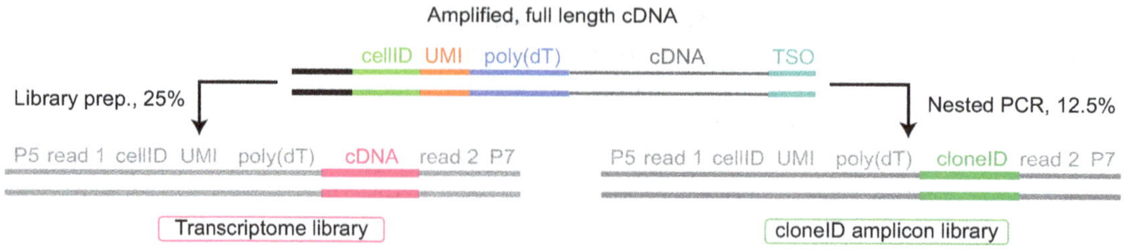

Fig. 6 Barcode recovery from transcriptome and amplicon libraries. To maximize cloneID recovery rates, we prepare two libraries for each sample: a transcriptome library (left) following the 10x Genomics 3' protocol and a cloneID amplicon library (right) using a targeted PCR approach described here. (Adapted from [9])

C. Transfer tube to DynaMag™ and incubate at room temperature until solution clears.

D. Discard supernatant and wash twice with 200 μL of 80% ethanol for 30 s.

E. Centrifuge briefly, place on DynaMag™, and remove remaining ethanol.

F. Air dry at room temperature for 2 min.

G. Remove from DynaMag™ and add 23 μL of water.

H. Vortex for 10 s and incubate for 2 min at room temperature.

I. Place on DynaMag™ until the solution clears and transfer 22 μL to a PCR tube.

J. Determine DNA concentration using Qubit 1x dsDNA HS Assay Kit and 2 μL of eluate.

K. Expected yield is 1–5 ng/μL.

To set up the second PCR, combine the following components in a tube on ice:

Reagent	Volume [μL]
ddH2O	10.5
2x Q5 master mix	25
DMSO	2.5
MRX1587 (10 μM)	1
MRX1588 (10 μM)	1
PCR1 product	10
Total	**50**

Gently mix tube contents, spin down briefly, transfer to thermocycler, and run PCR2 using the following protocol:

Temperature [°C]	Time	# of cycles
98	5 min	
98 69 72	20 s 20 s 20 s	10
72	5 min	
4	hold	

The resulting PCR product will have a size of 414 bp. Purify the PCR product using SPRIselect bead cleanup:

A. Add 40 μL of SPRIselect beads (0.8× ratio).

B. Mix by vortexing and incubate at room temperature for 5 min.

C. Transfer tube to DynaMag™ and incubate at room temperature until solution clears.

D. Discard supernatant and wash twice with 200 μL of 80% ethanol for 30 s.

E. Centrifuge briefly, place on DynaMag™, and remove remaining ethanol.

F. Air dry at room temperature for 2 min.

G. Remove from DynaMag™ and add 21 μL of water.

H. Vortex for 10 s and incubate for 2 min at room temperature.

I. Place on DynaMag™ until the solution clears and transfer 20 μL to a PCR tube.

J. Determine DNA concentration using Qubit 1x dsDNA HS Assay Kit and 2 μL of eluate.

K. Expected yield is 10–20 ng/μL.

Set up the indexing PCR by combining the following components in a tube on ice:

Reagent	Volume [μL]
ddH2O	18.8
2x Q5 master mix	25
DMSO	2.5
MRX1589 (10 μM)	1
I7 primer from 10x index plate	2.5
PCR2 product	0.2
Total	**50**

Gently mix tube contents, spin down briefly, transfer to thermocycler, and run PCR2 using the following protocol:

Temperature [°C]	Time	# of cycles
98	5 min	
98 65 72	20 s 20 s 20 s	10
72	5 min	
4	hold	

The final PCR product will have a size of 502 bp. Purify the indexed PCR product using SPRIselect bead cleanup:

A. Add 40 μL of SPRIselect beads (0.8× ratio).

B. Mix by vortexing and incubate at room temperature for 5 min.

C. Transfer tube to DynaMag™ and incubate at room temperature until solution clears.

D. Discard supernatant and wash twice with 200 μL of 80% ethanol for 30 s.

E. Centrifuge briefly, place on DynaMag™, and remove remaining ethanol.

F. Air dry at room temperature for 2 min.

G. Remove from DynaMag™ and add 23 μL of water.

H. Vortex for 10 s and incubate for 2 min at room temperature.

I. Place on DynaMag™ until the solution clears and transfer 22 μL to a PCR tube.

J. Determine DNA concentration using Qubit 1x dsDNA HS Assay Kit and 2 μL of eluate.

K. Expected yield is 1–10 ng/μL.

L. Load 1 μL of indexed library to check library quality via Bioanalyzer High Sensitivity DNA Analysis (Fig. 7).

The resulting library is sequenced on an Illumina sequencer (e.g., MiSeq, NextSeq550) using 124 cycles (read 2) and around 1000 reads/cell. For example, if an amplicon library has been prepared from 10,000 cells, aim for a total of 10 million reads as sequencing depth for the amplicon library.

3.5 Barcode Extraction and Clone Reconstruction from scRNA-Seq Data

An essential part of the TREX workflow is the extraction and assembly of cloneIDs from single-cell transcriptomes, the reconstruction of cells, and the grouping of cells into clones based on cloneIDs (Fig. 8).

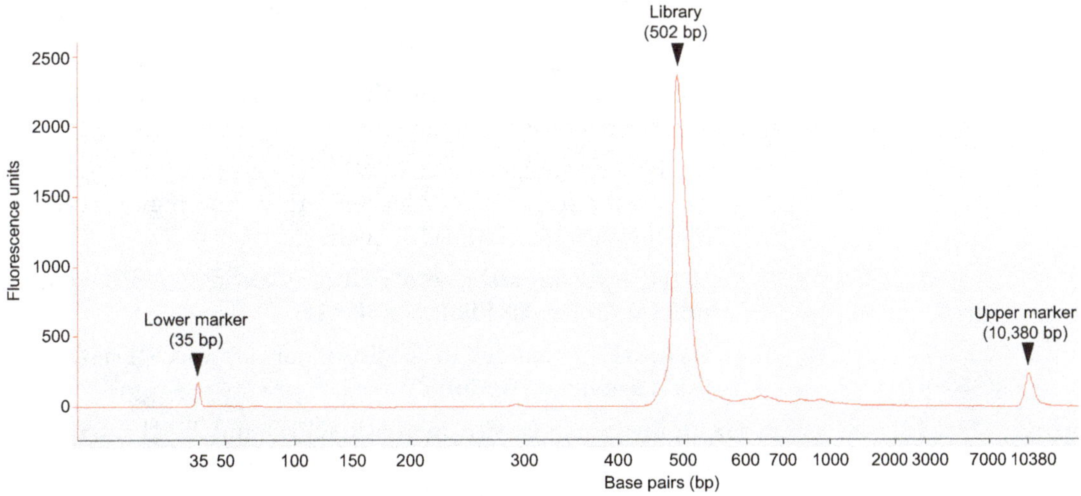

Fig. 7 Representative Bioanalyzer trace for indexed cloneID library (502 bp peak) prepared from full-length cDNA. A single peak at around 502 bp indicates a high-quality library that is ready for next-generation sequencing

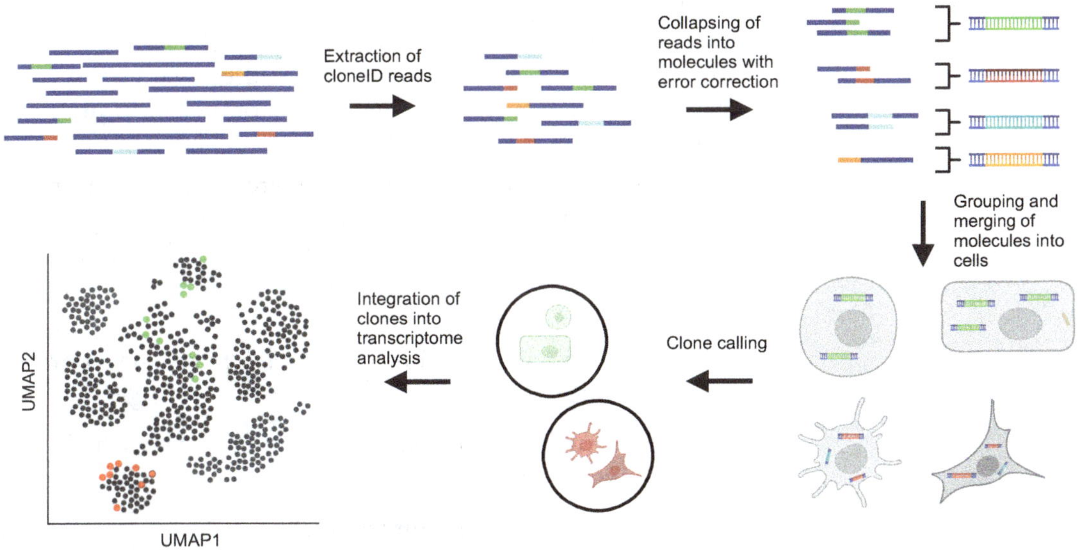

Fig. 8 TREX computational workflow. TREX software takes a file with mapped reads as input. It automatically detects the cloneID region and extracts all reads aligning to it. All reads with same UMI get collapsed to a consensus sequence that represents the original mRNA molecule. Molecules based on only one read get filtered out (e.g., orange molecule). Next, molecules are grouped into cells by their cellIDs. Molecules in the same cell with less than six differences are seen as copies of the same integrated barcode. Occasionally cells have more than one integrated barcode, those molecules show six or more differences in their sequence (see blue and red molecules in the two bottom row cells). Based on the ratio of overlapping barcodes between cells, cells are grouped into clones (black circles) by a graph-based clone calling process. The results can be integrated into existing single-cell transcriptomic datasets and clonally related cells highlighted in nonlinear dimensional reduction plots (red and green dots). Created with BioRender.com.

Our TREX software (https://github.com/frisen-lab/TREX/) performs all tasks with the following steps:

1. Retrieval of usable reads from the input BAM file: A usable read fulfills these requirements:
   - It aligns to the specified cloneID region or to the region that has been automatically identified to be the region containing the variable cloneID sequence.
   - It has both an associated cellular identifier (cellID) and a unique molecular identifier (UMI).
   - Its cellID is included in the list of permitted cellIDs if such a list is provided (optional).
2. Grouping of reads with identical cellID and UMI into molecules. The cloneID of the molecule is taken to be the consensus of all cloneIDs with same UMI.
3. Error-correction of cloneIDs of the molecules.
4. Grouping of molecules by cellIDs into cells. CloneIDs within a cell which have a Hamming distance below a defined threshold are merged.
5. Filtering of low-quality cloneIDs in each cell: Rare cloneIDs also found in another cell are considered contaminants. CloneIDs supported by only a single read (or a custom number) are removed.
6. Clustering of the cells into clones by creating a clone graph. Edges are drawn between cells that appear to belong to the same clone based on the set Jaccard index threshold. The connected components of the graph are considered to be clones.
7. Error-correction of the clone graph by removing spurious edges ("bridges").

If TREX is run on Smart-Seq2 (Ref. [21]) data which lack UMIs, it follows a similar pipeline with the following differences:

- A useable read does not require a UMI.
- Reads do not get grouped into molecules and their cloneIDs are not collapsed and error-corrected into consensus sequences.

In the following sections, we will discuss the required steps to prepare for a TREX run, perform it, and integrate the results.

3.5.1 Running scRNAseq Data Processing Software with Custom Chromosome

Generally, scRNAseq data processing can be performed following the instructions of the software used, for example, Cell Ranger (10x Genomics). However, to correctly map reads carrying (parts of) a cloneID, a custom reference transcriptome needs to be created.

1. Download and unpack the desired reference transcriptome and the corresponding general transfer format (GTF) file.
2. Keep the FASTA file with the cloneID and reporter gene sequence (e.g., EGFP_30N.fa) in the same folder or have the path available.

   The following **steps** (**3–6**) are specific for Cell Ranger (used to analyze Chromium Next GEM Single Cell 3′ data) and are not required for zUMIs (used to analyze Smart-seq3 data).
3. Create a GTF file in a text editor of choice with the following line in tab-separated values (TSV) format.

```
chrH2B-EGFP-30N    other exon  1     1451   .     +     .     gene_id

"chrH2B-EGFP-30N"; transcript_id "chrH2B-EGFP-30N";
```

   Replace the `chrH2B-EGFP-30N` chromosome name with a desired one. Replace `1451` with the correct position of the last base in the sequence.
4. Save the file with the same name as the FASTA file, for example, EGFP_30N.gtf, and in the same location.
5. Run the following command to combine the original reference with the additional chromosome as well as the original GTF file with the additional one. Modify the command to match with correct file names:

```
cat path/to/transcriptome_ref.fa path/to/chrH2B-EGFP-3N.fa > ref_H2B-

EGFP-30N_genome.fa

cat  path/to/genes.gtf  path/to/chrH2B-EGFP-30N.gtf  >  ref_H2B-EGFP-

30N_genes.gtf
```

6. A custom reference transcriptome needs to be created as follows. Modify the command as required:

```
cellranger   mkref   --genome=ref_H2B-EGFP-30N   --fasta=ref_H2B-EGFP-

30N_genome.fa   --genes=ref_H2B-EGFP-30N_genes.gtf   >   ref_H2B-EGFP-

30N.out
```

Be aware that the spaces between the words are not a group of whitespaces but a tab character denoted by *\t*. This step might take about ≈1 h.

7. For zUMIs, the additional chromosome is not added to the reference but to the configuration file as an additional file (*See* **Note 10**).
8. Run the scRNAseq data processing software as preferred, using the customized reference transcriptome. Be aware that this step might be time intensive.

3.5.2 Installing TREX

1. Open terminal on your computer.
2. Install TREX by copying the following lines into your local terminal:

```
python3 -m venv trex-venv
trex-venv/bin/pip install git+https://github.com/frisen-lab/TREX.git
```

It creates the virtual environment *trex-venv* and installs TREX within this environment.

If a different package management software such as *conda* is used, run the following command within the virtual environment:

```
pip install git+https://github.com/frisen-lab/TREX.git
```

3. Activate the virtual environment. If *venv* was used (as in **step 2**), run the following command every time the terminal is started:

```
source trex-venv/bin/activate
```

4. Test if the installation was successful by typing the following command:

```
trex --version
```

3.5.3 Preparations for Running TREX

It is highly recommended to not run TREX on all cells in the dataset, but to add a preceding filtering step removing low-quality cells and doublets which otherwise interfere with the clone calling process. Especially doublets may disturb assignment of cells to clones as they might contain two cloneIDs of two unrelated cells.

1. Follow a preferred single-cell transcriptomics analysis routine with tools such as Seurat [22] or Scanpy [23] including filtering of low-quality cells and doublets.
2. When all filtering steps have been performed, extract a list of cellIDs in TSV format as follows:

```
1    CellID1
2    CellID2
...
```

3.5.4 Running TREX

The essential part of running the TREX software for a given dataset is to build an individual command line. A TREX run usually takes a few minutes only and is not computationally heavy, but this depends on the input dataset, especially on the size of the input BAM file.

1. Currently, there are three versions of the TREX code available:
   (a) `run10x`: For executing TREX on 10x Genomics Chromium data processed with Cell Ranger (or 10x Visium data processed with Space Ranger software).
   (b) `smartseq2`: For executing TREX on Smart-seq2 data processed with zUMIs software. Since Smart-Seq2 data lack UMIs, this pipeline counts reads. It can also be run on Smart-Seq3 data if read instead of UMI counting is desired.
   (c) `smartseq3`: For executing TREX on Smart-Seq3 data processed with zUMIs software.

   Choose one of the three options. To run TREX on a Visium sample, use the `run10x` option and add the `--visium` flag. It might be desired to count reads instead of UMIs in Smart-Seq3 data due to low capture of UMIs (*See* **Note 11**)

   In the following examples, we will use `run10x`. The command line looks as follows so far: `trex run10x`.
2. In the following steps, we will go through the most important command line options to build an individual command line. To see all command line options, type `trex run10x -h` or check the documentation on GitHub. Replace `run10x` with the version of your choice to find all options specific for that version.
3. The only compulsory input is the path to the directory, which either contains (1) the Cell/Space Ranger `/outs` folder (10x Genomics data) or (2) a single BAM file with all reads (Smart-seq2/3) or (3) a folder with several BAM files (Smart-Seq2/3). The command line then looks as follows: `trex run10x path/to/folder/with/outs/`. This command is already sufficient to run TREX successfully, using default settings for each command line option. However, it is highly recommended to go through the next steps to adapt the command line further. To run TREX on multiple samples, *see* **Note 12**.
4. It is highly recommended to remove low-quality cells and doublets from the TREX run by adding a list of cellIDs to

keep. This can be done with the `--filter-cellids` flag. The resulting command is:

```
trex run10x --filter-cellids path/to/filtered_cells.tsv path/to/folder/with/outs/
```

Separate the two paths with a whitespace.

5. Most important for the outcome of clone calling with TREX are parameters that determine the filtering of cloneIDs and cells. For each parameter, we have performed extensive testing to find the most reasonable default setting. However, it is recommended to consider the default settings for each individual dataset and adjust when necessary. It might also be desired to compare loose and stringent settings to decide for the best outcome for the given dataset. In the following, three parameters are discussed.
   (a) The `--min-length` option is the minimum number of detected nucleotides required to keep a cloneID in the analysis and the default setting is 20. This has been calculated because the total length of the cloneID is 30 nucleotides and even with only 20 nucleotides a sufficient diversity is ensured to distinguish two cloneIDs from each other. However, if the cloneID has a different expected length or if a more loose/stringent exclusion of cloneIDs is desired, this parameter can be changed.
   (b) The `--max-hamming` parameter decides which number of differences in nucleotides between two cloneIDs of equal length are allowed to consider these two cloneIDs as identical and is meant to adjust for sequencing errors. The default of five is again based on the total length of cloneIDs and the likelihood of two different cloneIDs to appear identical due to sequencing errors. This number can be adjusted if more stringent/loose conditions are desired.
   (c) The Jaccard score is a similarity coefficient to estimate the overlap of two sets of elements, in our case the overlap of two sets of cloneIDs. It is calculated by dividing the number of overlapping, unique cloneIDs between cell A and cell B by the total number of unique cloneIDs in cell A and B. An index of 0 indicates no overlapping cloneIDs and an index of 1.0 a perfect match. The Jaccard threshold is the Jaccard index above which two cells are merged into one clone. It can be set with the `--jaccard-threshold` flag and is set to 0.7 by default. In our experience, this threshold ensures that most clones only contain truly

related cells. However, it could also result in splitting a single, large clone into two or more smaller clones. Therefore, it might be helpful to run TREX with various Jaccard thresholds and to compare the results.

After this step, an example command line might look like this:

```
trex run10x --min-length 22 --max-hamming 4 --jaccard-threshold 0.7 -
-filter-cellids path/to/filtered_cells.tsv path/to/folder/with/outs/
```

6. TREX is designed to automatically detect the positions of the cloneID sequence in the reference. This occurs by calculating the variability of nucleotides in each position. Positions containing a nucleotide of the cloneID are expected to be variable, with none of the four bases dominating, while all other positions are assumed to have one base dominating in up to 100% of all matching reads. However, this part of the pipeline might take some time and does not always yield correct results (this can be checked in the log.txt output file). To avoid this risk, the `--start` and `--end` flags can be used to indicate the first and last position of the cloneID sequence in the custom reference chromosome. After following all described steps, the command line could, for example, read as follows:

```
trex run10x --min-length 22 --max-hamming 4 --jaccard-threshold 0.7 --
start 2330 --end 2359 --filter-cellids path/to/filtered_cells.tsv
path/to/folder/with/outs/
```

3.5.5 Interpreting and Integrating TREX Output

TREX yields, depending on the added command line options, a range of output files, which are all described in the documentation. We will briefly discuss the most important ones:

- Most importantly, *clones.txt* contains the assignment of each cell to one clone, denoted with a clone number. As can be seen in clone_sequences.txt, each clone number is connected to the most common cloneID in a clone. We recommend adding clone numbers to a Seurat or AnnData (Scanpy) object as metadata. We then highlight single clones in nonlinear dimensionality reduction plots. Based on the frequency of shared cloneIDs amongst cell types of different developmental origin (e.g., microglia and neurons in the mouse brain) we assess the accuracy of clone calling.
- *log.txt* is a useful output when performing troubleshooting as it records the number of cells and cloneIDs remaining after each

step and additional information. It also contains a clone size histogram. We recommend inspecting this histogram and considering whether especially the largest clone sizes are in line with expectations and knowledge of the cells of interest. Keep in mind that clone sizes are expected to be much smaller when using scRNAseq for clonal tracing compared to classical approaches that utilize fluorescence imaging of tissue sections to reconstruct clones after sparse labeling of progenitor cells [9].

- If it is desired to understand how cells were assigned to clones in more depth, the other output files are of value. For example, *cells.txt* lists all cells and their respective cloneIDs together with a read/UMI count for each cloneID. *Umi_* or *read_count_matrix.csv* (with cells in columns, cloneIDs in rows, and UMI/read count in cells) files are the input data for clone calling and can be used to perform an own clone calling on a custom algorithm if desired. To receive such a matrix as output the flags `--umi-matrix` or `--read-matrix` need to be used, respectively.

4 Notes

1. *Low-Cycle PCR to Reduce Amplification Bias*

The ultramer template is a pool of single-stranded DNA molecules each containing a random, 30 bp long barcode. Technically, $>10^{18}$ different barcodes can be present in the template pool. As PCR leads to exponential amplification of DNA templates, any amplification bias should increase with the number of PCR cycles. Therefore, to mitigate amplification bias of a smaller number of barcodes, we reduce the number of PCR cycles.

2. *Increase Number of PCRs*

Increase the number of PCRs in case of low yields of DNA (<20 ng/μL) after low-cycle PCR. For example, four 50 μL low-cycle PCRs can be run in parallel followed by combining the reaction products for MinElute purification and elution in 12 μL water. This should yield enough amplicon DNA for the next step.

3. *Ratio of Insert to Backbone*

We routinely use a molar insert:vector ratio of 7:1 for a typical library assembly reaction. This yields at least tenfold more bacterial colonies containing a barcode compared to background colonies without barcode. Different ratios can be calculated easily for any pair of vector and insert using tools such as the NEBioCalculator: https://nebiocalculator.neb.com/#!/ligation

4. *Incubation Time of Plates*

Discrete bacterial colonies should be observed on both the small and large plates, which will enrich transformed bacterial cells that survived ampicillin selection. Limit the growth of bacterial cells to ≤16 h since too much growth may lead to competition between colonies and a skewed barcoded plasmid distribution.

5. *Next-Generation Sequencing of Barcoded Plasmid Versus Lentivirus Library*

Above we described our approach for next-generation sequencing of barcoded lentivirus preparations. In our experience, it is not necessary to sequence the barcoded plasmid library before lentivirus production since the estimated library size is similar between counting the number of colonies (Subheading 3.1.5) and next-generation sequencing of lentivirus preparations (Subheading 3.1.7). We therefore conclude that there are no major bottlenecks during lentivirus production that would diminish barcode library size or skew barcode distribution. If it is necessary to sequence the initial barcode plasmid library, a similar protocol as outlined in Subheading 3.1.7 can be used by skipping the "isolation and reverse transcription of lentiviral RNA" step and moving directly to "PCR amplification and indexing" of the barcode region.

6. *Double-Sided Size Selection and Bead Cleanup*

Reverse transcription of the lentiviral RNA genome will generate cDNA with a length of about 8.3 kb. Amplification of the barcode locus will result in a PCR product of 158 bp. Double-sided solid-phase reversible immobilization (SPRI) selection will lead to removal of larger DNA fragments such as partially amplified or full-length cDNA with a size ≥400 bp (first SPRI) and smaller fragments including primer dimers with a size ≤100 bp (second SPRI).

7. *Cell Isolation from 2-Week-Old Mouse Brains*

We isolate single cells from mouse brains between postnatal day 10–14, because all cell types of the brain have been generated and many neuronal types are mature in terms of molecular identity [24]. At the same time, the generation of myelin-producing oligodendrocytes has just begun [25] implying that removal of contaminating myelin from single-cell suspensions can be skipped in later steps. This is advantageous, because common cleanup methods for myelin removal from cell suspensions (e.g., Myelin removal beads) require large sample input and/or lead to significant cell loss.

8. *Number of Barcoded Cells*

In this chapter, we assume that barcoded cells are isolated from large tissue volumes such as the entire mouse neocortex, hippocampus, or the striatum [9]. This will yield a large number (tens to hundreds of thousands of cells) of barcoded cells per region which can be processed with droplet-based single-cell RNA-seq. If smaller brain regions are of interest, it should be expected that fewer barcoded cells will be isolated (hundreds to thousands). Because it is recommended for droplet-based assays to load at least ~1600 cells per reaction (10x Genomics), sequencing fewer barcoded cells will require a plate-based assay such as Smart-seq3 (Ref. [26]). This allows the preparation of single-cell RNA-seq libraries from individual cells in a 96- or 384-well plate format. We have successfully isolated ~1400 barcoded cells from the somatosensory cortex of a single mouse brain using the Smart-seq3 assay [10].

9. *Cell Viability After Enzymatic Tissue Dissociation*

The fraction of viable cells is ≥90% ideally and ≥70% is acceptable. However, a low number of viable cells will proportionally diminish the number of barcoded cells that can be isolated which contributes to incomplete clones. To maximize cell viability two factors are critical:

- ACSF: Prepare fresh on the day of the experiment and saturate with carbogen prior to use on ice. This ensures stable pH buffering (pH 7.3–7.4) and adequate oxygenation.
- Dissociation time: Cell viability decreases after prolonged treatment with papain, while a minimum time is required to dissociate tissue into single cells. Adjust dissociation times such that both viability and yield of single cells in suspension are maximized.

10. *Loose Versus Stringent Configuration in zUMIs*

When running zUMIs to align and process the raw reads, we recommend running it twice with stringent and loose alignment parameters. The stringent run is performed on the parameters commonly used (follow instructions from the zUMIs manual) and the output files are input for single-cell sequencing analysis, for example, for identifying cell types in Seurat. The loose run is aimed specifically at yielding better alignments of cloneID reads to the reference. As the barcode is a random sequence not aligning to its reference, more relaxed parameters improve the ratio of mapped cloneID reads. The output files of this loose run serve as input for TREX and should not be part of the single-cell analysis.

To run the loose and stringent parameters, we recommend to use a YAML configuration file as provided on the zUMIs GitHub (https://github.com/sdparekh/zUMIs/blob/main/zUMIs.yaml). The only difference between the files should be in the reference setup.

For stringent runs, use the following: `additional_STAR_-params: --clip3pAdapterSeq CTGTCTCTTATACACATCT`

For loose runs:

```
additional_STAR_params: --outFilterMismatchNmax 40 --
outFilterMismatchNoverLmax
0.4 --outFilterScoreMinOverLread 0.30 --outFilterMatchNminOverLread 0.30 --
clip3pAdapterSeq CTGTCTCTTATACACATCT
```

The `--clip3pAdapterSeq CTGTCTCTTATACACATCT` part is for adapter trimming (*see* https://knowledge.illumina.com/library-preparation/general/library-preparation-general-reference_material-list/000001314 for more information).

In both cases, indicate the path to the cloneID and reporter gene sequence with the `additional_files` option in the reference setup.

11. *UMIs Versus Read Counting*

Due to a lower capture of UMIs in Smart-seq3 datasets [26] and the resulting loss of cells with a cloneID, it is reasonable to run such a dataset twice, once with the `smartseq3` and once with the `smartseq2` option. In the latter case, TREX will ignore UMIs and count reads. The result is that cloneIDs without UMI are kept instead of being removed. Be aware that filtering based on reads only, and not on the existence of a UMI, will lead to more false positives. To increase accuracy of cloneID and cell filtering, adjust the minimum number of reads required for a cloneID to be kept in the analysis by using the `--readcount-threshold` flag. Compare the results of the two runs to balance cell drop out and accuracy.

12. *Running TREX on Multiple Samples and CloneID Amplicon Libraries*

It is not uncommon that transcriptome datasets of one animal are split across samples. Since members of one clone might also be split across samples, it is preferable to perform clone calling on all cells and cloneIDs at once. Adding several samples to one run is achieved by providing sample names and the respective paths to each samples `/outs` folder. Sample names can be added to each cellID as suffix with the `--sample` flag:

```
trex run10x --sample sample1,sample2 path/to/folder/with/outs1/
path/to/folder/with/outs2/
```

Additionally, we generate a cloneID amplicon library for each sample to increase cloneID capture rates (*see* Subheading 3.4). To integrate amplicon data into a TREX run, use the `--amplicon` flag and provide the directories in the same order as the transcriptome directories:

```
trex run10x --amplicon path/to/folder/with/outs/amplicon1/ path/to/folder/with/outs/amplicon2/ --sample sample1,sample2 path/to/folder/with/outs1/ path/to/folder/with/outs2/
```

Acknowledgment

We would like to thank Michael Hagemann-Jensen with help to configure the zUMI runs.

References

1. Kretzschmar K, Watt FM (2012) Lineage tracing. Cell 148:33–45
2. Ma J, Shen Z, Yu Y-C et al (2018) Neural lineage tracing in the mammalian brain. Curr Opin Neurobiol 50:7–16
3. Beattie R, Postiglione MP, Burnett LE et al (2017) Mosaic analysis with double markers reveals distinct sequential functions of Lgl1 in neural stem cells. Neuron 94:517–533.e3
4. Llorca A, Ciceri G, Beattie R et al (2019) A stochastic framework of neurogenesis underlies the assembly of neocortical cytoarchitecture. Elife 8:e51381
5. Cadwell CR, Scala F, Fahey PG et al (2020) Cell type composition and circuit organization of clonally related excitatory neurons in the juvenile mouse neocortex. Elife 9:e52951
6. Gao P, Postiglione MP, Krieger TG et al (2014) Deterministic progenitor behavior and unitary production of neurons in the neocortex. Cell 159:775–788
7. Bonaguidi MA, Wheeler MA, Shapiro JS et al (2011) In Vivo Clonal Analysis Reveals Self-Renewing and Multipotent Adult Neural Stem Cell Characteristics. Cell 145:1142–1155
8. Yuste R, Hawrylycz M, Aalling N et al (2020) A community-based transcriptomics classification and nomenclature of neocortical cell types. Nat Neurosci 23:1456–1468
9. Ratz M, von Berlin L, Larsson L et al (2022) Clonal relations in the mouse brain revealed by single-cell and spatial transcriptomics. Nat Neurosci 25:285–294
10. Mold JE, Weissman MH, Ratz M, et al (2024) Clonally heritable gene expression imparts a layer of diversity within cell types. Cell Syst 15:149–165.e10
11. Kameneva P, Artemov AV, Kastriti ME et al (2021) Single-cell transcriptomics of human embryos identifies multiple sympathoblast lineages with potential implications for neuroblastoma origin. Nat Genet 53:694–706
12. Gibson DG, Young L, Chuang R-Y et al (2009) Enzymatic assembly of DNA molecules up to several hundred kilobases. Nat Methods 6:343–345
13. Baekelandt V, Eggermont K, Michiels M et al (2003) Optimized lentiviral vector production and purification procedure prevents immune response after transduction of mouse brain. Gene Ther 10:1933–1940
14. Li H, Durbin R (2009) Fast and accurate short read alignment with Burrows–Wheeler transform. Bioinformatics 25:1754–1760
15. Li H, Handsaker B, Wysoker A et al (2009) The sequence alignment/map format and SAMtools. Bioinformatics 25:2078–2079
16. Mangold K, Mašek J, He J et al (2021) Highly efficient manipulation of nervous system gene expression with NEPTUNE. Cell Rep Methods 1:100043
17. Mangold K, He J, Stokman S et al (2022) Murine Neural Plate Targeting by In Utero

Nano-Injection (NEPTUNE) at embryonic day 7.5. JoVE J Vis Exp 180:e63148

18. Pierfelice TJ, Gaiano N (2010) Ultrasound-guided microinjection into the Mouse Forebrain In Utero at E9.5. JoVE J Vis Exp 45: e2047
19. Paxinos G, Franklin KBJ (2019) Paxinos and Franklin's The mouse brain in stereotaxic coordinates. Academic Press, an imprint of Elsevier London, London
20. Hagemann-Jensen M, Ziegenhain C, Chen P, et al (2020) Smart-seq3 protocol
21. Picelli S, Faridani OR, Björklund ÅK et al (2014) Full-length RNA-seq from single cells using Smart-seq2. Nat Protoc 9:171–181
22. Stuart T, Butler A, Hoffman P et al (2019) Comprehensive integration of single-cell data. Cell 177:1888–1902.e21
23. Wolf FA, Angerer P, Theis FJ (2018) SCANPY: large-scale single-cell gene expression data analysis. Genome Biol 19:15
24. Zeisel A, Hochgerner H, Lönnerberg P et al (2018) Molecular architecture of the Mouse nervous system. Cell 174:999–1014.e22
25. Nishiyama A, Shimizu T, Sherafat A et al (2021) Life-long oligodendrocyte development and plasticity. Semin Cell Dev Biol 116: 25–37
26. Hagemann-Jensen M, Ziegenhain C, Chen P et al (2020) Single-cell RNA counting at allele and isoform resolution using Smart-seq3. Nat Biotechnol 38:708–714

Chapter 7

Probing Cell-Type Specificity of Mutant Phenotype at Transcriptomic Level Using Mosaic Analysis with Double Markers (MADM)

Giselle Cheung, Florian M. Pauler, and Simon Hippenmeyer

Abstract

Mosaic Analysis with Double Markers (MADM) represents a mouse genetic approach coupling differential fluorescent labeling to genetic manipulations in dividing cells and their lineages. MADM uniquely enables the generation and visualization of individual control or homozygous mutant cells in a heterozygous genetic environment. Among its diverse applications, MADM has been used to dissect cell-autonomous gene functions important for cortical development and neural development in general. The high cellular resolution offered by MADM also permits the analysis of transcriptomic changes of individual cells upon genetic manipulations. In this chapter, we describe an experimental protocol combining the generation and isolation of MADM-labeled cells with downstream single-cell RNA-sequencing technologies to probe cell-type specific phenotypes due to genetic mutations at single-cell resolution.

Key words Mosaic Analysis with Double Markers (MADM), Genetic mosaics, Single-cell RNA sequencing, Cell-type identity, Cell-type diversity, Lineage, Development

1 Introduction

Mosaic Analysis with Double Markers (MADM) is a powerful genetic technology in mice that allows concomitant fluorescent labeling and genetic mutations of cells and their lineages at single-cell resolution [1, 2]. MADM relies on *Cre/loxP*-mediated interchromosomal recombination events to generate sparsely and differentially labeled homozygous mutant cells and control cells in an otherwise heterozygous environment (Fig. 1) [2]. Recently, MADM has been expanded to all 19 mouse autosomes allowing genetic mosaic manipulations of >96% of the mouse genome [3].

Over the last decade, MADM technology has greatly advanced our understanding of brain development and in particular, allowed

Giselle Cheung and Florian M. Pauler contributed equally with all other contributors.

Jorge García-Marqués and Tzumin Lee (eds.), *Lineage Tracing: Methods and Protocols*, Methods in Molecular Biology, vol. 2886, https://doi.org/10.1007/978-1-0716-4310-5_7,

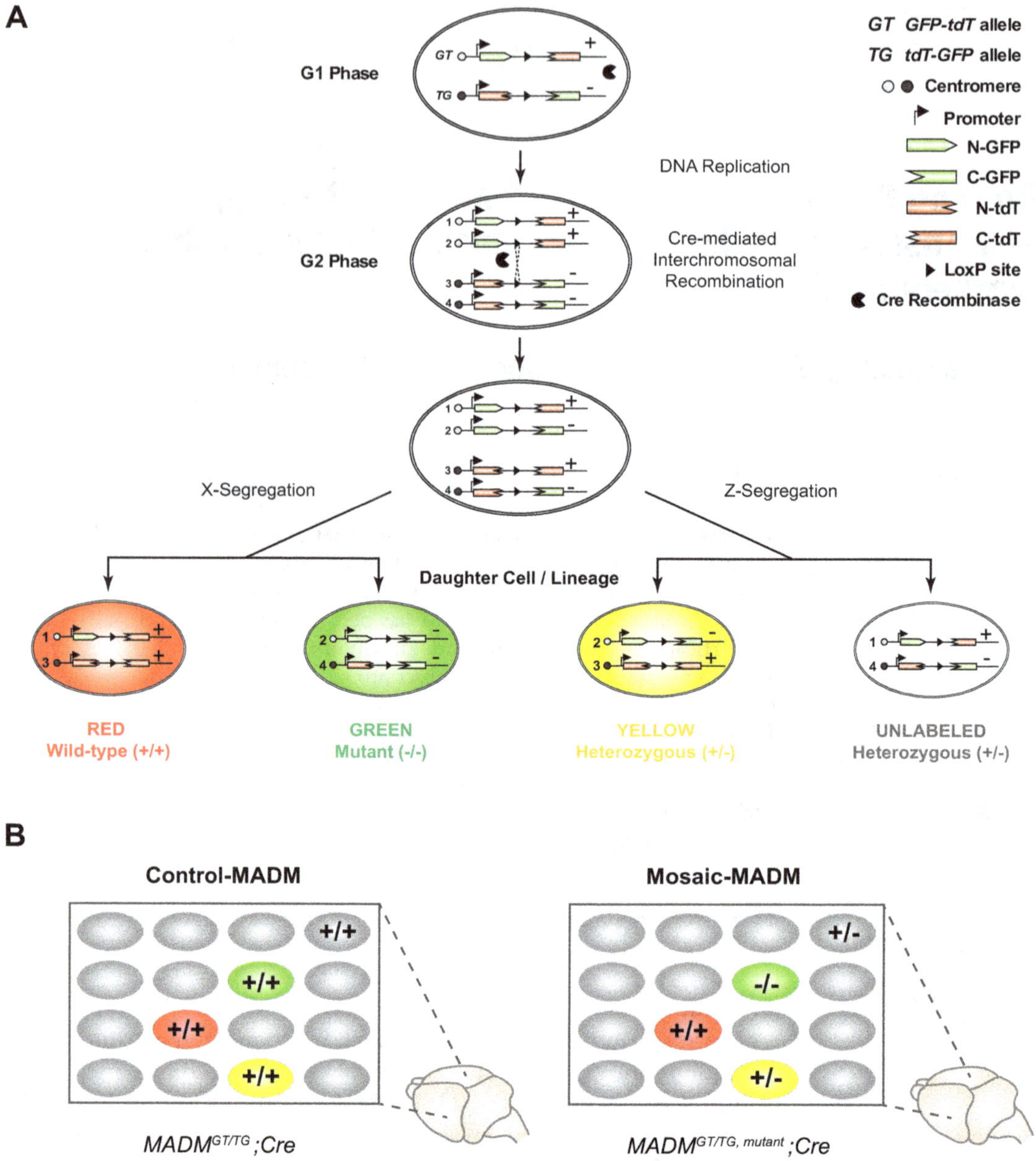

Fig. 1 Principle of MADM mutant analysis. (**a**) Illustration showing MADM principle generating differentially labeled MADM cells with distinct genotypes for mutant gene analysis. (Reproduced and adapted with permission from Contreras et al. [3]). (**b**) Schematics illustrating respective cellular genotypes in Control-MADM and Mosaic-MADM brains

us to dissect stem cell lineage progression in the cerebral cortex [4]. Sparse MADM mosaics have been used to dissect a variety of cell-autonomous gene functions, regulating cortical development [5–8]. The precise spatial and temporal resolution offered by MADM has permitted systematic lineage tracing studies in the neocortex [6, 9] and other brain regions [10–15]. We recommend consulting detailed protocols specific for the generation of MADM transgenic mice [16], isolation of MADM-labeled cells [17, 18], and performing MADM clonal analysis [19]. While MADM dual-labeling offers phenotypic assessment of both control and mutant populations at single-cell resolution, any less obvious changes like gene expression, cannot be easily detected. Furthermore, whether a mutation specifically affects subtypes of cells is even more difficult to study without the ability to robustly identify and isolate specific cell types. In order to enable such analysis, we present here single-cell RNA-sequencing (scRNA-seq), a high-throughput and unbiased method to identify and probe specific cell types, in combination with single-cell MADM labeling/genetic manipulation paradigm.

ScRNA-seq has become a standard lab technique due to easy-to-use and cost effective commercial solutions. Popular examples include droplet microfluidics (i.e., 10X Genomics) or split barcoding SPLiT-Seq [20] which provide high-throughput readout of the transcriptomes of tens of thousands of cells in one run. Such data allows high-resolution identification of cell types using different freely available analysis tools, like Seurat v4 [21] and SCANPY [22], that cluster cells based on shared transcriptomic features. Recent years have seen an explosion of scRNA-seq datasets that cataloged cellular atlases of different regions of the brain in mouse and human tissues [23–28]. This development has been made possible by advanced methods which allow us to selectively collect samples from specific tissues using micro-dissection and/or genetic cellular labeling. Such experimental set up, combined with sample collection at different developmental time points allows powerful multi-dimensional (i.e., space and time) investigations of cell-type diversity and lineages [29]. Dedicated analysis tools even reconstruct developmental trajectories, giving a high-resolution picture of the emergence and diversification of diverse cell types during development [23, 24]. Due to the highly quantitative nature of scRNA-seq data, relative changes in cell-type abundance as well as transcriptomic changes between control and mutant single cells can be investigated [30]. In this chapter, we describe a protocol in which we take full advantage of the single-cell readout of MADM and combine MADM mosaic genetic manipulations with scRNA-seq to uniquely probe differential cell-type specific functions of candidate genes.

2 Materials

2.1 Experimental Mice

1. Transgenic mice carrying a mutant/floxed allele (e.g., gene knockout) of a candidate gene of interest.

2. Transgenic mice carrying a *MADM-TG* cassette located on the same chromosome as the candidate gene of interest (*see* **Note 1**).
3. Transgenic mice carrying *MADM-GT* cassettes located on the homologous chromosome as the *MADM-TG* cassette (*see* **Note 1**).
4. Transgenic mice with CRE expression driven by a cell-type specific promoter of choice (*see* **Note 2**).

2.2 Drugs and Solutions

1. Papain.
2. DNase.
3. Ovomucoid Inhibitor.
4. RNase-free Earle's Balanced Salt Solution (EBSS).
5. Dulbecco's Modified Eagle Medium F12 (DMEM/F12).
6. Heat-inactivated Fetal Bovine Serum (FBS).
7. Horse Serum (HS).
8. Papain-EBSS solution: Papain (20 units/mL) in 5 mL EBSS.
9. DNase-EBSS solution: DNase (166.7 units/mL) in 500 μL EBSS.
10. Papain-DNase solution: 5 mL of Papain-EBSS + 250 μL of DNase-EBSS solutions.
11. Solution 2: 250 μL of DNase-EBSS solution + 400 μL of Ovomucoid Inhibitor + 4.35 mL with EBSS.

2.3 Tools and Equipment

1. Clean dissection tools for brain extraction.
2. Shaking waterbath.
3. Centrifuge.
4. Fluorescence-activated Cell Sorting (FACS) cell sorter.
5. Proprietary commercial equipment or standard laboratory equipment for capturing and barcoding single cells (*see* **Note 3**).
6. RNA sequencer.

2.4 Analysis Software and Packages

1. Programing software (*see* **Note 4**).
2. RNA-sequencing data analysis tools (*see* **Note 5**).

3 Methods

This protocol follows four main steps outlined in Fig. 2. Step 1 is the generation of MADM mice containing relevant genetic mutations. Step 2 is the preparation of single-cell suspension consisting

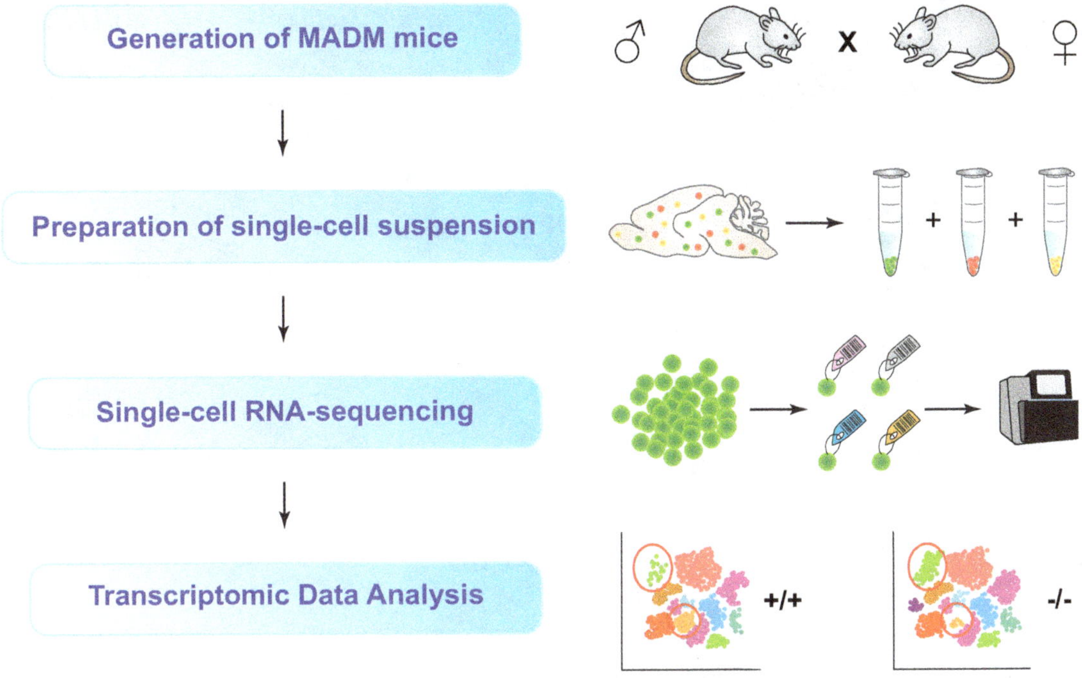

Fig. 2 General workflow. Flowchart and illustration showing the general workflow for scRNA-seq of MADM-labeled cells to analyze cell-type specific mutant gene functions at single cell resolution

of fluorescently labeled cells of interest at the time of sampling. Step 3 is to process samples immediately for scRNA-seq. Finally, in Step 4, transcriptomic data analyses are carried out to compare cell-type abundance, gene expression, and developmental trajectory between control and mutant cells. Each of these steps is detailed in the following sections.

3.1 Generation of MADM Mice

The following simplified protocol describes basic procedures to generate MADM-labeled brains with sparse genetic mutations. A simplified breeding scheme is illustrated in Fig. 3. We advise readers to consult previously published MADM library resource [3] and protocol [16] for further details and adaptations including chromosome-specific considerations.

1. Cross transgenic mice with genetic mutation of interest to one of *MADM-GT* or *MADM-TG* mouse lines of the same chromosome. If mutation is coupled to *MADM-TG* cassette, the final MADM-labeled homozygous mutant cells will be green (GFP^+); if coupled to *MADM-GT* cassette, homozygous mutant cells will be red (tdT^+). *See* **Note 6**.
2. Cross transgenic mice with *Cre*-driver of interest to the other MADM cassette. *See* **Note 7**.

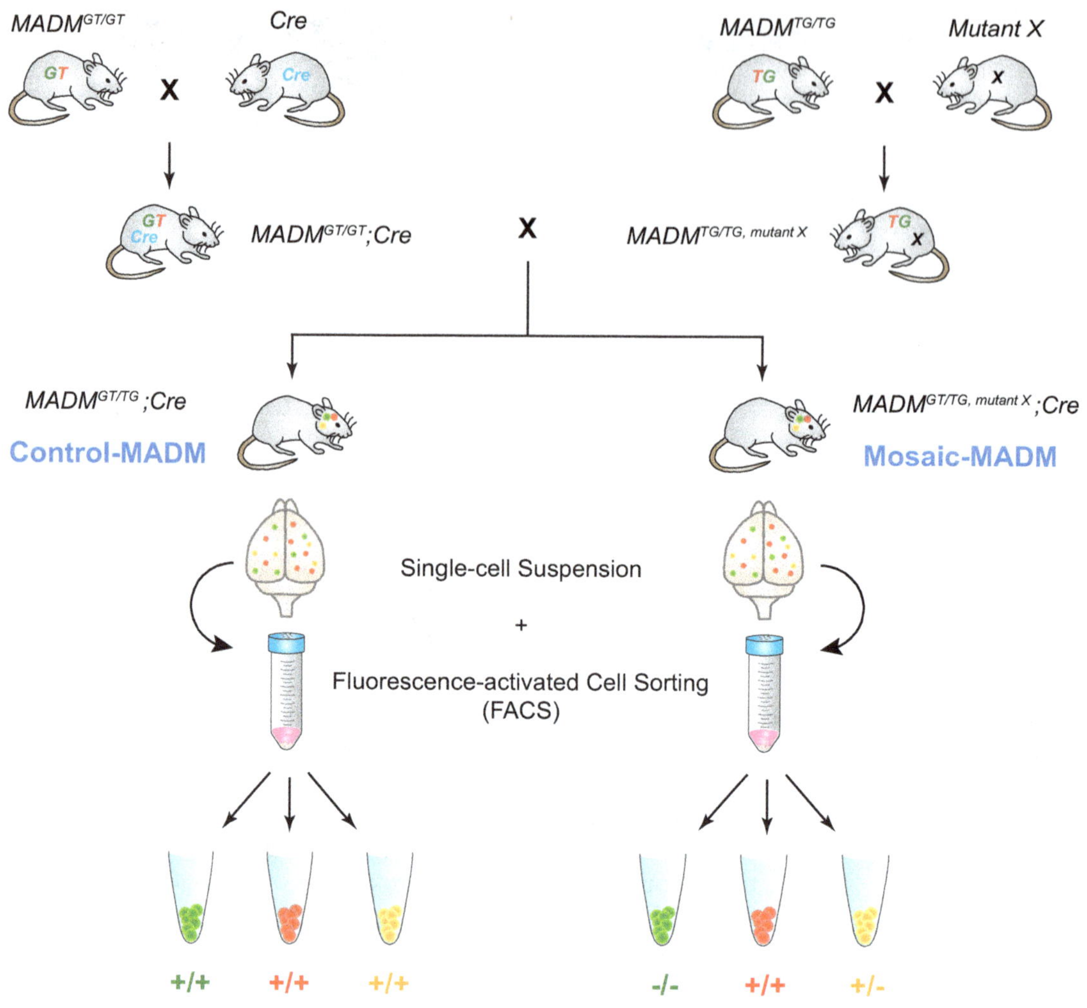

Fig. 3 Generation of MADM mice and single-cell suspension of MADM-labeled cells. Schematics showing the main steps in generating Control-MADM and Mosaic-MADM mice followed by the isolation of different fluorescently labeled populations from respective brains to obtain differential genotypes for gene X

3. After selecting for offspring with correct genotypes, cross resulting mice from 1 and 2.
4. The resulting litter will contain both Control-MADM and Mosaic-MADM mice for experiments. Perform necessary genotyping to identify Control-MADM and Mosaic-MADM mice. *See* also [3, 16].

3.2 Preparation of Single-Cell Suspension

The following simplified protocol describes basic procedures to obtain fresh single-cell suspension of fluorescently labeled MADM cells from young mouse brains aged up to P7. It is recommended to refer to previously published protocol [16, 18] for details and specific considerations for older mice.

1. On the day of sample collection, sacrifice MADM mice by lethal doses of anesthesia or by decapitation.
2. Carefully dissect whole brains or specific regions of interest on ice (Fig. 3).
3. Cut tissue into smaller pieces with clean razor blades and transfer pieces into separate 50 mL bioreactor tubes containing Papain-DNase solution. Separate experimental conditions (i.e., control vs. mosaic) in different tubes. *See* **Notes 8** and **9**.
4. Incubate tubes at 37 °C, shaking at 150 rpm for 30 min in a water bath protected from light.
5. Add Solution 2 to suspension and gently dissociate cell suspension by pipetting up and down with a P-1000 pipette. *See* **Note 9**.
6. Collect cell pellets by centrifugation at 1000 rpm for 10 min at room temperature.
7. Resuspend pellet in Solution 2 and gently dissociate cell suspension by pipetting up and down with a P-1000 pipette. *See* **Note 9**.
8. Collect cell pellets by centrifugation at 1500 rpm for 10 min at room temperature.
9. Resuspend pellet in DMEM containing 10% HS and 10% FBS and keep on ice (Fig. 3). *See* **Note 9.**
10. Using a FACS sorter, collect fluorescently labeled cells in DMEM containing 10% Serum in separate 1.5 mL Eppendorfs at 4 °C. Depending on the breeding schemes used, green, red, and yellow cells in Control-MADM are all control (+/+); while in Mosaic-MADM, green cells are homozygous mutant (−/−), red cells are +/+, and yellow cells are heterozygous (+/−). *See* Fig. 3 and **Note 10**.
11. Proceed immediately to Subheading 3.3.

3.3 Single-Cell RNA-Sequencing

1. Depending on the choice of sequencing method, follow vendor's protocol for capturing single cells and library construction (Fig. 4). For example, for 10X Genomics, sorted cells are loaded into a 10X Chromium controller.
2. Sequencing of libraries should be carried out via a dedicated facility. For example, a commonly used platform is the Illumina NovaSeq platform. *See* **Note 11**.

3.4 Single-Cell RNA-Sequencing Data Analysis

In principle, downstream analyses can be performed to compare between any two or among multiple conditions. For the sake of simplicity, we will focus on comparing green +/+ cells from Control-MADM with green −/− cells from Mosaic-MADM. *See* **Note 6**.

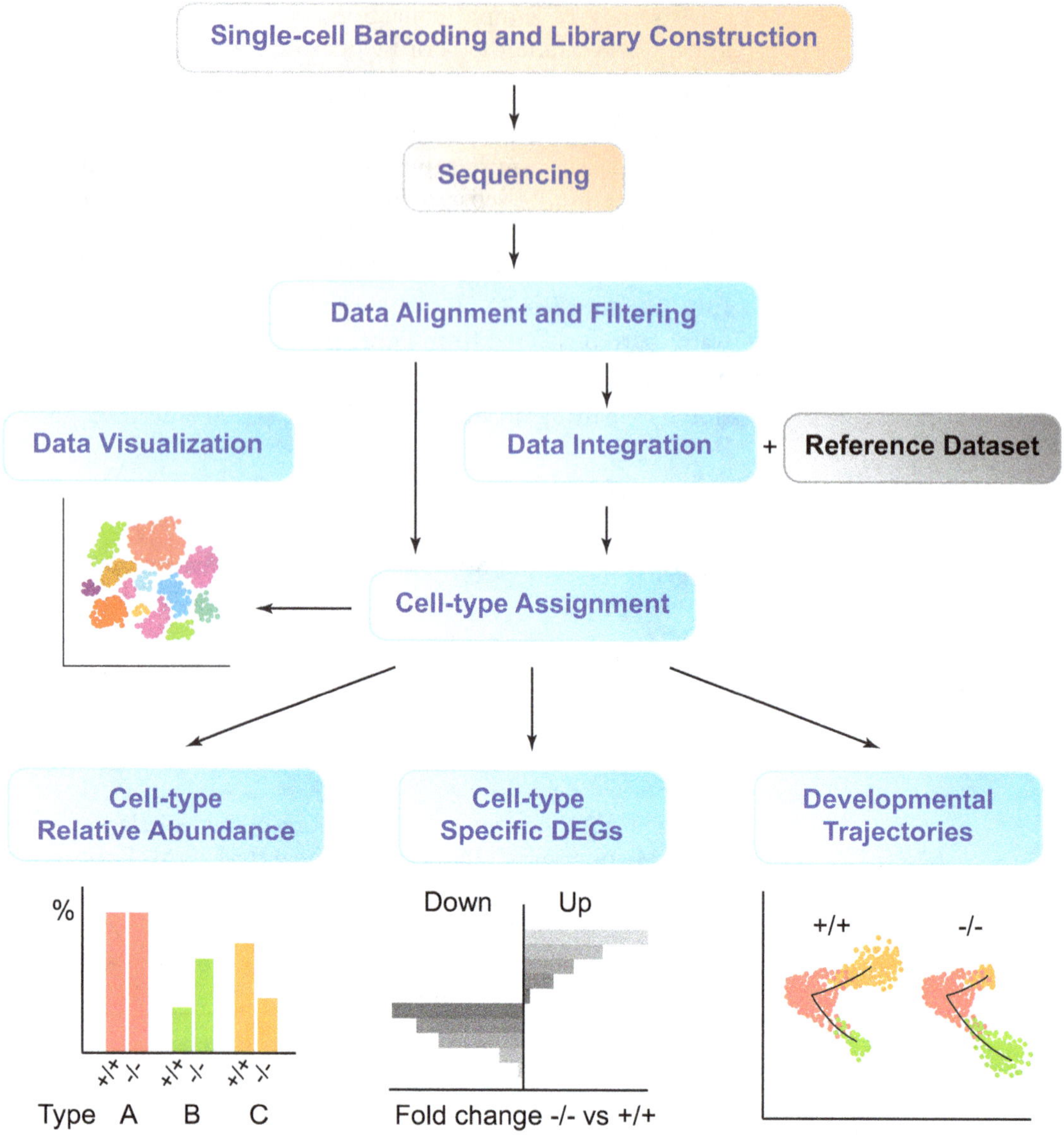

Fig. 4 Single-cell RNA-sequencing and data analysis. Flowchart illustrating the main steps involved in single-cell RNA-sequencing and subsequent data processing and analysis steps. Hypothetical plots of three types of downstream analyses to compare cell-type specificity of mutant genes are shown. DEG differentially expressed gene

Sequencing data analyses should be tailored to the specific features of the scRNA-seq dataset and the research question of interest. Good analysis packages offer detailed tutorials that explain step-by-step data analysis process, for example, https://satijalab.org/seurat/ or https://scanpy.readthedocs.io/en/stable/tutorials.html. Here we broadly describe possible analysis directions (Fig. 4).

1. *Data alignment/gene expression:* In the first step of data analysis, sequencing reads are assigned to specific cells via barcodes introduced during library generation. Using sequence alignment to a reference genome, expression of annotated genes is determined by counting sequencing reads assigned to annotated genes. Typically, specific analysis tools tailored for the specific scRNA-seq technology are used at this step (e.g., Cell-Ranger pipeline for 10X Genomics). These tools produce a matrix linking each cell to its gene expression values, which is the basis for all downstream analyses.
2. *Filtering of high quality cells:* Filtering out low quality cells is the second step of data analysis. Low quality cells usually show a high proportion of reads from mitochondrial genes and have a low number of genes with detectable expression. Exact filtering thresholds depend on scRNA-seq technology, sequencing depth and cell type. *See* **Note 12**.
3. *Cell-type assignment by known markers:* Cells can be first clustered using methods provided in the analysis package. In a second step, specificity of expression of known marker genes in each cluster is determined. Alternatively, *see* **step 4** below.
4. *Cell-type assignment based on a reference:* As an alternative to **step 3**, cell-type assignment can also be carried out based on published reference atlas. Different solutions are possible and depend on the analysis package used. For example, integrating datasets allows combined analysis of reference and scRNA-seq data, offering the possibility of analyzing increased cell numbers. Alternatively, cluster labels from reference cell clusters can be used to label cell types in the scRNA-seq data.
5. *Visualization using tSNE/UMAPs:* In order to visualize high dimensional transcriptomic data, dimensionality reduction workflows are used. Many different tools exist for this task, with *t*-distributed stochastic neighbor embedding (*t*-SNE) and uniform manifold approximation and projection (UMAP) being commonly used for scRNA-seq. Dimensionality reduction is a field of active research and needs careful consideration of parameters [31, 32]. *See* **Note 13**.
6. *Compare relative cell-type abundance between control and ko:* After assigning cell-type identity to individual single-cells using either **step 3** or **4**, the relative abundance of the same cell type across different experiment conditions can be compared. Depending on the availability of biological replicates, different statistical tests can be performed to test significance of difference in cell numbers between clusters of similar cells in different sample groups (i.e., control and mutant).
7. *Compare cell-type specific differential gene expression between control and ko:* Next, transcriptional changes in groups of

individual cells of a specific transcriptional subtype can also be compared across different experiment conditions. We typically compare transcriptomic changes in terms of the number of differential expressed genes (DEGs) as well as strength of differential expression across conditions. A wide range of statistical tests are available in different analysis packages to calculate statistical significance of gene expression differences between sample groups (i.e., control and mutant). The optimal test depends on the specific features of the scRNA-seq dataset and the research question. *See* **Note 6**.

8. *Compare developmental trajectories between control and ko:* As a result of genetic manipulations, the developmental potential of different cell types may be changed. This can be studied by comparing the trajectories of temporal changes in gene expression. A wide range of algorithms are available to infer developmental trajectories. The choice of the optimal algorithm depends on the specific features of the scRNA-seq dataset and the research question.

4 Notes

1. All 19 pairs of MADM mice are publicly available at either The Jackson Laboratory and/or European Mouse Mutant Archive EMMA [3].
2. For MADM to work, it is crucial that CRE is expressed in dividing cells.
3. For example, 10X Genomics Chromium Controller as proprietary equipment or standard laboratory equipment for SPLiT-Seq.
4. For example, R and Python.
5. For example, Seurat and SCANPY.
6. Recombination of MADM cassettes followed by X-segregation results in homozygosing of the parental alleles, called Uniparental Disomy (UPD). Depending on the parental origin of the MADM cassettes, green and red cells are strictly linked to either paternal or maternal UPD that can cause transcriptional effects [8, 17]. To control for UPD effects, it is important to compare control and mutant MADM with the same UPD.
7. Both constitutively expressed and inducible *Cre/ER*-drivers can be used in principle. However, to obtain sufficient number of MADM-labeled cells, we recommend using constitutively expressed *Cre*-drivers where possible.
8. Multiple brains can be pooled into one tube to control for individual variations. This is particularly important to achieve

good yield when a sparse MADM line and/or a small brain region is used.

9. Exact volume of solutions to be used depends on the amount of tissue to be dissociated. Refer to previous protocol [16, 18] for recommendations.
10. To preserve optimal RNA quality, FACS sorting should be performed at 4 °C. Sorted cells should be kept on ice and processed immediately for scRNA-seq. High levels of viable cells ensure high number of high-quality single-cell transcriptomes.
11. Choice of sequencing platform depends on read type and read depth, which in turn depend on the specific library. Multiple possibilities for sequencing platforms exist and change frequently, so it is advisable to get in touch with relevant sequencing facility prior to starting the experiment.
12. Whereas thresholds for detected genes largely vary, since they are specific to scRNA-seq technology and sequencing depth, commonly used thresholds for % mitochondrial genes are between 5% and10%.
13. Dimensionality reduction and visualization can be a powerful quality control tool. Samples containing a variety of cell types should produce a number of well-defined clusters of cells. Low quality cells tend to cluster together, independent of cell type.

Acknowledgments

We thank all Hippenmeyer lab members for support and discussions. Experimental steps described were optimized with support provided by the Imaging & Optics Facility (IOF) and Preclinical Facility (PCF) at ISTA, Vienna BioCenter Core Facilities (VBCF), and Christoph Bock lab at Center for Molecular Medicine (CeMM). G.C. received funding from European Commission (IST plus postdoctoral fellowship). This work was supported by ISTA institutional funds: The Austrian Science Fund Special Research Programmes (FWF SFB F78 Neuro Stem Modulation) to S.H.

Competing Interests The authors declare no competing interests.

Author Contributions Project Management: S.H.; Conceptualization and Design: S.H., G.C., and F.M.P.; Manuscript Drafting: G.C. and F.M.P.; Critical Revision of the Manuscript: All authors; Final version was read by all authors and approved to be published.

References

1. Hippenmeyer S (2013) Dissection of gene function at clonal level using mosaic analysis with double markers. Front Biol 8(6): 557–568. https://doi.org/10.1007/s11515-013-1279-6
2. Zong H, Espinosa JS, Su HH, Muzumdar MD, Luo L (2005) Mosaic analysis with double markers in mice. Cell 121(3):479–492. https://doi.org/10.1016/j.cell.2005.02.012
3. Contreras X, Amberg N, Davaatseren A, Hansen AH, Sonntag J, Andersen L, Bernthaler T, Streicher C, Heger A, Johnson RL, Schwarz LA, Luo L, Rulicke T, Hippenmeyer S (2021) A genome-wide library of MADM mice for single-cell genetic mosaic analysis. Cell Rep 35(12):109274. https://doi.org/10.1016/j.celrep.2021.109274
4. Hippenmeyer S (2023) Principles of neural stem cell lineage progression: insights from developing cerebral cortex. Curr Opin Neurobiol 79:102695. https://doi.org/10.1016/j.conb.2023.102695
5. Amberg N, Pauler FM, Streicher C, Hippenmeyer S (2022) Tissue-wide genetic and cellular landscape shapes the execution of sequential PRC2 functions in neural stem cell lineage progression. Sci Adv 8(44):eabq1263. https://doi.org/10.1126/sciadv.abq1263
6. Beattie R, Postiglione MP, Burnett LE, Laukoter S, Streicher C, Pauler FM, Xiao G, Klezovitch O, Vasioukhin V, Ghashghaei TH, Hippenmeyer S (2017) Mosaic analysis with double markers reveals distinct sequential functions of Lgl1 in neural stem cells. Neuron 94(3):517–533 e513. https://doi.org/10.1016/j.neuron.2017.04.012
7. Hansen AH, Pauler FM, Riedl M, Streicher C, Heger A, Laukoter S, Sommer C, Nicolas A, Hof B, Tsai LH, Rülicke T, Hippenmeyer S (2022) Tissue-wide effects override cell-intrinsic gene function in radial neuron migration. Oxford Open Neurosci 1. https://doi.org/10.1093/oons/kvac009
8. Laukoter S, Pauler FM, Beattie R, Amberg N, Hansen AH, Streicher C, Penz T, Bock C, Hippenmeyer S (2020) Cell-type specificity of genomic imprinting in cerebral cortex. Neuron 107(6):1160–1179 e1169. https://doi.org/10.1016/j.neuron.2020.06.031
9. Gao P, Postiglione MP, Krieger TG, Hernandez L, Wang C, Han Z, Streicher C, Papusheva E, Insolera R, Chugh K, Kodish O, Huang K, Simons BD, Luo L, Hippenmeyer S, Shi SH (2014) Deterministic progenitor behavior and unitary production of neurons in the neocortex. Cell 159(4):775–788. https://doi.org/10.1016/j.cell.2014.10.027
10. Ma T, Wong SZH, Lee B, Ming GL, Song H (2021) Decoding neuronal composition and ontogeny of individual hypothalamic nuclei. Neuron 109(7):1150–1167. e1156. https://doi.org/10.1016/j.neuron.2021.01.026
11. Shi W, Xianyu A, Han Z, Tang X, Li Z, Zhong H, Mao T, Huang K, Shi SH (2017) Ontogenetic establishment of order-specific nuclear organization in the mammalian thalamus. Nat Neurosci 20(4):516–528. https://doi.org/10.1038/nn.4519
12. Wong SZH, Scott EP, Mu W, Guo X, Borgenheimer E, Freeman M, Ming GL, Wu QF, Song H, Nakagawa Y (2018) In vivo clonal analysis reveals spatiotemporal regulation of thalamic nucleogenesis. PLoS Biol 16(4): e2005211. https://doi.org/10.1371/journal.pbio.2005211
13. Zhang T, Liu T, Mora N, Guegan J, Bertrand M, Contreras X, Hansen AH, Streicher C, Anderle M, Danda N, Tiberi L, Hippenmeyer S, Hassan BA (2021) Generation of excitatory and inhibitory neurons from common progenitors via Notch signaling in the cerebellum. Cell Rep 35(10):109208. https://doi.org/10.1016/j.celrep.2021.109208
14. Zhang YH, Xu M, Shi X, Sun XL, Mu W, Wu H, Wang J, Li S, Su P, Gong L, He M, Yao M, Wu QF (2021) Cascade diversification directs generation of neuronal diversity in the hypothalamus. Cell Stem Cell 28(8): 1483–1499. e1488. https://doi.org/10.1016/j.stem.2021.03.020
15. Cheung G, Pauler FM, Koppensteiner P, Krausgruber T, Streicher C, Schrammel M, Gutmann-Ozgen N, Ivec AE, Bock C, Shigemoto R, Hippenmeyer S (2024) Multipotent progenitors instruct ontogeny of the superior colliculus. Neuron 112(2):230–246 e211. https://doi.org/10.1016/j.neuron.2023.11.009
16. Amberg N, Hippenmeyer S (2021) Genetic mosaic dissection of candidate genes in mice using mosaic analysis with double markers. STAR Protoc 2(4):100939. https://doi.org/10.1016/j.xpro.2021.100939
17. Laukoter S, Amberg N, Pauler FM, Hippenmeyer S (2020) Generation and isolation of single cells from mouse brain with mosaic analysis with double markers-induced uniparental chromosome disomy. STAR Protoc 1(3): 100215. https://doi.org/10.1016/j.xpro.2020.100215

18. Amberg N, Cheung G, Hippenmeyer S (2024) Protocol for sorting cells from mouse brains labeled with mosaic analysis with double markers by flow cytometry. STAR Protoc 5(1): 102771. https://doi.org/10.1016/j.xpro.2023.102771

19. Beattie R, Streicher C, Amberg N, Cheung G, Contreras X, Hansen AH, Hippenmeyer S (2020) Lineage tracing and clonal analysis in developing cerebral cortex using mosaic analysis with double markers (MADM). J Vis Exp 159:e61147. https://doi.org/10.3791/61147

20. Rosenberg AB, Roco CM, Muscat RA, Kuchina A, Sample P, Yao Z, Graybuck LT, Peeler DJ, Mukherjee S, Chen W, Pun SH, Sellers DL, Tasic B, Seelig G (2018) Single-cell profiling of the developing mouse brain and spinal cord with split-pool barcoding. Science 360(6385):176–182. https://doi.org/10.1126/science.aam8999

21. Hao Y, Hao S, Andersen-Nissen E, Mauck WM 3rd, Zheng S, Butler A, Lee MJ, Wilk AJ, Darby C, Zager M, Hoffman P, Stoeckius M, Papalexi E, Mimitou EP, Jain J, Srivastava A, Stuart T, Fleming LM, Yeung B, Rogers AJ, McElrath JM, Blish CA, Gottardo R, Smibert P, Satija R (2021) Integrated analysis of multimodal single-cell data. Cell 184(13): 3573–3587 e3529. https://doi.org/10.1016/j.cell.2021.04.048

22. Wolf FA, Angerer P, Theis FJ (2018) SCANPY: large-scale single-cell gene expression data analysis. Genome Biol 19(1):15. https://doi.org/10.1186/s13059-017-1382-0

23. Di Bella DJ, Habibi E, Stickels RR, Scalia G, Brown J, Yadollahpour P, Yang SM, Abbate C, Biancalani T, Macosko EZ, Chen F, Regev A, Arlotta P (2021) Molecular logic of cellular diversification in the mouse cerebral cortex. Nature 595(7868):554–559. https://doi.org/10.1038/s41586-021-03670-5

24. La Manno G, Siletti K, Furlan A, Gyllborg D, Vinsland E, Mossi Albiach A, Mattsson Langseth C, Khven I, Lederer AR, Dratva LM, Johnsson A, Nilsson M, Lonnerberg P, Linnarsson S (2021) Molecular architecture of the developing mouse brain. Nature 596(7870):92–96. https://doi.org/10.1038/s41586-021-03775-x

25. Zeisel A, Hochgerner H, Lonnerberg P, Johnsson A, Memic F, van der Zwan J, Haring M, Braun E, Borm LE, La Manno G, Codeluppi S, Furlan A, Lee K, Skene N, Harris KD, Hjerling-Leffler J, Arenas E, Ernfors P, Marklund U, Linnarsson S (2018) Molecular architecture of the mouse nervous system. Cell 174(4):999–1014. e1022. https://doi.org/10.1016/j.cell.2018.06.021

26. Siletti K, Hodge R, Albiach AM, Hu L, Lee KW, Lönnerberg P, Bakken T, Ding S-L, Clark M, Casper T, Dee N, Gloe J, Keene CD, Nyhus J, Tung H, Yanny AM, Arenas E, Lein ES, Linnarsson S (2022) Transcriptomic diversity of cell types across the adult human brain. bioRxiv 2022.2010.2012.511898. https://doi.org/10.1101/2022.10.12.511898

27. Tasic B, Menon V, Nguyen TN, Kim TK, Jarsky T, Yao Z, Levi B, Gray LT, Sorensen SA, Dolbeare T, Bertagnolli D, Goldy J, Shapovalova N, Parry S, Lee C, Smith K, Bernard A, Madisen L, Sunkin SM, Hawrylycz M, Koch C, Zeng H (2016) Adult mouse cortical cell taxonomy revealed by single cell transcriptomics. Nat Neurosci 19(2): 335–346. https://doi.org/10.1038/nn.4216

28. Eze UC, Bhaduri A, Haeussler M, Nowakowski TJ, Kriegstein AR (2021) Single-cell atlas of early human brain development highlights heterogeneity of human neuroepithelial cells and early radial glia. Nat Neurosci 24(4): 584–594. https://doi.org/10.1038/s41593-020-00794-1

29. Svensson V, Vento-Tormo R, Teichmann SA (2018) Exponential scaling of single-cell RNA-seq in the past decade. Nat Protoc 13(4):599–604. https://doi.org/10.1038/nprot.2017.149

30. Heumos L, Schaar AC, Lance C, Litinetskaya A, Drost F, Zappia L, Lucken MD, Strobl DC, Henao J, Curion F, Single-cell Best Practices C, Schiller HB, Theis FJ (2023) Best practices for single-cell analysis across modalities. Nat Rev Genet 24(8): 550–572. https://doi.org/10.1038/s41576-023-00586-w

31. Becht E, McInnes L, Healy J, Dutertre CA, Kwok IWH, Ng LG, Ginhoux F, Newell EW (2018) Dimensionality reduction for visualizing single-cell data using UMAP. Nat Biotechnol 37:38. https://doi.org/10.1038/nbt.4314

32. Kobak D, Linderman GC (2021) Initialization is critical for preserving global data structure in both t-SNE and UMAP. Nat Biotechnol 39(2): 156–157. https://doi.org/10.1038/s41587-020-00809-z

Chapter 8

Lineage Tracing by Light-Sheet Microscopy and Computational Reconstruction

Maria Kalogeridi, Ioannis Liaskas, John Rallis, and Anastasios Pavlopoulos

Abstract

Lineage tracing based on modern live imaging approaches enables to visualize, reconstruct, and analyze the developmental history, fate, and dynamic behaviors of cells in vivo in a direct, comprehensive, and quantitative manner. Light-sheet fluorescence microscopy (LSFM) has greatly boosted lineage tracing efforts, because fluorescently labeled specimens can be imaged in their entirety, over long periods of time, with high spatiotemporal resolution and minimal photodamage. In addition, an increasing arsenal of commercial and open-source software solutions for cell and nuclei segmentation and tracking can be employed to convert data from pixel-based to object-based representations, and to reconstruct the lineages of cells in their native context as they organize in tissues, organs, and whole organisms. This chapter describes the preparation of LSFM image datasets and the use of three freely available platforms, namely, the Fiji/ImageJ plugins Massive Multiview Tracker (MaMuT), Mastodon and TrackMate, for small-scale and large-scale lineage tracing purposes using manual, semi-automated, and fully automated pipelines for nuclei or cell tracking. Lineage tracing with these tools is described on LSFM image datasets of fluorescently labeled embryos from the crustacean model *Parhyale hawaiensis* that lends itself to multi-scale investigations of development and regeneration.

Key words Light-sheet fluorescence microscopy, Cell lineage, Segmentation, Tracking, Fiji, ImageJ, MaMuT, Mastodon, TrackMate, Crustaceans, *Parhyale hawaiensis*

1 Introduction

There is a longstanding interest in biological and biomedical research in creating cellular and molecular atlases based on microscopy or multi-omics datasets of increasing scope and resolution [1]. Of interest here are lineage tracing methodologies employed by biologists for more than a century, in order to identify the fate of cells in developing multicellular organisms [2] [3]. In theory, direct

Authors Maria Kalogeridi, Ioannis Liaskas and John Rallis have contributed equally to this work and are listed in alphabetical order.

Jorge García-Marqués and Tzumin Lee (eds.), *Lineage Tracing: Methods and Protocols*, Methods in Molecular Biology, vol. 2886, https://doi.org/10.1007/978-1-0716-4310-5_8,

observation can reveal how a fertilized egg gives rise to the different tissues, organs, and cell types of the adult form. In practice, this endeavor had been hindered for many decades by myriad technical and biological complications that had limited comprehensive lineage tracing to relatively simple embryos composed of few cells like the nematode *Caenorhabditis elegans* [4].

Fluorescence microscopy has become an indispensable, noninvasive tool to observe complex and dynamic processes at multiple levels of biological organization [5]. Recent and ongoing advances in functional genetics and genomics, strategies for labeling cellular components in vivo, microscopy techniques for live imaging, and computational image analysis tools are increasingly enabling life scientists to record and reconstruct cell lineages in diverse experimental model organisms, in four dimensions (three-dimensional space over time), and with high spatial and temporal resolution [6]. Imaging-based lineage tracing can resolve the developmental history and fate of many or even all cells in the imaged specimen. Furthermore, it can provide quantitative information on the spatiotemporal dynamics of cellular behaviors, such as cell division, cell death, cell shape changes and cell rearrangements, shaping tissues, organs, and whole organisms [7]. Combined with other ongoing advances in single-cell and spatial omics approaches, these high-resolution atlases of development can make the link between anatomy, cell lineages, cell dynamics, gene regulatory states, and fates of cells in their native in vivo context [8].

Compared to more conventional imaging modalities, like wide-field, point scanning, or spinning disc confocal fluorescence microscopy, light-sheet fluorescence microscopy (LSFM) is currently revolutionizing multi-scale, in toto imaging, spanning from the (sub)micrometer-size (sub)cellular level to the centimeter-size tissue or even whole organism level [9, 10]. LSFM is used to image fixed, cleared and expanded samples, but it really excels in meso-scale, live imaging applications, which are essential for lineage tracing. Although there is no one-size-fits-all implementation, the LSFM principle strikes the best balance between several key requirements that need to be met for imaging live samples with cellular resolution and observe processes in their native context: sample health, long-term imaging with low phototoxicity and photobleaching, high spatiotemporal resolution, high signal-to-noise ratio, broad dynamic range of pixel intensities, and reconstruction of entire specimen from multiple angles (multi-view LSFM; mvLSFM) with almost isotropic resolution [11, 12].

The benefits of mvLSFM come at the cost of demanding computations necessary to reconstruct and visualize imaged volumes over time. We will only briefly describe here LSFM image acquisition and processing and refer readers to other publications for more details [13–15]. For lineage tracing purposes, cell or nuclei segmentation and tracking in these multi-terabyte, multi-dimensional

(4D, multi-channel, multi-view) image datasets is essential to reconstruct cell lineage trees. Over the last years, an increasing number of commercial and freely available software solutions are enabling lineage reconstruction from LSFM datasets [6]. In this chapter, we cover extensively three open-source software packages available as Fiji/ImageJ plugins [16]:

(i) Massive Multiview Tracker (MaMuT), which is suitable for relatively small-scale lineage tracing applications using manual or semi-automated cell/nuclei tracking [17].

(ii) Mastodon, which is suitable for large-scale, systems-level lineage reconstructions [18], offering algorithms for the fully automated detection and tracking of nuclei.

(iii) TrackMate 7, which is a versatile platform for lineage tracing in 3D (2D over time) datasets, as it integrates external, machine learning and deep learning detectors for the automated segmentation and tracking of nuclei and cell outlines, coupled with interactive tools for track editing [19, 20]. Among these tools, we describe here the use of the cellpose detector for cell tracking [21], as well as the Image Surface Analysis Environment (ImSAnE) [22], and ilastik [23] toolkits for making 2D cartographic map projections from 3D LSFM images.

We demonstrate lineage tracing with these Fiji plugins on mvLSFM image datasets acquired from fluorescently labeled embryos from the crustacean model organism *Parhyale hawaiensis* [24, 25]. *Parhyale* is a genetically and optically tractable experimental model well suited for embryological and regenerative studies that have motivated the development of some of the lineage tracing tools described here and elsewhere [6, 17, 26].

2 Materials

1. *Parhyale* mvLSFM dataset: A *Parhyale* mvLSFM image dataset can be downloaded from http://bds.mpi-cbg.de/mamut-example/. It consists of the image file *MaMuT_Parhyale_-demo.h5* in HDF5 format and the text file *MaMuT_Parhyale_-demo.xml* in XML format (see point 3 below). The demo dataset consists of 10 time-points from a time-lapse LSFM recording on a Zeiss LightSheet Z1 microscope [17]. A 3-day old transgenic *Parhyale* embryo with fluorescently labeled nuclei was imaged every 7.5 min at 26 °C from the ventral-left side (Angle 235°), the ventral side (Angle 280°), and the ventral-right side (Angle 325°). The dataset contains a fourth Angle 0° corresponding to the fused z-stack from the 3 input z-stacks after spatiotemporal registration and multi-view deconvolution (Fig. 1).

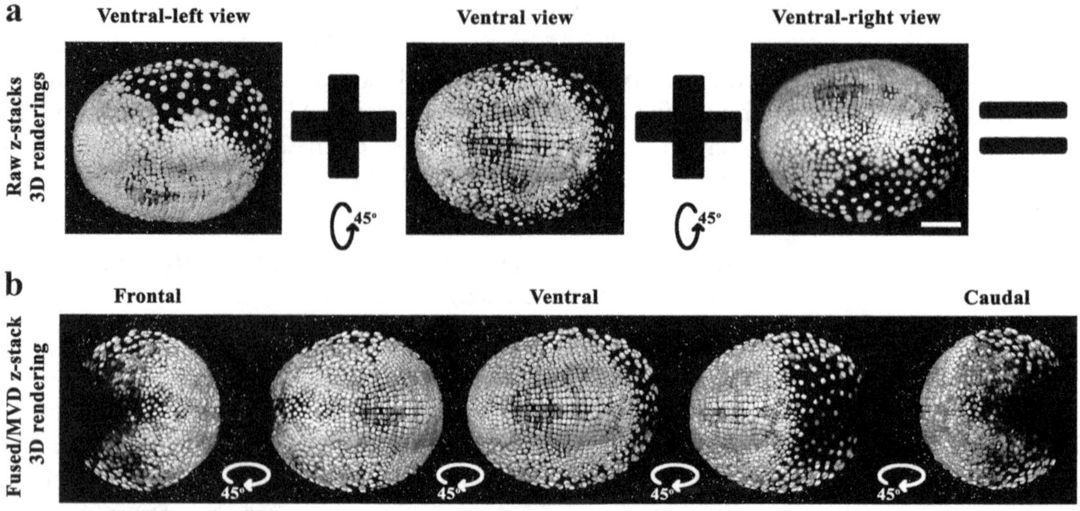

Fig. 1 Multi-view light-sheet fluorescence microscopy of *Parhyale* embryogenesis. (**a**) 3D renderings of a 3-day old transgenic *Parhyale* embryo with H2B-mRFP-labeled nuclei. The embryo was imaged from the three indicated views (ventral-left, ventral, ventral-right) with 45° physical rotation around the anterior-posterior body axis. Anterior toward the left in all panels. (**b**) Raw views were registered using the fluorescent beads surrounding the embryo and were fused with multi-view deconvolution into a single output 3D image, computationally rendered in different positions with 45° rotation around the left-right axis. Scale bar 100 μm, same in all panels

2. Fiji: Download and unpack Fiji (https://fiji.sc/).
3. Image processing: Process raw LSFM images to prepare them for lineage tracing using the collection of MATLAB scripts from the Keller group [14] (https://www.janelia.org/open-science/efficient-processing-and-analysis-of-large-scale-light-sheet-microscopy-data and https://www.janelia.org/open-science/high-throughput-multiview-image-registration-for-simview-microscopy) or the Fiji plugins Multiview Reconstruction [27, 28] (https://imagej.net/plugins/multiview-reconstruction) and its successor BigStitcher [29] (https://imagej.net/plugins/bigstitcher/). We only give an overview of LSFM image processing steps in Subheading 3.1 and refer readers to these other sources for detailed implementation.
4. MaMuT: Enable MaMuT (https://imagej.net/plugins/mamut/) from the Fiji updater, update Fiji, restart it, and access MaMuT from the *Plugins* dropdown menu. MaMuT comes with a tutorial (https://imagej.net/plugins/mamut/getting-started) describing how it combines the functionalities of the Fiji plugins BigDataViewer (https://imagej.net/plugins/bdv/) and TrackMate (https://imagej.net/plugins/trackmate/). BigDataViewer (BDV) is a browser for multi-terabyte, multi-view image datasets that uses the XML file to store image metadata, and the HDF5 file to store the image

volumes in tiles and at different resolutions [30]. This XML/HDF5 architecture of BDV is used by MaMuT and Mastodon for seamless and interactive navigation in all connected views of mvLSFM datasets sliced in any desired orientation. TrackMate is the annotation engine for tracking objects in fluorescence microscopy images offering visualization, editing and analysis tools for tracked cells/nuclei [19].

5. Mastodon: Enable Mastodon (https://imagej.net/plugins/mastodon) from the Fiji updater, update Fiji, restart it, and access Mastodon from the *Plugins* dropdown menu. Mastodon is under active development and all its features are updated and documented (https://mastodon.readthedocs.io/en/latest/docs/partA/getting_started.html#getting-mastodon).
6. ImSAnE: To download the ImSAnE toolbox for tissue cartography, unzip the folder imsane-tutorial_v1.0.0.zip from https://zenodo.org/records/7628300. Open the imsane-tutorial.html that contains the requirements and the instructions on how to unwrap 3D surfaces on 2D images [31]. ImSAnE is implemented in MATLAB (https://www.mathworks.com/products/matlab.html) and uses ilastik (https://www.ilastik.org) for pixel classification and surface segmentation. This tutorial was created by Tomancak lab members and colleagues to make cartographic projections at different depths (onion layers) and export them to Fiji for downstream analysis [32, 33].
7. TrackMate: TrackMate 7 is included in the main Fiji installation and can be launched from the *Plugins* › *Tracking* dropdown menu. ImageJ wikis provide extensive documentation of TrackMate (https://imagej.net/plugins/trackmate/) and the external state-of-the-art detectors (https://imagej.net/plugins/trackmate/trackmate-v7-detectors) for segmenting cells and nuclei.
8. Cellpose: To use the cellpose detector in TrackMate to segment cell outlines in the 2D cartographic projection, install cellpose, written in Python, following the instructions in the cellpose github page (https://github.com/MouseLand/cellpose) and the TrackMate-Cellpose module from the Fiji updater (https://imagej.net/plugins/trackmate/detectors/trackmate-cellpose#installing-cellpose).

3 Methods

3.1 Processing of LSFM Image Datasets for Lineage Tracing

1. Image file preprocessing: During a multi-view LSFM time-lapse recording, each acquired time-point is composed of multiple z-stacks corresponding to the different light-sheets, views

and/or channels (Fig. 1, top). The first task involves opening the raw TIFF files, applying some filters for noise reduction and resaving often in a new file format, such as the Keller Lab Block (KLB), HDF5 or others, offering lossless compression and flexible block architectures for fast read/write speeds and seamless data browsing.

2. Spatial Registration: In each time-point, the acquired views are first aligned (registered) in 3D space with respect to each other. Generalized spatial registration methods are based on blob-like features inside the sample (e.g., fluorescent nuclei) or surrounding the sample (embedded fluorescent beads) that can be reliably detected and matched between the different views in each time-point. Alternatively, the orthogonal geometry of LSFM microscopes can be taken into consideration to register 2 opposite views or 4 orthogonal views in a content-based manner.
3. Temporal Registration: Registration is then extended across time-points to stabilize the image series, compensating for drifts of the specimen during the long LSFM time-lapse recordings.
4. Fusion: Once the entire dataset is aligned, the individual views can be fused together (using the transformation parameters from previous steps) into a single output 3D image per time-point and channel. Voxel intensities in the fused image can be calculated by simple or weighted averaging between input views or by the more computationally intensive multi-view deconvolution for a better signal-to-noise ratio and more isotropic resolution (Fig. 1, bottom). Deconvolution requires the knowledge of point spread functions, which can be extracted from the fluorescent beads or nuclei. Users can also proceed directly after registration to data visualization and tracking without fusion.

3.2 Small-Scale Nuclei Tracking with MaMuT in LSFM Datasets

MaMuT is the first-generation cell lineaging tool in the Fiji/ImageJ ecosystem that was developed specifically for tracking cells synergistically and comprehensively from multiple views available in LSFM datasets [17]. The MaMuT architecture is suitable for tracking a relatively small number of cells or nuclei manually or semi-automatically (one nucleus at a time) or import cropped annotations from fully automated nuclei trackers, as described below. Please refer to the MaMuT documentation (https://imagej.net/plugins/mamut/getting-started) for more information.

1. Start a new MaMuT project by browsing to *Plugins › MaMuT › New MaMuT annotation* in Fiji. Select the BDV XML file *MaMuT_Parhyale_demo.xml* that accompanies the HDF5

image source *MaMuT_Parhyale_demo.h5*. The main GUI window of MaMuT opens (Fig. 2a, left).

2. Save the newly created MaMuT project any time by pressing the *Save* button in the main GUI. This MaMuT annotation file is a new XML typically named **-mamut.xml* (e.g.,*MaMuT_-Parhyale_demo-mamut.xml*) that can be reopened to continue lineage tracing by browsing to *Plugins* › *MaMuT* › *Open MaMuT annotation*.
3. Press the *MaMuT viewer* button in the main GUI to open the image data in a BDV *Viewer* window (Fig. 2b). Use the mouse and the keyboard to navigate through the image data in space and time. Press *Help* in the *Viewer* window to see the default key-bindings (*see* **Note 1**).
4. Select the desired raw or fused LSFM z-stack to visualize in the *Viewer* by pressing numeric keys (*1* to *4* in the demo dataset): *1* shows the fused volume (Angle 0°), *2* the ventral-right raw view (Angle 325°), *3* the ventral raw view (Angle 280°), and *4* the ventral-left raw view (Angle 235°). Change the brightness and color by selecting *Settings* › *Brightness & Color* in the *Viewer* (or press the *S* key) and switch with the *I* key between the nearest-neighbor and tri-linear methods to interpolate pixel values (the latter preferred for smoother display).

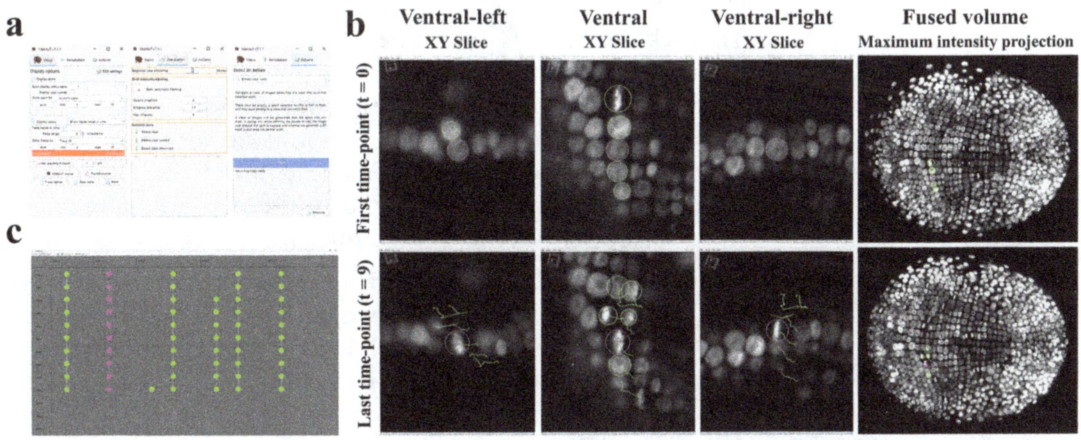

Fig. 2 Small-scale nuclear tracking with MaMuT. (**a**) The three tabs of the MaMuT GUI: Views tab (left), Annotation tab (middle), and Actions tab (right). (**b**) The MaMuT Viewer windows displaying raw image data (XY planes) from the ventral-left view (1st column), ventral view (2nd column) and ventral-right view (3rd column). Maximum intensity projections of the entire embryo after fusion of the registered raw views by multi-view deconvolution (MVD) are shown in the 4th column. Five nuclei marked with circles (in view) or dots (out of view) and their tracks (lines) were followed from the first time-point (top row) until the last time-point (bottom row) of the *Parhyale* demo dataset. All viewers are centered on the active selection marked in magenta. (**c**) The MaMuT TrackScheme window displaying the five nuclei (spots) and their tracks (vertical lines) over ten time-points. The selected nucleus in panels b is also highlighted in magenta in TrackScheme

5. Users have the flexibility to display z-stacks individually or in combination and slice the data in any desired position, orientation, zoom, and time-point. Desired positions and orientations can be saved as bookmarks for easy retrieval.
6. Press the *MaMuT viewer* button in the main GUI again and again to open as many *Viewer* windows as required for tracking purposes. All *Viewers* are synchronized and centered always on the same selected/activated spot (see next step) and time-point during lineage tracing (Fig. 2b).
7. In a *Viewer* window, move the mouse pointer over a cell (a nucleus to be more precise) in the demo dataset and double-click (or press *A* key) to create the first annotation that appears as a magenta circle (spot). To change its position, put the mouse pointer inside the spot, press and hold the *Spacebar* key, move the mouse pointer to the new position and the spot will follow it. Adjust spot radius with the *Q* and *E* keys to match the outline of the nucleus. The spot appears as a circle when the nucleus is in focus. It becomes progressively narrower and eventually appears as a dot when the nucleus moves out of focus (*see* **Note 2**).
8. Select the same spot in different *Viewers* to position it accurately in space. This can be easily done by having the XY slices of one z-stack (e.g., Angle 280°) in one *Viewer* and the orthogonal YZ or XZ slices of the same z-stack (Angle 280°) in another *Viewer*.
9. To track the nucleus over time, select it in a *Viewer* in the first time-point, activate the auto-linking mode of MaMuT with the *Shift* + *L* keys, move to a next time-point, double-click on the same nucleus, and, if needed, position the spot accurately as described in **steps** 7 and **8**. A line denotes the link (track) connecting the two spots created in the two time-points and the displacement of the tracked nucleus over time (Fig. 2b).
10. The *Viewers* can change the time in consecutive time-points or in multiples of 5 time-points, which is fine for relatively slow cells like in the demo dataset. For faster/slower moving cells, users can set the time interval to smaller/bigger values under the *Annotation* tab of the main GUI in the *Stepwise time browsing* field (Fig. 2a, middle).
11. In the case of mitotic divisions, in order to connect a mother cell to the two daughter cells, perform **steps** 7 to **9** to link the mother cell to the first daughter cell, then return to the first time-point and select the mother cell again, advance to the next time-point and create a second spot for the second daughter cell. A bifurcated track will appear connecting the two daughter cells (Fig. 2b).

12. Repeat **steps 7** to **11** to complete manual tracking for the desired number of cells and number of time-points according to the needs of the research question.
13. Press the *TrackScheme* button in the main GUI to open another window type in MaMuT, the *TrackScheme*. This window displays the reconstructed cell lineage tree with spots tracked over time depicted as nodes and tracks depicted as lines/branches connecting them (Fig. 2c). Each tracked cell is arranged in a separate column with the first time-point at the top and the last time-point at the bottom.
14. The *TrackScheme* is synced with the *Viewers*. The selected cell (s) in the *Viewers* are highlighted in the *TrackScheme* and vice versa. The mouse and keyboard can be used to browse the lineage tree in *TrackScheme* (*see* **Note 3**). Users are provided with different options on how to display spots and decorate and export the *TrackScheme* window.
15. Use the *TrackScheme* window to edit the lineage information, such as link or delete spots, rename or reorder sublineages.
16. Besides manual tracking, MaMuT offers the option of semi-automated tracking, where a single selected nucleus is tracked by the software over time. Select a nucleus in time-point t as described in **step 7** and press the *Semi-automatic tracking* button in the *Annotation* tab of the main GUI (or press *Shift* + *A*) to identify and track it over time-points t + 1, t + 2, etc., in an iterative manner.
17. Optimize the accuracy of semi-automated tracking for a particular dataset by adjusting the following three parameters in the *Semi-automatic tracking* field of the *Annotation* tab (Fig. 2a, middle):
    (a) *Quality Threshold* is an intensity-based measurement to detect and score candidate nuclei using a Laplacian of Gaussian filter.
    (b) *Distance tolerance* defines the search region for candidate nuclei in time-point t + 1 and link the one with the highest quality value to the initial nucleus in time-point t.
    (c) *Max nFrames* defines the number of iterations, i.e., the number of time-points over which the initial nucleus will be tracked (*see* **Note 4**).
18. For visual exploration and interpretation of the data, color-code spots, links, and tracks in the *Viewer* and *TrackScheme* windows based on primary or derived numerical features extracted from the data, such as the x, y, z, t coordinates of cells, the number of cell divisions, the duration of cell cycle, the displacement and velocity of cell movement, in silico clones, and others.

19. Create images or movies of tracked cells/nuclei using either the *Record Movie* (for slices) and *Record Max-Projection Movie* (for maximum intensity projections) commands under the *Tools* menu in the *Viewer* window or the *Extract track stack* command under the *Actions* tab of the main GUI (Fig. 2a, right).
20. Export the reconstructed lineage information from MaMuT in the XML file for customized analyses in other platforms (*see* **Note 5**).
21. MaMuT offers the option to import fully automated annotations generated by the Tracking with Gaussian Mixture Models software [14]. Follow instructions in [34] on how to install and run the TGMM 2.0 framework for nuclei segmentation and tracking. Browse to *Plugins › MaMuT › Import TGMM results in MaMuT*, import the TGMM annotations and the image data, and use the steps described above for manual curation and error correction (*see* **Note 6**).

3.3 Large-Scale Nuclei Tracking with Mastodon in LSFM Datasets

Mastodon is the second-generation cell lineaging tool in the Fiji/ImageJ ecosystem [18]. Mastodon is still unpublished in beta testing, but it is fully functional and continuously improved with the addition of extra features (*see* **Note 7**). It builds on the functionalities of MaMuT for the visualization, reconstruction and analysis of lineage information from mvLSFM datasets, but with a high-performance, memory-efficient architecture capable of dealing with many more annotations. Thus, Mastodon is suitable for large-scale lineage reconstructions based on semi-automated and fully automated tracking pipelines that can be implemented as described below. Please refer to the extended and regularly updated Mastodon documentation (https://mastodon.readthedocs.io/en/latest/index.html) for more information.

3.3.1 Basic Operations in Mastodon

1. Start a new Mastodon project by browsing to *Plugins › Mastodon* in Fiji and pressing the *new Mastodon project* button in the *Mastodon launcher* window (Fig. 3a, left-top). Alternatively, continue an existing Mastodon or MaMuT project by pressing the corresponding buttons. The main GUI of Mastodon opens (Fig. 3a, left-bottom).
2. Mastodon projects have the **.mastodon* suffix and can be saved by pressing the *save* or *save as* buttons in the main GUI or by assigning a key-binding. Press the *Control* + *Comma* keys to open the *Preferences* window, where you can access the default key-bindings and customize them in the *Keymap* menu.
3. Press the *bdv* button in the main GUI to open windows displaying the image data (Fig. 3b). These are identical with the MaMuT *Viewer* windows (both based on BigDataViewer for

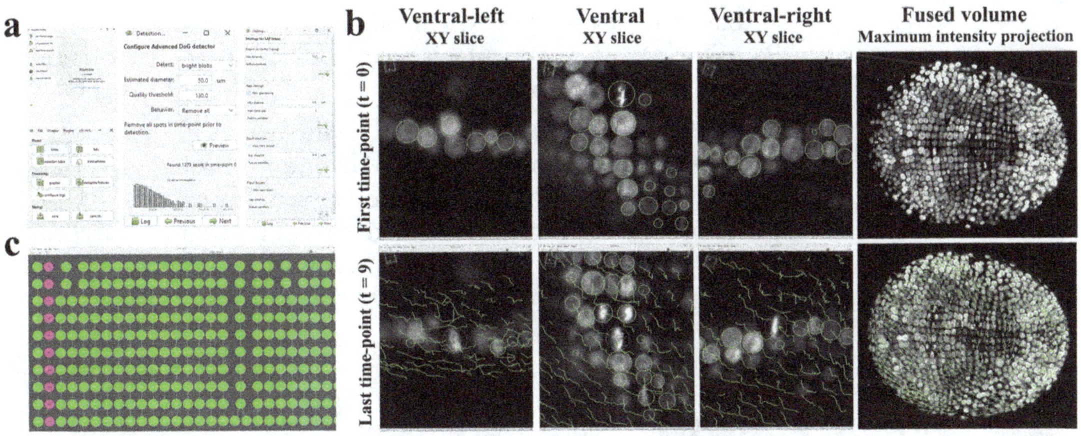

Fig. 3 Large-scale nuclear tracking with Mastodon. (**a**) Four Mastodon control panels: Launcher (left-top), Main GUI (left-bottom), Detection panel (middle), and Linker panel (right). (**b**) The Mastodon bdv windows displaying the same raw and fused image data as in Fig. 2b. All nuclei marked with circles (in view) or dots (out of view) and their tracks (lines) were followed from the first time-point (top row) until the last time-point (bottom row) of the *Parhyale* demo dataset. For reference, the Mastodon bdv windows are centered on the same active selection (marked in magenta) like the MaMuT Viewer windows in Fig. 2b. (**c**) The Mastodon TrackScheme window displaying a subset of nuclei (spots) and their tracks (vertical lines) over 10 time-points. The selected nucleus in panels b is also highlighted in magenta in TrackScheme

the interactive visualization and navigation of LSFM image datasets) with the same functionalities as described in Subheading 3.2, **steps 3** to **8**.

4. Create new spots for cells/nuclei of interest (pressing the *A* key by default), position them accurately and track them manually over time by following **steps 7** to **12** in Subheading 3.2.

5. Press the *trackscheme* button in the main GUI to open the *TrackScheme* window (Fig. 3c). As described in **steps 13** to **15** in Subheading 3.2, use *TrackScheme* to visualize and edit the lineage information (*see* **Note 8**). Dedicated key-bindings and mouse operations allow to easily navigate lineages in the *TrackScheme* and *bdv* windows.

6. Press the *table* and *selection table* buttons in the main GUI to open tabular representations of numerical features computed for the total number or a selected subset of created spots and links, respectively. Exported table data for spots and links in CSV format by browsing to *File* › *Export to CSV* in the table window.

7. The Mastodon *bdv*, *TrackScheme*, *table*, and *selection table* can be linked enabling the interactive visualization and editing of individual spots or groups of spots using the complementary information provided by the different window types. Users

have the flexibility to synchronize up to 3 groups of windows using a 3-lock system in their toolbars (e.g., lock number 1 activated in Fig. 3b, c.

8. Press the *configure tags* button in the main GUI to access the very useful tagging system in Mastodon. In the *Configure Tag Sets* window that opens, create tag-sets (categories) and tags (subcategories) that will label groups of tracked cells according to any desired quality (e.g., tag-set "germ layer" with tags "ectoderm," "mesoderm," and "endoderm").

3.3.2 Semi-automated Nuclei Tracking with Mastodon

1. Similar to MaMuT, Mastodon enables to follow automatically individual nuclei, an operation referred to as semi-automated tracking. To do so, create or select a spot and press the *Control* + *T* default keys for the software to track it.
2. Configure the semi-automated tracker by browsing to *Plugins* › *Tracking* › *Configure semi-automatic tracker* in any Mastodon window. In the *Preferences* window that opens, users can adjust parameters in the *Semi-automatic tracking* menu for the optimization of semi-automated tracking, which are similar to those parameters described in Subheading 3.2, **step 17**.
3. In the same menu, users can choose to track *Forward in time* or *Backward in time*, to allow or block the linking of newly created spots to existing ones, and to use the lock system to synchronize the semi-automated tracker with the *bdv* and *TrackScheme* windows to observe the tracker in real time (*see* **Note 9**).
4. Press the *Apply* button in the *Semi-automatic tracking* menu. Start the semi-automated tracking and follow a single nucleus at a time for a specified number of time-points. Return to the starting time-point, select another nucleus and repeat the procedure for all or a subset of nuclei. Proofread and manually correct the position or links of spots if necessary.
5. Repeat **step 4** for all tracked nuclei to extend the lineage forward or backward over the next or previous time-points, respectively. If needed, reconfigure the semi-automatic tracking parameters for optimal accuracy, as the tracked nuclei change their properties like their size, speed, and density over time.
6. Depending on experience, users can track more than one nucleus at a time. Simply select multiple nuclei and run the semi-automated tracker to speed up the full lineage reconstruction (*see* **Note 10**).

3.3.3 Fully Automated Nuclei Tracking with Mastodon

1. To start a fully automated tracking project, navigate to *Plugins* › *Tracking* › *Detection* from any Mastodon window.

2. In the *Detection* window that opens, select from the dropdown menu the source image dataset (channel and view) on which the analysis will be run and press *Next*. In our demo dataset, we selected the *Ventral* view (Fig. 3b, 2nd column).
3. In the *Region of interest* window, users have the option to operate on the entire dataset or enable the *Process only a ROI* box to define a spatiotemporal Region of Interest (ROI) interactively in a *bdv* window.
4. Press Next, the *Detector selection* window opens. Choose between the *Difference of Gaussian (DoG) detector*, *Advanced DoG detector* or *Laplacian of Gaussian (LoG) detector* of blob-like objects from the dropdown menu. We describe below the configuration options for the *Advanced DoG detector* (Fig. 3a, middle):
   (a) *Detect*: Choose between bright blobs or dark blobs.
   (b) *Estimated diameter*: Estimate the radius of blobs/nuclei.
   (c) *Quality threshold*: Set a minimum quality value for detected objects. These values are calculated by applying a filter to detect local maxima in the image and scoring candidate detections based on intensity and volume.
   (d) *Behavior*: Choose whether the detector will remove, keep, or replace (based on proximity) existing spots with new ones.
5. Press *Next*, the *Configure Advanced DoG detector* window opens. Set the parameters and press the *Preview* button. A new *bdv* window that pops up displays the detected objects on the source image with green circles, and the *Quality histogram* at the bottom of the *Configure Advanced DoG detector* window shows the frequency distribution of quality values (Fig. 3a, middle).
6. Modify the *Estimated diameter* and *Quality threshold* in the *Configure Advanced DoG detector* window and press the *Preview* button to optimize parameters for nuclei detection. In our demo dataset, we chose an *Estimated diameter* of 50 μm and a *Quality threshold* of 130 (Fig. 3a, middle).
7. Repeat **steps 5** to **6** to estimate optimal parameters for nuclei detection in other time-points across the entire image dataset by moving the time slider at the bottom of the *bdv* window and pressing the *Preview* button in the *Configure Advanced DoG detector* window.
8. Press the *Next* button in the *Configure Advanced DoG detector* window. Monitor the progress of detection in the *Log and status* window that opens and press the *Finish* button when completed (*see* **Note 11**).

9. Open a *bdv* window from the main Mastodon GUI to inspect the detection results (Fig. 3b). If satisfied, continue with the next step to link each detected nucleus across time-points.
10. Navigate to *Plugins* › *Tracking* › *Linking* from any Mastodon window.
11. In the *Linker target* window that opens, select to perform linking on all spots for all time-points. Users have also the option to perform linking for a subset of nuclei and time-points selected in the *bdv* window.
12. Press Next, the *Linker selection* window opens. Choose between the *LAP* (*Linear Assignment Problem*) *linker*, *Simple LAP linker* or *Linear Motion Kalman linker* from the drop-down menu (*see* **Note 12**). We describe below the configuration options for the *LAP linker* to track dividing nuclei (Fig. 3a, right):
    (a) *Frame to frame linking, Max distance*: Specify the maximum distance that a nucleus will travel between consecutive time-points.
    (b) *Gap closing, Max distance*: In case of missing detections (e.g., if the segmentation/detection of a nucleus fails for a few time-points or if the nucleus moves transiently out of focus), specify the maximum distance it can travel.
    (c) *Gap closing, Max frame gap*: Specify the maximum number of time-points for missing detections.
    (d) *Track division, Max distance*: Specify the maximum distance of splitting events, i.e., the maximum distance of daughter cells during division.
    (e) *Track fusion, Max distance*: Specify the maximum distance of fusion events (e.g., fusion of myoblast cells).
13. Press *Next*, the *Settings for LAP linker* window opens. Set the parameters described above and press *Next* (*see* **Note 13**). Monitor the progress of linking in the *Log and status* window that opens and press the *Finish* button when completed.
14. Open a *TrackScheme* window (Fig. 3c) and one or more *bdv* windows from the main Mastodon GUI to inspect the linking results. In *bdv* windows, segmented nuclei are indicated with green circles and their tracks with green lines (Fig. 3b).
15. Manually curate the fully automated annotations using both window types (*see* **Note 14**).

3.4 Cell Tracking with TrackMate in Cartographic Projections of LSFM Datasets

This section describes a pipeline for the segmentation and tracking of cell outlines (cell tracking for short) with TrackMate, which is a modular, user-friendly Fiji/ImageJ plugin for cell lineaging in 2D images over time [20]. For more detailed information, please refer to the extended TrackMate documentation https://imagej.net/

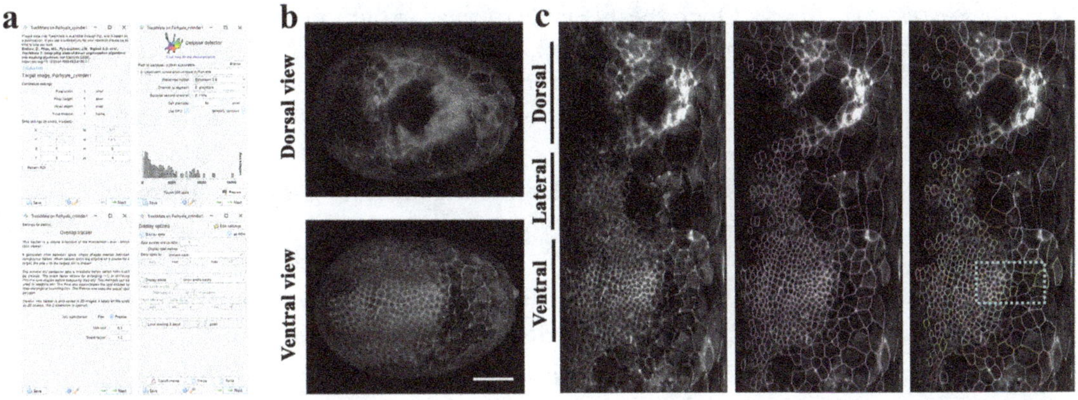

Fig. 4 Cell segmentation and tracking with TrackMate in cartographic projections of LSFM datasets. (**a**) TrackMate windows: Start window (left-top), Cellpose detector window (right-top), Overlap tracker window (left-bottom), and Display options window (right-bottom). (**b**) 3D renderings of the dorsal (top) and ventral (bottom) views of a *Parhyale* embryo labeled with LifeAct-mEmerald from a single time-point of a mvLSFM recording. Scale bar is 100 μm. Raw views were fused and the 3D surface of the entire embryo was unwrapped in the 2D cartographic map projection shown in c. (**c**) Unlabeled cartographic projection (left), segmented cells colored uniformly (middle), and segmented cells colored differentially by track index (right). The rectangular box shows the region magnified in Fig. 5c

plugins/trackmate/#documentation-and-tutorials. We use a multi-view LSFM recording of a *Parhyale* embryo expressing LifeAct-mEmerald, which labels the circumferential belt of actin cytoskeleton in the apical cortex of cells (apical cell outlines) (Fig. 4b). In each time-point, the fused 3D volume is first unwrapped on a 2D cartographic projection, a process which is analogous to making a map of earth, using the ImSAnE toolbox for tissue cartography (Heemskerk & Streichan, 2015). This step reduces the spatial dimensions of the data from three to two and simplifies the downstream, quantitative analyses of all cells covering the entire, curved surface of the embryo (Fig. 4c) (*see* **Note 15**). Users dealing with a flat region of interest may simply transform a 3D specimen to 2D with maximum intensity projections and can skip Subheading 3.4.1 that briefly outlines the basic tools and steps for tissue cartography with ImSAnE.

3.4.1 Generation of Cartographic Map Projections Using ImSAnE

Use a workflow established for fly embryos [31] to generate cartographic projections (*see* **Note 16**). In brief:

1. Load the data in MATLAB and run the ImSAnE code.
2. Segment the embryo surface using ilastik [23].
3. Project the embryo surface at different depths on 2D maps (onion layers).
4. Create the cartographic projections from maximum intensity projections of desired layers for all time-points and save them as a TIFF file.

3.4.2 Cell Segmentation and Tracking

1. Open the TIFF file of the cartographic (or maximum intensity) projections in Fiji. If required, enhance the signal-to-noise ratio with standard or more advanced denoising methods available in Fiji [35] to improve the efficiency of automated cell segmentation.
2. Launch TrackMate by browsing to *Plugins* › *Tracking* › *TrackMate*. The start window that pops up displays the properties of the TIFF file (Fig. 4a, left-top) (*see* **Note 17**).
3. Press *Next*. In the new *Select detector* window, choose the *Cellpose detector* from the dropdown menu.
4. Press *Next*. In the *Cellpose detector* window (Fig. 4a, right-top) set the slider to a desired time-point in the dataset to optimize the following parameters for segmentation (*see* **Note 18**):
   (a) *Path to cellpose/python executable*: Specify the file path to the cellpose executable. There are several ways of installing cellpose (https://github.com/MouseLand/cellpose). If cellpose was installed via the precompiled executables, enter the path to the executable (e.g., *C:\Users\user\Downloads\cellpose.exe*). If Cellpose was installed via Conda, enter the path to the python executable of the Conda environment created for cellpose (e.g., *C:\Users\user\.conda\envs\cellpose\python.exe*).
   (b) *Pretrained model*: Choose between three pretrained models in cellpose: The *Cytoplasm* and the *Cytoplasm 2.0* models segment cytoplasm or membranes and the *Nuclei* model segments nuclei. A fourth option is the *Custom* model that users have either trained or downloaded (*see* **Note 19**).
   (c) *Channel to segment*: Specify the channel containing the objects to be segmented. Choose *0* for a grayscale combination of all channels or select between channels *1*, *2*, or *3* in multi-channel acquisitions (or the red, green, and blue channels in RGB images).
   (d) *Optional second channel*: Cytoplasm/membrane segmentation with the *Cytoplasm* or *Cytoplasm 2.0* models is more robust in datasets with a nuclear channel for seeding cell detection. Select *0* for datasets without nuclei or *1*, *2*, *3* to select the nuclear channel if present.
   (e) *Cell diameter*: Cellpose takes into account the cell diameter for segmentation. Specify the average cell diameter in physical units (e.g., in μm if the source image pixel size is given in μm). Users can either use the ImageJ ROI tools to measure the cell diameter or set diameter to *0* for cellpose to calculate it automatically (*see* **Note 20**).

(f) *Use GPU*: Check this box if cellpose is installed with required libraries and hardware for GPU usage. If the box is left unchecked or GPU support is not available, cellpose will run on the CPU.

(g) *Simplify contours*: Check this box to smoothen the cell outlines generated by the segmentation (*see* **Note 21**).

5. After setting these parameters, press the *Preview* button to see the segmentation results for the chosen time-point. The number of objects/spots detected will be displayed in the *Area histogram* field (Fig. 4a, right-top). Change the parameters and press *Preview* again until users are happy with the results. For reference, in our single-channel LifeAct-mEmerald images (Fig. 4c, left), we picked the pretrained model *Cytoplasm 2.0* for membrane segmentation, selected *0 Channel to segment* and *0 Optional second channel*, estimated the cell diameter with the ImageJ ROI tools at 50 pixels, and checked both the *Use GPU* and *Simplify contours* boxes for cell segmentation (Fig. 4a, right-top and Fig. 4c, middle).
6. Press *Next* to start the object detection and segmentation in all selected time-points. Monitor the process in the progress bar. A *Detection done* notice indicates that segmentation has finished.
7. Press *Next*. The *Initial thresholding* window opens. Filter detected objects based on their size in pixel units. For example, set a cut-off threshold to exclude small detections not corresponding to cells.
8. Press *Next*. In the *Spot filter* window that opens, filter detections based on several features such as area, location, signal intensity, and others.
9. Press *Next* to choose the algorithm for linking the segmented (and filtered) cells between time-points in the *Select a tracker* window. Please refer to the TrackMate documentation and tutorials (https://imagej.net/plugins/trackmate/) for available tracker options and their configuration. For example, the Linear Assignment Problem (LAP) and the Overlap trackers enable the detection of cell divisions. We chose the *Overlap tracker* for our dataset setting the parameters as described in the TrackMate-Cellpose documentation (https://imagej.net/plugins/trackmate/detectors/trackmate-cellpose) (Fig. 4a, left-bottom).
10. Press *Next* to initiate the tracking process that can be monitored in the progress bar. A *Tracking done* notice indicates that tracking has finished.

11. Press *Next*. The *Track filter* window opens. Similar to spots, tracks can be filtered according to their features, such as length, speed, and location.
12. Press *Next*. The *Display options* window opens (Fig. 4a, right-bottom). The segmented cells and/or tracks can be color-coded based on various features that can be selected from the respective dropdown menus (Fig. 4c, middle and right, and Fig. 5a). Press the *Edit settings* button at the top of the window to change visualization options in the configuration window that opens. Press the *Tracks* and *Spots* buttons at the bottom of the window to export all TrackMate results as a CSV file for customized analyses.
13. Press the *TrackScheme* button in the *Display options* window to open the *TrackScheme* window for the visualization, manual editing, and analysis of the lineage information (similar to

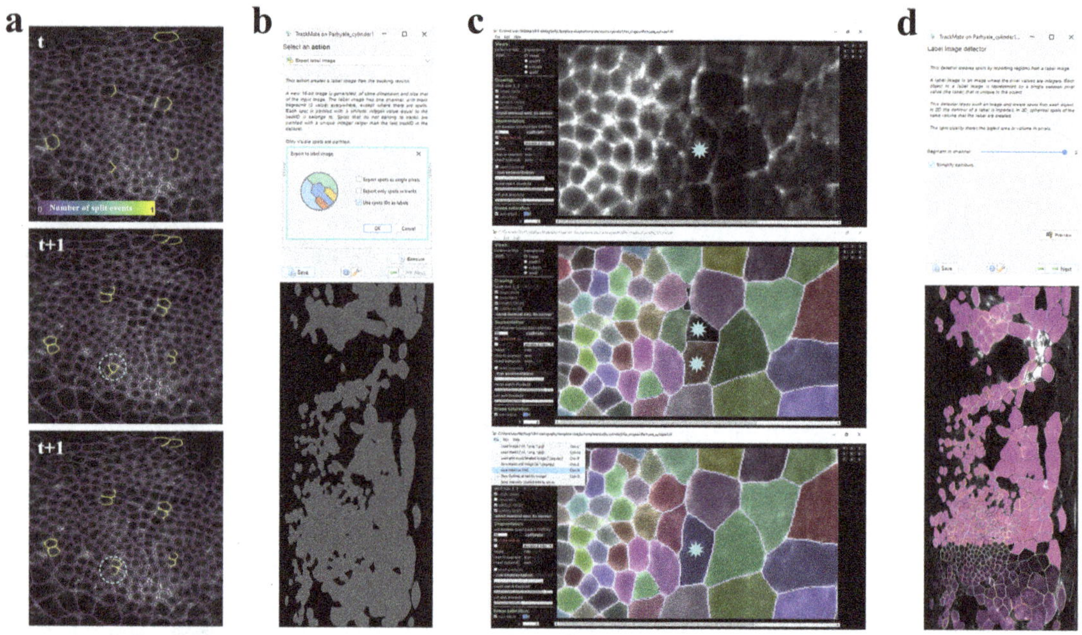

Fig. 5 Editing of segmented cells and tracks. (**a**) Close-ups of a region of the cartographic projection at two consecutive time-points, before (t, top) and after (t + 1, middle and bottom) division of cells with yellow outlines. All other nondividing cells have purple outlines. A tracking error linking a daughter cell to the wrong mother cell (middle, cyan circle) is manually corrected in TrackMate (bottom, cyan circle). (**b**) Export of the label image from the Actions window in TrackMate (top) with the mask of segmented cell outlines (bottom). (**c**) Correction of cell segmentation labels in the region of the cartographic projection indicated by the rectangular box in Fig. 4c. The source image data are displayed without (top) or with (middle and bottom) the segmentation labels. Two falsely segmented cells are merged together (asterisks) in cellpose. (**d**) The Label image detector window (top) is used to segment the source image data (grayscale) based on the corrected mask (magenta) in TrackMate (bottom)

MaMuT and Mastodon) following the detailed instructions in TrackMate documentation.

14. As an example, we colored segmented cells by the *Number of split events* in our demo dataset to label each tracked cell (sub-lineage) according to the number of cell divisions (Fig. 5a). By inspecting the color-coded cells for any mistakes in segmentation and tracking, we discovered that a daughter cell was linked to a wrong mother cell (Fig. 5a, middle). We edited the respective track in the *TrackScheme* window and corrected this mistake (Fig. 5a, bottom) (*see* **Note 22**).
15. Press *Next*. In the *Plot features* window that opens, plot and quantitatively analyze spot or track features. Choose between the *All*, *Selection*, or *Tracks of Selection* options at the bottom of the window to include all or part of the data in the graphs. Users can configure the plot and export it as an image file or as a table in CSV format.
16. Press *Next* to open the last *Actions* window. Select among different actions (e.g., see Subheading 3.4.3) described in the dropdown menu and press the *Execute* button at the bottom of the window.

3.4.3 Editing of Cell Segmentation

Segmentation results can be manually edited (e.g., merge over-segmented cells or split under-segmented cells) using cellpose and the corrected labels can be loaded again in TrackMate using the pipeline described below:

1. Browse to the last *Actions* window in TrackMate, select the *Export label image* action from the dropdown menu and press *Execute* (Fig. 5b, top).
2. Check the *Use the spots IDs as labels* box in the *Export to label image* field to create separate masks for daughter cells and press *OK* (Fig. 5b, top). The labels (masks) of the segmented cells will be generated (Fig. 5b, bottom). Save the label image as a TIFF file (*see* **Note 23**).
3. Load the source image data and the label image in cellpose.
4. Manually curate the segmentation according to guidelines in cellpose (https://cellpose.readthedocs.io/en/latest/index.html). In our example, we merged two falsely segmented cells in one (Fig. 5c).
5. Save the correct label image under *File › Save masks as PNG* (Fig. 5c, bottom). The filename ends in **_cp_masks* (e.g., *Parhyale_cylinder1_cp_masks.png*).
6. Load the source image data and the corrected label image in Fiji.

7. In Fiji, browse to *Image › Color › Merge channels* to load the masks as an extra channel in the source images and save the new 2-channel image data. In our example, source image data are in channel 1 in grayscale and the labels in channel 2 in magenta (Fig. 5d, bottom).
8. Start a new project in TrackMate and proceed to the *Select detector* window as described in Subheading 3.4.2, **steps 2** and **3**. Select the *Label image detector* (Fig. 5d, top) and specify the channel containing the labels (channel 2 in our example).
9. Follow the rest steps for cell tracking and analysis with TrackMate as described in Subheading 3.4.2.

4 Notes

1. The key-bindings used to navigate the MaMuT/BDV *Viewers* can be customized through a *mamut.properties* text file as described here (https://imagej.net/plugins/mamut/example-properties-file).
2. Spots in MaMuT indicate objects of interest with a sphere. Users can change the position, the radius, and the time-point of the spot.
3. The key-bindings used to navigate the MaMuT *TrackScheme* cannot be customized.
4. The quality of the image dataset, including the signal-to-noise ratio and the distance of cell displacement between consecutive time-points, will guide users in the selection of semi-automated tracking (and the corresponding optimal parameters) versus manual tracking. For good datasets, the semi-automated method works well for tracking nuclei during interphase (between mitoses) aided by manual tracking around mitoses to accurately link mother cells and daughter cells.
5. Earlier versions of MaMuT provided a third window type, called *3D Viewer*, with interactive animations of tracked objects depicted as spheres [17]. Although the *3D Viewer* is not available in more recent versions, the MaMuT XML file can be imported into other software tools for the visualization of cell lineages with 4D animations, such as CeLaVi [36].
6. The graph data structure of MaMuT can handle efficiently up to about one hundred thousand annotations, which is a reasonable number of annotations generated with manual or semi-automated lineage tracing. Fully automated frameworks like TGMM result in much higher numbers of annotations that can be easily handled with Mastodon, but not with MaMuT.

As a trade-off, users are provided with the option to crop the imported TGMM annotations in space and time to make them compatible with MaMuT.

7. Mastodon provides an open-source, extensible platform accommodating an increasing number of new features, plugins, and imported data from other software solutions for automated tracking, custom lineage analyses, and other applications. Of note are the capacity for collaborative distributed annotations of large LSFM datasets, the import of automated annotations generated with TGMM, the fitting of ellipsoids around spots, and other analysis tools developed by the Mastodon Deep Lineage consortium.
8. Compared to MaMuT, the TrackScheme in Mastodon provides huge improvements in performance, allowing users to browse seamlessly millions of annotations.
9. The extra features for semi-automated tracking in Mastodon are very useful for general-purpose lineage tracing, as well as for more specialized applications, like backtracking the origin of cells of interest or filling the gaps in existing lineages. Mastodon developers are also proud of the Undo (*Control* + *Z*) and Redo (*Control* + *Shift* + *Z*) functions as extremely important, time-saving operations.
10. The synchronization of the different Mastodon window types is crucial to supervise the accuracy of semi-automated tracking and spot errors in real time. Having at least two synced *bdv* windows with orthogonal slices of the image data helps to ensure the proper localization of spots in space. Backtracking is preferred to correctly link daughter cells to mother cells, and the proofreading of all divisions is strongly recommended. Erroneous branching patterns observed in the *TrackScheme* window may also uncover mistakes in semi-automated tracking.
11. If optimal parameters for nuclei detection vary over time (or for different regions), run **steps 1** to **8** in Subheading 3.3.3 for different spatiotemporal ROIs enabling the *Process only a ROI* box in the *Region of interest* window.
12. Select the *Linear motion Kalman linker* to track objects moving with a roughly constant velocity vector.
13. Users can enable or disable the linking parameters in the *Settings for LAP linker* window and constrain spot linking based on various features (like detection quality, number of links, or position) in the *Feature penalties* fields.
14. Even a very low tracking error rate results in the breaking of most lineages within a few time-points [37]. Depending on the research question, different error levels in lineage tracing can

be tolerated, thus, different investments are required for manual curation.

15. Projecting curved three-dimensional surfaces to two-dimensional maps during tissue cartography introduces distortions that need to be corrected before measuring quantities such as size, shape, angle, and direction in the maps [22]. Check the ImSAnE tutorial on how to perform quantitative measurements on cartographic projections [31].
16. ImSAnE requires cylindrical shapes to work properly. Users can generate cartographic projections of cylindrical-like embryos from fused LSFM volumes covering the entire embryo. To generate cartographic projections from individual LSFM views covering part of the embryo, mirror the dataset to imitate a cylindrical shape [31].
17. Users can crop their dataset to select a ROI in their images and/or specific time-points for segmentation in the *Crop settings* field and by pressing the *Refresh ROI* button.
18. Extra parameters in cellpose segmentation include the *flow threshold* and the *probability threshold*. The default settings are used in TrackMate-cellpose and cannot be changed.
19. Cellpose 2.0 [38] allows the fast and easy training of new custom models for improved segmentation, as described in Github page https://github.com/MouseLand/cellpose.
20. Cellpose may over-segment cells if the diameter is set to lower values and under-segment cells if it is set to higher values than real ones.
21. Cellpose and other detectors in TrackMate can define the shape of detected objects to perform morphological analyses. Check the *Simplify contour* for more precise measurement of cell shape descriptors. If the contours are not simplified, the cell perimeter is overestimated, affecting the measurement of morphological features that depend on it.
22. If a large number of mistakes are detected, improve the results by altering the tracking parameters in TrackMate as described in Subheading 3.4.2, **step 9**.
23. Cellpose can load masks as TIFF, PNG, or JPEG files.

Acknowledgements

We are grateful to Philipp Keller (Howard Hughes Medical Institute Janelia Research Campus), Pavel Tomancak (Max Planck Institute of Molecular Cell Biology and Genetics) and their groups for providing access to the SiMView and Zeiss light-sheet microscopes, respectively. We particularly thank William Lemon, Raghav

Chhetri, Akanksha Jain, Marina Cuenca, Bruno Vellutini, Jean-Yves Tinevez, Vladimir Ulman, and Jan Peychl for their help with different aspects of light-sheet image acquisition and analysis. This work was supported by the HFSP Research Grant RGP022/2023, the Janelia's Visiting Scientist Program to J.R. and A.P., an Erasmus+ traineeship to M.K., and a Hellenic Foundation for Research & Innovation PhD scholarship to J.R.

References

1. Quake SR (2022) A decade of molecular cell atlases. Trends Genet 38:805–810
2. Buckingham ME, Meilhac SM (2011) Tracing cells for tracking cell lineage and clonal behavior. Dev Cell 21:394–409
3. Kretzschmar K, Watt FM (2012) Lineage tracing. Cell 148:33–45
4. Sulston JE, Schierenberg E, White JG et al (1983) The embryonic cell lineage of the nematode Caenorhabditis elegans. Dev Biol 100: 64–119
5. Wu Y, Shroff H (2022) Multiscale fluorescence imaging of living samples. Histochem Cell Biol 158:301–323
6. Wolf S, Wan Y, McDole K (2021) Current approaches to fate mapping and lineage tracing using image data. Development 148: dev198994
7. Lemke SB, Nelson CM (2021) Dynamic changes in epithelial cell packing during tissue morphogenesis. Curr Biol 31:R1098–R1110
8. Liu Z, Keller PJ (2016) Emerging imaging and genomic tools for developmental systems biology. Dev Cell 36:597–610
9. Power RM, Huisken J (2017) A guide to light-sheet fluorescence microscopy for multiscale imaging. Nat Methods 14:360–373
10. Lemon WC, McDole K (2020) Live-cell imaging in the era of too many microscopes. Curr Opin Cell Biol 66:34–42
11. Wan Y, McDole K, Keller PJ (2019) Light-sheet microscopy and its potential for understanding developmental processes. Annu Rev Cell Dev Biol 35:655–681
12. Stelzer EHK, Strobl F, Chang B-J et al (2021) Light sheet fluorescence microscopy. Nat Rev Methods Primers 1:1–25
13. Schmied C, Stamataki E, Tomancak P (2014) Open-source solutions for SPIMage processing. Methods Cell Biol 123:505–529
14. Amat F, Höckendorf B, Wan Y et al (2015) Efficient processing and analysis of large-scale light-sheet microscopy data. Nat Protoc 10: 1679–1696
15. Jain A, Ulman V, Krumnikl M et al (2024) Image processing and analysis of light sheet microscopy data. In: Light sheet fluorescence microscopy. Wiley, pp 203–229
16. Schindelin J, Arganda-Carreras I, Frise E et al (2012) Fiji: an open-source platform for biological-image analysis. Nat Methods 9: 676–682
17. Wolff C, Tinevez J-Y, Pietzsch T et al (2018) Multi-view light-sheet imaging and tracking with the MaMuT software reveals the cell lineage of a direct developing arthropod limb. Elife 7
18. https://github.com/mastodon-sc/mastodon
19. Tinevez J-Y, Perry N, Schindelin J et al (2017) TrackMate: an open and extensible platform for single-particle tracking. Methods 115:80–90
20. Ershov D, Phan M-S, Pylvänäinen JW et al (2022) TrackMate 7: integrating state-of-the-art segmentation algorithms into tracking pipelines. Nat Methods 19:829–832
21. Stringer C, Wang T, Michaelos M et al (2021) Cellpose: a generalist algorithm for cellular segmentation. Nat Methods 18:100–106
22. Heemskerk I, Streichan SJ (2015) Tissue cartography: compressing bio-image data by dimensional reduction. Nat Methods 12: 1139–1142
23. Berg S, Kutra D, Kroeger T et al (2019) ilastik: interactive machine learning for (bio)image analysis. Nat Methods 16:1226–1232
24. Rallis J, Kapai G, Pavlopoulos A (2021) Chapter 16 – Parhyale hawaiensis, Crustacea. In: Boutet A, Schierwater B (eds) Handbook of marine model organisms in experimental biology: established and emerging. CRC Press
25. Paris M, Wolff C, Patel NH et al (2022) The crustacean model Parhyale hawaiensis. Curr Top Dev Biol 147:199–230
26. Sugawara K, Çevrim Ç, Averof M (2022) Tracking cell lineages in 3D by incremental deep learning. Elife 11:e69380

27. Preibisch S, Saalfeld S, Schindelin J et al (2010) Software for bead-based registration of selective plane illumination microscopy data. Nat Methods 7:418–419
28. Preibisch S, Amat F, Stamataki E et al (2014) Efficient Bayesian-based multiview deconvolution. Nat Methods 11:645–648
29. Hörl D, Rojas Rusak F, Preusser F et al (2019) BigStitcher: reconstructing high-resolution image datasets of cleared and expanded samples. Nat Methods 16:870–874
30. Pietzsch T, Saalfeld S, Preibisch S et al (2015) BigDataViewer: visualization and processing for large image data sets. Nat Methods 12: 481–483
31. Vellutini BC (2022) How to make cartographic projections using ImSAnE. https://zenodo.org/records/7628300
32. Jain A, Ulman V, Mukherjee A et al (2020) Regionalized tissue fluidization is required for epithelial gap closure during insect gastrulation. Nat Commun 11:5604
33. Vellutini BC, Cuenca MB, Krishna A et al (2023) Patterned embryonic invagination evolved in response to mechanical instability. https://www.biorxiv.org/content/10.1101/2023.03.30.534554v2
34. McDole K, Guignard L, Amat F et al (2018) In Toto imaging and reconstruction of post-implantation mouse development at the single-cell level. Cell 175:859–876.e33
35. Shroff H, Testa I, Jug F et al (2024) Live-cell imaging powered by computation. Nat Rev Mol Cell Biol
36. Salvador-Martínez I, Grillo M, Averof M et al (2021) CeLaVi: an interactive cell lineage visualization tool. Nucleic Acids Res 49:W80–W85
37. Hirsch P, Epstein L, Guignard L (2022) Chapter 20 – Mathematical and bioinformatic tools for cell tracking. In: Schnoor M, Yin L-M, Sun SX (eds) Cell movement in health and disease. Academic, pp 341–361
38. Pachitariu M, Stringer C (2022) Cellpose 2.0: how to train your own model. Nat Methods 19:1634–1641

Chapter 9

Reconstructing Progenitor State Hierarchy and Dynamics Using Lineage Barcoding Data

Weixiang Fang, Yi Yang, Hongkai Ji, and Reza Kalhor

Abstract

Measurements of cell phylogeny based on natural or induced mutations, known as lineage barcodes, in conjunction with molecular phenotype have become increasingly feasible for a large number of single cells. In this chapter, we delve into Quantitative Fate Mapping (QFM) and its computational pipeline, which enables the interrogation of the dynamics of progenitor cells and their fate restriction during development. The methods described here include inferring cell phylogeny with the Phylotime model, and reconstructing progenitor state hierarchy, commitment time, population size, and commitment bias with the ICE-FASE algorithm. Evaluation of adequate sampling based on progenitor state coverage statistics is emphasized for interpreting the QFM results. Overall, this chapter describes a general framework for characterizing the dynamics of cell fate changes using lineage barcoding data.

Key words Lineage tracing, Lineage barcoding pipeline, Quantitative fate mapping (QFM), Progenitor state dynamics, Phylotime, ICE-FASE, Time-scaled cell phylogeny

1 Introduction

Development happens gradually over time as cells transform from relatively undifferentiated states into specialized cell types, all while undergoing the process of cell division. Historically, tracking cell lineage has been an important technique for researchers to trace the origins and fates of embryonic cells during development [1]. One way to track lineage is using lineage barcodes, which are heritable DNA sequences that uniquely mark each progenitor cell (Fig. 1a). A lineage barcode can be static or cumulative. Static barcodes are integrated into the genome at a single time point [2–4]. Cumulative or evolving lineage barcodes entail heritable genomic mutations that accumulate over a period of time in development. With static lineage barcodes, clones of progenitor cells can be identified. With cumulative lineage barcodes, a *cell phylogeny* can be inferred, which refers to a phylogenetic tree that represents the complete division

Jorge García-Marqués and Tzumin Lee (eds.), *Lineage Tracing: Methods and Protocols*, Methods in Molecular Biology, vol. 2886, https://doi.org/10.1007/978-1-0716-4310-5_9,

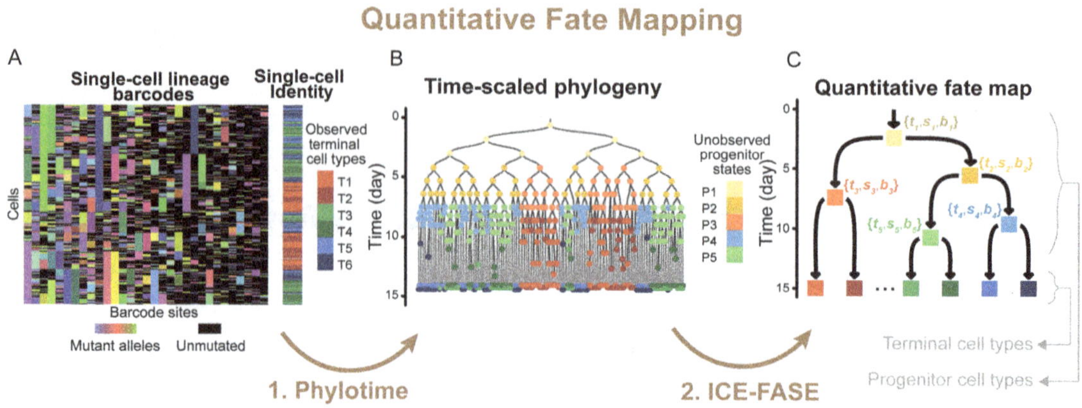

Fig. 1 Workflow for applying the quantitative fate mapping (QFM) pipeline. (**a**) Example of lineage barcoding experiment data, rows are cells and columns are barcoding sites. **Step 1** applies Phylotime to single cell lineage barcodes to infer time-scaled phylogeny. (**b**) Example of a time scaled phylogeny with progenitor states assigned to the nodes. **Step 2** applies ICE-FASE to single cell phylogeny to obtain fate map topology and cell state parameters. (**c**) Example of a quantitative fate map

history of a population of cells (Fig. 1b). In this chapter, we focus on evolving lineage barcodes, which can be engineered or naturally occurring. Engineered barcodes include the recently developed CRISPR-editing based barcoding schemes [5–12]. In these systems, a panel of guide RNAs guides the cas9 protein to induce double stranded break(s) and repair at an array of independent target sites, generating random mutations. Examples of the naturally occurring lineage barcode include somatic indels and point mutations [13–16]. We refer to the CRISPR target site or the location where somatic mutation occurs as the barcoding site (Fig. 1a). Advancements in single cell sequencing have made it possible to sequence these mutations in a large number of single cells [17]. The readout of single cell lineage barcodes has enabled high throughput retrospective experiments to reconstruct the developmental history. In such a design, the cell identities are profiled alongside with their lineage barcodes [18]. As a result, one can simultaneously learn the cell phylogeny with the cell types (Fig. 1a). We refer to these observed cell types as *terminal cell types*. With such experiment design, it is possible to investigate the progression of cell fate restriction during development. In this chapter, we describe a method called quantitative fate mapping (QFM) and the corresponding computational pipeline to reconstruct the progenitor state dynamics. QFM reconstructs progenitor state dynamics as quantitative fate maps. A *fate map* refers to a collection of cell states organized in a hierarchy (Fig. 1c) called the fate map topology. A fate map is quantitative in the sense that it further defines the following three parameters for each progenitor state: (i) *commitment time*: the time at which cells of a progenitor state commit to downstream states; (ii) *population size*: the number of progenitor cells before they commit to downstream states; and (iii) *commitment bias*: the proportions of cells of a progenitor state committing to each of the downstream states.

Reconstructing a quantitative fate map involves the following steps, which are discussed in detail in the ensuing sections. First, a cell phylogeny needs to be inferred from single cell lineage barcodes. In Subheading 2.1, we describe a general mutagenesis model that covers somatic mutation as well as distributed mutations created by a CRISPR-cas9 system, and a method called Phylotime (PHYlogeny inference using Likelihood Of TIME) to infer the phylogeny. Oftentimes, the lineage barcode readout has a possibility of dropout (failure to detect single cell mutation at certain barcoding site). To address that, an imputation strategy is discussed in Subheading 2.2. After a single cell phylogeny is inferred, one can infer the progenitor state changes along the cell phylogeny. For this, we discuss the ICE-FASE algorithm. The ICE-FASE algorithm consists of two main steps. The first step (FASE step) reconstructs the fate map topology based on fate separations (*see* Subheading 2.5). The second step recovers progenitor state dynamics parameters (*see* Subheading 2.6). In Subheading 2.7, we discuss that successful application of QFM depends on adequate sampling of the progenitor population, and address how to evaluate the level of sampling with progenitor state coverage (PSCov) statistics. Throughout this chapter, we accompany each of these components with the corresponding R code, facilitating the execution of the entire pipeline. To install the QFM R package, the user can refer to the instructions available at https://github.com/Kalhor-Lab/QFM. Additionally, the necessary files for running the provided example can be accessed at https://github.com/Kalhor-Lab/QFM/tree/main/data/example. The content in this chapter is largely based on the original QFM publication [19], although details of ICE-FASE algorithm, particularly the definition and identification of inferred commitment events have been updated for improved clarity and concision.

2 Methods

2.1 Single Cell Phylogeny Reconstruction with Phylotime

The first step in processing evolving lineage barcoding data is to infer the cell phylogeny. Much like phylogeny of species is inferred from their DNA sequences, cell phylogenies can be inferred from the mutations that comprise their lineage barcodes. First, as the goal is to infer the dynamics of cell state change, the branch lengths in the cell phylogeny should reflect actual time in between cell divisions. Maximum parsimony methods for inferring cell phylogeny, for example, have branch length in the unit of the number of character changes [20]. Maximum likelihood methods, on the other hand, are based on well-defined statistical models and do infer time-scaled phylogenies [21]. However, optimizing the parsimony or likelihood objective involves searching through a large tree space, and can become computationally infeasible for large datasets. Lineage barcoding experiments can involve thousands of single cells, and traditionally, phylogenetic inference methods deal with only hundreds of tips [22].

In this section, we describe the Phylotime model, which computes the maximum likelihood estimator for the temporal distance between pairs of single cells. By applying UPGMA (a distance-based tree reconstruction method to infer the final phylogenetic tree) to these estimators, Phylotime ensures that the method is scalable while producing a time-scaled phylogeny. The likelihood used is based on a general mutagenesis model that applies to engineered lineage barcodes with multiple independent integrations as well as somatic mutations.

We will review the mutagenesis model first. The mutagenesis model assumes independent barcoding sites that accumulate mutations according to a Poisson point process with a constant rate. Each mutation converts an active copy of the site into one of many possible inactive mutated alleles, each with a distinct emergence probability (Fig. 2). In the case of engineered mutations by the CRISPR-cas9 scheme, each site is a different target site and the mutant alleles are different outcomes of DNA double stranded break repair that are not subject to targeting by the same guide RNA. In the case of somatic mutation, each site is a different locus where point mutations or indel can occur. Because the chances of the same site mutating multiple times are considered small, the mutations are considered final in these cases. In summary, Phylotime is applicable as long as the lineage barcode is generated at independent sites, each with the potential for irreversible mutation.

Phylotime employs the above mutagenesis model to estimate the temporal distances between cells using a maximum likelihood approach. The estimation is focused on the branch length

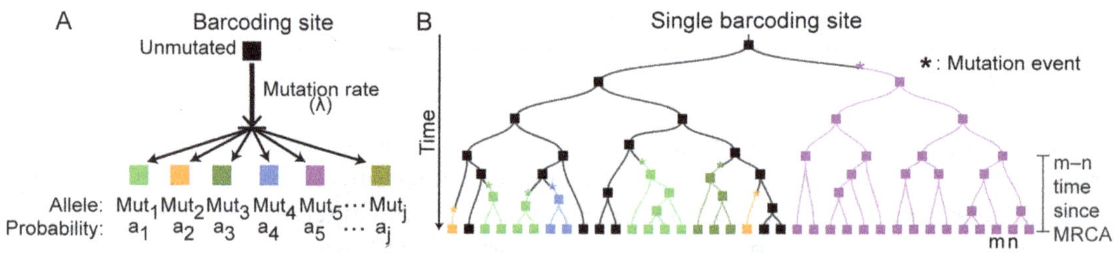

Fig. 2 (**a**) Model of a barcoding site which mutates with a constant rate (λ) to convert to one of many possible mutant alleles (Mut_1, ..., Mut_j), each with an emergence probability (a_1, ..., a_j). (**b**) Example time-scaled phylogeny showing the inheritance of a single barcoding site. Rectangles represent the site; their colors represent their allelic states according to (**a**). Asterisks denote mutation events. Only the unmutated allele (black) can change; mutated alleles (colored) are inherited from a cell to all its descendants. The time since MRCA is shown for two terminal cells, m and n

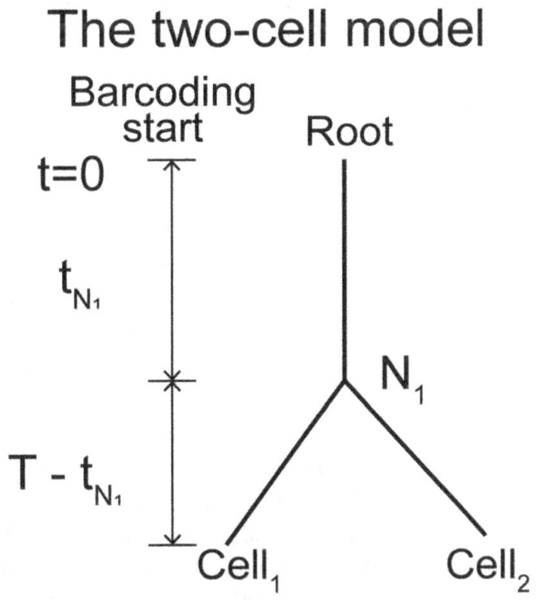

Fig. 3 The two-cell model. An illustration of a phylogenetic model with two terminal cells used to estimate their distance since their most recent common ancestor (MRCA). N_1 is the MRCA of the two cells. T is time from root to sampling. t_{N_1} is the time from root to MRCA

parameter t_{N_1}, which is the time since the most recent common ancestor (MRCA) of the two cells, as illustrated in Fig. 3.

In this two-cell model, let's consider a barcoding site labeled i with a mutation rate of λ_i and $a_i = (a_1, a_2, \ldots, a_J)_i$ of mutating into alleles $(A_1, A_2, \ldots, A_J)_i$, the likelihood of observing a given allele at this site in two cells (c_1, c_2) is the sum of two terms, as in Eq. 1.

$$p_i = p_{i,\mathrm{N}_1,1} + p_{i,\mathrm{N}_1,0}\left(p_{i,\mathrm{C}_1} p_{i,\mathrm{C}_2}\right) \tag{1}$$

The first term, $p_{i,N_{1,1}}$ represents the probability of a mutation occurring before the MRCA, leading to identical alleles in both cells. The second term considers the probability of observing the same allele in each terminal cell (p_{i,C_1} and p_{i,C_2}) assuming no mutation occurred before the MRCA ($p_{i,N_1,0}$).

For the first term, we define the term $p_{i,N_1,A_{ij}}$ for each mutant allele A_{ij}, where $p_{i,N_1,A_{ij}}$ is calculated based on whether both C_1 and C_2 carry allele A_{ij} for the site. If yes, then $p_{i,N_1,A_{ij}} = \left(1 - e^{-\lambda_i t_{N_1}}\right) a_{ij}$ and $p_{i,N_1,A_{ij}} = 0$ otherwise. Then the first term is calculated as in Eq. 2.

$$p_{i,\mathrm{N}_1,1} = \sum\nolimits_{j=1}^{J_i} p_{i,\mathrm{N}_1,A_{ij}} \tag{2}$$

which is the sum over all probable mutations, where at most one term can be non-zero.

For the second term, we have Eqs. 3 and 4.

$$p_{i,N_{1,1}} = e^{-\lambda_t t_{N_1}} \tag{3}$$

$$p_{i,c} = \left(1 - e^{-\lambda_i\left(T - t_{N_1}\right)}\right)^{1-\text{Ind}(M_{ic}=\text{``0''})} \Pi_{j=1}^{J_i} a_{ij}^{\text{Ind}\left(M_{ic}=A_{ij}\right)} \left(e^{-\lambda_i\left(T - t_{N_1}\right)}\right)^{\text{Ind}(M_{ic}=\text{``0''})} \tag{4}$$

where M_{ic} denotes the allele observed for $c = C_1$, C_2 and "0" denotes an unmutated allele, and Ind(.) is the indicator function.

Given the assumption of independent barcoding sites, the likelihood for the set of alleles observed in all barcoding sites is the product of their individual likelihoods in Eq. 5.

$$p(\{\lambda_i, a_i\}_i, t_{N_1}) = \Pi_i p_i \tag{5}$$

To get estimates of pairwise distance, or equivalently, time until MRCA between two cells (t_{N_1}), we plug in estimates of mutation rates and allele emergence probabilities. In this context, estimates of mutation rates can be calculated as $\widehat{\lambda} = -\frac{1}{T}\log(1 - F)$, based on the mutated fraction (F) of the site, defined as the fraction of cells with a mutant copy and the total time (T) from the start of the experiment to sample collection. Mutant allele emergence probabilities are set according to a uniform prior by default, that is $a_j = 1/J$ for $j = 1, \ldots, J$. Mutant allele emergence probabilities can be set according to a uniform prior by default, but better estimates [23] may be used to enhance Phylotime's performance. For example, CRISPR-cas9 induced double stranded break repair outcomes follow certain patterns and may be predicted from a computational model [24].

To find the optimal value of t_{N_1}, we solve the score equation in Eq. 6 using the Newton-Raphson method. The distance between the two cells is then $2\ (T - t_{N_1})$.

$$\frac{d\log(p(t_{N_1}))}{dt_{N_1}} = 0 \tag{6}$$

Once pairwise distances are computed, they can be organized into a distance matrix (Fig. 4b), and UPGMA hierarchical clustering [25] can be applied to derive a phylogenetic tree where branch lengths correspond to actual time (Fig. 4c). UPGMA is chosen because it yields an ultrametric phylogeny, where the distance from the root to all tips is the same, as expected in a lineage barcoding experiment where all cells are collected at the same time.

2.2 Imputing Missing Lineage Barcodes

To apply a phylogeny inference method such as Phylotime, mutations must be present for all sites for each cell. However, in actual experimental data, lineage barcodes may be missing for some sites due to failure in readout. In this case, one strategy is to first impute the missing mutations. Here, we describe one such imputation method inspired by multiple imputation with chained equations

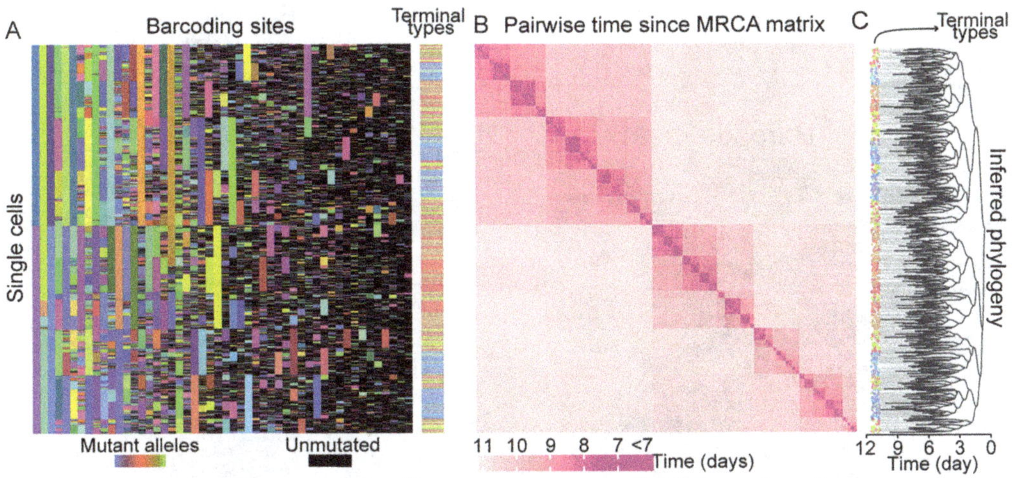

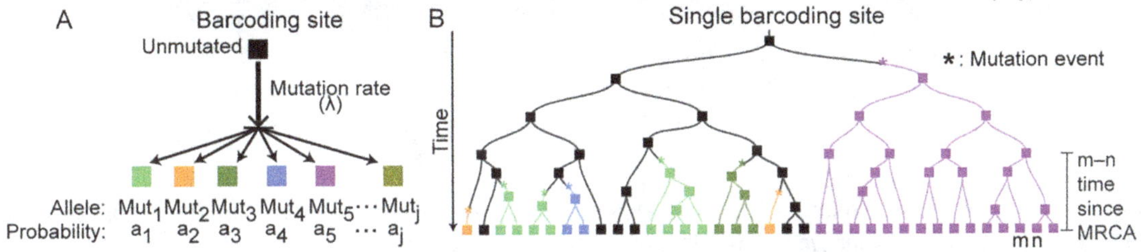

Fig. 4 Phylotime infers single cell phylogeny from lineage barcodes. (**a**) Lineage barcodes along with terminal cell types for an example lineage barcoding experiment. Each barcoding site is a column and each cell is a row. Unmutated alleles are in black; mutant alleles are in other colors. Color bar on the right shows distinct cell types. (**b**) Heatmap showing the estimated pairwise time since MRCA for all cells in **a**. (**c**) Time-scaled phylogenetic tree inferred by applying Phylotime to **a**. Colors of terminal nodes signify cell type

[26]. The method iterates through each site with missing values and builds a model to estimate the missing values based on observed values from other variables. First, the missing percentages of all barcoding sites are computed. Next, each site is imputed one by one going from the one missing in the fewest cells to the one missing in the most cells (Fig. 5).

In each iteration, a site with missing values is treated as the dependent variable, and all other sites are treated as dependent variables. Each site is treated as a categorical variable where the categories are the different mutant alleles. The design matrix of the model is constructed by one-hot encoding each site. Missing independent variables are treated as unmutated, and all independent variables are column concatenated. Here, a choice of model that can fit a categorical dependent variable may be used. In the implementation of the QFM package, XGboost is chosen as the method for model fitting and prediction. XGboost is a machine learning library based on gradient-boosted decision trees [27]. By default, QFM package uses the multinomial softmax objective for model fitting.

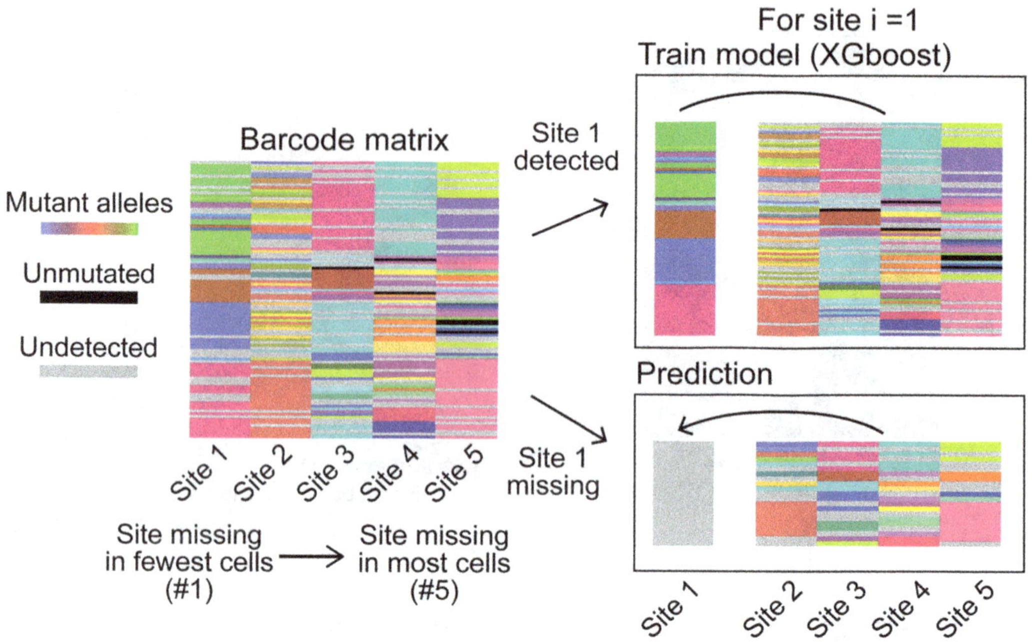

Fig. 5 Strategy for imputing missing mutations. Different sites are imputed from least missing to most missing. For each site, all the cells for which the site has observed data is used as training

2.3 Preparing Phylotime Input Data

To infer a time-scaled phylogeny using Phylotime, a lineage barcode matrix in the form of a character matrix should be prepared as the input. Optionally, mutation rate estimates and mutation emergence probabilities can be provided if known. If not specified, these parameters will be estimated by default from the data.

A character matrix encodes the lineage barcodes of a dataset. A character matrix specifies the mutations of cells at different barcoding sites. The rows correspond to single cells and columns correspond to barcoding sites. Example of a character matrix is shown in Table 1. The header line indicates the name of the barcoding site. This character matrix is used as input to Phylotime.

Mutation rates, if known from other sources, can be specified in the mutation rate table. Mutation rate table specifies the rate of mutation of different sites. An example mutation rate table is shown in Table 2. Each barcoding site should have a unique site name that is matched to the site name in the character matrix.

A mutation emergence probability table specifies the probability of a specific mutation occurring for each barcoding site. An example mutation emergence probability table is shown in Table 3. Site names should match the header in the character matrix.

Table 1
An example of a character matrix table

Cell	Site_1	Site_2	Site_3	Site_4	Site_5
Typ1_Cell1	Site1_Mut22	0	Site3_Mut28	0	Site5_Mut5
Type1_Cell2	Site1_Mut102	0	Site3_Mut1	0	Site5_Mut6
Type1_Cell3	Site1_Mut2	0	Site3_Mut1	0	Site5_Mut8
Type1_Cell4	0	0	Site3_Mut1	0	Site5_Mut8
Type1_Cell5	0	0	Site3_Mut84	0	Site5_Mut8
Type1_Cell6	0	0	Site3_Mut1	Site4_Mut7	Site5_Mut8
Type1_Cell7	0	0	Site3_Mut8	0	Site5_Mut8
Type1_Cell8	Site1_Mut4	0	Site3_Mut13	0	Site5_Mut8
Type1_Cell9	Site1_Mut22	0	Site3_Mut28	0	Site5_Mut35

For example, "Site1_Mut22" is a mutation that occurred on barcoding site 1. "0"s are used to code unmutated copies

Table 2
An example of a mutation rate table

Site	Mutation rate (Muts/day)
Site_1	0.93676574
Site_2	0.429635091
Site_3	0.640923333
Site_4	0.429635091
Site_5	0.93676574
Site_6	0.429635091
Site_7	0.394114734
Site_8	0.268539625
Site_9	0.936485602
Site_10	0.533760199

2.4 Running Phylotime

We go through the code for running Phylotime in this section. For the content of this subsection and later section with R code, we will load the QFM R package:

```
library(qfm)
```

First, we can read in a character matrix in the format of Table 1.

```
cell_allele_tb = readr::read_csv("mutant_allele_matrix.csv")
```

Table 3
An example of mutation emergency probability table

Site	Mutation	Probability
Site_1	Site1_Mut1	0.07561821
Site_1	Site1_Mut2	0.062085832
Site_1	Site1_Mut3	0.060753932
Site_1	Site1_Mut4	0.048325995
Site_1	Site1_Mut5	0.04545283

Next, we convert it to a matrix whose row names are set to the cell labels. The column names should be set to the site names.

```
chr_mat = as.matrix(cell_allele_tb[-1])

rownames(chr_mat) = cell_allele_tb$cell
```

We can now run Phylotime. By default, Phylotime runs in parallel mode via the `future` package, the number of workers can be set with the `plan` function. The user can set `parallel=F` If sequential mode is preferred. The argument `t_total` should be set to the duration from the root cell division to the time of sample collection. If not known or the absolute scale of time does not matter, `t_total` can be set to 1.0 by default.

```
library(furrr)
plan(multisession, workers = 8)
tr = phylotime(chr_mat,
               t_total = 15. - 0.6)
```

To visualize the inferred tree with the lineage barcodes:

```
plot_barcodes(chr_mat,
              tr = tr,
              show_column_names = F)
```

This produces a heatmap in the format of Fig. 1a.

To plot the inferred tree in a larger format:

```
plot_tr(tr)
```

This produces a dendrogram in the format of Fig. 1b.

The distance matrix can also be returned by itself if the user wants to apply other distance based phylogeny inference method:

```
dmat = phylotime(chr_mat,
                 t_total = 15. - 0.6,
                 return_dist = T)
```

If the mutation rate and/or the allele emergence probabilities are known, they can be provided to the phylotime function:

```
mut_rate_vec = read_csv('mut_p_mut_rate.csv')
recur_vec_tb = read_csv('mut_p_recur_vec.csv')
tr = phylotime(chr_mat,
               t_total = 15. - 0.6,
               mut_rate = mut_rate,
               recur_vec = recur_vec)
```

If the input character matrix contains missing data, the missing values should be coded as NA in R, and the `impute_characters` function will apply the XGboost based imputation to the data.

```
chr_mat_imputed = impute_characters(chr_mat)
```

2.5 Fate Map Topology Reconstruction Using Fate Separations (FASE)

The fate map inference starts with identifying a topology connecting the terminal cell types. Most often the topology comes in the form of a branching tree, either bifurcating or multifurcating. Within this tree, internal nodes represent progenitor states that existed during the development. It is important to clarify that while "cell state" often refers to molecular measurements of cells, here, "progenitor state" should be understood in terms of their developmental potency. Each progenitor state is defined by the set of terminal cell types that it produces. Ideally, there should be a direct correspondence between a progenitor cell's molecular state and its potency level; however, it is possible that a progenitor cell's molecular state has the potential to generate a broader range of terminal fates than it does in the embryo.

To extract the fate map topology from time-scaled phylogenies, it is natural to focus on the potency changes of internal nodes of the cell phylogeny. Although phylogeny inference with Phylotime is presented in the previous subsection, the algorithm presented hereafter—ICE-FASE—is a general method that can be applied to any time-scaled cell phylogeny regardless of how it was generated. The ICE-FASE algorithm relies on the timings of observed fate separations between terminal cell types. Initially, each node within the phylogenetic tree is annotated with its observed fate, representing the cell types of its observed descendants (Fig. 6). For example,

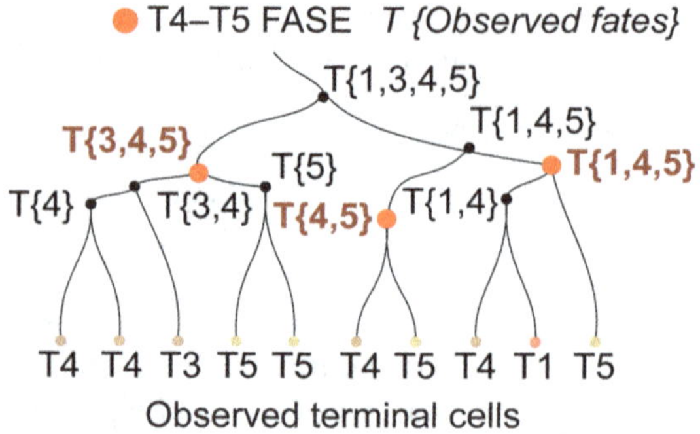

Fig. 6 Example phylogenetic subtree. Each internal node is labeled by its observed fate. T4–T5 FASEs are colored orange

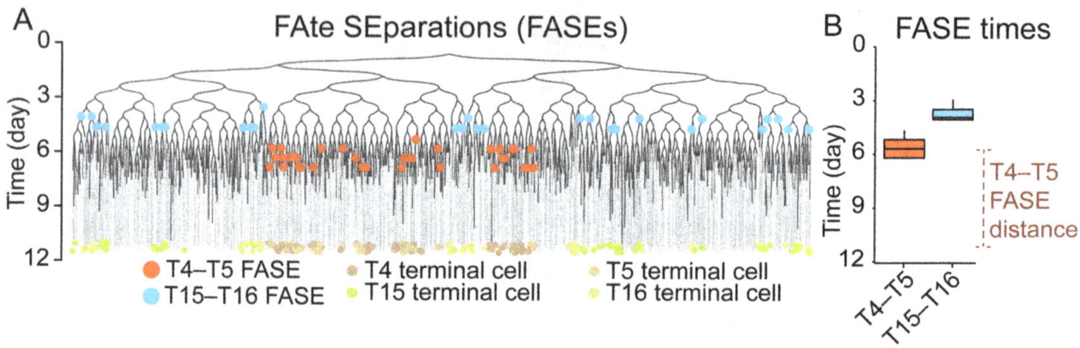

Fig. 7 (**a**) FASEs between T4–T5 (orange) and T15–T16 (blue) are highlighted on the phylogeny. T4, T5, T15, and T16 are colored according to the key on the bottom. (**b**) Boxplots showing the temporal distribution of T4–T5 and T15–T16 FASEs in the tree in (**a**)

the topmost node in Fig. 6 corresponds to a clone with observed cell types {T1, T3, T5, T6}. Subsequently, nodes whose observed fates are more potent than that of both their daughter nodes are identified. For example, the left red node in Fig. 6 led to terminal cell types {T3, T4, T5}, but only {T3, T4} were observed in one branch, and {T5} was exclusively observed in the other branch, this node signified a Fate Separation (FASE) between T3 and T5, as well as between T4 and T5. FASEs are identified across the entire phylogenetic tree. For each FASE, all pairs of terminal cell types that the FASE separates are recorded. The average time elapsed since FASE events between two terminal cell types, referred to as FASE distance, serves as a measure of their developmental separation: Greater FASE distances suggest an earlier divergence in development, while shorter FASE distances indicate a more recent divergence. For example, Fig. 7 compares early FASEs between T15 and T16 and late FASEs between T4 and T5. If no FASE event exists for a pair of terminal types, the FASE time is assumed to be zero. As an example, all FASE between T4 and T5 are annotated as red in Fig. 6. Consequently, a matrix of FASE distances between

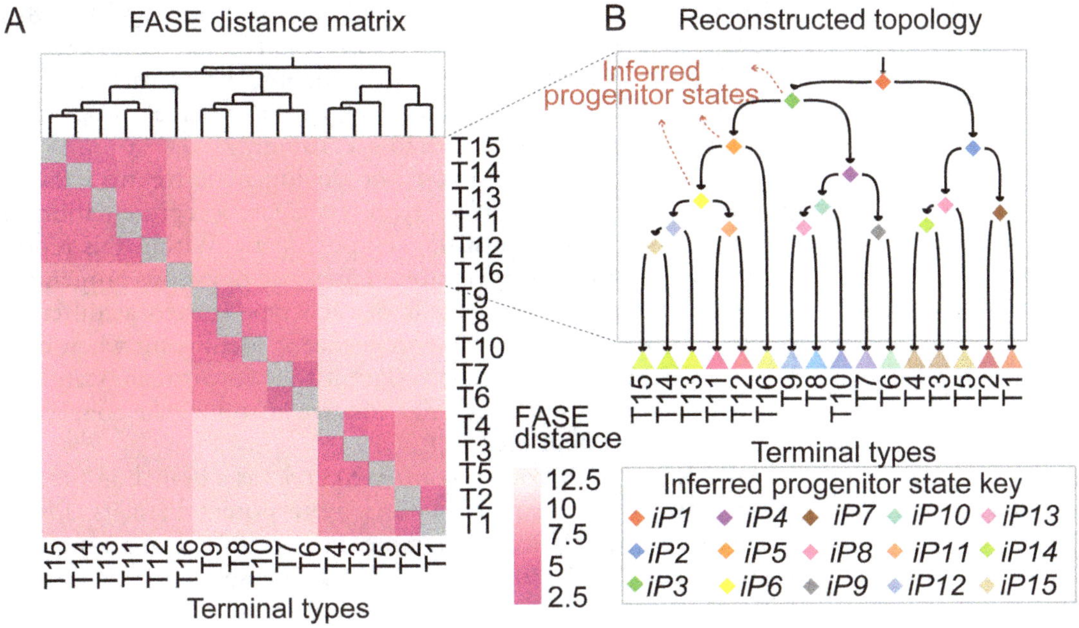

Fig. 8 (**a**) Heatmap showing the FASE distance matrix for all pairs of terminal types in the tree in Fig. 7a. Dendrogram shows hierarchical clustering results. (**b**) The fate map topology reconstructed by clustering the FASE distance matrix in **a**. Triangles, observed terminal types; diamonds, inferred progenitor states (*iPs*)

all terminal cell types (Fig. 8) can be constructed. To reconstruct the fate map topology, the UPGMA hierarchical clustering algorithm is applied. This fate map topology establishes a hierarchy of increasingly diverse but decreasingly potent "inferred" progenitor states (labeled with "*iP*"s followed by numerals) that give rise to the terminal cell types (Fig. 8).

2.6 Progenitor State Dynamics Reconstruction via Inferred Commitment Events (ICEs)

Given the now reconstructed fate map topology and the progenitor states identified within it, we can characterize fate restriction in terms of changes in progenitor states. The defining trait of a progenitor state is its potency, specifically the range of potential end states it can ultimately lead to. Likewise, each node within the time-scaled phylogeny can be assigned an observed potency, which is determined by the assortment of cell types covered by its descendants. The determination of the progenitor state ascribed to a node in the phylogeny relies on its potency. Essentially, a node is ascribed an inferred state sharing the same level of capacity as itself. In cases where such a state is not present in the reconstructed fate map, the node will be assigned the least potent state from the fate map that surpassed the node's potency but still encompassed all the observed cell types associated with the node's potency. For example, in Fig. 9b, the rightmost blue node has an observed fate of {T2, T3, T4}, but it is assigned a more potent inferred state of *iP2* capable of

{T1 to T5} as the now-reconstructed fate map topology (Fig. 8b) indicates that *iP2* differentiates into fates {T1, T2} and {T3, T4, T5} (Fig. 9). Figure 9a shows a phylogeny, where each internal node in the phylogeny has been assigned an inferred progenitor state or terminal type and colored accordingly.

By attributing each node in the phylogenetic tree to either a progenitor state or a terminal type, we create a representation of progenitor states on top of the cell phylogeny. Within this representation, we can track the temporal presence of various progenitor states, revealing how many cells of each type existed at different points in time. Of particular significance is identifying when cells lose certain potency and hence commit to a downstream state. To delve deeper into this characterization, we introduce the concept of inferred commitment events (ICEs).

We define an *inferred commitment event* as a branch in the cell phylogeny whose upstream state is more potent than its downstream state (Fig. 10). For example, in Fig. 9b, when an internal node assigned to *iP2* (capable of {T1 to T5}) splits into two nodes with assigned states of T2 and *iP14* (capable of {T3, T4}), the two branches are counted as ICEs. With ICE defined, we need to next associate ICEs with progenitor states. To do that, we further define the concept of a *state path* for each ICE. The state path for a branch includes all the inferred progenitor states in the fate map topology between the state of the incoming node (closer to the root of the phylogeny) and the state of the outgoing node (closer to tips of the phylogeny) of the branch. For example, the state path for the branch with an incoming state of iP2 and an outgoing state of T2 has a state path of (*iP2* -> *iP7* -> *T2*). We can then associate ICEs to progenitor state(s). For each progenitor state, a branch is associated with it if its state path covers both the state itself and at least one of its immediate downstream states. For example, an ICE is associated with *iP2* if its state path covers {*iP2*, *iP7*} or {*iP2*, *iP8*}. Intuitively, ICEs that are associated with a progenitor state are where progenitor cells have transitioned from the state to a less potent state. ICEs improve on FASEs by leveraging the reconstructed fate map topology to identify a more confident set of edges that represent state

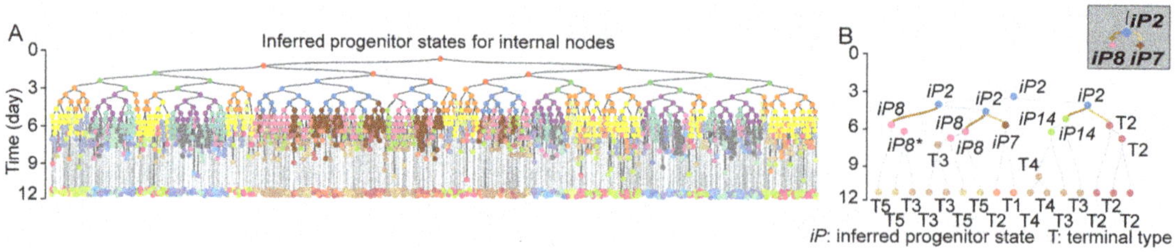

Fig. 9 (**a**) Phylogeny in Fig. 7a with internal nodes colored according to their inferred progenitor state. Nodes are colored according to Fig. 8b. (**b**) A subtree from **a** where each internal node is labeled by its inferred progenitor state or terminal type. Asterisks signify ICE nodes

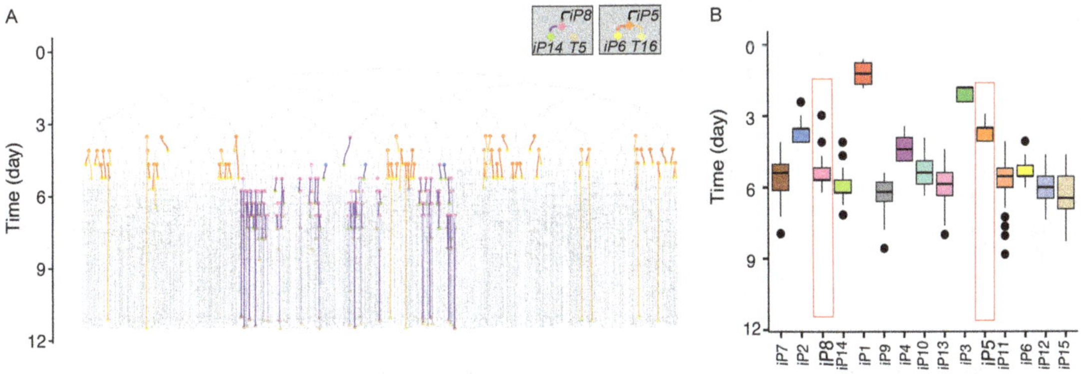

Fig. 10 (a) The phylogeny in Fig. 9a with ICEs associated with *iP5* and *iP8* labeled. (**b**) Boxplots showing the distribution of ICE times for all inferred progenitor states in the tree in (**a**), representing their commitment times. The ICE times for *iP5* and *iP8* are boxed in red

transitions. Figure 10a highlights the ICEs identified for *iP5* and *iP8*. Given the steps so far, the commitment time of a progenitor state can now be estimated by the distribution of incoming times of the ICE branches (Fig. 10b).

Analyzing ICEs across the entire cell phylogeny yields a picture of the population dynamics before and after the commitment events. Specifically, ICEs' incoming nodes offer a glimpse into the cell population preceding commitment, and we can estimate the population size of the progenitor state by counting the unique incoming nodes associated with the ICEs. Conversely, the outgoing nodes linked to the ICEs directly relate to the population sizes post-commitment. It's worth noting that each ICE tied to a progenitor state exclusively contains one of its downstream states, allowing for their classification as such. To determine the founding population size of each downstream state, we tally the outgoing nodes attributed to each specific downstream state. The fraction of outgoing nodes directed toward a particular downstream state relative to all others provides an estimate of the commitment bias of the progenitor state (Fig. 11c). This approach offers a comprehensive view of cell population dynamics surrounding commitment events.

2.7 The Effect of Sampling and Progenitor State Coverage (PSCov)

An important condition for successful retrospective fate mapping is that the progenitor population should be well represented by the sample collected. This can be best understood via the concept of sampling fraction. The sampling fraction is the proportion of the progenitor state population whose progeny are represented among sampled terminal cells. When the progenitor states are being sufficiently sampled, their observed progenies faithfully represent their

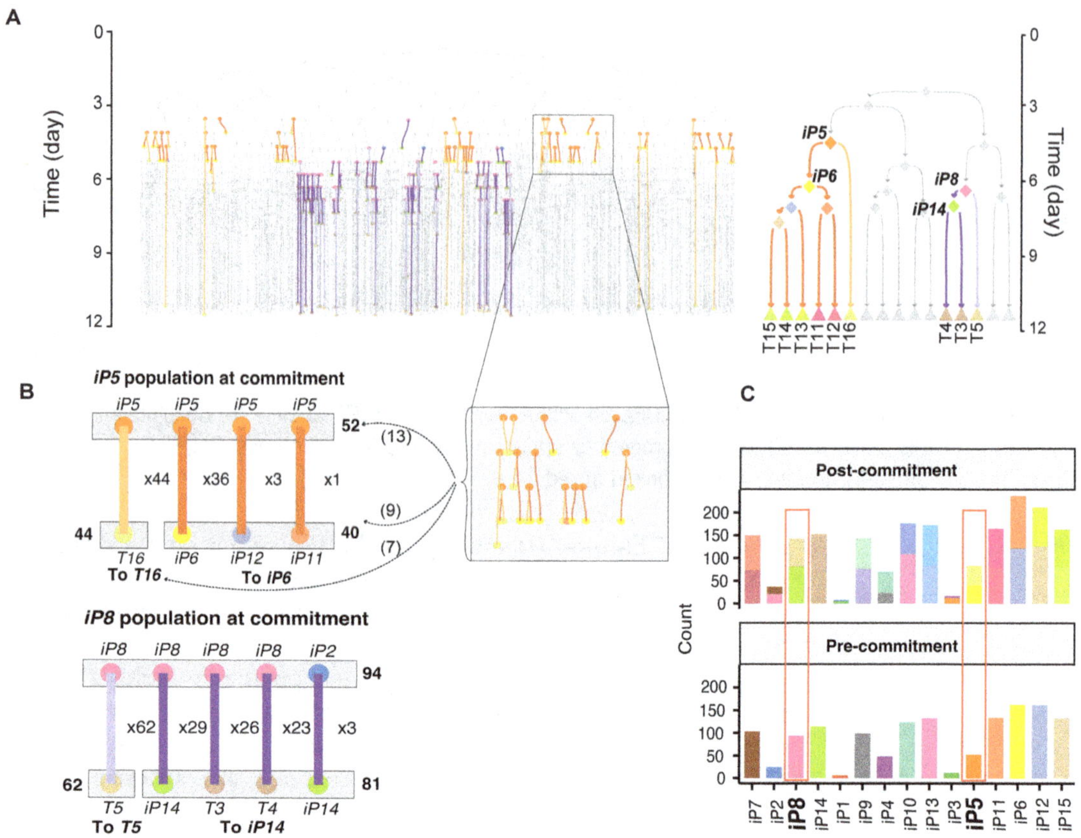

Fig. 11 (**a**) The ICEs associated with iP5 and iP8 are highlighted (left) and the relevant state changes are highlighted depending on the downstream states (right). The ICEs in the phylogeny are classified and colored based on the downstream states they commit to, nodes based on their inferred progenitor state or terminal type, with the fate map serving as the color key. (**b**) Tally of edges and pre- and post-commitment nodes associated with *iP5* (left) and *iP8* (right) from the phylogeny in (**a**), a part of which is magnified for added clarity. The number next to each edge indicates how many times the combination of incoming and outgoing progenitor state or terminal type is observed. Gray boxes on top indicate the nodes that are counted toward progenitor state population size; gray boxes on the bottom indicate post-commitment nodes counted for commitment bias. The number next to each gray box indicates the total count of nodes in that box. Arrows from (**b**) inset show in parentheses counts being added to relevant tallies from the inset. (**c**) Barplots showing the estimated population size of each inferred progenitor state. Red boxes mark *iP5* and *iP8* estimates from **b**

developmental potentials, and the fate restriction patterns can thus be reconstructed accurately. If a progenitor state is not being sampled adequately, its associated estimates will be biased. For example, in Fig. 12a, true FASEs in green are missed because the sampled progenies do not represent all terminal cell types the progenitor is capable of. Similarly, in Fig. 12b, because T3 and T4 progenies of the progenitor cell are not being sampled, the *iP2* cell is mislabeled as *iP8* cell, which causes wrong ICEs to be identified. Overall, undersampling a progenitor state will cause commitment time

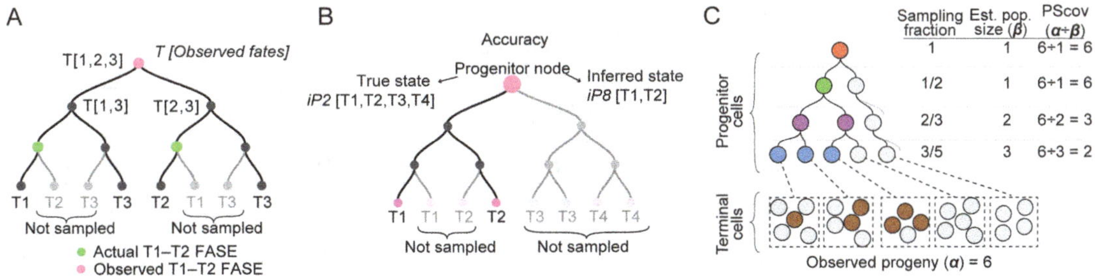

Fig. 12 (**a**) Example phylogenetic subtree, showing how undersampling can lead to FASEs appearing earlier in the tree (pink circles) than they really are (green circles). Gray circles: unsampled cells. (**b**) Example phylogenetic subtree, showing how undersampling can lead to a node being assigned a progenitor state with less potency than its true state. (**c**) Schematic showing how PScov is calculated. Each circle represents a cell; colors represent different progenitor states the cells have been assigned to; gray circles represent unsampled progenitor and terminal cells. *Est* estimate, *pop* population

estimates to be biased toward earlier, population size estimates to be biased toward smaller and commitment bias estimates to be biased toward 0.5.

The sampling fraction, although intuitive, is not directly observed in the experiment, so an alternative statistic to gauge how undersampled each progenitor state is of practical use. One such statistic is the estimated progenitor state coverage (PScov). PScov is defined as the number of observed terminal progeny for a progenitor state divided by its estimated population size (Fig. 12a). Intuitively, this statistic indicates how many terminal descendants are being sampled per inferred progenitor cell. Figure 12c illustrates the idea of PScov, compared to the blue stage where there are six cells covering a progenitor population of three, the green stage where there are six cells covering a population of one is considered more trustworthy.

2.8 Preparing Input Data for ICE-FASE

The ICE-FASE algorithm takes a cell phylogeny alongside with the terminal cell types as input. A cell phylogeny can be represented in different ways. In the QFM package, we use the `phylo` class from the ape package as the standard data structure for cell phylogeny. As input to QFM, we describe the edge table format and the newick format. Table 4 shows an example of an edge table. Each row of the edge table specifies the incoming node (From column), outgoing node (To column), incoming time (From_time column), and outgoing time (To_time column).

The Newick standard represents a tree using nested parentheses. Briefly, each tip or node is followed by a column and then the branch length leading to it. Tips and nodes descending from the same upstream node are separated by comma in one parenthesis and the immediate upstream node is outside the parentheses. The

Table 4
Example of an edge table specifying a time-scaled phylogeny

From	To	From_time	To_time
Node-1	Node-2	0	0.716611
Node-2	Node-8	0.716611	1.701454
Node-8	Node-34	1.701454	2.771955
Node-34	Node-238	2.771955	3.180666
Node-238	Node-385	3.180666	3.98387
Node-385	Node-807	3.98387	4.652447
Node-807	Type5_Cell39	4.652447	14.30797
Node-807	Node-1116	4.652447	4.801977
Node-1116	Type5_Cell28	4.801977	14.30797

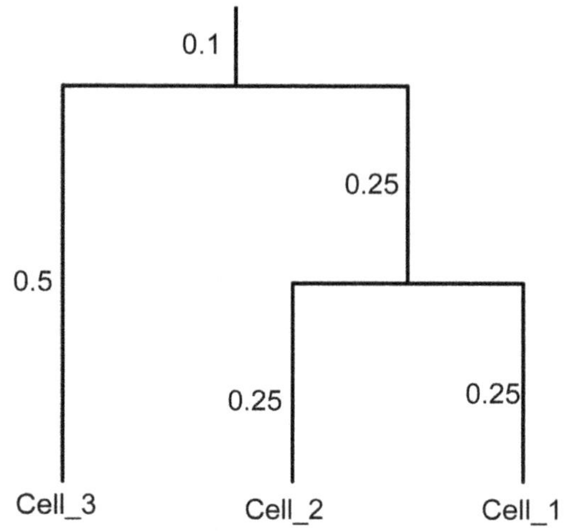

Newick format:
(Cell_3:0.5,(Cell_2:0.25,Cell_1:0.25):0.25):0.1;

Fig. 13 Example of a Newick string encoding a cell phylogeny

Newick string ends with a semicolon. The details could be seen in the website: https://evolution.genetics.washington.edu/phylip/newicktree.html. Figure 13 shows an example of a tree encoded in the Newick format.

The terminal cell type table specifies the cell types of the terminal cells in the cell phylogeny. Table 5 displays an example of the terminal cell type table.

2.9 Running ICE-FASE Algorithm

To apply ICE-FASE, we need the cell type classifications in addition to the phylogeny. From the terminal cell type table, we construct a named vector where the names of the vector are the cell names.

```
cell_type_tb = read_csv('cell_type.csv')
sc_celltypes = cell_type_tb $type
names(sc_celltypes) = cell_type_tb$cell
```

Instead of using the Phylotime output, the user may also choose to import a phylogeny obtained elsewhere. For the Newick format:

```
tr = ape::read.tree("tr.newick")
```

For the edge table format:

Table 5
Example of a terminal cell type table

Cell	Type
Type1_Cell1	Type1
Type1_Cell2	Type1
Type1_Cell3	Type1
Type2_Cell4	Type2
Type2_Cell5	Type2
Type2_Cell6	Type2

```
tr = read_edge_table('edge_table.csv')
```

Then, the lineage barcode can be visualized alongside the observed cell types.

```
plot_barcodes(chr_mat,
              tr = tr,
              tip_celltype = sc_celltypes,
              show_column_names = F)
```

The above command will produce a heatmap in the format of Fig. 4a.

The following code runs the ICE-FASE algorithm, `root_time` is the time until barcode activation, and `total_time` is the time until sample collection.

```
res = ice_fase(tr,
               sc_celltypes,
               total_time = 15 - 0.6,
               root_time = 0.6)
```

The user can also choose to reconstruct the fate map topology only:

```
gr = reconstruct_graph(tr,
                       sc_celltypes,
                       total_time = 15 - 0.6)
```

With ICE-FASE applied, we can plot the reconstructed fate map topology.

```
plot_topology(gr = res$gr,
              total_time = 15,
              gr_node_time = res$gr_trans_time,
              type_col = res$col_pal,
              show_node_label = T)
```

The above command will produce a plot similar to Fig. 8b.

To visualize the node assignment on the cell phylogeny:

```
plot_tr(res$tr,
        node_types = res$tr_node_assign,
        type_col = res$col_pal)
```

The above command will produce a plot similar to Fig. 9a.

To visualize the distribution of commitment times:

```
plot_ice_times(res)
```

The above command will produce a plot similar to Fig. 10b.

To visualize the pre- and post- commitment population sizes:

```
plot_node_sizes(res)
```

The above command will produce a plot similar to Fig. 11c.

To output a summary table of progenitor state parameters:

```
output_estimates(res)
```

An example output is shown in Table 6.

Acknowledgements

This work was supported by grants from the Simons Foundation (SFARI 606178, R.K.), the National Institutes of Health (NIH) (U01HL156056, R.K.; R01HG012357, R.K.; R01HG010889, H.J.), and the David & Lucile Packard Foundation (2020-71380, R.K.). The authors would also like to acknowledge the Advanced Research Computing at Hopkins (ARCH), which is supported by the National Science Foundation (NSF) (OAC1920103).

Table 6
Example output of progenitor state parameter estimates

Progenitor state	Commit time	Population size	Downstream state1	Downstream size1	Downstream state2	Downstream size2	Commitment bias1	Commitment bias2	Ps cov
Node-1	2.2468	10	Node-2	8	Node-3	8	0.5	0.5	160
Node-2	4.61528020	181	Node-9	115	Node-10	138	0.454545454	0.545454545	2.2099
Node-3	2.70564999	17	Node-4	14	Node-5	11	0.56	0.44	70.5882
Node-4	4.58481316	117	Type6	83	Type5	81	0.506097560	0.493902439	1.7094
Node-5	3.64055915	46	Node-6	27	Node-7	33	0.45	0.55	21.7391

References

1. Garcia-Marques J, Espinosa-Medina I, Lee T (2021) The art of lineage tracing: from worm to human. Prog Neurobiol 199:101966
2. Weinreb C, Rodriguez-Fraticelli A, Camargo FD, Klein AM (2020) Lineage tracing on transcriptional landscapes links state to fate during differentiation. Science 367. https://doi.org/10.1126/science.aaw3381
3. Wagner DE, Weinreb C, Collins ZM et al (2018) Single-cell mapping of gene expression landscapes and lineage in the zebrafish embryo. Science 360:981–987
4. Biddy BA, Kong W, Kamimoto K et al (2018) Single-cell mapping of lineage and identity in direct reprogramming. Nature 564:219–224
5. Kalhor R, Kalhor K, Mejia L et al (2018) Developmental barcoding of whole mouse via homing CRISPR. Science 361. https://doi.org/10.1126/science.aat9804
6. McKenna A, Findlay GM, Gagnon JA et al (2016) Whole-organism lineage tracing by combinatorial and cumulative genome editing. Science 353:aaf7907
7. Spanjaard B, Hu B, Mitic N et al (2018) Simultaneous lineage tracing and cell-type identification using CRISPR-Cas9-induced genetic scars. Nat Biotechnol 36:469–473
8. Liu X, Weng W, He L, Zhou B (2023) Genetic recording of in vivo cell proliferation by ProTracer. Nat Protoc 18:2349–2373
9. Bowling S, Sritharan D, Osorio FG et al (2020) An engineered CRISPR-Cas9 mouse line for simultaneous readout of lineage histories and gene expression profiles in single cells. Cell 181:1693–1694
10. Kalhor R, Mali P, Church GM (2017) Rapidly evolving homing CRISPR barcodes. Nat Methods 14:195–200
11. Alemany A, Florescu M, Baron CS et al (2018) Whole-organism clone tracing using single-cell sequencing. Nature 556:108–112
12. Espinosa-Medina I, Feliciano D, Belmonte-Mateos C et al (2023) TEMPO enables sequential genetic labeling and manipulation of vertebrate cell lineages. Neuron 111:345–361.e10
13. Bizzotto S, Dou Y, Ganz J et al (2021) Landmarks of human embryonic development inscribed in somatic mutations. Science 371:1249–1253
14. Spencer Chapman M, Ranzoni AM, Myers B et al (2021) Lineage tracing of human development through somatic mutations. Nature 595:85–90
15. Coorens THH, Moore L, Robinson PS et al (2021) Extensive phylogenies of human development inferred from somatic mutations. Nature 597:387–392
16. Salipante SJ, Horwitz MS (2006) Phylogenetic fate mapping. Proc Natl Acad Sci U S A 103:5448–5453
17. Baron CS, van Oudenaarden A (2019) Unravelling cellular relationships during development and regeneration using genetic lineage tracing. Nat Rev Mol Cell Biol 20:753–765
18. Wagner DE, Klein AM (2020) Lineage tracing meets single-cell omics: opportunities and challenges. Nat Rev Genet 21:410–427
19. Fang W, Bell CM, Sapirstein A et al (2022) Quantitative fate mapping: a general framework for analyzing progenitor state dynamics via retrospective lineage barcoding. Cell 185:4604–4620.e32
20. Fitch WM, Margoliash E (1967) Construction of phylogenetic trees. Science 155:279–284
21. Felsenstein J (1981) Evolutionary trees from DNA sequences: a maximum likelihood approach. J Mol Evol 17:368–376
22. Yang Z (2014) Molecular evolution: a statistical approach. Oxford University Press
23. Leeper K, Kalhor K, Vernet A et al (2021) Lineage barcoding in mice with homing CRISPR. Nat Protoc 16:2088–2108
24. Shen MW, Arbab M, Hsu JY et al (2018) Predictable and precise template-free CRISPR editing of pathogenic variants. Nature 563:646–651
25. Sokal RR, Michener CD (1958) A statistical method for evaluating systematic relationships. University of Kansas
26. Rubin DB (2009) Multiple imputation for nonresponse in surveys. Wiley
27. Chen T, Guestrin C (2016) XGBoost: a scalable tree boosting system. In: Proceedings of the 22nd ACM SIGKDD international conference on knowledge discovery and data mining. Association for Computing Machinery, New York, pp 785–794

Chapter 10

Backtracking Cell Phylogenies in the Human Brain with Somatic Mosaic Variants

Sara Bizzotto

Abstract

Somatic mosaic variants, and especially somatic single nucleotide variants (sSNVs), occur in progenitor cells in the developing human brain frequently enough to provide permanent, unique, and cumulative markers of cell divisions and clones. Here, we describe an experimental workflow to perform lineage studies in the human brain using somatic variants. The workflow consists in two major steps: (1) sSNV calling through whole-genome sequencing (WGS) of bulk (non-single-cell) DNA extracted from human fresh-frozen tissue biopsies, and (2) sSNV validation and cell phylogeny deciphering through single nuclei whole-genome amplification (WGA) followed by targeted sequencing of sSNV loci.

Key words Somatic mosaic variants, Human tissue, Deep whole-genome sequencing, Single-nuclei sorting, Whole-genome amplification, Single-cell DNA sequencing, Amplicon sequencing, Bioinformatic analyses, Cell phylogenies

1 Introduction

Somatic variants consist in changes in the DNA sequence that start occurring in the human genome with the first postzygotic cell division and continue accumulating during the entire life. In normal non-pathological human tissues, somatic variants occur at rates that vary based on the type of variant but also depending on the tissue and cell type [1–3]. Somatic variants can be due to both endogenous processes (e.g., DNA replication, transcription, erroneous DNA repair) and exposure to exogenous agents (e.g., tobacco smoking, UV light etc.) [3–6]. Somatic single nucleotide variants (sSNVs) are by far the most frequent variant type, and occur at a median rate of 2.8×10^{-7} per base pair (bp) as calculated in primary human dermal fibroblasts [7]. In the human brain, sSNV rates have been studied in neural progenitors, and in the context of aging, in neurons and oligodendrocytes. In neural progenitors, one study estimated ~8-9 sSNVs per cell cycle [8, 9], while another

Jorge García-Marqués and Tzumin Lee (eds.), *Lineage Tracing: Methods and Protocols*, Methods in Molecular Biology, vol. 2886, https://doi.org/10.1007/978-1-0716-4310-5_10,

study suggested ~1.4 variants per division in brain development [10]. Aging neurons and oligodendrocytes accumulate ~16 and ~29 sSNVs per genome per year, respectively [2, 3, 11].

sSNVs occurring in progenitor cells are inherited by daughter cells, providing permanent and cumulative markers of cell divisions and clones [12]. Furthermore, if we exclude recurrent SNVs that are common in the general population and mutation hotspots, we can assume that the likelihood of the same sSNV occurring twice in the same individual is almost null. Thus, sSNVs occurring in progenitor cells mark permanently, cumulatively, and uniquely the genome within a cell lineage, providing a natural barcoding system that is forever embedded in each cell and that allows to reconstruct its developmental history. For this reason, sSNVs have recently been used to reconstruct and track cell phylogenies through lineage tracing, providing novel insights into human development [8, 12–18].

Few aspects and limitations need to be discussed in association with lineage tracing in the human brain using somatic variants. Since the great majority of sSNVs in normal human tissues are found in intronic and intergenic non-transcribed regions [12], whole-genome sequencing (WGS) is required in order to identify them with good sensitivity. Although we know very little about the clonal architecture of the human brain—defined as the number of neural progenitor lineages contributing to a specific region of interest (ROI)—current knowledge suggests that the human brain may be highly polyclonal [12, 15, 19]. This implies that sensitivity in somatic variant detection is one of the main aspects to consider when performing lineage tracing. Indeed, high levels of clonality mean that each neural progenitor would contribute a small fraction of the cells in an ROI. Findingi the somatic variants that mark such small clones can be extremely challenging when starting from bulk (non-single cell) DNA, even when performing very deep (>200×) WGS.

Although the ideal strategy to overcome sensitivity issues would be to identify somatic variants through single-cell (sc)WGS of thousands of cells, this is unpractical and too costly with current technologies. For this reason, to retrospectively build cell phylogenies we favor a two-step approach that is summarized in Fig. 1 and that is made of: (1) a "calling phase" performed through deep bulk WGS, and (2) a "validation phase" performed through single-cell targeted sequencing, where cells are obtained from a separate sample that is spatially adjacent to the biopsy used in step 1. There are two main implications associated with this approach. First, the final resolution of the obtained cell phylogeny is highly dependent on the sensitivity of somatic variant discovery, which in turn is dependent on the amount of starting material used to extract the bulk DNA and on the sequencing depth. Second, only those

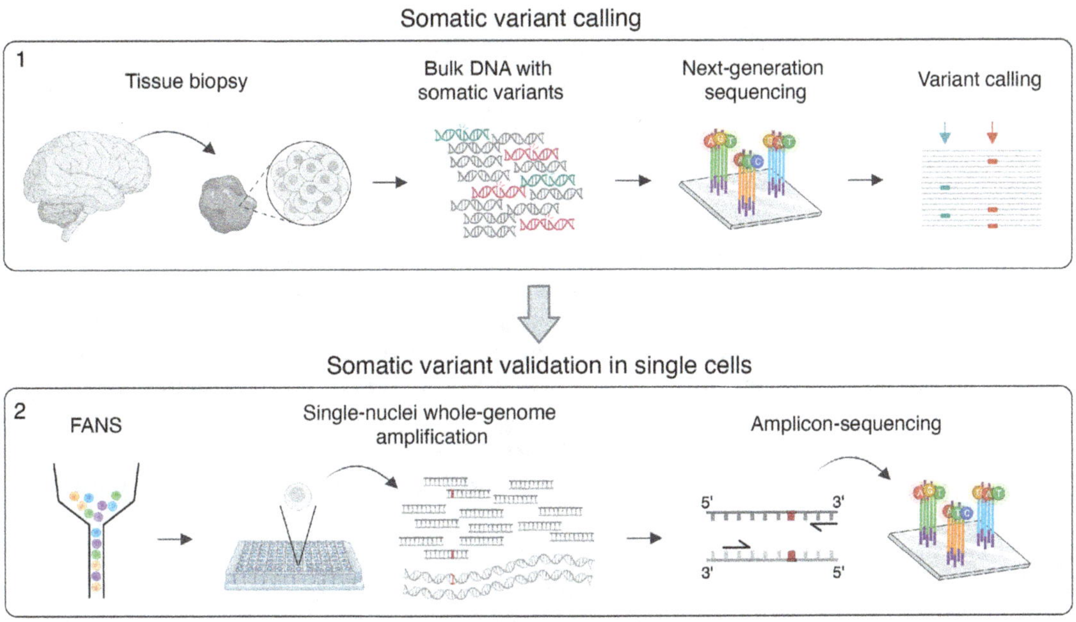

Fig. 1 Schematic of the workflow presented in this chapter. Step 1 (calling phase): a fresh-frozen brain tissue biopsy is dissected and subjected to bulk DNA extraction, which is then sent for deep WGS with Illumina next-generation sequencing (NGS). Deep WGS data are then processed and analyzed for somatic variant calling. Somatic variants (in green and red) are discerned from germline variants as those that are carried by <50% of the sequencing reads. As part of Step 1, somatic variants are validated through amplicon sequencing (not shown) using an aliquot of the same bulk DNA sample initially used for deep WGS, and additional samples if variant assessment in other tissues or tissue regions is required by the experimental design. Step 2 (validation phase): single nuclei are isolated through fluorescence activated nuclear sorting (FANS) from a biopsy obtained from a tissue region adjacent to that used in Step 1. Cell types of interest (e.g. neurons) could be enriched during this step. Following FANS, single-nuclei whole-genome amplification (WGA) is performed. Single-cell amplified genomes are then used as input DNA for amplicon sequencing and variant validation. (Figure created with BioRender.com)

lineages that are present in both the initial calling sample (step 1) and the validation sample (step 2) can be placed in the phylogeny. Thus, if a lineage is restricted in space such that it is only present in the calling sample, this will be excluded from the final phylogeny. For this reason, we strongly advise to use two spatially adjacent samples for the calling and validation phases.

Another limitation currently associated with lineage tracing in humans is the poor ability in obtaining the cell type composition of each lineage [9]. In editable model systems, this is done by high-throughput single-cell (sc)RNA-seq, which simultaneously captures the full cell transcriptome as well as the region (barcode) where nucleotide changes (scars) analogous to somatic variants are artificially introduced [20, 21]. Since we cannot control where somatic variants occur in the human genome, and since the majority of sSNVs are found outside of transcribed regions, our ability to get the genotyping of the variant loci as well as the cell type identity (e.g. from single-cell transcriptomes) at high-throughput remains significantly limited. Indeed, existing technologies such as G&T-seq (Genome and Transcriptome sequencing), PRDD-seq (Parallel RNA and DNA analysis after Deep-sequencing), and GoT (Genotyping of Transcriptome) are either low-throughput or limited to the detection of very few variant loci [19, 22, 23]. In this chapter however, we are not addressing this specific aspect of lineage tracing in humans, but rather focus on how to backtrack cell divisions and lineages.

Finally, studies performed in humans are limited by tissue access and choice of sample(s). Although some conclusions inferred from a restricted region can to a certain extent be generalized to the entire tissue, what we are able to effectively reconstruct remains limited to the available tissue. In spite of this, the big advantage associated with this strategy is the ability of studying cell phylogenies directly in humans thus, providing access to information that may be difficult to extrapolate from animal or in vitro models. With the progressive reduction of sequencing costs and the methodologies described here becoming more efficient and widely accessible, future studies based on somatic variants will progressively increase our understanding of developmental cell phylogenies in humans.

In this chapter, we focus on sSNVs as a tool for lineage tracing in the human brain, and present a basic workflow that can be used and modified/adapted to perform such studies. The objective is to resolve the developmental process into a cell phylogeny that describes the formation of the different (sub)lineages (or (sub)-clones) that contributed to a specific brain region.As part of the methodology, we also provide a general basic description of how to perform bioinformatic analyses of both WGS and targeted sequencing data, and on how to build a cell phylogeny starting from collected data. However, we must clarify that this chapter does not provide an exhaustive coverage of the bioinformatic analyses that can be performed with such data.

2 Materials

To perform this protocol, accessibility to a computer cluster with capacity to store big data and perform advanced bioinformatic analyses is necessary, as well as access to an NGS facility equipped with an Illumina NovaSeq6000 sequencer, or similar.

Start from fresh-frozen human brain specimens (post-mortem or post-surgical). All experiments on human tissues must have ethics approval in accordance with governmental and institutional regulations, and informed consent must be obtained for all human samples. Store human tissue specimens at −80 °C. When removed from −80 °C, tissue must be maintained in dry-ice or in a −20 °C environment (e.g., cryostat chamber). Lysed tissue and DNA should be kept on ice during the entire procedure, unless specified, and at −20 °C for overnight or longer storage. Similarly, all buffers and reagents should be kept on ice as much as possible, unless specified. If possible, avoid too many freeze-thaw cycles for reagents and DNA samples.

Manipulate human specimens in a level 2 (L2) confinement laboratory, when possible under a type II laminar flow hood equipped with a High Efficiency Particulate Air (HEPA) filters. Wear a lab coat, gloves, and possibly a face mask covering mouth and nose during manipulation. All experimenters must be screened for immunization prior to the experiment, in accordance with local institutional guidelines.

Before starting the experiment, make sure that all surfaces and tools are free of residual nucleic acids and nucleic-acid-degrading enzymes. This is critical when working with somatic variants since nucleic acid contamination could lead to the generation of artifacts in the sequencing data. Clean all surfaces and materials with DNAZap and RNaseZap. In addition, perform a second decontamination using 70% (vol/vol) ethanol.

Below, find the materials required for each step described in the method section.

2.1 Bulk DNA Deep WGS and Somatic Variant Calling

2.1.1 Tissue Dissection

1. L2 cryostat, sterile cryostat blades, sterile scalpels, sterile plastic tweezers, sterile paintbrush.
2. 70% vol/vol ethanol.
3. Low DNA binding 1.5 mL tubes.

2.1.2 DNA Extraction

1. Lysis Buffer ATL (Qiagen).
2. Proteinase K solution (Thermo Fisher Scientific).
3. Phenol-chloroform.
4. Phase-lock (phasemaker) tubes.
5. Low DNA binding 1.5 mL tubes.
6. Pre-cool 100% ethanol for molecular biology at −20 °C.
7. 3M Sodium Acetate, pH 5.5, RNase-free.
8. 70% vol/vol ethanol.
9. Nuclease-free (NF) ultra-pure water.
10. Isopropanol (for molecular biology, DNAse, RNAse, and Protease free).
11. PicoGreen dsDNA Assay Kit and dsDNA Reagents with 96-well black plates (Thermo Fisher Scientific).
12. Vortex.
13. Thermomixer for 1.5 mL tubes.
14. Benchtop centrifuge with 1.5 mL tube rotor.

15. Nanodrop instrument.
16. Bioanalyzer or Tape Station instrument for DNA quantification and quality control (QC).

2.1.3 Somatic Variant Validation Through Amplicon Sequencing

1. 96-well thermal cycler for Polymerase Chain Reaction (PCR).
2. Primer pools targeting each candidate locus, designed using the Ion AmpliSeq custom designer tool from Thermo Fisher Scientific (https://www.thermofisher.com/fr/fr/home/life-science/sequencing/next-generation-sequencing/ion-torrent-next-generation-sequencing-workflow/ion-torrent-next-generation-sequencing-select-targets/ampliseq-target-selection/custom-panels.html) and order 2× panels (*see* **Note 1**). We advise including in the panel few (around 10) germline loci as positive controls.
3. 2× Phusion U Mastermix (Thermo Fisher Scientific).
4. AMPure XP Bead-Based Reagent (Beckman Coulter).
5. NF ultra-pure water.
6. 10× Tango Buffer (Thermo Fisher Scientific).
7. 100 mM ATP (Thermo Fisher Scientific).
8. 10 U/μL T4 Polynucleotide Kinase (Thermo Fisher Scientific).
9. 10 mM dNTPs (Thermo Fisher Scientific).
10. 5 U/μL T4 DNA Polymerase (Thermo Fisher Scientific).
11. 5 M NaCl.
12. 1 M Tris-HCl.
13. UltraPure 0.5 M Ethylendiaminetetraacetic acid (EDTA), pH 8.0 (Thermo Fisher Scientific).
14. High Performance Liquid Chromatography (HPLC-Grade Water) decontaminated with a 30 min UV irradiation before use.
15. 10× Oligo Hybridization Buffer (1 mL): 100 μL 5 M NaCl, 10 μL 1 M Tris-HCl, 2 μL 0.5 M EDTA, 888 μL UV HPLC-water. Irradiate the buffer with UV for 30 min with an open lid, then store at −20 °C.
16. Hybridization mix for adapter P5 (200 μM): 40 μL 500 μM IS1-adapter.P5 (A*C*A*C* TCTTTCCCTACAC GACGCTCTTCCG *A*T*C*T*N*N*N*N*N*N*N), 40 μL 500 μM IS3-adapter.P5+P7 (N*N*N*N*N*N*N*A*-G*A*T*CGGAA*G*A*G*C), 10 μL 10× Oligo Hybridization Buffer and 10 μL UV HPLC-water. Store at −20 °C.
17. Hybridization mix for adapter P7 (200 μM): 40 μL 500 μM IS1-adapter.P7 (G*T*G*A* CTGGAGTTCA GACGTGTGCTCTTCCG *A*T*C*T*N*N*N*N*N*N

*N), 40 μL 500 μM IS3-adapter.P5+P7, 10 μL 10× Oligo Hybridization Buffer and 10 μL UV HPLC-water. Store at −20 °C.

18. Adapter mix (100 μM) with unique molecular identifiers (UMI): Mix 100 μL of hybridization mix for adapter P5 and 100 μL of hybridization mix for adapter P7, and incubate the reactions in a PCR thermal cycler for 10 s at 95 °C, followed by a ramp from 95 °C to 12 °C at a rate of 0.1 °C/s. Combine both reactions to obtain the final 100 μM adapter mix. Store at −20 °C.
19. 10× T4 DNA Ligase Buffer (Thermo Fisher Scientific).
20. 50% Polyethylene glycol (PEG)-4000 (Thermo Fisher Scientific).
21. 5 U/μL T4 DNA Ligase (Thermo Fisher Scientific).
22. 10× ThermoPol Reaction Buffer (New England BioLabs).
23. 8 U/μL Bst Polymerase Large Fragment (New England BioLabs).
24. 5× Phusion U HF Buffer (Thermo Fisher Scientific).
25. 1 U Phusion U Hot Start DNA Polymerase (Thermo Fisher Scientific).
26. i5 and i7 Illumina Indexing Primers.

2.2 Somatic Variant Validation in Single Cells and Cell Phylogeny Reconstruction

2.2.1 Fluorescence-Activated Nuclear Sorting (FANS)

1. See **item 1** in Subheading 2.1 for tissue dissection materials.
2. Ultracentrifuge with swinging bucket rotor.
3. Ultracentrifuge rotor-adapted plastic tubes.
4. Dounce homogenizers.
5. UltraPure Sucrose (Thermo Fisher Scientific).
6. Prepare 1 M $CaCl_2$ and store at 4 °C.
7. Prepare 0.5 M $MgAc_2$ and store at 4 °C.
8. UltraPure 0.5 M EDTA, pH 8.0 (Thermo Fisher Scientific).
9. Prepare 1 M Tris-HCl pH 8 and store at 4 °C.
10. NF ultra-pure water.
11. DTT, dithiothreitol (Thermo Fisher Scientific).
12. Prepare 1 M $MgCl_2$ and store at 4 °C.
13. PBS 10× (Thermo Fisher Scientific).
14. BSA Bovine Serum Albumine (Sigma-Aldrich).
15. 40 μm cell strainers.
16. Low DNA binding 1.5 mL tubes.
17. Antibodies for nuclei staining (see methods for more details).
18. DAPI solution.

19. 96-well plates with plate covers.
20. Cell sorter adapted for 96-well plates.
21. Benchtop centrifuge refrigerated with rotor adapted to 1.5 mL tubes.
22. Rotary shaker adapted to 1.5 mL tubes placed in a cold room.

2.2.2 Single-Cell WGA

1. Alkaline nuclear Lysis Buffer: 200 mM KOH, 5 mM EDTA, 40 mM DTT. Store at −20 °C.
2. Neutralization Buffer: 400 mM HCl, 600 mM Tris-HCl, pH 7.5. Store at −20 °C.
3. REPLI-g single-cell 96 reaction kit (Qiagen).
4. NF ultra-pure water.
5. PicoGreen dsDNA Assay Kit and dsDNA Reagents with 96-well black plates (Thermo Fisher Scientific).
6. 5× Phusion U HF Buffer (Thermo Fisher Scientific).
7. 2 U/μL Phusion Hot Start II DNA Polymerase (Thermo Fisher Scientific).
8. 10 mM dNTP mix.
9. Custom primer mix targeting at least 4 random DNA loci.
10. 96-well thermal cycler for PCR.
11. Capillary electrophoresis device such as QIAxcel Connect System.
12. PCR plate spinner.

3 Methods

3.1 Bulk DNA Deep WGS and Somatic Variant Calling

3.1.1 Tissue Dissection

Before starting, decontaminate an L2 confinement laboratory cryostat using UV light for about 30 min, and pre-cool the cryostat chamber at −20 °C. Clean all components of the cryostat, including blades, scalpels, tweezers, and paintbrush using DNAZap, RNaseZap and 70% vol/vol ethanol. Cut frozen human brain tissue specimens using a clean scalpel. Alternatively, if the shape of the biopsy allows it, cut several ~100 μm-thick slices using the cryostat blade. Dissect at least 10–30 mg of tissue per sample (*see* **Note 2**) and collect it in a low DNA binding 1.5 mL tube. To avoid any sample cross-contamination, decontaminate everything with 70% vol/vol ethanol and change gloves, blade, scalpel, and tweezers when starting with a new tissue specimen.

3.1.2 DNA Extraction

We follow a phenol-chloroform extraction and ethanol plus isopropanol precipitation protocol that gives high DNA yields and quality, and is relatively cheap compared to other commercially available kits. Work in a pre-PCR amplification space. All the DNA extraction steps can be performed at room temperature (RT) unless otherwise specified. When approximate volumes are indicated, adjust the volumes to each sample. Before starting, equilibrate 70% vol/vol ethanol and isopropanol at RT.

1. Add to the tube containing the tissue biopsy obtained from Subheading 3.1.1 180 μL Lysis Buffer ATL and 20 μL Proteinase K solution. Lyse and dissolve the tissue in a thermomixer at 1,400 rpm and 56 °C for at least 2 h by vortexing every 30 min. Wait until a homogeneous solution appears, with no visible tissue chunks, then add to the tube 1 volume (~250 μL) of phenol-chloroform and vortex for a few seconds.
2. Prepare one empty phase lock (phasemaker) tube per sample by centrifuging for 1 min at 15,000 rcf, then transfer the sample mixture to the phase lock tube and perform phase separation by centrifuging for 5 min at 14,000 rcf. Transfer the top phase (~200 μL) containing the DNA to a clean 1.5 mL low DNA binding tube.
3. Add to each 1.5 mL tube containing the DNA solution 2.5 volumes (~500 μL) of cold (−20 °C) 100% vol/vol ethanol and 0.1 volumes (~20 μL) of 3 M sodium acetate. Flick the tubes and incubate at least 30 min (ideally overnight) at −80 °C. After the incubation, centrifuge 30 min at 15,500 rcf, then remove and discard the supernatant and leave the pellet in the tube without removing the mucous substance around it. At this point, add 500 μL of RT 70% vol/vol ethanol and dislocate the pellet by pipetting and vortexing. Centrifuge again for 15 min at 15,500 rcf, then remove and discard the supernatant and leave the tubes open at RT to dry the pellet (be careful not to over-dry). Once all the residual ethanol has evaporated, resuspend the pellet in 100 μL NF ultra-pure water and incubate at 37 °C to dissolve for at least 1 h by vortexing from time to time.
4. Add 1 volume (100 μL) of RT isopropanol and gently flick the tube. At this point, a white precipitate should appear. Centrifuge at 14,000 rcf for 2 min, then remove and discard the supernatant. Add 500 μL RT 70% vol/vol ethanol and dislocate the pellet by vortexing for a few seconds. Centrifuge again at 14,000 rcf for 2 min then discard the supernatant. Repeat the ethanol wash one more time, then dry the pellet by leaving the tube open at RT or at 37 °C for 2–5 min being careful not to over-dry the pellet. Re-suspend the pellet in 100 μL NF ultra-pure water (or less if you started from a small biopsy and anticipate a low DNA yield) and dissolve the pellet by incubating at 37 °C for 1 h (vortex from time to time). Before proceeding to DNA QC and quantification (**step 5** in Subheading 3.1.2), we advise to leave the samples at 4 °C overnight in order to make sure that the DNA is well resuspended.

5. DNA QC and quantification: to verify that DNA samples do not contain traces of residual ethanol or isopropanol, run 1 μL on a Nanodrop instrument, and check that the A_{260}/A_{280} ratio is >2.0. Ratios lower than 2.0 may indicate contamination. In case of contamination, you can perform an additional ethanol wash however, this may cause some DNA loss and decrease the DNA extraction yield. Once verified that there is no ethanol/isopropanol contamination, proceed to DNA quantification with PicoGreen assay for dsDNA or an alternative fluorescence-based method. Alternatively, you can run the sample through a tape station or bioanalyzer, which will provide more accurate information regarding abundance, integrity, distribution, and impurities. A minimum of ~500 ng DNA are required to proceed with library prep for Illumina NGS.

3.1.3 NGS

For better performance, we advise making library preparations for WGS through a sequencing facility, but libraries can also be prepared in-house by following manufacturer instructions. When possible, use of PCR-free library protocols (*see* **Note 3**), such as Illumina TrueSeq DNA PCR free, is strongly recommended in order to avoid introduction of PCR artifacts. Each sample can be split into several libraries and across different sequencing lanes to reduce library- and lane-associated biases. For DNA samples prepared as indicated above, which start from 10–30 mg of tissue, we recommend targeting an average sequencing depth of >200×. Perform Illumina 150 bp paired-end sequencing on a NovaSeq6000 (300 cycles). The number and size of flow-cells will depend on the number of samples and the desired sequencing depth. Rely on the sequencing facility expertise for an optimal sequencing design.

3.1.4 Somatic Variant Calling from Bulk DNA Deep WGS Data

For optimal sSNV calling, we advise following the Brain Somatic Mosaicism Network (BSMN) best-practices [24] available at https://zenodo.org/record/4321679 and https://github.com/bsmn/bsmn-pipeline (*see* **Note 4**). Briefly, the workflow consists into a first step of read mapping to the reference genome. This is done starting from FASTQ files and performing: (1) read alignment using bwa [25]; (2) duplicate read (reads that come from the same original DNA fragment) marking using PICARD (http://broadinstitute.github.io/picard/); (3) insertion and deletion (indel) realignment and recalibration using GATK [26]. These steps produce Binary Alignment Map (BAM) files. A first step of candidate sSNV calling is then performed using GATK Mutect2 [27]. On this set of raw candidates, a first filter is applied by eliminating inaccessible genomic regions as defined by the 1000 Genome Strict Mask [24]. Accessible candidate sSNVs are further filtered to eliminate germline variants by (1) removing all variants that have a population frequency >0.001 as defined by the Genome Aggregation Database [28] and (2) eliminating sSNVs having a variant allele frequency (VAF, defined as the number of reads showing the variant over the total number of reads covering the variant locus) equal to 0.5. On the list of candidate sSNVs further apply quality filters by (1) keeping only sSNVs supported by at least five high quality independent non-duplicated reads, and (2) eliminating false-positive calls due to technical artifacts as identified in a panel of normal (PON) based on the data from the 1000 Genomes Project (https://www.internationalgenome.org/home) [29]. These steps give a list of high-quality candidate sSNVs that we advise to further filter using machine learning algorithms such as MosaicForecast [30] or DeepMosaic [31].

3.1.5 Somatic Variant Validation Through Amplicon Sequencing

Although the somatic variant calling approach described in **step 3.1.4** was shown to present low (<10%) false-discovery rates [24], it is best practice to validate at least a subset of the called somatic variants, especially when performing this type of experiments and analyses for the first time. Furthermore, this protocol can be used to assess the presence of the variants in other tissues or tissue regions of interest [12]. sSNV validation/assessment is done by re-sequencing the candidate variant loci using a targeted-sequencing approach. The DNA sample used for validation should ideally be an aliquot of the same sample used initially for WGS (DNA extraction in **step 3.1.2**). However, if not enough DNA is available, a DNA sample extracted from an adjacent region can alternatively be used, although this could give lower validation rates, especially for somatic variants called at very low VAFs in the original sample. We also advise to include a couple of negative control samples, e.g., bulk DNA extracted from other non-related individuals. These samples will help setting a threshold for false positive reads. Here, we present a validation approach that we previously used in one of our studies [12] that consists in amplicon sequencing and library preparation for Illumina NGS (*see* **Note 5**). The indicated volumes refer to a single reaction. Library adapter preparation for dual-index double stranded DNA NGS refers to the protocol described in Meyer and Kircher [32].

1. Before starting, quilibrate AMPure XP Bead-Based Reagent at RT.
2. Dilute the input DNA at 20 ng/μL and prepare the amplicon enrichment mix (19 μL/sample) with 10 μL 2× Phusion U Mastermix and 9 μL 2× AmpliSeq Primer Pool. Mix 1 μL (20 ng) DNA with 19 μL amplicon enrichment mix. Run the PCR as follows: (1) 30 s at 98 °C; (2) 10 s at 98 °C; (3) 30 s at 62 °C; (4) 30 s at 72 °C; (5) repeat **steps 2–4** per 18 cycles; (6) 10 min at 72 °C; (7) hold at 4 °C. If multiple primer pools are used with the same DNA, at this point pool each PCR

reaction obtained for the same DNA sample in a single tube. Put aside 5 μL reaction to run a gel and verify the size of the product (amplicons should be of the size of ~270 bp), then proceed to DNA purification with 1.8× AMPure XP Bead-Based Reagent, and finally elute in 55 μL NF water.

3. For library prep and Illumina NGS (*see* **Note 6**), prepare blunt-end repair mix (20 μL/sample) with 6.7 μL NF water, 7 μL 10× Tango Buffer, 0.7 μL 100 mM ATP, 3.5 μL 10 U/μL T4 Polynucleotide Kinase, 0.7 μL 10 mM dNTPs, and 1.4 μL 5 U/μL T4 DNA Polymerase. Mix 50 μL of purified amplicon enrichment with 20 μL blunt-end repair mix. Incubate the reaction in a PCR thermal cycler for 15 min at 25 °C, followed by 5 min at 12 °C, then place the tubes on ice. Perform a purification step with 1.8× AMPure XP Bead-Based Reagent and elute in 30 μL NF water. Prepare the adapter ligation mix (10 μL/sample) with 0.8 μL NF water, 4 μL 10× T4 DNA Ligase Buffer, 4 μL 50% PEG-4000, 0.2 μL 100 μM each adapter mix, and 1 μL 5 U/μL T4 DNA Ligase. Proceed to adapter ligation by mixing 30 μL of the previously purified reaction with 10 μL of adapter ligation mix. Incubate in a PCR thermal cycler for 30 min at 22 °C, then purify with 1.8× AMPure XP Bead-Based Reagent and elute in 30 μL NF water. Prepare the adapter fill-in reaction mix (10 μL/sample) with 3.5 μL NF water, 4 μL 10× ThermoPol Reaction Buffer, 1 μL 10 mM dNTPs, and 1.5 μL 8 U/μL Bst Polymerase Large Fragment. Mix the elution with 10 μL (total volume 40 μL) adapter fill-in reaction mix. Incubate in a PCR thermal cycler at 37 °C for 20 min, then purify with 1.8× AMPure XP Bead-Based Reagent, and resuspend in 30 μL NF water. Prepare the indexing mix (18 μL/sample) with 10 μL 5× Phusion U HF Buffer, 1 μL 10 mM dNTPs, 0.5 μL 1U Phusion U Hot Start DNA Polymerase, and 6.5 μL NF water. Finally, proceed to indexing by mixing 18 μL indexing reaction mix, 30 μL elution from the previous step, 1 μL 10 μM i5 Indexing Primer and 1 μL 10 μM i7 Indexing Primer. Run the PCR reaction as follows: (1) 30 s at 98 °C; (2) 10 s at 98 °C; (3) 20 s at 60 ° C; (4) 20 s at 72 °C; (5) repeat 10–15 times **steps 2–4**; (6) 10 min at 72 °C; (7) hold at 4 °C. Once indexing is completed, multiple samples can be pooled in a single tube before AMPure XP Bead-Based Reagent purification, which is performed in two steps: (1) at 0.5× to remove fragments larger than 400–500 bp, from this step keep the supernatant and discard the beads; and (2) at 1.4× to remove shorter fragments (<150 bp). Finally, resuspend the purified library into 100 μL NF water. At this point, proceed to Illumina 150 bp paired-end sequencing. The average depth of the sequencing can vary depending on the VAF of the somatic variants targeted for

validation. We prefer to have at least 5 alternate reads supporting the variant so, in order to validate somatic variants with VAF<1% we suggest an average amplicon sequencing depth >500×.

4. Amplicon sequencing data analysis: for variant calling in amplicon sequencing data, we suggest using Bcftools [33], which will output a VCF file with the genotyping likelihoods for each variant site in each sample.

3.2 Somatic Variant Validation in Single Cells and Cell Phylogeny Reconstruction

3.2.1 Fluorescence-Activated Nuclear Sorting (FANS)

Perform all the steps until sorting in an L2 confinement laboratory under a well-decontaminated laminar flow hood. Before starting this procedure, prepare tissue biopsies by cutting or slicing (~100 μm thickness) fresh-frozen brain specimens using an L2 cryostat as described in **Step 3.1.1**. For more efficient variant validation, use a tissue region immediately adjacent to the one used in **Step 3.1.1**. Always keep the tissue on dry-ice until lysis. Keep all the samples and buffers on ice for the entire procedure.

1. Before starting, pre-cool on ice plastic tubes adapted for an ultracentrifuge swinging bucket rotor (1 per tissue sample) and Dounce homogenizers (one per tissue sample). Pre-cool the ultracentrifuge, including the rotor (e.g., Thermo Scientific TH3) at 4 °C and set up the parameters as follows: duration 1 h 30 min, 13,000 rpm, max acceleration, and max deceleration.
2. Prepare fresh Lysis Buffer (6 mL/sample) with 660 mg UltraPure Sucrose, 30 μL 1 M $CaCl_2$, 36 μL 0.5 M $MgAc_2$, 1.2 μL 0.5 M EDTA pH 8, 60 μL 1 M Tris-HCl pH 8, and NF water until volume completion. Note that the volume should be adjusted after the sucrose has completely dissolved. Add to the Lysis Buffer 6 μL fresh DTT (final concentration 1 mM). Pre-cool the Lysis Buffer on ice before starting. Prepare fresh Sucrose Buffer (10 mL/sample) with 6.3 g UltraPure Sucrose, 36 μL 0.5 M $MgAc_2$, 60 μL 1 M Tris-HCl pH 8, NF water to adjust the volume. Similar to the Lysis Buffer, add 10 μL fresh DTT (final concentration 1 mM) only once sucrose has completely dissolved and the volume of the buffer has been adjusted. Pre-cool the Sucrose Buffer on ice before starting. Transfer 5 mL Lysis Buffer in each Dounce homogenizer, and 9 mL Sucrose Buffer in each ultracentrifuge plastic tube. Transfer the tissue into the Dounce homogenizer containing Lysis Buffer and homogenize until no big chunks of tissue are visible (10–20 times). Lay the lysed tissue homogenate on top of the Sucrose Buffer into the ultracentrifuge plastic tube taking care not to mix the two phases, then proceed to ultracentrifugation for ~1 h and 30 min. While the centrifuge is running, prepare Resuspension Solution (3 mM $MgCl_2$, PBS 1×) and Blocking

Solution (3 mM $MgCl_2$, 3% BSA, PBS 1×). The volume of each solution per sample depends on the number of desired staining conditions for the sorting (*see* **Note 7**). Calculate 120 μL/condition and 2 mL/condition of Resuspension and Blocking solutions, respectively. Once ready, pre-cool the solutions on ice. Once the ultracentrifugation is over, carefully remove the plastic tubes from their supports and place them on ice. Remove and discard the supernatant being careful not to touch the walls of the tube with the pipette tip and without touching the bottom center of the tube, where the invisible nuclei pellet is located. Add dropwise the desired amount of Resuspension Solution on top of the pellet and leave 10 min on ice without resuspending, then carefully resuspend the nuclei (~10 times) and transfer the suspension into a 1.5 mL tube by filtering through a 40 μm cell strainer.

3. Transfer 100 μL of filtered nuclei suspension from **Step 1** in Subheading 3.2.1 into a new 1.5 mL tube containing 900 μL Blocking Solution, and incubate 30 min at 4 °C on a tube rotary shaker (this step is optional, in case of limited time we can directly proceed to antibody staining). Add the antibody (ies) at the desired concentration and incubate 30 min at 4 °C on a tube rotary shaker. At this point, we can add an additional optional step which consists in a short centrifugation of 5 min at 500 rcf and 4 °C, removal of the supernatant, and re-suspension of the nuclei pellet in 1 mL Blocking Solution. This step helps cleaning the nuclei suspension by removing the residual antibody however, it is not strictly required and has the downside of causing some nuclei loss so, in case of limited starting material, we advise not to perform this additional centrifugation. Add DAPI to the stained nuclei and proceed to FANS.
4. Sort single nuclei into a 96-well plate (one nucleus per well), and directly into previously prepared Lysis Buffer for WGA (see next **Step 3.2.2**), then proceed to next step. The total number of sorted nuclei depends on the desired depth of the cell phylogeny. The higher the cell number, the deeper the phylogeny.

3.2.2 Single-Cell WGA

Here, we describe the protocol to follow in order to perform WGA through multiple-displacement amplification (MDA). We favor this protocol since it is more cost-effective compared to other methodologies, and it provides a quality of DNA amplification that is sufficient for our purpose of validating somatic variants. However, alternative WGA methods are available that could replace MDA (*see* **Note 8**).

All steps described below, until MDA, should be performed under a decontaminated pre-PCR hood to avoid sample contamination with external sources of DNA. Before proceeding to nuclei sorting, prepare 96-well plates containing 2.8 μL/well of ice-cold previously prepared (*see* **Step 2.2.2**) alkaline nuclear Lysis Buffer. After sorting, spin down the plates (1–2 min at 300 rcf) in order to make sure that Lysis Buffer and nuclei reach the bottom of the well. To note, plate spin down should be performed every time we need to make sure all reaction volumes are located at the bottom of the tube. Perform alkaline lysis by placing the 96-well plate on ice for 30 min, then add 1.4 μL/well previously-prepared (*see* Step **Step 2.2.2**) Neutralization Buffer. Prepare MDA master mix (15.8 μL/sample) with 2.7 μL NF water, 0.11 μL DTT, 12.2 μL REPLI-g sc Reaction Buffer, and 0.84 μL REPLI-g sc DNA Polymerase (all ingredients are included in the REPLI-g single-cell kit from Qiagen). Add 15.8 μL/well MDA master mix, then spin down and proceed to MDA by incubating in a PCR thermal cycler for 2 h at 30 °C, followed by a 3 min step at 65 °C to inactivate the polymerase, then hold at 4 °C. At this point, plates can be stored at −20 °C until next steps are performed.

A WGA quantification and QC step is performed before proceeding to the next step. Although QC is not strictly required, it helps selecting good quality single-cell DNAs, and it is advised in order to avoid wasting reagents on low-quality DNA. Under a post-PCR hood, prepare 1:50 dilutions for each MDA reaction by mixing 1 μL MDA product with 49 μL NF water in a new 96-well plate. At this point, quantify the DNA using a PicoGreen assay for dsDNA (expected yield for 2 hours MDA should be ~10 μg). For QC, we perform a multiplex-PCR reaction targeting 4 randomly selected genomic loci of choice. As negative and positive controls, include at least one no-DNA well, and one well where the reaction is performed on a high-quality bulk human DNA, respectively. Each PCR reaction is prepared as follows: 5 μL 5× Phusion U HF Buffer, 0.25 μL 2 U/μL Phusion Hot Start II DNA Polymerase, 16.25 μL NF water, 0.5 μL 10 mM dNTP mix, 1 μL primer mix (custom design), and 2 μL template DNA. For easier PCR product verification, we recommend using a capillary electrophoresis device such as QIAxcel Connect System (Qiagen), which allows automate reading of one 96-well plate per run. For downstream experiments, we advise using single-cell DNA where at least three out of four loci were successfully amplified.

3.2.3 Somatic Variant Validation Through Amplicon Sequencing

Follow the same protocol described in **step 3.1.5**. Here, we suggest targeting an average sequencing depth per amplicon per cell of 30-50×, which is sufficient to detect somatic variants that are heterozygous in single cells and thus should be found in half (50%) of the reads.

3.2.4 Data Analysis and Lineage Reconstruction

As for **step 3** in Subheading 3.1.5, for variant calling we suggest using Bcftools [33]. To account for allelic imbalances introduced by MDA, we recommend using genotyping posterior probabilities to derive the genotyping of each variant in each single cell [12]. Our objective is then to use this information to reconstruct a rooted lineage tree consisting of observed "leaves," represented by each single cell, and unobserved "ancestral" internal nodes defined by variant events (Fig. 2). To account for allelic dropouts, each cell is assigned to each clade using maximum likelihood estimates [12].

4 Notes

1. Primer design can fail for some difficult to amplify loci that will be excluded from validation.

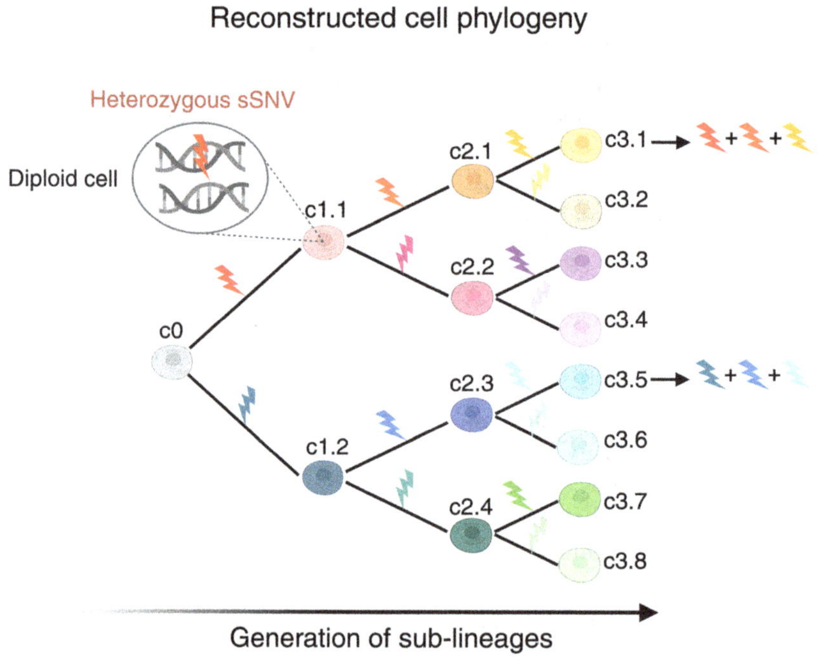

Fig. 2 Example of a reconstructed cell phylogeny. Each branch (sub-lineage) is defined by a somatic variant event that is shared by a fraction of the cells and excluded from the remaining fraction. The leaves of the tree (c3.1–c3.8) correspond to observed (sequenced) cells, while internal nodes (c0–c2.4) are unobserved ancestral nodes that are inferred from the variants found in observed leaves. For example, the observed cell c3.1 carries the yellow somatic variant, which is not found in any other observed cell. Observed cell c3.1 carries in addition an orange variant that is also carried by the observed cell c3.2, and a red variant that is also carried by observed cells c3.2, c3.3, and c3.4. Thus, the orange variant defines the c2.1 internal node connecting c3.1 and c3.2, while the red variant defines the c1.1 internal node connecting c3.1, c3.2, c3.3, and c3.4. (Figure created with BioRender.com)

2. The amount of dissected material, as well as the number and type(s) of tissue regions dissected per individual, largely depend on the study design and the downstream protocol used for DNA extraction, library preparation, and sequencing. For the purpose of this chapter, we describe a standard procedure that starts from 10–30 mg of tissue, which corresponds to a yield of ~1–4 μg of DNA. Other protocols allow to perform bulk low-input sequencing starting from as low as 100 cells (0.5–1 ng DNA), and generally involve tissue fixation and laser-capture microdissection (LCM) [34]. The advantage of such protocols is that they facilitate the detection of low frequency variants without requiring to increase the sequencing depth. The limitation is however, that the detection of somatic variants with such low-input protocols is restricted to a very small tissue area. Thus, a decision must be made at the time of study design, and a trade-off must be found between the desired somatic variant frequency detection limit and the extension of the tissue area where somatic variants are called.
3. DNA amplification through PCR introduces artifact variants as a consequence of DNA polymerase errors, even when using high fidelity polymerases. These errors can then be confounded with true somatic variants at the time of analysis. Thus, it is important when possible, to use PCR-free protocols for library prep. If the starting amount of DNA is too low to avoid PCR, we recommend limiting PCR cycles at the minimum required in order to reach DNA yields sufficient for sequencing.
4. Somatic variant calling for the purpose of lineage tracing studies presents some challenges compared to more diffuse approaches of somatic variant calling in cancer studies. When calling somatic variants in cancer, experimenters favor a tumor-control mode, which consists in running the tumor sample against a matched control that represents non-cancerous tissue from the same patient (e.g., blood or healthy tissue sampled from the area adjacent to the tumor). This allows to exclude variants that are found in both tumor and control samples since these are either germline or not relevant for cancer. However, some of these "shared" variants can represent true somatic events that are relevant for lineage tracing. For this reason, matched control samples cannot be used in lineage tracing studies, and the tumor-control strategy of somatic variant calling cannot be applied. This has led to generate somatic variant callers that are specifically designed to identify somatic variants in single samples without the need for a matched control, and without sacrifying sensitivity and accuracy [30, 31, 35]. These analytical tools start from raw candidate calls obtained with more classical tools such as Mutect2, which present

high sensitivity but also high false-positive rates, especially when used in single-mode, and identify true somatic variants by applying machine learning trained on a set of validated calls.

5. The protocol presented here is ideal to validate a high number of candidate somatic variants across multiple samples, though it can be applied also to a single DNA. Alternative validation methods could also be used such as IonTorrent [36] sequencing or droplet digital (dd)PCR. Due to the elevated costs, we advise using ddPCR only when targeting for validation a small number (<10) of candidate somatic variants.

6. This protocol optimizes the DNA input and number of PCR cycles, and keeps both at a minimum to avoid introducing variant artifacts (*see* **Note 4**).

7. Sorting conditions vary depending on the experimental strategy and objective. To reconstruct a phylogeny that includes all cell types, sort exclusively for DAPI staining. In this case, prepare an unstained sample and a sample containing DAPI, and skip the blocking and antibody incubation steps. To reconstruct a cell phylogeny including only neuronal cells, perform DAPI + NeuN staining. In this case, prepare the following conditions: (1) unstained, (2) DAPI, (3) NeuN, (4) DAPI + NeuN. We favor the use of conjugated antibodies for optimal results however, if this is not possible, we can couple with a secondary antibody. In this case, conditions to prepare are the following: (1) unstained, (2) DAPI, (3) secondary antibody, (4) DAPI + NeuN + secondary antibody. Ideally, if a secondary antibody is used, incubate the primary first, then pellet the nuclei by centrifuging for 5 min at 500 rcf and 4 °C, remove the supernatant, and re-suspend the nuclei pellet in 1 mL Blocking Solution before adding the secondary antibody for another 30 min incubation. However, in case of time constraint, primary and secondary antibodies can be added at the same time for a single incubation step.

8. Several methods for single-cell WGA exist, among which Multiple Annealing and Looping Based Amplification Cycles (MALBAC), MDA and Primary Template-directed Amplification (PTA) [37–40]. The most recent method, PTA, performs a quasi-linear amplification, and has been shown to perform better than MDA in terms of genome amplification coverage and uniformity, allelic skewing and polymerase error propagation [40]. The reason why we suggest using MDA in this protocol is that at this time it still consists in the less expensive option. Furthermore, although MDA does present limitations such as allelic skewing and lower coverage and uniformity in genome amplification, for our purpose, which is of validating somatic variants rather than performing *de novo* calling in scWGS, MDA is usually good enough.

Acknowledgments

We thank the team "Genetics and Pathophysiology of Epilepsy" at Paris Brain Institute (ICM). Sara Bizzotto was supported by the Horizon2020 Research and Innovation Program Marie Skłodowska-Curie Actions (MSCA) Individual Fellowship (grant agreement no. 101026484—CODICES), and is now supported by a European Research Council (ERC) starting grant (grant agreement project no. 101115984—LINMOS).

References

1. Martincorena I (2019) Somatic mutation and clonal expansions in human tissues. Genome Med 11(1):35
2. Luquette LJ, Miller MB, Zhou Z et al (2022) Single-cell genome sequencing of human neurons identifies somatic point mutation and indel enrichment in regulatory elements. Nat Genet 54(10):1564–1571
3. Ganz J, Luquette LJ, Bizzotto S et al (2024) Contrasting somatic mutation patterns in aging human neurons and oligodendrocytes. Cell 187(8):1955–1970
4. Lodato MA, Rodin RE, Bohrson CL et al (2018) Aging and neurodegeneration are associated with increased mutations in single human neurons. Science 359(6375):555–559
5. Yoshida K, Gowers KHC, Lee-Six H et al (2020) Tobacco smoking and somatic mutations in human bronchial epithelium. Nature 578(7794):266–272
6. Martincorena I, Roshan A, Gerstung M et al (2015) Tumor evolution. High burden and pervasive positive selection of somatic mutations in normal human skin. Science 348(6237):880–886
7. Milholland B, Dong X, Zhang L et al (2017) Differences between germline and somatic mutation rates in humans and mice. Nat Commun 8(1):15183
8. Bae T, Tomasini L, Mariani J et al (2018) Different mutational rates and mechanisms in human cells at pregastrulation and neurogenesis. Science 359(6375):550–555
9. Bizzotto S, Walsh CA (2022) Genetic mosaicism in the human brain: from lineage tracing to neuropsychiatric disorders. Nat Rev Neurosci 23:275–286
10. Werner B, Case J, Williams MJ et al (2020) Measuring single cell divisions in human tissues from multi-region sequencing data. Nat Commun 11:1035
11. Bizzotto S (2023) The human brain through the lens of somatic mosaicism. Front Neurosci 17
12. Bizzotto S, Dou Y, Ganz J et al (2021) Landmarks of human embryonic development inscribed in somatic mutations. Science 371(6535):1249–1253
13. Lodato MA, Woodworth MB, Lee S et al (2015) Somatic mutation in single human neurons tracks developmental and transcriptional history. Science 350(6256):94–98
14. Fasching L, Jang Y, Tomasi S et al (2021) Early developmental asymmetries in cell lineage trees in living individuals. Science 371(6535):1245–1248
15. Breuss MW, Yang X, Schlachetzki JCM et al (2022) Somatic mosaicism reveals clonal distributions of neocortical development. Nature 604(7907):689–696
16. Coorens THH, Moore L, Robinson PS et al (2021) Extensive phylogenies of human development inferred from somatic mutations. Nature 597:387–392
17. Spencer Chapman M, Ranzoni AM, Myers B et al (2021) Lineage tracing of human development through somatic mutations. Nature 595:85–90
18. Park S, Mali NM, Kim R et al (2021) Clonal dynamics in early human embryogenesis inferred from somatic mutation. Nature 597(7876):393–397
19. Huang AY, Li P, Rodin RE et al (2020) Parallel RNA and DNA analysis after deep sequencing (PRDD-seq) reveals cell type-specific lineage patterns in human brain. Proc Natl Acad Sci U S A 117(25):13886–13895
20. Alemany A, Florescu M, Baron CS et al (2018) Whole-organism clone tracing using single-cell sequencing. Nature 556(7699):108–112

21. Chan MM, Smith ZD, Grosswendt S et al (2019) Molecular recording of mammalian embryogenesis. Nature 570(7759):77–82
22. Nam AS, Kim KT, Chaligne R et al (2019) Somatic mutations and cell identity linked by genotyping of transcriptomes. Nature 571(7765):355–360
23. Macaulay IC, Haerty W, Kumar P et al (2015) G&T-seq: parallel sequencing of single-cell genomes and transcriptomes. Nat Methods 12(6):519–522
24. Wang Y, Bae T, Thorpe J et al (2021) Comprehensive identification of somatic nucleotide variants in human brain tissue. Genome Biol 22(1):92
25. Li H, Durbin R (2009) Fast and accurate short read alignment with Burrows-Wheeler transform. Bioinformatics 25(14):1754–1760
26. McKenna A, Hanna M, Banks E et al (2010) The genome analysis toolkit: a MapReduce framework for analyzing next-generation DNA sequencing data. Genome Res 20(9): 1297–1303
27. Benjamin D, Sato T, Cibulskis K et al (2019) Calling somatic SNVs and indels with Mutect2. bioRxiv:861054
28. Chen S, Francioli LC, Goodrich JK et al (2024) A genomic mutational constraint map using variation in 76,156 human genomes. Nature 625(7993):92–100
29. Auton A, Abecasis GR, Altshuler DM et al (2015) A global reference for human genetic variation. Nature 526(7571):68–74
30. Dou Y, Kwon M, Rodin RE et al (2020) Accurate detection of mosaic variants in sequencing data without matched controls. Nat Biotechnol 38(3):314–319
31. Yang X, Xu X, Breuss MW et al (2023) Control-independent mosaic single nucleotide variant detection with DeepMosaic. Nat Biotechnol 41:870–877
32. Meyer M, Kircher M (2010) Illumina sequencing library preparation for highly multiplexed target capture and sequencing. Cold Spring Harb Protoc
33. Danecek P, Bonfield JK, Liddle J et al (2021) Twelve years of SAMtools and BCFtools. GigaScience 10(2)
34. Ellis P, Moore L, Sanders MA et al (2021) Reliable detection of somatic mutations in solid tissues by laser-capture microdissection and low-input DNA sequencing. Nat Protoc 16(2):841–871
35. Huang AY, Zhang Z, Ye AY et al (2017) MosaicHunter: accurate detection of postzygotic single-nucleotide mosaicism through next-generation sequencing of unpaired, trio, and paired samples. Nucleic Acids Res 45(10):e76
36. Doan RN, Miller MB, Kim SN et al (2021) MIPP-Seq: ultra-sensitive rapid detection and validation of low-frequency mosaic mutations. BMC Med Genomics 14(1):47
37. Wang X, Liu Y, Liu H et al (2022) Recent advances and application of whole genome amplification in molecular diagnosis and medicine. MedComm 3(1):e116
38. Dean FB, Hosono S, Fang L et al (2002) Comprehensive human genome amplification using multiple displacement amplification. Proc Natl Acad Sci U S A 99(8):5261–5266
39. Zong C, Lu S, Chapman AR et al (2012) Genome-wide detection of single-nucleotide and copy-number variations of a single human cell. Science 338(6114):1622–1626
40. Gonzalez-Pena V, Natarajan S, Xia Y et al (2021) Accurate genomic variant detection in single cells with primary template-directed amplification. Proc Natl Acad Sci U S A 118(24):e2024176118

Chapter 11

Paired Single-Cell Transcriptome and DNA Barcode Detection in Zebrafish Using ScarTrace

Chloé S. Baron and Anna Alemany

Abstract

ScarTrace is a CRISPR/Cas9-based genetic lineage tracing method that allows for uniquely barcoding the DNA of single cells at a target GFP sequence during developing zebrafish embryos. Single cells from barcoded adult zebrafish can be isolated from various tissues (e.g., marrow, brain, eyes, fins), and their transcriptome and barcode sequences are captured by single-cell cDNA amplification and genomic DNA nested PCR, respectively. Computationally, cell type and barcode identification permit clone tracing and lineage tree reconstruction of tissues to unravel fate decisions during embryogenesis.

Key words CRISPR/Cas9, Genetic lineage tracing, Zebrafish development, Single-cell transcriptomics, Single-cell barcode capture, Next-generation sequencing, Clone tracing, Lineage tree reconstruction

1 Introduction

Genetic lineage tracing aims to introduce permanent barcodes in the DNA of single cells. Those barcodes, retrievable by sequencing, are inherited by the cellular offspring and allow to retrospectively reconstruct the lineage relationships between all the cell types present in an organism with single-cell resolution. Genetic lineage tracing can be achieved using retroviral barcode libraries or more sophisticated genome editing techniques. In this chapter, we present ScarTrace, a genetic lineage tracing technique that benefits from the ability of CRISPR/Cas9 to introduce pseudo-random insertions and deletions (indels) in the desired genomic loci by means of a single guide RNA (sgRNA). These indels, also known as scars, are permanent and inherited by the progeny of every cell, making them ideal lineage barcodes. In the original ScarTrace, a sgRNA targeting GFP was used in a zebrafish transgenic line containing eight copies of the GFP transgene in its genome. This permits the dynamic introduction/accumulation of scars in the

Jorge García-Marqués and Tzumin Lee (eds.), *Lineage Tracing: Methods and Protocols*, Methods in Molecular Biology, vol. 2886, https://doi.org/10.1007/978-1-0716-4310-5_11,

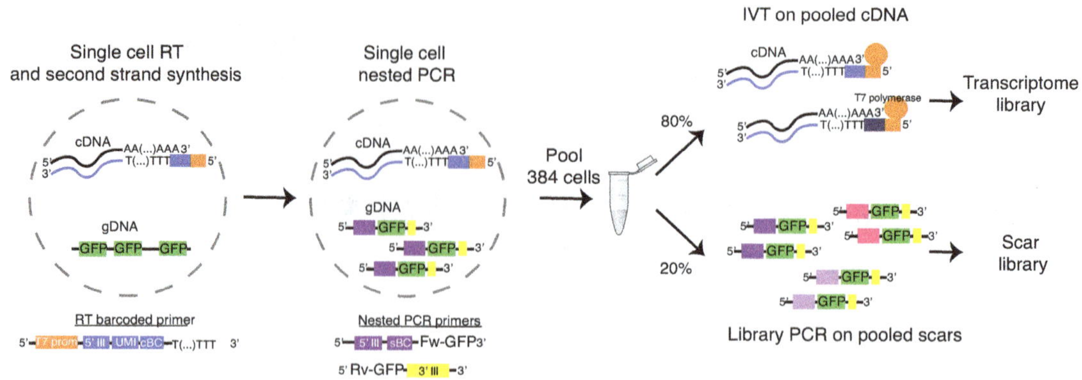

Fig. 1 Overview of the ScarTrace protocol from single-cell lysis to transcriptome and scar libr ary preparation

different copies of the GFP transgene during the first hours of embryonic development without perturbing the genomic landscape of the growing zygote.

Diverse tissues from adult scared zebrafish are collected for single-cell dissociation, fluorescent activated cell sorting followed by single-cell transcriptome and ScarTrace library construction (Fig. 1). Sequencing of both libraries allows for cell type and scar identification with single-cell resolution for clone tracing and lineage tree reconstruction. Crucially, the ScarTrace protocol allows for scar capture from the DNA using a nested PCR performed at the level of single cells. This approach allows for a higher scar capture efficiency as transcription of the GFP transgene can be affected by the presence of indels. Here, we present a detailed protocol for ScarTrace, from zebrafish embryo injection to bioinformatic analysis of clonal composition and lineage relationship between cells of zebrafish adult organs.

2 Materials

2.1 Zebrafish Husbandry

1. Heterozygous zebrafish from the transgenic line Tg(h2afva: GFP)^{kca66} (*see* **Note 1**).

2.2 Laboratory Equipment

1. Thermomixer.
2. Flow cytometer.
3. Centrifuge.
4. Mosquito LV liquid handler (TTP lab tech).
5. Nanodrop II liquid handling platform (GC Biotech).
6. 384-well plate PCR machine.
7. 0.5-mL tube PCR machine.

8. Qubit fluorometer.
9. Agilent bioanalyzer or tape station.

2.3 Zebrafish Embryo Injections

1. Cas9 RNA generated from linearized pCS2-mCas9n vector (Addgene plasmid #47929) using the mMESSAGE mMACHINE SP6 Transcription Kit.
2. Cas9 protein.
3. GFP sgRNA (spacer sequence: GGTGTTCTGCTGGTAGTGGT, Fig. 2) template synthesized with T4 DNA polymerase (New England Biolabs) by partially annealing two single-stranded DNA oligonucleotides containing the T7 promotor and the GFP-binding sequence, and the tracrRNA sequence. The template was in vitro transcribed using the MEGAscript T7 Transcription Kit (Thermo Scientific).

2.4 Single-Cell Dissociation of Organs

1. DPBS.
2. FBS.
3. Heparin.
4. Viability dye (such as DRAQ7 or DAPI).
5. DMEM/F12.
6. Penicillin/Streptomycin.
7. Papain.
8. L-Cysteine.

5'- ATG GTG AGC AAG GGC GAG GAG CTG TTC ACC GGG GTG GTG CCC
ATC CTG GTC GAG CTG GAC GGC GAC GTA AAC GGC CAC AAG TTC AGC
GTG TCC GGC GAG GGC GAG GGC GAT GCC ACC TAC GGC AAG CTG ACC
CTG AAG TTC ATC TGC ACC ACC GGC AAG CTG CCC GTG CCC TGG CCC
ACC CTC GTG ACC ACC CTG ACC TAC GGC GTG CAG TGC TTC AGC CGC
TAC CCC GAC CAC ATG AAG CAG CAC GAC TT→C TTC AAG TCC GCC ATG
CCC GAA GGC TAC GTC CAG GAG CGC ACC ATC TTC TTC AAG GAC GAC
GGC AAC TAC AAG→ ACC CGC GCC GAG GTG AAG TTC GAG GGC GAC ACC
CTG GTG AAC CGC ATC GAG CTG AAG GGC ATC GAC TTC AAG GAG GAC
GGC AAC ATC CTG GGG CAC AAG CTG GAG TAC AAC TAC AAC AGC CAC
AAC GTC TAT ATC ATG GCC GAC AAG CAG AAG AAC GGC ATC AAG GTG
AAC TTC AAG ATC CGC CAC AAC ATC GAG GAC GGC AGC GTG CAG CTC
G CC G AC CAC TAC CAG CAG AAC ACC CCC ATC GGC GAC ←GGC CCC
GTG CTG CTG CCC GAC AA←C CAC TAC CTG AGC ACC CAG TCC GCC CTG
AGC AAA GAC CCC AAC GAG AAG CGC GAT CAC ATG GTC CTG CTG GAG
TTC GTG ACC GCC GCC GGG ATC ACT CTC GGC ATG GAC GAG CTG TAC
AAG TAA -3'

Fig. 2 GFP sequence with the PAM and the sgRNA indicated in red and cyan, respectively, and sequence of the forward and reverse primers used for the nested PCR indicated by underlying their sequence

9. DNAseI.
10. Glucose.
11. HEPES 1 M.
12. HANKS' solution.
13. Protease XIV.
14. Collagenase A.
15. Trypsin.

2.5 Single-Cell Sorting and ScarTrace

1. 384 hard well plate.
2. CEL-Seq2 primers: 5′- GCCGGTAATACGACTCACTA TAGGCCTTGGCACCCGAGAATTCCANNNNNN [8 bp cell barcode]TTTTTTTTTTTTTTTTTTTTTTTTVN-3′.
3. Aluminum foil seals.
4. RNAse-free PCR strips.
5. Vapor lock.
6. Nuclease-free water.
7. DNA polymerase I.
8. SuperScript II.
9. RNAseOUT.
10. Second-strand buffer.
11. 0.1 M DTT.
12. *Escherichia coli* DNA ligase.
13. RNase H.
14. dNTP 10 mM.
15. Spike-In.
16. Proteinase K (20 mg/mL).
17. 2× High-fidelity PCR Master Mix.
18. eGFP_232 forward primer (nested PCR I, Fig. 2): 5′-CACAT GAAGCAGCACGACTT-3′.
19. eGFP_597 reverse primer (nested PCR I, Fig. 2): 5′-ACTGG GTGCTCAGGTAGTGG-3′.
20. eGFP uniquely barcoded forward primer (nested PCR II, Fig. 2), containing the 5′ TruSeq small RNA kit adapter: 5′-G CCGGTAATACGACTCACTATAGGGAGTTCTACAGTCC GACGATCNNN -[8 bp cell barcode]-AGGACGTCGGCAACTACAAG-3′.
21. eGFP-scar reverse primer (nested PCR II, Fig. 2), containing the 3′ TruSeq small RNA kit adapter: 5′-GTCGGGCAGCAG CACGGGGCC TTCCTTGGCACCCGAGAATTCCA-3′.
22. AMPure XP beads.

23. Bead binding buffer, 20% PEG8000, 2.5 M NaCl.
24. EtOH.
25. In Vitro Transcription (IVT) Kit.
26. EXO-SAP.
27. Fragmentation buffer.
28. Fragmentation stop buffer.
29. RNAClean XP beads.
30. First-strand buffer.
31. Random hexamer.
32. Blood buffer, 19 mL DPBS (no Ca^{2+}, no Mg^{2+}), 1 mL FBS, 20 mL Heparin (1000×).
33. Dissection medium, 9.8 mL DMEM/F12, 200 μL penicillin/streptomycin (100×).
34. L-Cysteine solution: 10 mL nuclease-free water, 120 mg L-cysteine; this solution can be prepared in advance and stored at 4 °C for up to 2 weeks or aliquoted and stored at −20 °C for up to a year.
35. DNAseI solution: 1 mL nuclease-free water, 10 mg DNAseI; this solution can be prepared in advance and stored at 4 °C for up to 2 months.
36. Papain solution: 5 mL DMEM/F12, 100 μL of papain (140 units total), 200 μL L-cystine solution; this solution needs to be prepared fresh and sterilized with a 0.22-μM filter before use.
37. Brainwashing solution: 93.85 mL DPBS, 650 μL 45% glucose, 500 μL HEPES 1 M, 5 mL FBS; this solution needs to be sterilized with a 0.22-μM filter before use and can be stored at 4 °C for up to 2 months.
38. Caudal fin dissociation solution: 480 μL HANKS' solution, 10 μL Protease XIV (15 mg/mL), 10 μL Collagenase A (100 mg/mL).
39. Caudal fin wash solution: 9.6 μL HANKS' solution, 400 μL FBS.

3 Methods

3.1 Zebrafish Embryo Injections to Induce Scars

1. Incross heterozygous zebrafish from the transgenic line Tg (h2afva:GFP)^{kca66} the evening before injection.
2. Inject single-cell stage embryos with 1 nL of Cas9 protein (1590 ng/μL) or Cas9 mRNA (300 ng/μL) and GFP sgRNA (25 ng/μL). Cas9 protein injections are performed in the cell, and Cas9 RNA injections are performed in the yolk (*see* **Note 2**).

3. Viability and healthy development of embryos are assessed at 24 h post fertilization (hpf) and 48 hpf. Healthy embryos are raised to adulthood.

3.2 Single-Cell Suspension Preparation

3.2.1 Kidney Marrow

1. Dissect the kidney as described in [1] and collect 300 μL of blood buffer.
2. Pipet up and down with a p1000 20 times to release hematopoietic cells.
3. Filter cell suspension over a 40-μL cell strainer.
4. Add 1 mL of blood buffer.
5. Spin down the sample at 400 g for 5 min at 4 °C.
6. Remove the supernatant and resuspend the pellet in 400 μL of blood buffer.
7. Add 1 μL of viability dye (such as DRAQ7 or DAPI).

3.2.2 Brain and Eyes

1. Dissect and transfer the brain tissues or eyes into a 1.5-mL tube containing 800 μL of brain dissection medium. Remove the dissection medium by pipetting, without touching the brain tissue.
2. Add 500 μL of papain solution and digest the brain tissue at 37 °C for 10 min.
3. Transfer the brain tissue with 500 μL of the papain solution into a 15-mL conical tube using a 1000-μL pipet tip cut at ~0.25 inches from the bottom. Gently dissociate the brain tissue by slowly pipetting up and down 10 times with the same tip.
4. Incubate the brain tissue again at 37 °C for 10 min followed by pipetting up and down 10–13 times using an uncut 1000-μL regular tip.
5. Stop the enzymatic digestion by adding 2 mL of brainwashing solution and centrifuge the cell suspension at 800 g for 5 min at room temperature.
6. Carefully remove the supernatant and resuspend the pellet in the remaining solution by tapping the tube vigorously with a finger and then by pipetting up and down with a 1000-μL regular tip. Next, add 2 mL of washing solution.
7. Centrifuge the cell suspension again at 800 g for 5 min at room temperature. Remove the supernatant and resuspend the pellet in 400-μL dissection medium.
8. Add 1 μL of viability dye (such as DRAQ7 or DAPI).

3.2.3 Caudal Fin

1. Cut the caudal fin with a razor blade and collect in 250 μL of caudal fin dissociation solution.

2. Shake at 400 rpm at 30 °C for 50 min in a thermomixer and pipet up and down every 10 min with a p200 pipet.
3. Spin down cells at 400 g for 4 min at 4 °C.
4. Remove supernatant and resuspend in 2.5 μL of PBS.
5. Add 2.5 μL of 5% Trypsin and incubate at room temperature for 5 min.
6. Add 500 μL of PBS.
7. Spin down cells at 400 g for 4 min at 4 °C.
8. Remove supernatant and resuspend in 250 μL of caudal fin wash solution.
9. Spin down cells at 400 g for 4 min at 4 °C.
10. Remove the supernatant and resuspend in 400 μL of wash solution.
11. Add 1 μL of viability dye (such as DRAQ7 or DAPI).

3.3 Single-Cell Sorting

1. Prepare 384-well plates for single-cell sorting by adding 5 μL of vapor lock per well using a multichannel pipette.
2. Dispense in each well 50 nL of 7.5 ng/μL uniquely barcoded CEL-Seq2 primers using a Mosquito LV liquid handler.
3. Spin down plates at 1000 g for 1 min at 4 °C (*see* **Note 3**).
4. Prepare the following lysis mix:

Lysis mix	For 1 well (μL)	For 2 plates (μL)
Spike (1:50,000)	0.02	44.4
dNTP 10 mM	0.01	22.2
SUPERase in	0.01	22.2
H_2O	0.01	22.2

5. Dispense 50 nL of the lysis mix into each well using a Nanodrop II liquid handling platform.
6. Spin down plates at 1000 g for 1 min at 4 °C.
7. Maintain plates on ice until single-cell sorting is performed.
8. Sort one cell per well using a flow cytometer. Keep six wells of choice empty (zero cell sorted) to serve as technical negative controls (*see* **Note 4**).
9. Spin down plates at 1000 g for 1 min at 4 °C.
10. Snap freeze plates on dry ice.
11. Store at −80 °C for up to 6 months for transcriptome and ScarTrace library preparation.

3.4 Amplified RNA Generation from Single Cells

1. Spin down plates at 1000 g for 1 min at 4 °C.
2. Lyse cells by incubating plates at 65 °C for 5 min in a 384-well PCR machine.
3. Spin down plates at 1000 g for 1 min at 4 °C.
4. Cool plates on ice.
5. Prepare the following reverse transcription mix:

Reverse transcription mix	For 1 well (μL)	For 2 plates (μL)
5× First-strand buffer	0.035	60.48
0.1 M DTT	0.0175	30.24
H_2O	0.005	8.64
RNaseOUT	0.00875	15.12
Superscript II	0.00875	15.12

6. Dispense 75 nL of the reverse transcription mix into each well using a Nanodrop II liquid handling platform.
7. Spin down plates at 1000 g for 1 min at 4 °C.
8. Incubate at 42 °C for 1 h, 4 °C for 5 min, 70 °C for 10 min, 4 ° C hold.
9. Spin down plates at 1000 g for 1 min at 4 °C.
10. Cool plates on ice.
11. Prepare the following second-strand synthesis mix:

Second-strand synthesis mix	For 1 well (μL)	For 2 plates (μL)
Second-strand buffer 5×	0.21875	184.8
dNTP 10 mM	0.021875	18.5
E. coli ligase	0.007875	6.65
E. coli DNA polymerase I	0.030625	25.9
RNAse H	0.007875	6.65
H_2O	0.67375	569.3

12. Dispense 960 nL of the second-strand synthesis mix into each well using a Nanodrop II liquid handling platform.
13. Spin down plates at 1000 g for 1 min at 4 °C.
14. Incubate at 16 °C for 2 h, 4 °C hold.
15. Spin down plates at 1000 g for 1 min at 4 °C.
16. Prepare the following Proteinase K mix:

Proteinase K mix	For 1 well (μL)	For 2 plates (μL)
Proteinase K	0.04	69.12
H_2O	0.025	43.2

17. Dispense 65 nL of the Proteinase K mix into each well using a Nanodrop II liquid handling platform.
18. Spin down plates at 1000 g for 1 min at 4 °C.
19. Incubate at 80 °C for 10 min, 4 °C hold.
20. Spin down plates at 1000 g for 1 min at 4 °C.
21. Prepare GFP Nested PCR I mix:

GFP Nested PCR I mix	For 1 well (μL)	For 2 plates (μL)
eGFP_232 forward primer	0.025	20.6
eGFP_597 reverse primer	0.025	20.6
2× PCR Mastermix	1.25	1030

22. Dispense 1300 nL of the GFP Nested PCR I mix into each well using a Nanodrop II liquid handling platform.
23. Spin down plates at 1000 g for 1 min at 4 °C.
24. Perform the following PCR cycle (*see* **Note 5**):
    - (a) 98 °C, 1 min
    - (b) 10–14 cycles of:
        - (i) 98 °C, 10 s
        - (ii) 58 °C, 30 s
        - (iii) 72 °C, 30 s
    - (c) 72 °C, 10 min
    - (d) 4 °C, hold
25. Spin down plates at 1000 g for 1 min at 4 °C.
26. Dispense in each well 75 nL of 20 μM uniquely barcoded ScarTrace primer using a Mosquito LV liquid handler.
27. Prepare GFP Nested PCR II mix:

GFP Nested PCR II mix	For 1 well (μL)	For 2 plates (μL)
eGFP-scar reverse primer	0.075	70.39
2× PCR Mastermix	0.25	234.63
H20	0.01	93.85

28. Dispense 425 nL of the GFP Nested PCR II mix into each well using a Nanodrop II liquid handling platform.
29. Spin down plates at 1000 g for 1 min at 4 °C.
30. Perform the following PCR cycle (*see* **Note 5**):
    (a) 98 °C, 1 min
    (b) 10–14 cycles of:
        (i) 98 °C, 10 s
        (ii) 64 °C, 30 s
        (iii) 72 °C, 30 s
    (c) 72 °C, 10 min
    (d) 4 °C, hold
31. Spin down plates at 1000 g for 1 min at 4 °C.
32. Warm AMPure XP beads to room temperature.
33. Pool the content of all wells from a plate by spinning the plate upside down in a collection funnel at 200 g for 1 min.
34. Collect the entire aqueous phase and transfer to a 1.5 mL collection tube.
35. Measure the volume recovered (expected volume: 1152 μL).
36. Vortex AMPure XP beads for 30 s.
37. Add 1× volume of AMPure beads diluted 1 in 5 in bead binding buffer.
38. Incubate for 15 min at room temperature.
39. Place on magnet stand for 5 min or until liquid clears.
40. Remove and discard supernatant without disrupting beads.
41. Wash the bead pellet carefully with 1 mL of freshly prepared 80% EtOH.
42. Incubate for 1 min.
43. Remove as much 80% EtOH as possible.
44. Wash the bead pellet carefully with 1 mL of freshly prepared 80% EtOH.
45. Remove as much 80% EtOH as possible.
46. Dry beads for 10 min at room temperature.

47. Add 10 μL of nuclease-free water and resuspend beads.
48. Place on magnet stand for 1 min or until liquid clears.
49. Collect 8 μL of supernatant and transfer to a clean 0.5 mL collection tube.
50. Split the sample for transcriptome library preparation (6.4 μL, 80% of the material, *see* Subheading 3.5) and ScarTrace library preparation (1.6 μL, 20% of the material, *see* Subheading 3.6).

3.5 Transcriptome Library Preparation

1. Prepare the following IVT mix:

IVT mix	For 1 sample (μL)
A (75 mM)	1.6
G (75 mM)	1.6
C (75 mM)	1.6
U (75 mM)	1.6
T7 10× reaction buffer	1.6
T7 enzyme mix	1.6

2. Add 9.6 μL of IVT mix to each sample (6.4 μL, 80% of post-bead cleanup material).
3. Mix well by pipetting up and down and spin down.
4. Incubate at 37 °C for 14 h, 4 °C hold.
5. Amplified RNA is stable at 4 °C for a few hours and at −20 °C for up to 7 days.
6. Add 6 μL of EXO-SAP and 2 μL of TURBO DNAseI per sample.
7. Mix well by pipetting up and down and spin down.
8. Incubate at 37 °C for 15 min.
9. Spin down and cool on ice.
10. Add 5.5 μL of fragmentation buffer per sample.
11. Incubate at 94 °C for 3 min.
12. Immediately move to ice and add 2.75 μL of fragmentation stop buffer per sample.
13. Warm RNAClean XP beads to room temperature.
14. Vortex RNAClean XP beads for 30 s.
15. Add 1.8× volume of RNAClean beads (55 μL) per sample.
16. Incubate for 15 min at room temperature.

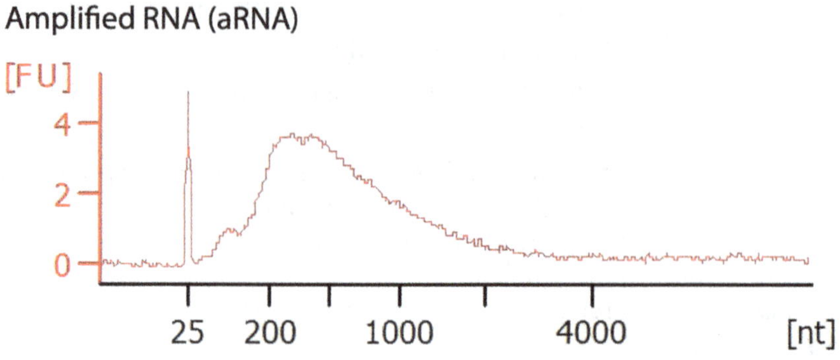

Fig. 3 Representative example of a bioanalyzer profile of amplified RNA (aRNA) from 384 single macrophages from a scared zebrafish kidney marrow

17. Place on magnet stand for 5 min or until liquid clears.
18. Remove and discard supernatant without disrupting beads.
19. Wash the bead pellet carefully with 200 μL of freshly prepared 80% EtOH.
20. Incubate for 1 min.
21. Remove as much 80% EtOH as possible.
22. Wash the bead pellet carefully with 200 μL of freshly prepared 80% EtOH.
23. Remove as much 80% EtOH as possible.
24. Dry beads for 10 min at room temperature.
25. Add 7 μL of nuclease-free water and resuspend beads.
26. Place on magnet stand for 1 min or until liquid clears.
27. Collect 7 μL of supernatant and transfer to a clean 0.5 mL collection tube.
28. Assess the quality of amplified RNA by running each sample on a bioanalyzer or tape station (high-sensitivity RNA reagents, Fig. 3).
29. Transfer 5 μL of amplified RNA to a clean 0.5 mL collection tube.
30. Add 1 μL of random hexamer primer and 0.5 μL of dNTP.
31. Incubate at 65 °C for 5 min.
32. Spin down and cool on ice.
33. Prepare the following reverse transcription mix:

Reverse transcription mix	For 1 sample (μL)
First-strand buffer	2
0.1 M DTT	1
RNAseOUT	0.5
SuperScript II	0.5

34. Incubate at 25 °C for 10 min, 42 °C for 1 h, 4 °C hold.
35. Prepare library PCR mix:

Library PCR mix	For 1 sample (μL)
2× PCR Mastermix	25
Illumina RNA PCR primer (RP1)	2
H2O	11

36. Add 38 μL of Library PCR mix to each sample.
37. Add 2 μL of uniquely indexed Illumina RNA PCR primer (RPIX).
38. Mix well by pipetting up and down and spin down.
39. Perform the following PCR cycle:
    (a) 98 °C, 1 min
    (b) 10–14 cycles of:
        (i) 98 °C, 10 s
        (ii) 60 °C, 30 s
        (iii) 72 °C, 30 s
    (c) 72 °C, 10 min
    (d) 4 °C, hold
40. Add 50 μL of AMPure beads to each sample (1×).
41. Incubate for 15 min at room temperature.
42. Place on magnet stand for 5 min or until liquid clears.
43. Remove and discard supernatant without disrupting beads.
44. Wash the bead pellet carefully with 200 μL of freshly prepared 80% EtOH.
45. Incubate for 1 min.
46. Remove as much 80% EtOH as possible.

47. Wash the bead pellet carefully with 200 μL of freshly prepared 80% EtOH.
48. Remove as much 80% EtOH as possible.
49. Dry beads for 10 min at room temperature.
50. Add 25 μL of nuclease-free water and resuspend beads.
51. Place on magnet stand for 1 min or until liquid clears.
52. Collect 25 μL of supernatant and transfer it to a clean 0.5-mL collection tube.
53. Perform second bead cleanup by adding 25 μL of AMPure beads to each sample (1×).
54. Place on magnet stand for 5 min or until liquid clears.
55. Remove and discard supernatant without disrupting beads.
56. Wash the bead pellet carefully with 200 μL of freshly prepared 80% EtOH.
57. Incubate for 1 min.
58. Remove as much 80% EtOH as possible.
59. Wash the bead pellet carefully with 200 μL of freshly prepared 80% EtOH.
60. Remove as much 80% EtOH as possible.
61. Dry beads for 10 min at room temperature.
62. Add 10 μL of nuclease-free water and resuspend beads.
63. Place on magnet stand for 1 min or until liquid clears.
64. Collect 10 μL of supernatant and transfer to a clean 0.5 mL collection tube.
65. Measure the library amount with a Qubit fluorometer (high-sensitivity DNA reagents). A concentration equal to or superior to 1 ng/μL is expected.
66. Assess library quality by running each sample on a bioanalyzer or tape station (high-sensitivity DNA reagents, Fig. 4).

3.6 ScarTrace Library Preparation

1. Warm AMPure XP beads to room temperature.
2. Add 8.4 μL of nuclease-free water to 1.6 μL of postbead cleanup material.
3. Add 2 μL of EXO-SAP.
4. Mix well by pipetting up and down and spin down.
5. Incubate at 37 °C for 15 min, 80 °C for 15 min.
6. Spin down and cool on ice.
7. Vortex AMPure XP beads for 30 s.
8. Add 1× volume of AMPure beads (12 μL).
9. Incubate for 15 min at room temperature.

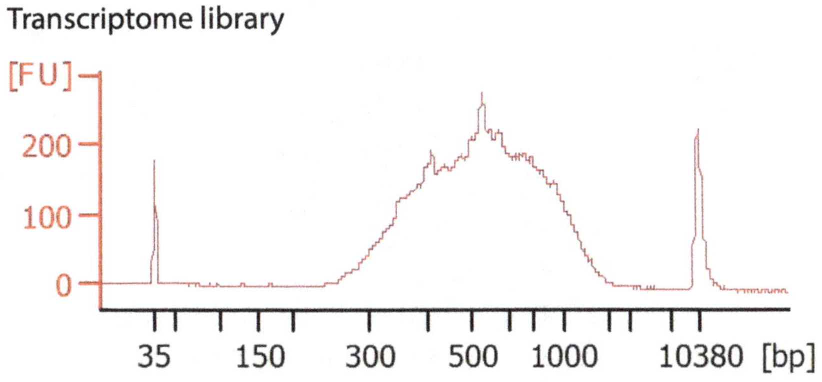

Fig. 4 Representative example of a bioanalyzer profile of a DNA transcriptome library from 384 single macrophages from a scared zebrafish kidney marrow

10. Place on a magnet stand for 5 min or until liquid clears.
11. Remove and discard supernatant without disrupting beads.
12. Wash the bead pellet carefully with 200 μL of freshly prepared 80% EtOH.
13. Incubate for 1 min.
14. Remove as much 80% EtOH as possible.
15. Wash the bead pellet carefully with 200 μL of freshly prepared 80% EtOH.
16. Remove as much 80% EtOH as possible.
17. Dry beads for 10 min at room temperature.
18. Add 10 μL of nuclease-free water and resuspend beads.
19. Place on a magnet stand for 1 min or until liquid clears.
20. Collect 10 μL of supernatant and transfer to a clean 0.5 mL collection tube.
21. Prepare library PCR mix:

Library PCR mix	For 1 sample (μL)
2× PCR Mastermix	25
Illumina RNA PCR primer (RP1)	2
H_2O	11

22. Add 38 μL of Library PCR mix to each sample.
23. Add 2 μL of uniquely indexed Illumina RNA PCR primer (RPIX).
24. Mix well by pipetting up and down and spin down.
25. Perform the following PCR cycle:

   (a) 98 °C, 1 min
   (b) 10–14 cycles of:
      (i) 98 °C, 10 s
      (ii) 60 °C, 30 s
      (iii) 72 °C, 30 s
   (c) 72 °C, 10 min
   (d) 4 °C, hold

26. Add 40 μL of AMPure beads to each sample (0.8×).
27. Incubate for 15 min at room temperature.
28. Place on magnet stand for 5 min or until liquid clears.
29. Remove and discard supernatant without disrupting beads.
30. Wash the bead pellet carefully with 200 μL of freshly prepared 80% EtOH.
31. Incubate for 1 min.
32. Remove as much 80% EtOH as possible.
33. Wash the bead pellet carefully with 200 μL of freshly prepared 80% EtOH.
34. Remove as much 80% EtOH as possible.
35. Dry beads for 10 min at room temperature.
36. Add 25 μL of nuclease-free water and resuspend beads.
37. Place on magnet stand for 1 min or until liquid clears.
38. Collect 25 μL of supernatant and transfer to a clean 0.5 mL collection tube.
39. Perform second bead cleanup by adding 22.5 μL of AMPure beads to each sample (0.9×).
40. Place on magnet stand for 5 min or until liquid clears.
41. Remove and discard supernatant without disrupting beads.
42. Wash the bead pellet carefully with 200 μL of freshly prepared 80% EtOH.
43. Incubate for 1 min.
44. Remove as much 80% EtOH as possible.
45. Wash the bead pellet carefully with 200 μL of freshly prepared 80% EtOH.
46. Remove as much 80% EtOH as possible.
47. Dry beads for 10 min at room temperature.
48. Add 10 μL of nuclease-free water and resuspend beads.
49. Place on magnet stand for 1 min or until liquid clears.
50. Collect 10 μL of supernatant and transfer to a clean 0.5 mL collection tube.

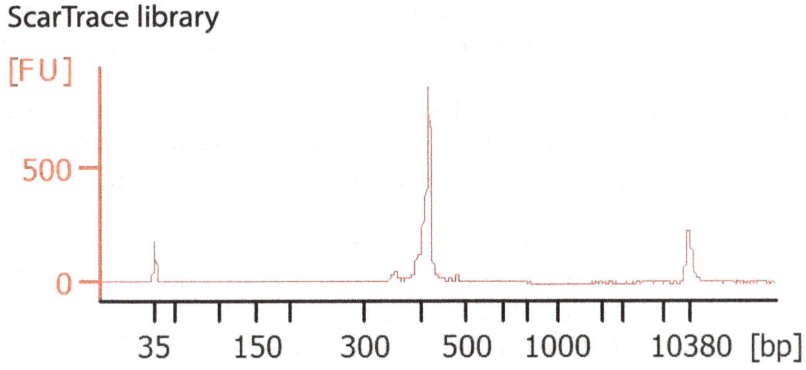

Fig. 5 Representative example of a bioanalyzer profile of a DNA ScarTrace library from 384 single macrophages from a scared zebrafish kidney marrow

51. Measure the library amount with a Qubit fluorometer (high-sensitivity DNA reagents). A concentration equal to or superior to 1 ng/μL is expected.
52. Assess library quality by running each sample on a bioanalyzer or tape station (high-sensitivity DNA reagents, Fig. 5).

3.7 Sequencing

Both transcriptomic and ScarTrace libraries can be sequenced in any Illumina platform using a pair-end approach. We recommend at least 2 and 0.5 million reads per plate for transcriptome and ScarTrace libraries, respectively.

3.8 Data Analysis

3.8.1 Transcriptome Libraries

Read 1 contains the cell barcode the unique molecular identifier (UMI) information, and the polyA of the sequenced transcript. Read 2 contains the sequence of the transcript. To analyze the transcriptome libraries, we recommend the following workflow:

1. Perform a demultiplexing step in which all Read 2 with a valid barcode extracted from their corresponding Read 1 are selected. Append the corresponding cell barcode and UMI sequences at the name of each selected Read 2. Read 1 is no longer used in downstream analysis.
2. Trim adapters from the selected reads, using TrimGalore [2] or any equivalent software.
3. Map trimmed reads to the reference, which can be achieved by mapping to the reference genome using STAR [3] or mapping to the reference transcriptome using bwa mem [4], among other options.
4. Generate a count table containing the number of observed reads and observed UMIs for each cell and each gene. This can be achieved using featureCounts [5], STAR, or custom-made code. To estimate the number of estimated transcripts, we recommend the use of the Poisson correction described by Grün and colleagues [6].

5. Analyze the data using standard single-cell RNA-seq analysis workflows such as RaceID3, Seurat, or scanpy [7–9].

3.8.2 ScarTrace Libraries

Read 1 contains the cell barcode (which is different from the one in transcriptome libraries) and a fragment of the GFP starting at the position of the forward primer (Fig. 1). Note that Read 1 starts with a UMI-like sequence that is used to improve cluster calling in the flow cell during sequencing. However, this UMI-like sequence is not used in downstream analysis since all amplification steps during scar library preparation are performed by PCR, defeating the purpose of the UMI. Read 2 contains a fragment of the GFP starting at the position of the reversed primer, which sits next to the sgRNA target site and therefore contains the scar. To analyze the ScarTrace libraries we recommend the following workflow:

1. Perform a demultiplexing step in which all Read 2 with a valid barcode extracted from their corresponding Read 1 are selected. Append the corresponding cell barcode sequence at the name of each selected Read 2. Read 1 can be ignored for downstream analysis.
2. Map reads to the reference GFP sequence using bwa mem.
3. Select reads mapped to the GFP with proper strand orientation, containing the primer sequences used for scar amplification with three or less mismatches.
4. Count how many times each read is seen in each cell. We recommend performing this step with read resolution to correct sequencing errors using raw counts and avoid false positives in scar calling.
5. Remap unique reads using pairwise2.align.globalms function from biopython with match, mismatch, open, and extend parameters set to 1, 0.25, −1, and −0.1, respectively.
6. Identify the CIGAR string for each scar and create a count table containing the number of reads observed for each cell and each CIGAR string. Create a dictionary of CIGARS and sequenced reads. If two or more CIGARS are very close by sequence, consider merging them to encode for the same scar. If a CIGAR contains different sequences with similar total raw counts, consider splitting them to encode for different scars.
7. Perform quality checks to filter out both cells and scars (i.e., CIGAR strings) with few counts:
   (a) The histogram of the logarithm of total number of counts per cell is expected to follow a bimodal distribution. The low peak contains cells with low quality which we recommend removing from downstream analysis (Fig. 6).
   (b) The logarithm of the histogram of the logarithm of total number of counts per scar is expected to follow a three-

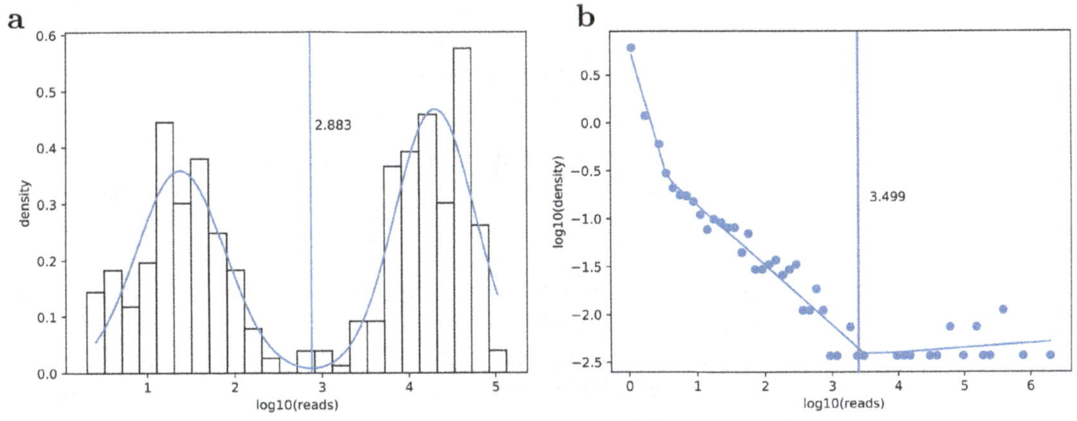

Fig. 6 Quality checks during analysis. (**a**) Normalized histogram of the number of reads per cell in log10 scale (black bins) and fit to double Gaussian distribution (blue curve). The vertical line and the text label 2.883 indicate the position of the minimum of the double Gaussian distribution function and the suggested threshold used to select cells from empty barcodes. (**b**) Normalized histogram of the number of reads per scar (blue circles), along with a fit to three-domain piecewise-defined linear function. The vertical line and the text label 3.499 indicate the starting position of the third domain and the suggested threshold to select real scars instead of sequencing errors for downstream analysis

domain piecewise-defined linear function (Fig. 6). The first domain corresponds to highly-detected scars, while the third domain corresponds to scars originating from sequencing errors that were not filtered out by the CIGAR string. We recommend removing from downstream analysis cigars that fall in this third domain of the histogram and keeping the ones in the first and second domains.

8. At this stage, data can be normalized by library size.
9. For each single cell, explore the scar content. It is common to observe scar bleed-through between cells (that is, scars present in some cells with a high-count number can be observed in very small count numbers in cells in which they are absent). This can easily be filtered by removing scars with a very low count from each cell.
10. In the original ScarTrace, we do not expect more than eight scars for each cell since there are only eight copies of the GFP sequence. If more than eight scars are seen, filtering thresholds should be revisited.
11. Cluster cells according to scar content to identify clones.
12. The copy number of each scar for each clonal family can be extracted using a noise model described in [10, 11], which assumes that the detected fraction of reads per scar, x_i, in a clone family follows the following analytical expression:

$$p(x_i|f_i) \simeq \sqrt{\frac{f_i^{1/2}}{4\pi b x_i^{3/2}}} \exp\left[-\frac{\left(\sqrt{x_i}-\sqrt{f_i}\right)^2}{b}\right]$$

where f_i is the expected fraction (i.e., copy number of a scar divided by total number of GFP copies) and b is proportional to the variance.

3.8.3 Integrative Analysis

Keep cells for which both transcriptome and scar pattern are detected (this is usually around 60% of the plate). To check whether a clone is enriched or depleted in a cell type, the Fisher exact test can be performed. Alternatively, one can compute the lineage coupling correlation [12, 13].

To build lineage trees, we used the maximum parsimony approach. Since scars are sequence-dependent, we recommend them to rank them by their probability of occurrence and take such probability into account for tree building. Be aware that by design, big deletions can take place between sequential GFP copies. These can result in the loss of information and an incorrect lineage tree.

4 Notes

1. We observed that homozygous zebrafish tend to have more WT hence we recommend using heterozygous lines.
2. In Cas9 protein injections, scars accumulate approximately during the first 3 h of embryo development. Conversely, in Cas9 mRNA injections, scars accumulate during the first 11 h of development.
3. Centrifugating plates after the addition of any new reagent in each well is essential for successful reactions.
4. Cell viability and population purity after flow cytometry sorting need to be evaluated in-house for every new sample type. We recommend sorting the desired population in a collection tube, adding a fresh viability marker, and analyzing the sorted population for viability and purity (expected purity >80%).
5. ScarTrace nested PCR cycle numbers need to be optimized for each sample type. We recommend starting with 12 cycles for PCR I and 12 cycles for PCR II. However, we have observed that 10 cycles for PCR I and 12 cycles for PCR II yielded less PCR duplicates for cell types such as blood cells. We have done as much as twice 16 cycles on sorted cell types from the fins and yielded good quality ScarTrace sequencing data.

Acknowledgments

C.B. is supported by the European Molecular Biology Organization (EMBO LT-290-2019) and Human Frontier Science Program (HFSP LT000494/2020-L) fellowships. A. A. is supported by the European Research Council Starting grant (ERC-StG 101042634-i-Signaltrace) and the Nederlandse Organisatie voor Wetenschappelijk Onderzoek (NWO) VIDI grant (VI.Vidi.203.050).

References

1. Stachura DL, Traver D (2011) Cellular dissection of zebrafish hematopoiesis. Methods Cell Biol 101:75–110
2. Krueger F (2021) TrimGalore. GitHub repository. https://doi.org/10.5281/zenodo.7598955
3. Dobin A et al (2013) STAR: ultrafast universal RNA-seq aligner. Bioinformatics 29(1):15–21
4. Li H, Durbin R (2009) Fast and accurate short read alignment with Burrows-Wheeler Transform. Bioinformatics 25:1754–1760
5. Liao Y, Smyth GK, Shi W (2014) featureCounts: an efficient general purpose program for assigning sequence reads to genomic features. Bioinformatics 30(7):923–930
6. Grün D, Kester L, Van Oudenaarden A (2014) Validation of noise models for single-cell transcriptomics. Nat Methods 11(6):637–640
7. Hao Y et al (2021) Integrated analysis of multimodal single-cell data. Cell 184(13): 3573–3587
8. Herman JS, Sagar N, Gruen D (2018) FateID infers cell fate bias in multipotent progenitors from single-cell RNA-seq data. Nat Methods 15(5):379–386
9. Wolf FA, Angerer P, Theis FJ (2018) SCANPY: large-scale single-cell gene expression data analysis. Genome Biol 19:1–5
10. Levy SF et al (2015) Quantitative evolutionary dynamics using high-resolution lineage tracking. Nature 519(7542):181–186
11. Alemany A, Florescu M, Baron CS, Peterson-Maduro J, Van Oudenaarden A (2018) Whole-organism clone tracing using single-cell sequencing. Nature 556(7699):108–112
12. Wagner DE et al (2018) Single-cell mapping of gene expression landscapes and lineage in the zebrafish embryo. Science 360(6392): 981–987
13. Bandler RC et al (2022) Single-cell delineation of lineage and genetic identity in the mouse brain. Nature 601(7893):404–409

Chapter 12

LINNAEUS: Simultaneous Single-Cell Lineage Tracing and Cell Type Identification

Bastiaan Spanjaard and Jan Philipp Junker

Abstract

A key goal of biology is to understand the origin of the many cell types that can be observed during diverse processes such as development, regeneration, and disease. Single-cell RNA-sequencing (scRNA-seq) is commonly used to identify cell types in a tissue or organ. However, organizing the resulting taxonomy of cell types into lineage trees to understand the origins of cell states and relationships between cells remains challenging. Here we present LINNAEUS (Spanjaard et al, Nat Biotechnol 36:469–473. https://doi.org/10.1038/nbt.4124, 2018; Hu et al, Nat Genet 54:1227–1237. https://doi.org/10.1038/s41588-022-01129-5, 2022) (LINeage tracing by Nuclease-Activated Editing of Ubiquitous Sequences)—a strategy for simultaneous lineage tracing and transcriptome profiling in thousands of single cells. By combining scRNA-seq with computational analysis of lineage barcodes, generated by genome editing of transgenic reporter genes, LINNAEUS can be used to reconstruct organism-wide single-cell lineage trees. LINNAEUS provides a systematic approach for tracing the origin of novel cell types, or known cell types under different conditions.

Key words Single-cell mRNA, Single-cell lineage tracing, Barcoding, CRISPR/Cas9, Tree building, Phylogenetics

1 Introduction

Measuring lineage relationships between cell types is important for understanding fundamental mechanisms of cell differentiation in development and disease [1, 2]. In early development and in adult systems with a constant turnover of cells, short-term lineage predictions can be computed directly on scRNA-seq data by ordering cells along pseudotemporal trajectories according to transcriptome similarity [3–9]. However, the developmental origin of cells in the adult body cannot be identified using these approaches alone. Several approaches for lineage tracing exist. Genetically encoded fluorescent proteins are widely used as lineage markers [10, 11], but due to limited spectral resolution, optical lineage tracing methods have mostly been restricted to relatively small numbers of cells.

Jorge García-Marqués and Tzumin Lee (eds.), *Lineage Tracing: Methods and Protocols*, Methods in Molecular Biology, vol. 2886, https://doi.org/10.1007/978-1-0716-4310-5_12,

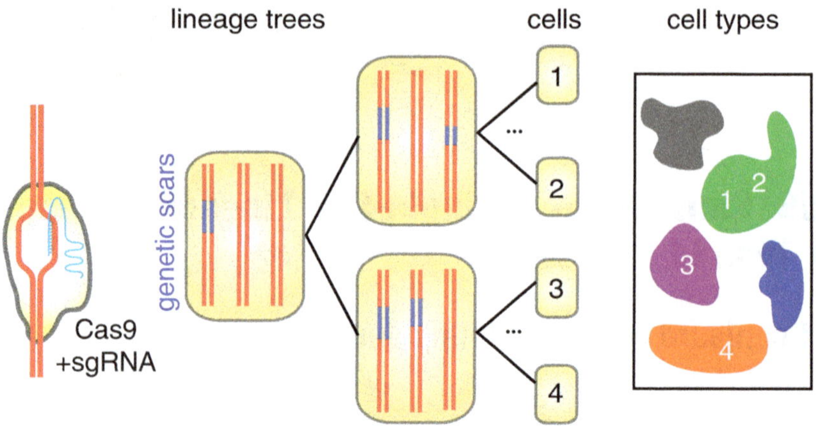

Fig. 1 LINNAEUS combines single-cell lineage tracing and transcriptomics. CRISPR/Cas9 creates heritable genetic "scars" (left) that can be used to build a lineage tree (middle) and combined with scRNA-seq (right). (Reproduced from the original publication, Ref. [22])

Pioneering studies based on viral barcoding [12, 13], transposon integration sites [14], microsatellite repeats [15], somatic mutations [16, 17], *Cre*-mediated recombination [18], and genome editing of reporter constructs [19–21] have used sequence information to increase the diversity of lineage labels. However, these methods have not been coupled with single-cell transcriptome sequencing and therefore do not provide any information on cell type.

Here we describe LINNAEUS for simultaneous measurement of single-cell transcriptomes and lineage markers in vivo [22, 23]. The approach is based on the observation that, in the absence of a template for homologous repair, Cas9 produces short insertions or deletions at its target sites, which are variable in their length and position [19, 20, 24]. These insertions or deletions (hereafter referred to as genetic "scars") constitute heritable cellular barcodes that can be used for lineage analysis and read out by scRNA-seq (Fig. 1). Similar methods have been published concurrently in zebrafish [25, 26], and later in mice.

In this chapter, we focus on the bioinformatic analysis and lineage tree reconstruction using the LINNAEUS algorithm. The figures shown here, and parts of the text, are sourced from the original publications [22, 23]. In the materials section, we introduce the necessary datasets and software and, in the methods section, we explain the steps and rationale of the lineage tree reconstruction. In the notes section, we provide additional detail on and guidance for parameter choices for the filtering and tree building algorithms and help users design experiments with the goal of producing high-resolution single-cell lineage tracing data.

2 Materials

As this chapter describes the bioinformatics pipeline of LINNAEUS, we here list the necessary datasets, computing environments, and scripts. Users will need two datasets. The first is the single-cell sequencing data itself, containing both a transcriptome and a lineage dataset (Fig. 2a). The experimental creation of such datasets is detailed in the original publication [22]. The exact format of the lineage dataset can vary between techniques. In LINNAEUS, there are multiple target integrations, each of which is read out separately. This means that scarring events do not influence each other, and integrations cannot be excised by nearby scarring events. There will be lineage information for a relatively high number of cells, but this information will almost always be incomplete. These characteristics strongly influenced the design of the lineage reconstruction algorithm. *See* **Note 1** for further considerations.

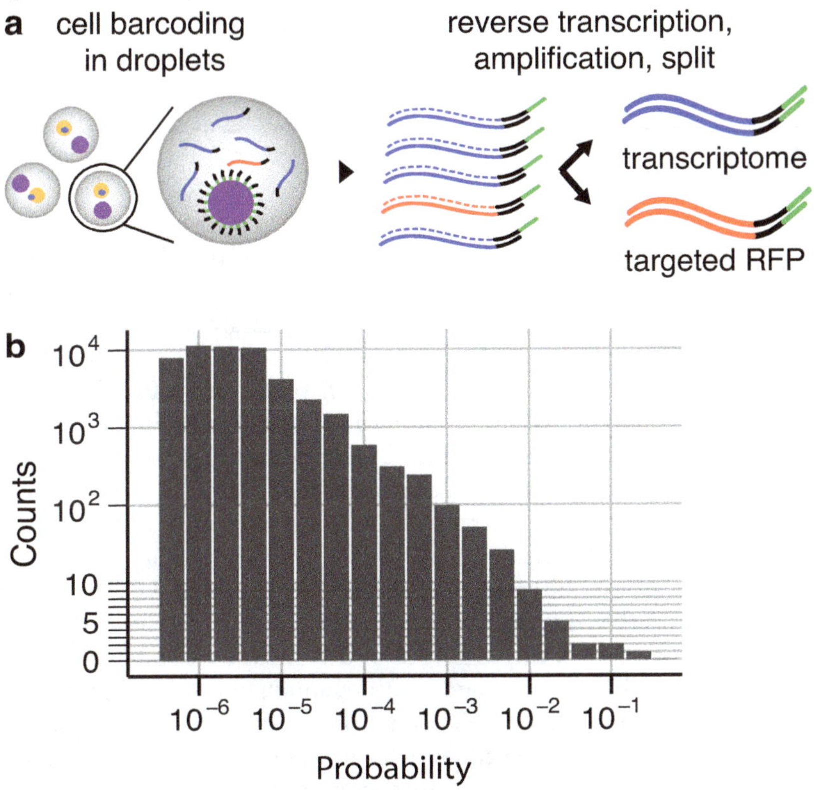

Fig. 2 Datasets needed for lineage tree building. (**a**) Approach for simultaneous detection of scars and transcriptome from single cells. Cells are captured by droplet microfluidics, followed by lysis, reverse transcription, and amplification. After amplification, the material is split and processed into a whole transcriptome library and a targeted library for scar detection. (**b**) Probability distribution of scars, shown as measured in bulk experiments on the DNA level. (Reproduced from the original publication, Ref. [22])

The second dataset needed is a set of scar probabilities (Fig. 2b). The scarring process can create more than 100,000 different scars. However, due to biases in the DNA repair machinery that is used during scar creation, some of these scars are created much more frequently than others. A highly probable scar can be created multiple times during development and is therefore not a reliable lineage marker. Scar probabilities can be inferred by bulk sequencing several CRISPR/Cas9-scarred organisms. Multiple samples are needed to generate a thorough covering of the scar space and to disentangle the intrinsic scar probability from the effects caused by earlier or later scar generation. After the removal of all unscarred reads from each organism and normalizing the scar contents to one, the scar probabilities can be calculated as the average ratio with which each scar is observed.

For computations, users will need a computing environment capable of running single-cell transcriptome alignment such as a high-performance computing cluster, as well as the option of running interactive R sessions, for example, through the RStudio IDE. Users will also need to clone the repository at https://github.com/Bastiaanspanjaard/LINNAEUS and follow the installation instructions for the collapsibleTree package.

3 Methods

The LINNAEUS algorithm follows four steps that we here briefly describe together before explaining them in detail below. Initially, the transcriptome and lineage data are both aligned to reference sequences: the transcriptome data are aligned to a reference transcriptome to obtain transcript counts for each cell barcode and are then used to determine cell types (**step 1**). The lineage data are aligned to the sequence of the transgene that is targeted by CRISPR/Cas9 (**step 2**). The lineage data are initially filtered to remove sequencing artifacts and chimeric reads [27] (**step 3**) and later filtered to remove doublets and other cell-specific effects (**step 4**). The reader may notice that the lineage data are subject to stricter filtering than the transcriptome data. The rationale for this is that lineage data are binary: a cell either has a scar or it does not, and both the structure of the lineage tree and the placement of the cell in the lineage tree are much more susceptible to incorrectly identified scars. This is not the case for transcriptome data: here, expression levels, rather than binary behavior, matter for cell type determination and further computations such as trajectory analysis. As a corollary, small errors in the number of transcripts are negligible and do not need to be filtered out. After filtering, bulk probabilities are appended to the observed scars (**step 5**) and the lineage tree is built, and the cells are placed in the lineage tree (**step 6**). We now describe these steps in further detail.

1. After the alignment of single-cell transcriptomes to a reference transcriptome, cell type determination can be done with a method of the reader's choosing. In our publications, we used Seurat [28], but scanpy [29] would be equally suitable, as well as other algorithms for cell type determination. In our implementation, this step is performed by the script R_scripts/Larvae_Seurat_mt.R. Importantly, this step creates a whitelist of cellular barcodes that have sufficient transcriptomic information. It also groups cells by cell type, information that will later be used to determine doublets. Our research suggests that every cell type has a different average amount of detected scars, likely caused by overall expression levels as well as a variable number of active promoters of the scar-expressing transgene, both of which can vary by cell type (Fig. 3). We can use these averages in a later stage to identify and filter out cells that have a higher number of scars than expected—cells that could potentially be doublets. The typical graphical output of cell type determination, a two-dimensional representation of the transcriptomes of all cells in the dataset, is not strictly necessary for LINNAEUS but can provide a useful overview of the dataset (Fig. 4).
2. As an initial step in the tree building, the lineage data are aligned to a reference sequence and the resulting data are filtered. These steps are performed by the script scar_extraction/extract_scar_[X].sh (where [X] is a library designation) after changing the name of the sequencing and reference files in the script, as well as the primer sequence in the call to scar_CIGAR_sc_10X_v2.pl. Scar reads have the same structure as transcript reads: they consist of a barcode, a Unique Molecular Identifier [30] (UMI), and a scar. Initially, barcodes with more than 500 scar reads are selected. The scar sequences are aligned using *bwa mem* [31] to a reference that represents the region targeted by CRISPR/Cas9. Reads that are unmapped, have an incorrect barcode, or do not start with the exact PCR primer used are removed. All scar sequences are truncated to 75 nucleotides and shorter sequences are filtered out. *See* **Note 2** for adjustable parameters in this step.
3. To mitigate the effect of sequencing errors, we implemented several rounds of scar filtering (Fig. 5). The filtering algorithm counts the number of times each molecule was sequenced. Sequencing errors will typically have fewer reads than the actual scars they originate from. As a first filtering step, all molecules only seen once are removed to reduce the complexity in the dataset for consecutive filtering steps.
4. The second filtering step, implemented in scar_filter.py, which is automatically called by extract_scar_[X].sh, aims to remove easily recognizable sequencing errors and chimeric reads

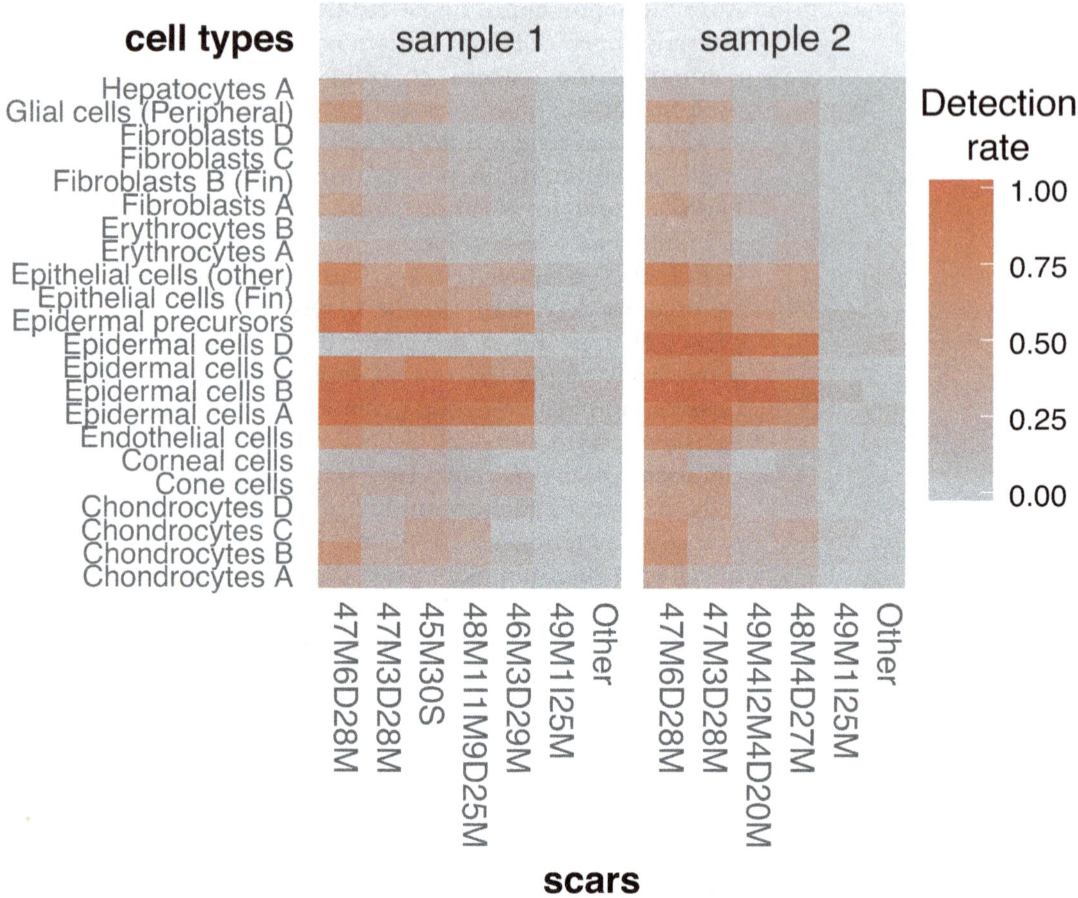

Fig. 3 Single-cell scar detection rates. Scar detection efficiency (here measured as average over cells in two organisms where every cell had the same scar profile) depends on cell type, which probably reflects differences in cell size or promoter strength. While most genomic integrations (i.e., different scar sequences) have similar detection efficiencies, one integration (CIGAR code 49M1I25M) has significantly lower expression. Such lowly expressed scars are not considered for tree reconstruction, since they carry very little lineage information and may be hard to distinguish from sequencing errors. (Reproduced from the original publication, Ref. [22])

[27]. To this end, the algorithm consecutively considers scar sequences that have the same cellular barcode and UMI, UMIs that have the same cellular barcode and scar sequence, and cellular barcodes that have the same UMI and scar sequence. In each step, the algorithm retains only the molecule with the highest number of reads. The rationale behind this is that it is very improbable to have two valid scar sequences in the same cell with the same UMI or to have a scar sequence with the same UMI appear in two different cells. The observation of two different UMIs for the same scar in the same cell is much more likely and corresponds to the detection of multiple transcripts

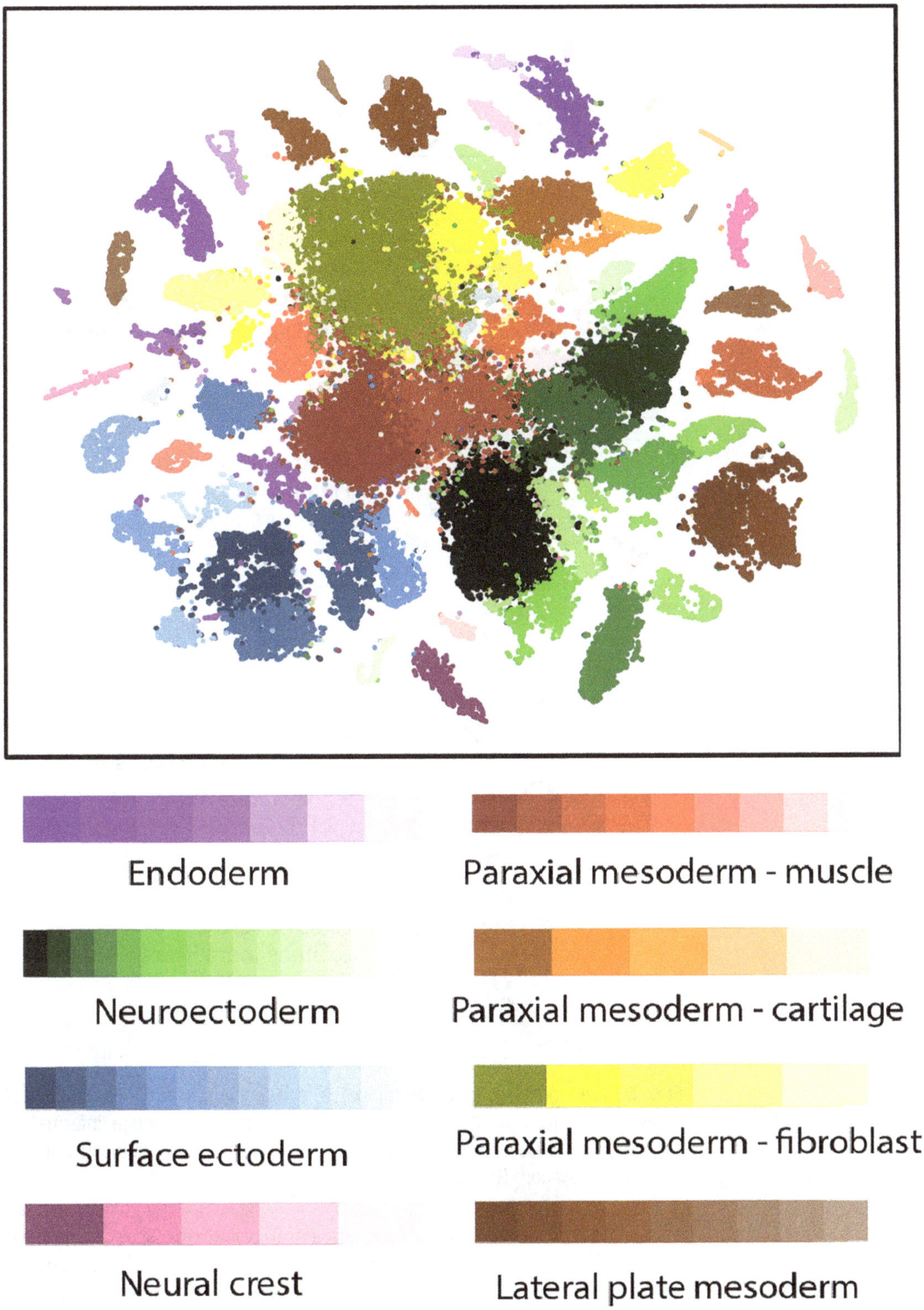

Fig. 4 t-SNE representation of scRNA-seq data and identified cell types for dissociated zebrafish larvae (5 dpf, $n = 7$ animals). Cell types are grouped into eight categories as indicated by the color code. (Reproduced from the original publication, Ref. [22])

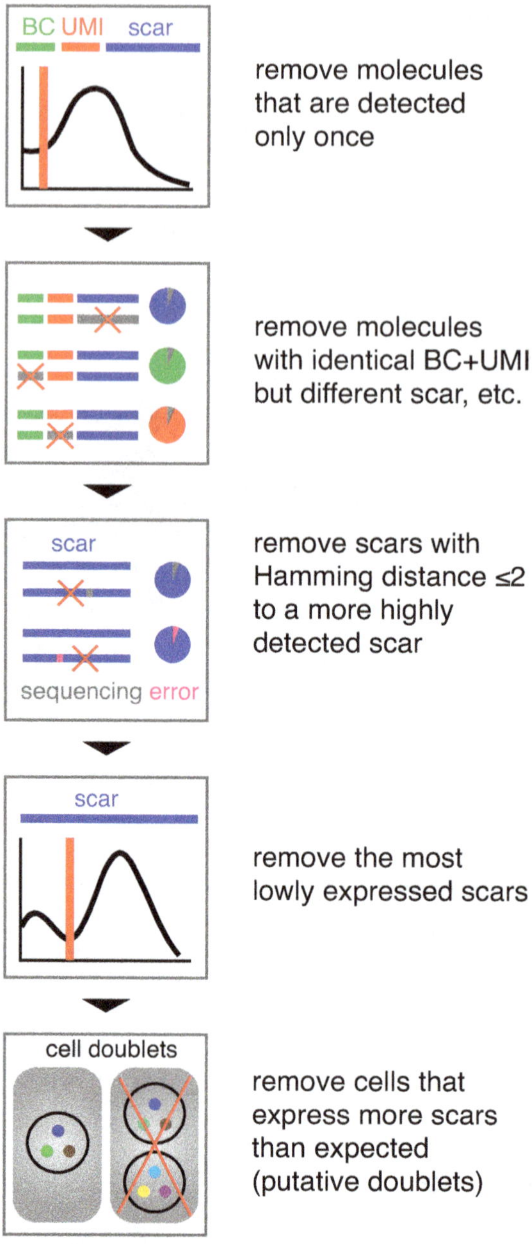

Fig. 5 Scar filtering pipeline mitigates sequencing errors. To distinguish sequencing errors in scars and barcodes from genuine lineage information, scar libraries are oversequenced and filtered as shown. (Reproduced from the original publication, Ref. [22])

from the same locus, but information about scar expression levels is not required in our downstream analysis. The name of extract_scar_[X].sh's output follows the naming convention [X]_scar_filtered_scars.csv. Adjustable parameters for the first two filtering steps are described in **Note 3**.

5. The third, fourth, and fifth filtering steps occur in an interactive R session and are coded in library_scar_filtering/Preprocess_-droplet_scar_[X]_allreads.R. Initially, valid cell barcodes are identified based on the single-cell transcriptome data (*see* **step 1**). The third filtering step specifically targets sequencing errors within each cell. Scar sequences found in the same cell are compared to each other. Sequences with a Hamming distance of 2 or less to another scar sequence in the same cell that occurred in at least eight times as many reads are removed. Scar sequences in the same cell that are 1 Hamming distance apart but that have a read ratio less than 8 are tested on three criteria if both occurred at least twice in the scar library:
   - Do both scars have more than one transcript?
   - Do both scars occur in cells independently from each other?
   - Do the UMIs of both scars have a Hamming distance of two or more?

   If two of these criteria are true, the scars are kept and the sequences are placed on a list of validated scars that, if they occurred in the same cell in another library, do not have to be tested anymore. If one or zero criteria are true, the scar that had only one transcript, or the scar that did not occur independently, is filtered out.

6. In the fourth filtering step, the algorithm determines the distribution of reads for the scars that were kept so far. Based on this distribution, the user sets a cutoff to filter out the scars that do not have at least this number of reads.

7. Then, for each cell type, the algorithm calculates the distribution of different scars seen per cell and the user sets a maximum number of scars a cell of that type can have. Cells with more than this maximum number are filtered out as possible doublets.

   The output of these filtering steps is named [X]_used_s-cars.csv. Parameters for these steps are described in **Note 4**.

8. Finally, in the script library_scar_filtering/Scar_comparison_-between_fish.R, the user adds scar probabilities to the observed scars. In the case of multiple organisms, this script can also be used to count the total libraries in which scars independently occur. This measure is correlated to the probability of creating a scar: scars with a high creation probability are more likely to occur in multiple libraries independently. While each scar is identified by its sequence, scars are labeled using their ranking in the bulk scar frequency distribution (e.g., "scar 77") or their CIGAR code (e.g., "47M6D28M") as a shorthand notation. Since scars cannot be modified once created, each scar is considered as a separate entity for lineage tracing independent of its sequence. The output of this script is named [X]_scars_compared.csv.

9. Our computational method for lineage tree reconstruction consists of two phases that are both coded in R_scripts/Iterative_tree_building.R. First, the algorithm derives the order of scarring events. To do so, it uses a scar network graph, a representation of all pairwise combinations of scars that are experimentally observed together in single cells (Fig. 6a). If all scar connections are detected, the scar that is created first has the highest degree of connections in the scar network graph, followed by scars that were created next, enabling lineage tree reconstruction in an iterative manner (Fig. 6b). Due to incomplete scar detection, we do not have full lineage information about every single cell. However, the structure of the scar network graph is robust toward scar dropouts since it is based on the collective information of thousands of single cells. **Note 5** discusses the tree building algorithm and its outputs. To ensure that lineage tree reconstruction is not affected by known experimental biases, the tree building algorithm also includes the following measures:
   - *Double scarring:* Some scars have a higher intrinsic probability than others. To minimize the chance of considering scars that may have been created twice or more in the same fish, by default all scars that have a probability higher than 0.1% are excluded. This threshold can be changed by the user, a lower threshold would be appropriate in systems with a high number of scars whereas systems with a lower amount of scars could benefit from a higher threshold. The relevant parameters are max.scar.p and max.larvae.
   - *Cell doublets:* Co-encapsulation of two cells in one droplet is a known limitation of scRNA-seq techniques that are based on droplet microfluidics. Cell doublets can lead to spurious connections between scars in the network graph. Incomplete tissue dissociation, limited barcode diversity, barcode sequencing errors, and free-floating RNA from cells burst in the microfluidic system may potentially have similar consequences. As a protection against this effect, the algorithm only accepts connections in the scar network graph that are more highly detected than expected by chance given a library-specific doublet rate (typically around 10%, depending on the experimental cell loading rate). The relevant parameter is doublet.rate.
   - *Missing connections:* In case of very low cell numbers or scar detection efficiencies, it is possible that a connection is missed in the scar network graph. To address this issue, a statistical test is performed for each scar to check whether the number of observed connections is compatible with the scar being on top of the current subbranch, given the number of cells and the observed scar dropout rates. In each

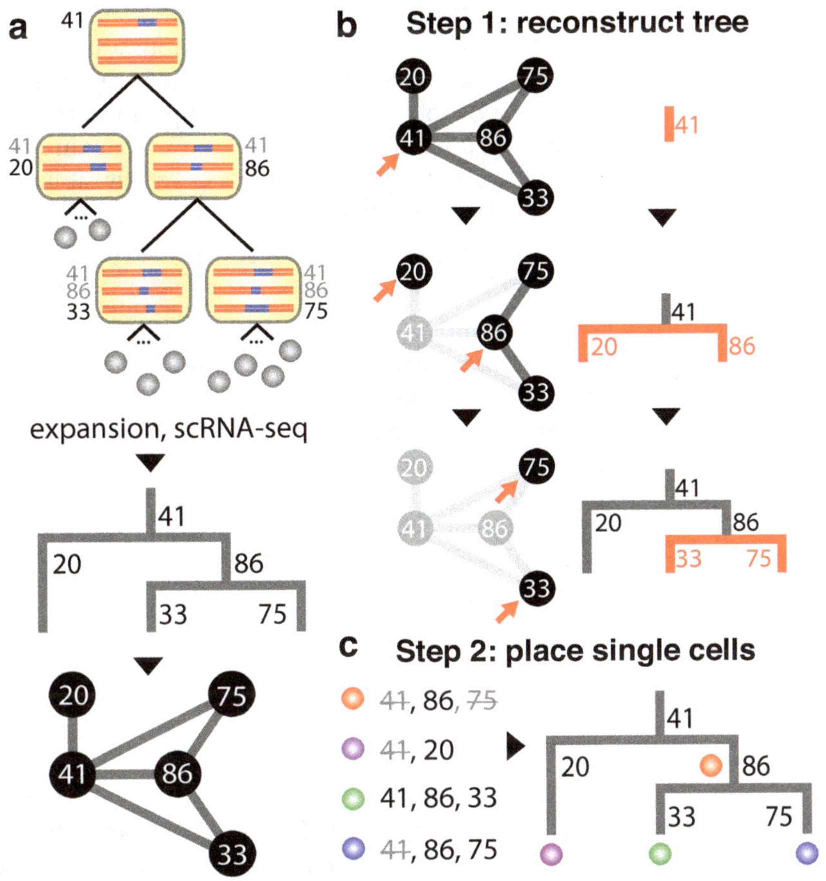

Fig. 6 Computational reconstruction of lineage trees on the single-cell level. (**a**) In a developmental lineage tree (top), each scar can be identified by a unique number corresponding to its ranking in the bulk scar frequency distribution (Fig. 2b). Newly created scars are indicated in black font. The resulting scar tree (middle), a reduced representation of the order of scarring events, can be represented as a network graph (bottom). In a scar network graph, each node corresponds to a different scar, and pairs of scars that are coexpressed in single cells are connected by gray lines. In LINNAEUS, scar network graphs are experimentally measured, and they form the base on which the underlying lineage tree is computationally reconstructed. (**b**) Cartoon of the computational approach. Network graphs allow reconstructing the order of scar creation events in an iterative approach. The first scar is determined as the one with the highest connectivity (red arrow). Upon removal of the first scar and its connections, the following scars are identified as the most highly connected ones in the reduced network. (**c**) After the scar tree has been built, all individual cells are placed in the tree according to their scar profile. Incomplete scar detection efficiency may lead to loss of information in single cells (black numbers: detected scars; gray crossed-out numbers: missed scars). As a consequence, some cells cannot be placed all the way down to the lowest branch of the tree (e.g., red cell, in which scars 41 and 75 were not detected). However, some missing scars can be reconstructed (e.g., blue cell, in which scar 41 can be inferred). (Reproduced from the original publication, Ref. [22])

iteration, the algorithm tests only those scars whose inferred detection rate (if placed on top of the corresponding sub-branch) is higher than 0.05. The relevant parameter is min. detection.rate.

- *Pruning the tree:* Especially for later, smaller branches, it is possible that not enough connections are observed to accurately place them in the lineage tree, resulting in the positioning of the branch too high up in the tree. The lineage tree is pruned for such branches by removing branches that have less than 12.5% of the cells their siblings have. The relevant parameter is the branch.size.ratio.

All parameters for the tree reconstruction are discussed further in **Note 6**.

10. In the second phase, we place all cells in the lineage tree according to their scar profile (Fig. 6c). Cells are placed as low in the tree as their scars allow. The cell type pie charts shown in every node count the cells that are placed in that node and lower; this means that the pie chart at the root node of the tree depicts all cells in the tree. The lineage trees can be visualized using the collapsibleTree package (Fig. 7), saved as html widgets or exported for further analysis as edge lists. *See* **Note 7** for parameter tuning and sanity checks on the validity of the created lineage trees.

4 Notes

1. LINNAEUS can be applied to any model system with transgenes whose expression does not influence the system. The number of branching events recorded for every cell is bounded by the number of transgene integrations: the maximum depth of the tree depends on the number of integrations. While we have created lineage trees for organisms with around five integrations, we recommend at least ten integrations for high-resolution lineage trees.

   A lineage tree can only be informative if the scarring occurs during the right timeframe. For zebrafish development, for example, a timeframe from 0 to 8hpf is very informative, but this timeframe will be very different for development in other organisms or for other processes than development. Whether scars are created during the correct timeframe can be measured through bulk sequencing of scars at several time points after scar creation is initiated (Fig. 8). These same experiments also show the scarring efficiency; the lineage tree resolution of systems with scarring efficiency lower than 50% is questionable in our experience.

2. Some of the parameters in the initial scar read selection are highly library specific. The first two lines of extract_scars_[X].sh count the reads for each barcode and select only those

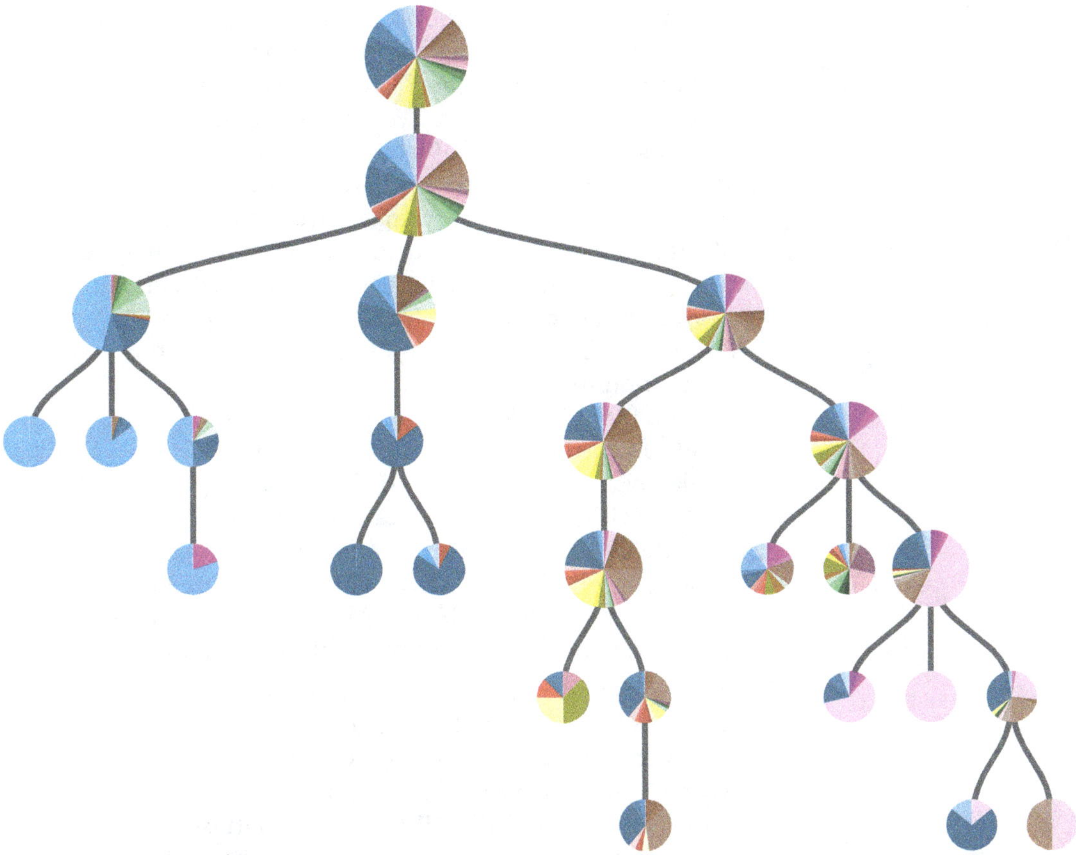

Fig. 7 Lineage tree for one 5 dpf larva. Color code for cell types as in Fig. 4. In general, developmental lineages separate well in the tree. However, since scarring ends at ~10 h post fertilization, the endpoints of the branches may still give rise to multiple cell types in multiple tissues. (Reproduced from the original publication, Ref. [22])

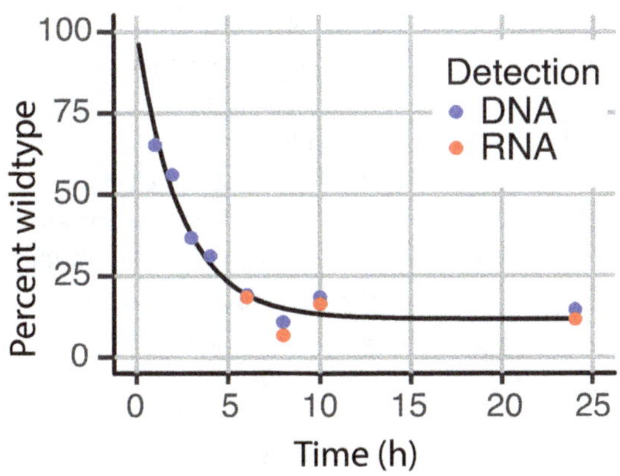

Fig. 8 Scarring dynamics as measured on the DNA and RNA levels, with exponential fit. (Reproduced from the original publication, Ref. [22])

barcodes with more than 500 reads. The snippet `substr($1, 1, 16)` in the first line extracts the barcodes which, in this case, are the first 16 bases of the R1. Users may wish to adjust this to their library specifications. For lower-sequenced libraries, users can reduce the cutoff of 500 to a lower number or skip this part altogether and initially select barcodes that are transcriptome-whitelisted, i.e., barcodes that have enough transcriptomic information to be included in the cell type analysis. In that case, users will have to later impose scar read cutoffs on barcodes to remove low-read barcodes, but this can be done in a library-specific manner, for example through the scar read distribution in a library.

The UMIBC file in the script contains the necessary information from the nongenomic part of the read: the UMI and cellular barcode concatenated. The mapping script scar_CIGAR_sc_10X_v2.pl will truncate scar reads to 75 nucleotides, align them to the provided reference (in the script this is lintrace_hGFP_RFP_ERCC92.fa), and output all aligned scar reads together with CIGAR, UMI and barcode in the file [op]_scars.txt, with the output prefix (op) defined by the user.

3. In our experience, there is a vast difference between scar reads that have been seen only once and scar reads that have been seen twice or more times. The complexity of scar reads that occur more than once is in the place where we expect it: the locus that is cut by CRISPR/Cas9. In contrast, the complexity of scar reads that occur only once is not tied to a specific location on the read but seems to be ubiquitous. Most of these reads, therefore, are sequencing errors, and they are removed. We would like to encourage the users to ascertain this by themselves to make sure that the cutoff of one read is correct for them. A secondary reason for removing scar reads that only occur once is that it strongly reduces the (spurious) complexity of the scar dataset, so that the subsequent steps can more easily be controlled.

   The next filtering step, in scar_filter.py which is called in the last line of extract_scars_[X].sh, does not have any parameters: it selects the dominant sequence (i.e., the sequence with most reads) for every barcode + UMI combination, then the dominant UMI for each barcode + sequence combination, and finally the dominant barcode for each UMI + sequence combination. The rationale behind this is that, due to the randomness of UMIs, it is very unlikely to observe the same UMI for two different scars in the same cell; if two scars in the same cell have the same UMI, this is very likely caused by sequence overwriting through chimeric reads. Chimeric reads occurred quite frequently in our data, but this could depend on experimental parameters. The script only retains the dominant

UMI for every barcode + sequence combination; this is not strictly necessary and removes information on scar transcript counts. There are certainly cases where this information can be useful, and users can consider skipping the relevant lines (56–70) in scar_filter.py.

4. The third, fourth, and fifth filtering steps are all executed by the interactive script library_scar_filtering/Preprocess_droplet_scar_[X]_allreads.R. The third filtering step, meant to remove scars with multiple errors (e.g., a chimeric read that also had an error early in PCR amplification) is coded in lines 40–207. Like the previous filtering step, we implemented this because we saw these scars occur frequently, but this could again depend on experimental parameters. Users can determine its necessity by looking at the relevant dataframes, scars.assess, and scars.assess.2. The main parameter is log2.cutoff with a default value of 3, to filter out errors that have been made after three PCR rounds. Note that this step also uses a dataframe of scars all.scars.g1; this is the output of the scar alignment after removing scars with only one read.

   The fourth filtering step is meant to remove any scar transcripts with an unexpectedly low number of reads and is coded in lines 209–219. The user plots a histogram of the number of reads of all remaining scar transcripts and selects (in line 218) the proper cutoff. In the case of oversequenced libraries, the distribution is clearly bimodal.

   The fifth filtering step is meant to remove doublets or at least cells with an unexpected number of scars. It is coded in lines 226–265. In this step, the algorithm will plot a distribution of the number of scars per cell for each cell type and request the user to determine the maximum amount of scars a cell of this type can have. The distributions are strongly cell type-dependent, as expected (Fig. 3), so cell type information is indispensable here. Given the strong possible effect of doublets on the final lineage tree, we advise a conservative approach in this filtering step.

5. The tree building algorithm is coded in Iterative_tree_building.R. As discussed above, the algorithm depicted in Fig. 6b is true for the idealized case where all scars are observed in every cell where they are present, and no scars are observed in the incorrect cells. Most of the coded tree building algorithms corrects for the fact that this idealized case is never experimentally attained. In particular, the tree building algorithm does not identify the maximally connected scar but the scar that would have been maximally connected if its detection rate had been 1. To do this, the algorithm computes the detection rate of every scar under the assumption that it is the earliest

created scar (in that case, it should have been present in every cell so the detection rate can be computed). The algorithm then tests the congruence between the observed connections and detection rate if the scar would have been the earliest created scar and selects the scar with the highest congruence.

Both in the ideal case and in the real case, tree building is an iterative process where the earliest scar is determined from a scar graph, removed from the scar graph, after which the earliest scar is determined from the resulting scar graph, and so forth. In the real case, an additional iterative structure is superimposed on this to detect "weak scars," scars that are below a detection threshold and whose lineage information is not supported by other scars (meaning the connections in the scar graph are also formed by other scars). After identifying one or more weak scars during tree building, the weak scars are removed from the dataset and the tree building is restarted using the scar graph made from the resulting dataset. In Iterative_tree_building.R, the tree building is coded in lines 273–434 and the resulting tree structure is encoded as edgelists in the dataframe tree.summary.

In lines 436–545, the position of the cells in the tree is determined and encoded in the dataframe correct.cell.placement.

In lines 548–649, the tree is collapsed, and the cells are assigned to the correct nodes in the collapsed tree. Multiple scars can be created in the same cell and in this step, the linear branch that is created by these scars is collapsed into a single node.

In lines 652–714, the number of cells, stratified by cell type, is calculated for every node in the tree.

Since scarring is a stochastic process, some cell divisions will not be marked by a scar, resulting in a lineage tree with nodes that can split into three or more branches. For correctness, the algorithm indicates which splits may have occurred after an unobserved cell division by counting the cells in every branch and comparing all subtrees that branch off from one node. Branches with lower-than-expected cell numbers are flagged as "off-main." This determination is done in lines 716–795.

In lines 797–858, edge lists for the final tree are created, and in lines 860–943, these edgelists together with the previously calculated cell type counts per node are used to visualize the full tree. In lines 945–1040, users can zoom into sets of cell types and plot the lineage tree for only these cell types, to visualize how, for example, certain lineages are distributed over the tree.

Finally, in lines 1042–1045, tree statistics are summarized, and in lines 1047–1158, the number of inferred scars, scars that have not been observed in a cell but can be inferred due to the tree placement of that cell, are calculated.

6. The tree building parameters are set in lines 25–46 of Iterative_tree_building.R. The parameters max.scar.p and max.larvae control the maximum scar creation probability of scars in the tree; as mentioned above, scars have a wide range of creation probabilities (Fig. 2b) and the more scars have been created, the more likely it becomes that high-probability scars are created more than once. Since the algorithm considers every scar as a unique event, a scar that is created more than once will lead to an incorrect tree and therefore scars with a high creation probability are filtered out. In our experience, setting max.scar.p to 0.001 (the default) for 100–300 scars, and to 0.01 for 50–100 scars works well; however, different scar probability distributions will influence the optimal parameter choice. The parameter max.larvae is another way to control high-probability scars; if users have multiple libraries that are integrated through Scar_comparison_between_fish.R, max.larvae is an upper bound for the amount of libraries a scar can appear in. In our experience, setting max.scar.p is sufficient, and users can verify this in their own datasets.

   Scars that, despite the filtering, are observed in a cell where they do not belong, will lead to incorrect connections in the scar graph and incorrect branches in the lineage tree. At the level of the scar graph, we implemented a final test for incorrect connections, controlled by the parameter doublet.rate (since doublets are a probable cause of incorrect connections). This test assumes that the connection strength between two scars in the scar graph (i.e., the number of times two scars are observed in the same cell) is solely due to the doublet rate and calculates the p-value of the observed connection strength under this assumption. The algorithm retains cars with a false discovery rate of less than 0.01, a parameter that is hardcoded in lines 240–242 and continues tree building without the information encoded in cells flagged as probable doublets. Note that the algorithm will still try to place these cells in the trees after the full tree structure has been determined; however, since the tree structure is built without these cells, it is possible that they are not uniquely placeable. In that case, these cells are flagged as definitive doublets and are not placed.

   The tree building algorithm revolves around determining the earliest created scar from the scar graph. The earliest created scar is the scar with the most connections after correcting for its detection rate. This correction works well for scars with high detection rates but breaks down for lower detection rates,

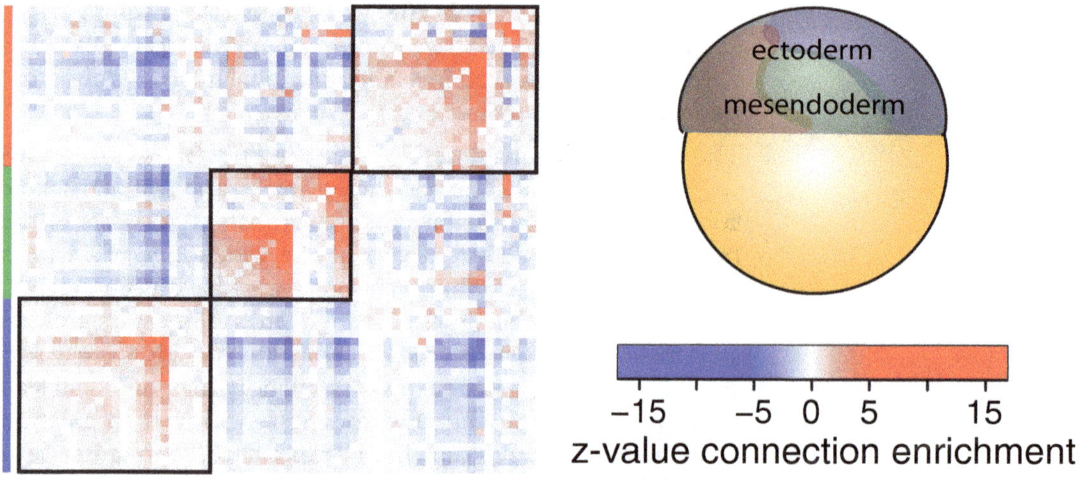

Fig. 9 Enriched connections form clusters on the fate map. We calculated the *z*-values of connection enrichments between cell types. Cell type distances were calculated based on these connection enrichments, and the cell types were clustered on these distances. Cartoon on the right: Cell type clusters determined based on scar connections form contiguous domains on the zebrafish fate map at the shield stage (yolk sac shown in orange). (Reproduced from the original publication, Ref. [22])

a cutoff set through the parameter min.detection.rate. We therefore flag scars that have too many connections for their observed detection rates and only consider them as the earliest scar if there are no scars with a detection rate higher than min.detection.rate. Scars with a lower detection rate are also flagged as "difficult scars" and if they are not supported by another scar, they are removed from the dataset as "weak scars." As a conservative measure, the algorithm restarts the tree building with these weak scars removed from the dataset.

After building the tree, the algorithm considers the different branches. Subtrees that have a cell ratio to their parent node that is less than min.branch.ratio are flagged as later splits and are plotted by dashed lines or left out in plotting.

7. As is apparent from the methods, scar filtering is a complex process. It corrects for a wide variety of artifacts that can depend on sequencing technology, single-cell encapsulation, and scar library creation, and as a result, there is no fixed prescription to set the filtering parameters. We offer two suggestions to help find the correct parameter settings for each library. The first is to run a bulk analysis of the lineage data after the first two scar filtering steps. This analysis (coded in R_scripts/Link_enrichment.R) determines which cell types have an enriched amount of lineage relationships. By clustering on the link enrichments, we were able to reproduce a coarse-grained view of zebrafish development to show that the lineage tracing data contained relevant information (Fig. 9). The

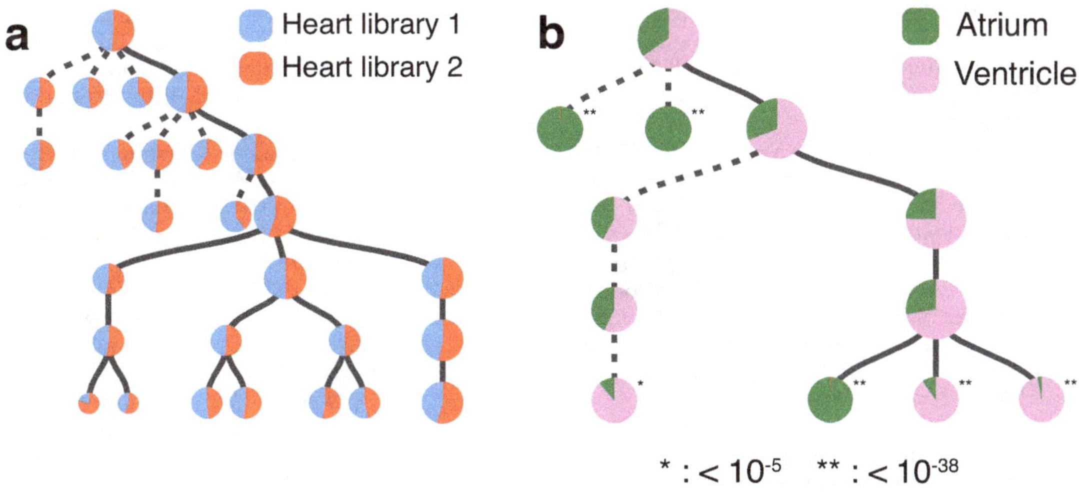

Fig. 10 Technical validation of lineage trees using split libraries. (**a**) Randomly split libraries of a single zebrafish heart show no (out of 14) leaf nodes significantly enriched for either library. (**b**) Atrial/ventricular split libraries of a single zebrafish heart show all six leaf nodes significantly enriched for either library. (Reproduced from the original publication, Ref. [23])

second suggestion is to build lineage trees for different parameter settings and assess the tree topology. Most of the filtering is designed to prevent "spurious links," a connection in the lineage tree that does not exist but is caused by an artifact such as a chimeric read or a doublet. These spurious links would strongly influence the tree topology, and this can be determined post hoc by reviewing the evidence for the scars that are highest in the tree. For example, a doublet that is not filtered out can lead to the merger of two separate branches.

Finally, users can perform sanity checks on lineage trees. One sanity check is that, in developmental lineage trees, immune cells from different body parts should be in the same lineage tree branch. Another sanity check can be implemented by splitting the library. Cells from random splits should be evenly divided over the lineage tree, whereas cells from cell type or location-based splits should show biases in lineage tree placement (Fig. 10).

References

1. Woodworth MB, Girskis KM, Walsh CA (2017) Building a lineage from single cells: genetic techniques for cell lineage tracking. Nat Rev Genet 18:230–244. https://doi.org/10.1038/nrg.2016.159
2. Spanjaard B, Junker JP (2017) Methods for lineage tracing on the organism-wide level. Curr Opin Cell Biol 49:16–21. https://doi.org/10.1016/j.ceb.2017.11.004
3. Trapnell C, Cacchiarelli D, Grimsby J et al (2014) The dynamics and regulators of cell fate decisions are revealed by pseudotemporal ordering of single cells. Nat Biotechnol 32:381–386. https://doi.org/10.1038/nbt.2859

4. Setty M, Tadmor MD, Reich-Zeliger S et al (2016) Wishbone identifies bifurcating developmental trajectories from single-cell data. Nat Biotechnol 34:637–645. https://doi.org/10.1038/nbt.3569
5. Haghverdi L, Büttner M, Wolf FA et al (2016) Diffusion pseudotime robustly reconstructs lineage branching. Nat Methods 13:845–848. https://doi.org/10.1038/nmeth.3971
6. Wolf FA, Hamey FK, Plass M et al (2019) PAGA: graph abstraction reconciles clustering with trajectory inference through a topology preserving map of single cells. Genome Biol 20:59. https://doi.org/10.1186/s13059-019-1663-x
7. Schiebinger G, Shu J, Tabaka M et al (2019) Optimal-transport analysis of single-cell gene expression identifies developmental trajectories in reprogramming. Cell 176:928–943.e22. https://doi.org/10.1016/j.cell.2019.01.006
8. Lange M, Bergen V, Klein M et al (2022) CellRank for directed single-cell fate mapping. Nat Methods 19:159–170. https://doi.org/10.1038/s41592-021-01346-6
9. Qiu X, Zhang Y, Martin-Rufino JD et al (2022) Mapping transcriptomic vector fields of single cells. Cell 185:690–711.e45. https://doi.org/10.1016/j.cell.2021.12.045
10. Barker N, van Es JH, Kuipers J et al (2007) Identification of stem cells in small intestine and colon by marker gene Lgr5. Nature 449: 1003–1007. https://doi.org/10.1038/nature06196
11. Livet J, Weissman TA, Kang H et al (2007) Transgenic strategies for combinatorial expression of fluorescent proteins in the nervous system. Nature 450:56–62. https://doi.org/10.1038/nature06293
12. Lu R, Neff NF, Quake SR, Weissman IL (2011) Tracking single hematopoietic stem cells in vivo using high-throughput sequencing in conjunction with viral genetic barcoding. Nat Biotechnol 29:928–933. https://doi.org/10.1038/nbt.1977
13. Naik SH, Perié L, Swart E et al (2013) Diverse and heritable lineage imprinting of early haematopoietic progenitors. Nature 496:229–232. https://doi.org/10.1038/nature12013
14. Sun J, Ramos A, Chapman B et al (2014) Clonal dynamics of native haematopoiesis. Nature 514:322–327. https://doi.org/10.1038/nature13824
15. Frumkin D, Wasserstrom A, Kaplan S et al (2005) Genomic variability within an organism exposes its cell lineage tree. PLoS Comput Biol 1:e50. https://doi.org/10.1371/journal.pcbi.0010050
16. Lodato MA, Woodworth MB, Lee S et al (2015) Somatic mutation in single human neurons tracks developmental and transcriptional history. Science 350:94–98. https://doi.org/10.1126/science.aab1785
17. Ju YS, Martincorena I, Gerstung M et al (2017) Somatic mutations reveal asymmetric cellular dynamics in the early human embryo. Nature 543:714–718. https://doi.org/10.1038/nature21703
18. Pei W, Feyerabend TB, Rössler J et al (2017) Polylox barcoding reveals haematopoietic stem cell fates realized in vivo. Nature 548:456–460. https://doi.org/10.1038/nature23653
19. McKenna A, Findlay GM, Gagnon JA et al (2016) Whole-organism lineage tracing by combinatorial and cumulative genome editing. Science 353:aaf7907. https://doi.org/10.1126/science.aaf7907
20. Junker JP, Spanjaard B, Peterson-Maduro J et al (2017) Massively parallel clonal analysis using CRISPR/Cas9 induced genetic scars. bioRxiv:056499
21. Frieda KL, Linton JM, Hormoz S et al (2017) Synthetic recording and in situ readout of lineage information in single cells. Nature 541: 107–111. https://doi.org/10.1038/nature20777
22. Spanjaard B, Hu B, Mitic N et al (2018) Simultaneous lineage tracing and cell-type identification using CRISPR–Cas9-induced genetic scars. Nat Biotechnol 36:469–473. https://doi.org/10.1038/nbt.4124
23. Hu B, Lelek S, Spanjaard B et al (2022) Origin and function of activated fibroblast states during zebrafish heart regeneration. Nat Genet 54: 1227–1237. https://doi.org/10.1038/s41588-022-01129-5
24. Schmidt ST, Zimmerman SM, Wang J et al (2017) Quantitative analysis of synthetic cell lineage tracing using nuclease barcoding. ACS Synth Biol 6:936–942. https://doi.org/10.1021/acssynbio.6b00309
25. Alemany A, Florescu M, Baron CS et al (2018) Whole-organism clone tracing using single-cell sequencing. Nature 556:108–112. https://doi.org/10.1038/nature25969
26. Raj B, Wagner DE, McKenna A et al (2018) Simultaneous single-cell profiling of lineages and cell types in the vertebrate brain. Nat Biotechnol 36:442–450. https://doi.org/10.1038/nbt.4103
27. Wang GC, Wang Y (1997) Frequency of formation of chimeric molecules as a consequence of PCR coamplification of 16S rRNA genes from mixed bacterial genomes. Appl Environ

Microbiol 63:4645–4650. https://doi.org/10.1128/aem.63.12.4645-4650.1997

28. Satija R, Farrell JA, Gennert D et al (2015) Spatial reconstruction of single-cell gene expression data. Nat Biotechnol 33:495–502. https://doi.org/10.1038/nbt.3192
29. Wolf FA, Angerer P, Theis FJ (2018) SCANPY: large-scale single-cell gene expression data analysis. Genome Biol 19:1–5
30. Kivioja T, Vähärautio A, Karlsson K et al (2012) Counting absolute numbers of molecules using unique molecular identifiers. Nat Methods 9:72–74. https://doi.org/10.1038/nmeth.1778
31. Li H (2013) Aligning sequence reads, clone sequences and assembly contigs with BWA-MEM

Chapter 13

In Vivo Clonal Analysis Using MADM with Spatiotemporal Specificity

Yuwei Peng, Efan Wang, Zhongfu Shen, Song-Hai Shi, and Jian Ma

Abstract

Mosaic analysis with double markers (MADM) is a powerful in vivo lineage tracing technique. It utilizes Cre recombinase-dependent interchromosomal recombination to restore the stable expression of two fluorescent proteins sparsely in individual dividing stem or progenitor cells and their progenies. Here, we describe the application of this technique for quantitative lineage analysis of radial glial progenitors in the developing mouse neocortex at the single-cell resolution.

Key words MADM, Lineage tracing, Clonal analysis, Interchromosomal recombination, Serial sectioning, 3D reconstruction

1 Introduction

Lineage analysis stands as a cornerstone in deciphering the developmental trajectories of diverse cells in living organisms, tracing their progenitor origins, and elucidating their eventual destinies. Insights gained from lineage analysis are critical for advancing our knowledge in developmental biology under both normal and disease conditions. Historically, genetic mosaics have been a valuable tool in dissecting the development and organization of multicellular organisms [1]. In Drosophila, genetic mosaics have been created through interchromosomal recombination in conjunction with a repressible cell marker (i.e., mosaic analysis with a repressive cell marker, MARCM) before cell division [2], a methodology not directly applicable in mammals due to different chromosome dynamics during mitosis [3, 4]. While the earlier strategies for generating genetic mosaics in mice primarily relied on Cre-loxP-mediated intrachromosomal recombination, the mosaic analysis with double markers (MADM) approach ingeniously took

Yuwei Peng and Efan Wang contributed equally with all other contributors.

Jorge García-Marqués and Tzumin Lee (eds.), *Lineage Tracing: Methods and Protocols*, Methods in Molecular Biology, vol. 2886, https://doi.org/10.1007/978-1-0716-4310-5_13,

advantage of Cre-mediated interchromosomal recombination, which results in genetic mosaics due to the relatively low recombination rate [5, 6].

MADM offers the advantage of simultaneously labeling the two daughter cells of dividing progenitor cells with distinct fluorescence. In particular, Cre recombinase-driven interchromosomal recombination in the G_2 phase of the dividing progenitor cells, followed by X chromosome segregation (G_2-X), reconstitutes one of the two fluorescent markers, enhanced green fluorescent protein (EGFP, green) or tandem dimer Tomato (tdTomato, red), in each of the two daughter cells, as well as their corresponding descendants for lineage tracing (*see* Fig. 1). With distinct fluorescent labeling of the two daughter cell lineages, it becomes possible to assess the division mode of labeled progenitor cells to be symmetric or asymmetric. In addition, upon G_2-Z recombination and segregation, G_1, or G_0 recombination events, green and red fluorescent proteins are restored concurrently in the same cell(s), resulting in double-labeled (i.e., yellow) cell(s).

By crossing the MADM mouse lines with specific *CreER*^{T2} strains, in which the inducible recombinase is expressed in specific stem or progenitor cells, one can achieve precise spatiotemporal control of the labeling of lineages of interest. Moreover, the labeling density can be tuned by adjusting the level of tamoxifen administration, thereby allowing lineage tracing of dividing progenitors at the single-cell resolution, i.e., clonal analysis [7–9]. The power of MADM has been demonstrated in numerous applications, such as lineage analyses of cerebellar granule cells [5] and neocortical and hippocampal neurons and glial cells [7, 9–11]. In addition, MADM allows to selectively ablate endogenous genes residing on the same chromosome as the MADM allele in specifically labeled cells for mosaic functional analysis, whereas other labeled cells serve as the wild-type (WT) control [10, 12, 13].

The application of MADM spans from developmental biology to stem cell biology to neurobiology, allowing researchers to trace cell lineages and understand the cellular dynamics of complex tissues at a fine level. The utilization of MADM for in vivo clonal analysis requires genetic crossing of MADM mouse strains with inducible *CreER*^{T2} lines. Subsequent steps involve tamoxifen administration, sample processing, fluorescence imaging, and three-dimensional (3D) reconstruction to recover and document the labeled cells at the desired time and condition. Here, we outline the use of MADM for a systematic and quantitative lineage analysis of individual radial glial progenitors (RGPs) in the developing mouse neocortex (*see* Fig. 2).

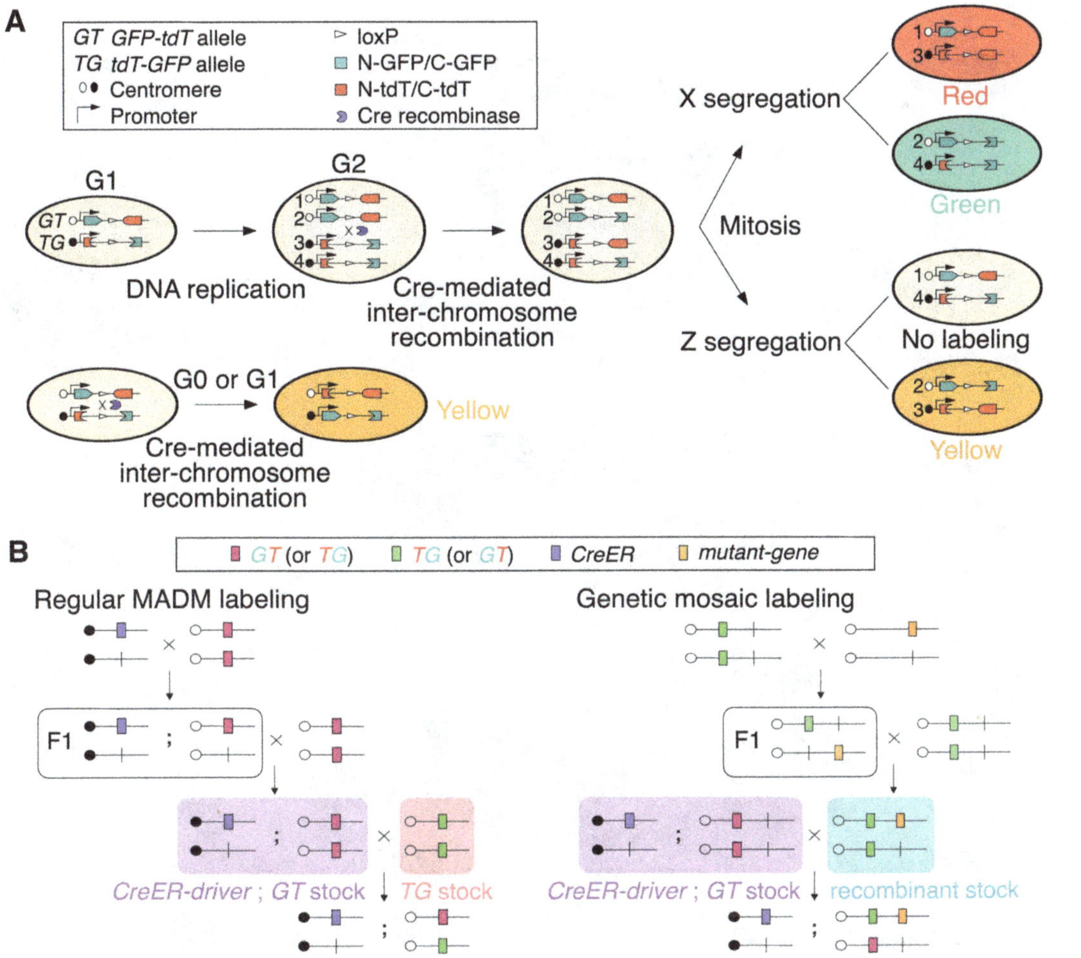

Fig. 1 Schematic of the MADM design and the mouse breeding strategy for experimental application. (**a**) The MADM cassettes are inserted into the same loci close to the centromere of a chromosome. During the G_2 phase, Cre recombinase mediates interchromosomal recombination of the duplicated chromosomes, leading to the reconstitution of the fluorescent markers EGFP and tdTomato. Following the X-segregation process, these markers are separated into the two new daughter cells. Consequently, each daughter cell and its subsequent descendant(s) are labeled with a permanent and distinct fluorescence (either EGFP/green or tdTomato/red). In comparison, during G_2-Z recombination and segregation events, or G_1 or G_0 recombination events, both EGFP and tdTomato are simultaneously reinstalled in the same cell, producing cells with a dual (EGFP/green and tdTomato/red as yellow) fluorescence. (**b**) For regular MADM labeling, a specific *CreER* line is crossed with MADM-*GT/GT* (or *MADM-TG/TG*) line to generate *CreER* and *GT* double heterozygous mice (*CreER/+;MADM-GT/+*) (F1). *CreER/+;MADM-GT/+* mice are crossed with *MADM-GT/GT* mice to generate *CreER/+*;MADM-*GT/GT* mice, which are then crossed with *MADM-TG/TG* mice to generate *CreER/+;MADM-GT/TG* mice for experimental use. For MADM genetic mosaic labeling, the mutant mouse line of the gene of interest is crossed with *MADM-TG/TG* (or *MADM-GT/GT*) to generate trans-heterozygous mice (F1). The trans-heterozygous mice are backcrossed to *MADM-TG/TG* (or *MADM-GT/GT*) to generate homozygous *MADM-TG/TG* cassette and, through meiotic recombination, the mutant allele of the gene of the interest is genetically linked to *MADM-TG/TG*. These recombinant mice are then crossed with *CreER/+;MADM-GT/GT* mice to generate the corresponding mice for experimental use

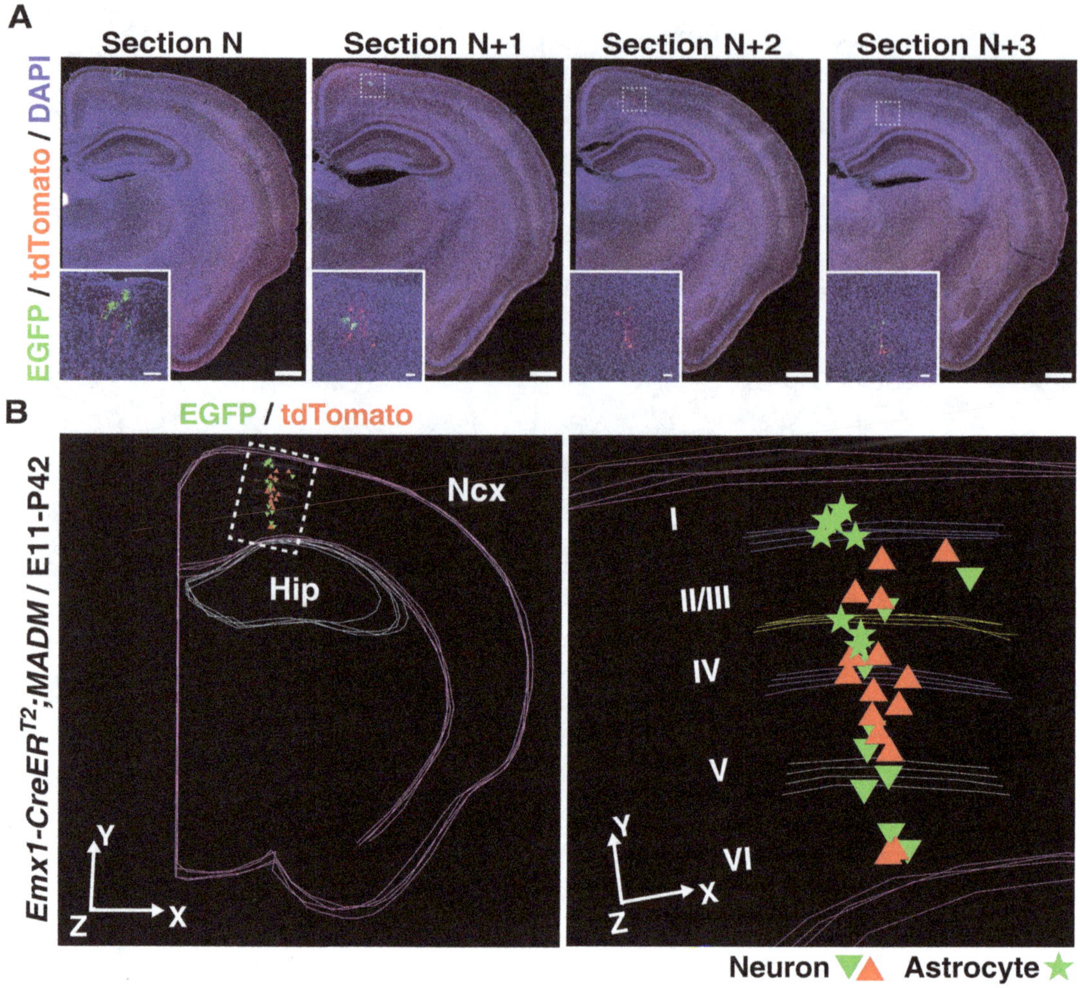

Fig. 2 Serial sectioning and 3D reconstruction of a MADM-labeled mouse neocortical clone. (**a**) Slide scanner images of four consecutive sections of the *Emx1-CreER*^{T2};*MADM*-labeled neocortex (TM administration at E11; tissue collection at P42) stained with antibodies against EGFP (green) and tdTomato (red), and with DAPI (blue). Scale bars: 500 and 50 μm. (**b**) 3D reconstruction image of the labeled cells representing a typical G_2-X clone in A using the Neurolucida software. Zoom-in image of the clone (broken line) is shown to the right. The magenta lines represent the contour of the neocortex, while the gray lines represent the contour of the hippocampus. Other colored lines indicate the layer boundaries. Neurons are represented by red or green triangle symbols, in which color indicates the type of fluorescent protein expressed. Astrocytes are represented by green star symbols. The x/y/z axes follow the spatial orientation of the clone with the y-axis parallel to the brain midline and extended dorsally. Hip hippocampus, Ncx neocortex

2 Materials

2.1 Mouse Lines

MADM-6^{GR} Jax mice stock# 006041 [5].

MADM-6^{RG} Jax mice stock# 006067 [5].

MADM-11^{GT} Jax mice stock# 013749 [6].

MADM-11^{TG} Jax mice stock# 013751 [6].

Emx1-CreER^{T2} Jax mice stock#027784 [14].

MADM alleles are also available on other chromosomes [15–17].

2.2 Chemicals and Antibodies

1. Tamoxifen (TM) corn oil stock solution (10 mg/mL): Dissolve 400 mg of TM in 40 mL corn oil (*see* **Note 1**). Shake at 37 °C and protect from light until TM is completely dissolved and the mixture becomes a translucent oil solution (*see* **Note 2**). Aliquot and store away from light at 4 °C (*see* **Note 3**) or freeze at −20 °C (*see* **Note 4**). The TM corn oil stock solution may be diluted as appropriate in experiments (*see* **Note 5**).
2. Avertin stock solution: Dissolve 10 mg of 2,2,2-tribromoethanol (avertin) in 10 mL 2-methyl-2-butanol. Shake at room temperature and protect from light until dissolved and store at 4 °C away from light (*see* **Note 6**).
3. Avertin working (2.5%) solution: Dilute Avertin stock solution in 1:40 with 1 mM Tris–HCl (pH 7.4), 0.25 mM EDTA (pH 7.4), and 0.8% NaCl.
4. Paraformaldehyde (PFA) solution (4%): Dissolve 4 g of PFA in phosphate-buffered saline (PBS) for a final volume of 100 mL. Place on a heating magnetic stirrer until the powder dissolves, then adjust pH to 7.4 after cooling to room temperature. Filter with a 0.22-μm filter set, and store at 4 °C.
5. Agarose gel (4%): Dissolve 2 g of agarose in PBS for a final volume of 50 mL. Heat in a microwave until dissolved and pour the gel into the cast.
6. Sucrose solution (30%): Dissolve 15 g of sucrose in PBS for a final volume of 50 mL. Store at 4 °C.
7. Blocking solution: 5% donkey serum, 0.5% Triton X-100 in PBS, and dissolve 0.5 g of sodium azide in the solution (*see* **Note 7**). After filtering with a 0.22-μm filter set, aliquot, and store at −20 °C.
8. Antibody solution: 1% donkey serum, 0.5% Triton X-100 in PBS, and dissolve 0.5 g sodium azide in the solution. After filtering with a 0.22-μm filter set, aliquot, and store at −20 °C.
9. PBS with Triton (PBST): 0.1% Triton X-100 in PBS.
10. PCR primers (*see* Table 1).
11. Antibodies and dyes (*see* Table 2).

2.3 Instruments

1. Peristaltic pump or perfusion syringes with a 3-way valve.
2. 24-well plate.
3. 500 mL 0.22-μm filter set.

Table 1
PCR primers

Gene	Primer sequence	Positive band	Negative band
GT&TG	TGGAGGAGGACAAACTGGTCAC TCAATGGGCGGGGGTCGTT TTCCCTTTCTGCTTCATCTTGC	230 bp	/
GT/TG	CTGTCCGCGGGGGGACGGCTGCCTTCG GAACTTCGGAGATCTTTGATACCAAC	240 bp (TG) 520 bp (GT)	/
CreER^{T2}	GCCTGGTCTGGACACAGTGCC CTGTCTGCCAGGTTGGTCAGTAAGC	398 bp	/

Table 2
Antibodies and dye

Antibody/dye	Source	Identifier	Recommended dilution ratio
Chicken anti-GFP	Aves	Cat# GFP-1020; RRID: AB_10000240	1:1000
Rabbit anti-RFP	Rockland	Cat# 600-401-379; RRID: AB_2209751	1:1000
Alexa flour 488-Donkey anti-chicken	Jackson ImmunoResearch Labs	Cat# 703-546-155; RRID: AB_162543	1:1000
Alexa flour 555-Donkey anti-rabbit	Thermo Fisher Scientific	Cat# A31572; RRID: AB_162543	1:1000
DAPI	Thermo Fisher Scientific	Cat# D1306; RRID: AB_2629482	1:1000

4. Horizontal shaker.
5. Vibratome or Cryostat.
6. Slide scanner.
7. Confocal microscope.
8. Neurolucida software (MBF Bioscience).

3 Methods

3.1 Mouse Breeding Strategy (See Fig. 1b)

For regular MADM labeling of stem/progenitors:

1. Cross the specific *CreER*^{T2} mouse strain (e.g., *Emx1-CreER*^{T2}) with MADM-*GT/GT* or MADM-*TG/TG* mouse line to

generate *CreER*^{T2}*;MADM-GT/GT* or *CreER*^{T2}*;MADM-TG/TG* mice (primers for genotyping, *see* Table 1).

2. Cross *CreER*^{T2}*;MADM-GT/GT* with *MADM-TG/TG* or *CreER*^{T2}*;MADM-TG/TG* with *MADM-GT/GT* to generate *CreER*^{T2}*;MADM-GT/TG* mice.

For generating genetic mosaic for functional analysis:

1. Choose the MADM alleles in the same chromosomes as the gene of interest. The probability of meiotic recombination can be estimated by determining the genetic distance between the chromosomal loci of the gene of interest and the MADM cassette on a particular chromosome.
2. Cross the mutant mouse line of the gene of interest with *MADM-TG/TG* (or *MADM-GT/GT*). This breeding results in offspring that are *TG* heterozygous, and also heterozygous for the mutant allele.
3. Following the initial cross, cross the above trans-heterozygote mice with *MADM-TG/TG* mice again to obtain the homozygous *MADM-TG/TG* offspring. Notably, through meiotic recombination, the mutation allele of the gene of interest is genetically linked to the *MADM-TG* cassette through meiotic recombination on the corresponding MADM chromosome, resulting in the *MADM-TG/TG;mutant gene/+* recombinant mice. As a result, the tdTomato-labeled cells are mutant cells, whereas the EGFP-labeled cells are control cells.
4. Generate the *Cre*-driver;*MADM-GT/GT* mouse line (as WT) as **step 1** in the above "For regular MADM labeling" breeding strategy.
5. Cross the *MADM-TG/TG;*mutant-*gene*/+ recombinant mouse line with the *Cre*-driver;*MADM-GT/GT* WT mouse line to generate the genetic mosaic mouse for experimental analysis (*see* **Note 8**).

3.2 Mouse-Timed Pregnancy

1. At 4–6 pm on the first day, pair adult male and female mice of the respective genotypes (e.g., *Emx1-CreER*$^{T2+/+}$*;MADM-GT/GT* and *MADM-TG/TG*; or *Emx1-CreER*$^{T2+/+}$*;MADM-TG/TG* and *MADM-GT/GT*) in individual cages according to a male to female ratio of 1:1 or 1:2.
2. Starting from the following day, and for the next 2–3 consecutive days (*see* **Note 9**), check the vaginal plugs of the females before 10 am daily (*see* **Note 10**). Transfer the females with vaginal plugs to separate cages and record the corresponding date as the embryonic day 0 (E0) (*see* **Note 11**).
3. For preparing the foster females, 2 days before the MADM mice are mated, CD-1 mice are mated as described above to obtain timed pregnant females (*see* **Note 12**).

3.3 TM Induction

At the desired embryonic or neonatal stages, TM is administrated to timed pregnant females or neonatal pups to induce interchromosomal recombination and labeling of dividing cells. TM administration is typically scheduled at 1 p.m. Different doses of TM are used to achieve the desired labeling density (e.g., the dose for sparse clonal labeling is described below).

1. Embryonic stage induction: Weigh-timed pregnant females. For E10–E14, TM corn oil solution is injected intraperitoneally at a dosage of 10–50 μg/g body weight [7, 11, 13]. For E15–E17, TM corn oil solution is administered via oral gavage at a dosage of 50–400 μg/g body weight [10] (*see* **Notes 13**, **14**, and **15**).
2. Neonatal stage induction: Weigh neonatal pups. TM corn oil is administered via intraperitoneal injection at a dosage of 10–100 μg/g body weight (*see* **Note 16**).

3.4 Cesarean Section and Fostering

To avoid dystocia, a cesarean section is performed at E18–E19 to recover the embryos, which are then fostered with nursing ICR (Institute of Cancer Research) female mice (*see* **Note 17**).

1. Euthanize E18–E19 pregnant dams by cervical dislocation and perform a midline incision to expose the uterus.
2. Incise the uterine wall carefully to extract the embryos one by one.
3. Dissect the embryo carefully from the yolk sac and amnion, and sever the umbilical cord.
4. Transfer the embryo to a clean, warm towel and gently swipe it with a cotton swab to remove amniotic fluid and secretions.
5. Wipe away any fluid from the nasal and oral areas while simultaneously stimulating respiration.
6. Wait for the pup to initiate spontaneous breathing. Indicators of spontaneous breathing include skin turning pink, vigorous movement, and the emission of squeaks when the tail is gently pinched.
7. Remove the foster mother from her original pups (*see* **Note 18**).
8. Gently rub the foster pups with bedding material from the foster mother, and put together the foster pups with the natural litter of the foster mother (if necessary, remove some of the natural litter pups).

3.5 Brain Collection

1. Anesthetize the animal by intraperitoneal injection of 2.5% Avertin solution at a dosage of 10–20 μL/g body weight.

Use forceps to gently pinch the paws of the mouse and proceed after confirming the absence of a pain response.

2. Secure the mouse limbs to the perfusion tray using tape. Make an incision in the mouse skin along the lower edge of the ribs and then excise the ribs on both sides.
3. Use hemostats to grasp and evert the cut sternum to expose the heart.
4. Rinse the perfusion hose with PBS and fill it with PBS.
5. Make an incision in the right atrial appendage and insert the injection needle connecting to the hose attaching to a peristaltic pump or a syringe filled with PBS into the left ventricle of the heart.
6. Turn on the perfusion pump (or perfuse manually using 10 mL syringes) and infuse ~10 mL PBS per 20 g body weight (the liver should turn pale) at a stable rate (~2 mL/min).
7. Replace the perfusion solution with precooled 4% PFA, and infuse ~30 mL PFA per 20 g body weight at a stable rate (~2 mL/min). Stop the perfusion once the tail and limbs of the mouse become stiff, and then remove the mouse from the perfusion tray.
8. Decapitate the mouse and carefully extract the brain from the skull.
9. Submerge the extracted brain in 4% PFA at 4 °C for 6–8 h with gentle agitation.
10. Wash the fixed brain three times with PBS at room temperature and then prepare for serial sectioning using a vibratome or cryostat.

3.6 Serial Sectioning Using a Vibratome

1. Prepare 4% agarose solution in PBS. Heat the solution in a microwave until completely dissolved and cool it at room temperature (RT) until it is not hot to touch.
2. Immerse the brain sample in the agarose solution in a Petri dish and cool in a 4 °C refrigerator.
3. Cut out the brain block carefully.
4. Fill the vibratome chamber with PBS and set the sectioning parameters (speed: 0.6 mm/s; amplitude: 0.8 mm). Carefully install the blade and adjust the angle.
5. Attach the brain/agarose block to the specimen disk using super glue.
6. Trim the block carefully and start the serial sectioning process to cut the brain into 60–100-μm sections. Collect the sections sequentially using a brush and place them into six consecutive wells of 24-well plates filled with PBS in order (*see* **Note 19**).

7. The sections can be stored in a 4 °C refrigerator. For long-term storage, include 0.01% sodium azide (NaN_3) in PBS.

3.7 Serial Sectioning Using a Cryostat

1. Immerse the PFA-fixed brain in 10% sucrose and agitate gently for about 4–8 h at 4 °C until it sinks to the bottom.
2. Replace the buffer with 20% sucrose and repeat **step 1**.
3. Replace the buffer with 30% sucrose and repeat **step 1**.
4. Replace the buffer with fresh 30% sucrose and repeat **step 1**.
5. Prepare Tissue-Tek Cryomold and add a small amount of optimal cutting temperature (OCT) compound to cover the bottom of the mold.
6. Remove the brain from 30% sucrose and dry it with absorbent paper. Lay it into the mold with the dorsal side up and the rostral side facing forward. Fill the mold with OCT compound to cover the brain.
7. Place the mold in a −80 °C freezer for a minimum of 15 min.
8. Extract the frozen brain block and place it into the cryostat chamber (*see* **Note 20**).
9. Label the frontal side of the brain block, remove the mold, and trim the block to make sure the bottom is flat.
10. Install the knife to the holder and make sure that the stage is stable.
11. Place the specimen disk inside the chamber to cool down and add the OCT compound to the surface. Once completely frozen, trim the OCT compound to create a flat and horizontal surface.
12. Add a small amount of OCT compound onto the flat surface, and rapidly mount the brain block vertically (frontal side up) onto the disk. Wait a few seconds and consolidate with an extra O.C.T. compound around it to ensure the block is stable enough during sectioning.
13. Load the specimen disk onto the object holder and set the feed-in to 60–100 μm.
14. Start the serial sectioning process, collect the sections sequentially using the forceps to lift the edge of the sections and place them into six consecutive wells of 24-well plates filled with PBS in order (*see* **Note 19**).
15. After washing out the OCT compound, the sections can be stored in a 4 °C refrigerator. For long-term storage, include 0.01% sodium azide (NaN_3) in PBS.

3.8 Immunostaining

1. After preparing the brain sections using either vibratome or cryostat, wash the sections carefully with PBS using a horizontal shaker at room temperature three times for 10 min each time (*see* **Note 21**).
2. Incubate the sections with blocking solution for 2 h at room temperature on a horizontal shaker.
3. Remove the blocking solution carefully and incubate the sections with the primary antibody solution at 4 °C on a horizontal shaker for 24 h.
4. Remove the primary antibody solution carefully and wash three times with PBST for 10 min each time at room temperature.
5. Remove PBST carefully and incubate the sections with the secondary antibody solution. Cover the 24-well plate with aluminum foil and incubate at room temperature on a horizontal shaker for 2 h.
6. Remove the secondary antibody solution carefully and wash three times with PBS for 10 min each time.

3.9 Serial Section Mounting

1. Prepare six 5-cm diameter Petri dishes filled with PBS. Using a brush, gently transfer all brain slices from the wells into Petri dishes, maintaining the order of wells of the 24-well plate (i.e., one well corresponds to one Petri dish).
2. Organize the brain sections collected at intervals in each Petri dish, arranging them from anterior to posterior.
3. Prepare a 15-cm Petri dish filled with PBS.
4. Transfer the anterior-most brain sections gently from each 5-cm Petri dish into the 15-cm Petri dish and arrange them sequentially.
5. Immerse one side of the glass slide into PBS, and use a brush to gently position the brain sections onto the appropriate location on the slide, maintaining the correct order.
6. Transfer the next set of sequentially arranged brain sections gently from each 5-cm Petri dish into the 15-cm Petri dish, organize them in order, and then transfer them onto the glass slide.
7. Repeat **steps 4–6** until the glass slides are fully occupied with brain sections.
8. Lift the glass slide from PBS carefully, use absorbent paper to remove the PBS solution from the back of the glass slide, and then place it diagonally on absorbent paper until it is semidry.
9. Based on the number of brain sections on the glass slide, apply an appropriate amount of mounting medium. Place the coverslip diagonally on one side of the glass slide carefully. Use a brush to lower it gradually to avoid air bubbles.

10. Label the glass slide with the corresponding brain and antibody information.
11. Repeat **steps 4–9** until all the collected serial sections of brain tissue are mounted.
12. Place the mounted slide flat on a drying rack. Wait until the mounting medium solidifies, and the coverslip no longer moves. Apply nail polish to the edge between the glass slide and the coverslip.
13. After the nail polish dries, organize the slides sequentially, insert them into a film box, and store them in a 4 °C refrigerator away from the light.

3.10 Fluorescence Imaging

1. Use a slide scanner to scan serially collected brain sections on the glass slides and acquire images in 405 nm (DAPI (4′,6-diamidino-2-phenylindole)for nuclei labeling), 488 nm (EGFP), and 555 nm (tdTomato) channels.
2. Based on the image data obtained with the scanner, record the serial number and specific location of the brain sections with labeled fluorescent cells and clones.
3. If necessary, employ confocal laser scanning microscopy to acquire high-magnification images of the specific regions of the brain sections with labeled fluorescent cells and clones (*see* **Note 22**).

3.11 3D Reconstruction

1. Fluorescent labeled cell tracing: Load the acquired images into the Neurolucida software. Establish the precise pixel dimensions for both the X and Y axes within the software workspace, congruent with the image's pixel dimensions.
2. Use Neurolucida software to trace labeled fluorescent cells and draw the contour of the respective brain region and subregions (e.g., neocortical layer boundaries). When drawing lines, zoom in and click the cursor to draw lines as smoothly and neatly as possible, allowing subsequent sequential overlaying of sections for 3D reconstruction.
3. Combine all the XML project files of the sections into a DAT project file. In the initial menu bar, select "open data file into serial section" under the file option and set the section cut thickness to the actual cutting thickness.
4. Choose "define automove" to create an operating space in the workplace. Use the midline of the brain region contour or the edge of the cortex as the axis line and move the brain sections to the custom operating space. Before dragging, select "move image and tracer" and select every aligned section simultaneously so that everything in the operating space can be dragged together. Finally, save the finished project files in DAT format.

5. Proceed by selecting the 3D viewer to transition into the 3D rendering interface. The visualization angle is adjustable, and user-defined actions and keyframes can be established to generate videos (*see* Fig. 2).

4 Notes

1. Weigh 400 mg of TM powder into a 50-mL Falcon tube and add 40 mL of corn oil. Before adding corn oil, shake the TM powder to disperse it along the wall of the tube; this helps the complete dissolution of TM.
2. It usually takes less than 24 h for TM to completely dissolve. If TM has absorbed moisture, the dissolution time significantly increases, which can be an indicator of TM degradation due to moisture.
3. TM corn oil solution stored at 4 °C can generally be used for 2–3 weeks. If crystal precipitation is observed, the drug concentration may have changed significantly, and a new solution should be made.
4. TM corn oil solution that has been frozen for a long time must be completely thawed to make sure that it is clear and precipitation-free before use.
5. When diluting, add fresh, appropriate amount of corn oil and mix thoroughly before use.
6. When preparing the avertin stock solution, make sure that it is completely dissolved and label the date. The stock solution can be stored at 4 °C away from the light for a period of time.
7. Make sure to follow safety guidelines and use appropriate personal protective equipment (PPE) when working with hazardous material sodium azide.
8. For maximal temporal efficiency in the generation of MADM genetic mosaic mice, the breeding steps rendering the MADM cassettes homozygous can be skipped and all the F1 mice (heterozygous for MADM cassettes, mutant gene, and Cre, respectively) are used directly to set up the experimental mating (*see* Fig. 1b).
9. It is advisable to separate the females for more than 3–4 days of inspection without an obvious plug from males for 2 weeks. Subsequent mating should only be conducted after confirming that the females are not pregnant.
10. Following each mating cycle, males need approximately 3 days of rest before the next mating cycle. If a female plugged with a male does not exhibit an actual pregnancy for two consecutive weeks (i.e., false plug), it is advisable to replace the male.

11. When housing individual animals in separate cages, ensure a conducive breeding environment.
12. The parturition of the foster dam should occur 1–2 days prior to the parturition of the MADM mouse to ensure that the age of the pups of the foster dam is similar to that of the MADM mouse pups, thereby enhancing the survival rate of the foster pups.
13. TM corn oil solution has a high viscosity and does not disperse rapidly from the injection site. Therefore, the needle should be withdrawn slowly postinjection to prevent the TM corn oil solution from leaking out.
14. Choice of tamoxifen (TM) or 4-hydroxy-tamoxifen. TM is metabolized into three intermediate and final metabolites, including N-desmethyl-tamoxifen, 4-hydroxy-tamoxifen, and endoxifen. Among them, 4-hydroxy-tamoxifen has a higher affinity for estrogen receptors and is rapidly metabolized. As a result, it can act more efficiently and quickly to activate CreER recombinase.
15. The TM induction dosage should be determined based on specific experimental requirements. The dosage ranges provided in this protocol have been tested using *Emx1-CreER* and *MADM11* mice and found to be appropriate for sparse labeling of radial glial progenitors in the developing neocortex for clonal analysis and lineage tracing at the single-cell resolution.
16. The abdominal cavity of neonatal pups is compact, and abdominal pressure is elevated, increasing the likelihood of fluid leakage. Therefore, the injection needle should be withdrawn slowly.
17. Monitor the condition of the pregnant females closely during the night of E18. If any signs such as curling up, tremors or bleeding are observed, it is imperative to perform a cesarean section immediately.
18. Select a foster mother from a strain known for its strong maternal instincts.
19. For example, place the 6n section into well 1, the $6n + 1$ section into well 2, the $6n + 2$ section into well 3, and so forth. This method simplifies the separation of two adjacent sections at a 300–500 μm interval within the same well.
20. If the tissue has been stored in the −80 °C freezer for more than 30 min, leave it in the chamber for a minimum of 20 min.
21. Exercise caution during all aspiration steps to avoid any loss or damage to the serial brain sections.
22. Prior to imaging, make sure that the designated imaging areas encompass the complete set of labeled cells and clone.

Acknowledgments

This work was supported by the STI2030-Major Projects (2021ZD0202300), the National Natural Science Foundation of China (32021002), Beijing Outstanding Young Scientist Program (BJJWZYJH01201910003012), Beijing Municipal Science & Technology Commission (Z221100003422011 and Z211100003321001), Chinese Institute for Brain Research (Beijing), and the New Cornerstone Investigator Program.

References

1. Zugates CT, Lee T (2004) Genetic mosaic analysis in the nervous system. Curr Opin Neurobiol 14(5):647–653
2. Lee T, Luo L (1999) Mosaic analysis with a repressible cell marker for studies of gene function in neuronal morphogenesis. Neuron 22(3):451–461
3. Koike H, Horie K, Fukuyama H et al (2002) Efficient biallelic mutagenesis with Cre/loxP-mediated inter-chromosomal recombination. EMBO Rep 3(5):433–437
4. Liu P, Jenkins NA, Copeland NG (2002) Efficient Cre-loxP-induced mitotic recombination in mouse embryonic stem cells. Nat Genet 30(1):66–72
5. Zong H, Espinosa JS, Su HH et al (2005) Mosaic analysis with double markers in mice. Cell 121(3):479–492
6. Hippenmeyer S, Youn YH, Moon HM et al (2010) Genetic mosaic dissection of Lis1 and Ndel1 in neuronal migration. Neuron 68(4): 695–709
7. Gao P, Postiglione MP, Krieger TG et al (2014) Deterministic progenitor behavior and unitary production of neurons in the neocortex. Cell 159(4):775–788
8. Beattie R, Postiglione MP, Burnett LE et al (2017) Mosaic analysis with double markers reveals distinct sequential functions of Lgl1 in neural stem cells. Neuron 94(3):517–533 e513
9. Bonaguidi MA, Wheeler MA, Shapiro JS et al (2011) In vivo clonal analysis reveals self-renewing and multipotent adult neural stem cell characteristics. Cell 145(7):1142–1155
10. Shen Z, Lin Y, Yang J et al (2021) Distinct progenitor behavior underlying neocortical gliogenesis related to tumorigenesis. Cell Rep 34(11):108853
11. Xu HT, Han Z, Gao P et al (2014) Distinct lineage-dependent structural and functional organization of the hippocampus. Cell 157(7):1552–1564
12. Liu C, Sage JC, Miller MR et al (2011) Mosaic analysis with double markers reveals tumor cell of origin in glioma. Cell 146(2):209–221
13. Lv X, Li S, Li J et al (2022) Patterned cPCDH expression regulates the fine organization of the neocortex. Nature 612(7940):503–511
14. Kessaris N, Fogarty M, Ianarelli P et al (2006) Competing waves of oligodendrocytes in the forebrain and postnatal elimination of an embryonic lineage. Nat Neurosci 9(2): 173–179
15. Contreras X, Amberg N, Davaatseren A et al (2021) A genome-wide library of MADM mice for single-cell genetic mosaic analysis. Cell Rep 35(12):109274
16. Tasic B, Miyamichi K, Hippenmeyer S et al (2012) Extensions of MADM (mosaic analysis with double markers) in mice. PLoS One 7(3): e33332
17. Hippenmeyer S, Johnson RL, Luo L (2013) Mosaic analysis with double markers reveals cell-type-specific paternal growth dominance. Cell Rep 3(3):960–967

Chapter 14

CARLIN: A Mouse Line for Simultaneous Readout of Lineage Histories and Gene Expression

Sarah Bowling and Fernando D. Camargo

Abstract

The CRISPR-activated repair lineage tracing (CARLIN) mouse line uses DNA barcoding to enable high-resolution tracing of cell lineages in vivo (Bowling et al, Cell 181, 1410–1422.e27, 2020). CARLIN mice contain expressed barcodes that allow simultaneous interrogation of lineage and gene expression information from single cells. Furthermore, barcode editing is fully inducible, resulting in cell lineage labeling that can be performed at any time point in development or adulthood. This chapter details the protocols followed for maintaining CARLIN mice, inducing barcoding, and amplifying the CARLIN barcode from DNA, RNA, and single-cell RNA-sequencing libraries for next-generation sequencing.

Key words DNA barcoding, Lineage tracing, CARLIN, Cas9-induced scarring, DNA barcode amplification protocol, Next-generation sequencing

1 Introduction

Tracing cell lineages in vivo has been a long-standing aim in biology. Until recently, lineage tracing was confined to single clones or populations of cells with specific gene expression patterns [1]. The development of next-generation sequencing (NGS) has opened up the possibility of using DNA itself as a molecular recorder, and a wealth of barcoding animal models have been developed in recent years [2, 3]. Inspired by the GESTALT and scGESTALT zebrafish lines that use CRISPR-Cas9 technology to induce DNA barcodes in a defined, transcribed genomic location [4, 5], we developed the CARLIN mouse line Bowling et al. [6]. Like GESTALT, CARLIN contains Cas9, gRNAs, and a barcode that comprises ten different guide RNA target sites. Following Cas9-mediated editing in the barcode region, error-prone DNA repair mechanisms generate a range of semi-random insertions and deletions (indels); this combination of indels in edited cells acts as unique and heritable markers of the cell and its progeny. CARLIN mice enable fully

Jorge García-Marqués and Tzumin Lee (eds.), *Lineage Tracing: Methods and Protocols*, Methods in Molecular Biology, vol. 2886, https://doi.org/10.1007/978-1-0716-4310-5_14,

inducible barcoding editing, and barcodes are expressed, enabling simultaneous interrogation of lineage and gene expression information from single cells. Furthermore, all transgenic elements in CARLIN mice are contained within defined genomic loci, enabling straightforward animal crossing into alternative genetic backgrounds.

Here, we provide protocols that can be used for maintaining CARLIN mice, inducing barcoding, and preparing NGS libraries from DNA, RNA, and 10× single-cell RNA-sequencing libraries. Not covered in this methods chapter are details of the analysis steps used to determine barcode sequences present in NGS data. Comprehensive instructions for running the analysis pipeline can be found at: https://gitlab.com/hormozlab/carlin.

Recently, we have described the generation of DARLIN mice, that use a Cas9-Dntt fusion protein to induce highly diverse DNA recombination sequences and contain two additional highly expressed barcode sequences [7]. Compared with CARLIN mice, DARLIN mice have (1) higher editing efficiency, (2) higher barcode diversity, and (3) higher barcode capture efficiency from single-cell RNA-sequencing libraries. While we do not provide protocols for library preparation of DARLIN mice here, the steps followed are similar, and method details, including primer sequence information, can be found in the DARLIN paper [7].

2 Materials (*See* Note 1)

1. Doxycycline hyclate.
2. 28 g insulin needles.
3. CARLIN mice (B6;129S4-Col1a1tm4(CAG-EGFP)Fcam/Mmjax—MMRRC Strain #067061-JAX).
4. iCas9; rtTA-M2 mice (B6;129S4-Gt(ROSA)26Sortm1(rtTA*M2)Jae Col1a1tm1(tetO-cas9)Sho/J—Jax strain 029415).
5. Primers (*see* Table 1 at the end for sequences).
6. Thermocycler.
7. Magnet for PCR tubes.
8. Buffer EB (Qiagen 19086).
9. KAPA library quantification kit (Kapa Biosystems KK4835).
10. MiSeq Reagent Kit v2—500 cycles (Illumina MS-102-2003).
11. Qubit fluorometer and reagents (Thermo Fisher Scientific).
12. Q5 polymerase (NEB M0491).
13. dNTPs (NEB N0447).
14. AMPure XP beads (Beckman Coulter A63880).
15. Trizol (Thermo Fisher 15596018).

Table 1
Primer sequences, as described in Bowling et al. [6]

Genotyping Col1a1 locus	
Col1a1_reverse	CCCTCCATGTGTGACCAAGG
Col1a1_fwd_WT	GCACAGCATTGCGGACATGC
Col1a1_fwd_mut	GCAGAAGCGCGGCCGTCTGG
Genotyping Rosa26 locus	
Rosa26_fwd	AAAGTCGCTCTGAGTTGTTAT
Rosa26_rev_WT	GGAGCGGGAGAAATGGATATG
Rosa26_rev_mut	GCGAAGAGTTTGTCCTCAACC
CARLIN amplification	
CARLIN_fwd1	GTCCTGCTGGAGTTCGTGAC
CARLIN_fwd2	GAGCTGTACAAGTAAGCGGC
CARLIN_rev	GCAACTAGAAGGCACAGTCG
Fragment analysis	
FAM_CARLIN_fwd	/56-FAM/GAGCTGTACAAGTAAGCGGC
Bulk RNA/DNA sequencing libraries	
RT_Bulk_CARLIN	CTACACGACGCTCTTCCGATCTNNNNNNNNNNNNNGCAAC TAGAAGGCACAGTCG
CARLIN_1	GTCCTGCTGGAGTTCGTGAC
NGS2_F	AATGATACGGCGACCACCGAGATCTACACTCTTTCCCTACACGACGC TCTTCCGATCT
CARLIN_2	GTGACTGGAGTTCAGACGTGTGCTCTTCCGATCTGAGCTGTACAAG TAAGCGGC
P5	AATGATACGGCGACCACCGA
NG2R_I#	CAAGCAGAAGACGGCATACGAGATXXXXXXGTGACTGGAGTTC XXXXXX = TruSeq Illumina indices
10× Chromium single -cell RNA-sequencing libraries	
10×-CARLIN_1	bio-GTCCTGCTGGAGTTCGTGAC
10×-CARLIN_2	GTGACTGGAGTTCAGACGTGTGCTCTTCCGATCTGAGCTGTACAAG TAAGCGGC
P5-PR1	AATGATACGGCGACCACCGAGATCTACACTCTTTCCCTACACGACGC TC
P5	AATGATACGGCGACCACCGA

16. GlycoBlue Co-precipitant (Thermo Fisher AM9516).
17. Superscript III Reverse Transcriptase (Thermo Fisher 18080085).
18. RNaseOUT (Thermo Fisher 10777-019).
19. Q5 polymerase (NEB M0491).
20. dNTPs (NEB N0447).
21. Whole tissue or cells from Fluorescence Activated Cell Sorting/ Magnetic Activated Cell Sorting (FACS)/MACS purification.
22. Single-cell RNA-sequencing library product from Chromium Single Cell 3' Reagent Kits v3 (follow steps for library generation in manufacturer guide until step 2.3—after cDNA amplification and cleaning—or equivalent step).
23. KAPA HiFi HotStart ReadyMix polymerase (Roche 07958935001), *see* **Note 2**.
24. Dynabeads (Thermo Fisher 60101).

3 Methods

3.1 CARLIN Mice Maintenance and Barcoding Induction

1. *Maintaining lines:* CARLIN and Cas9/rtTA-M2 mice are maintained as separate homozygous lines to avoid background editing of the CARLIN array. Lines can be purchased from Jax/ MMRRC and have no phenotype and no special housing requirements.
2. *Crossing mice:* When experimental mice are desired, CARLIN and Cas9/rtTA-M2 mice are crossed to generate $Col1a1^{CARLIN/iCas9}$;$Rosa26^{rtTA\text{-}M2/+}$ mice (*see* **Note 3**). For embryonic experiments, timed pregnancies are performed so that pregnant dams can be induced with doxycycline at specific embryonic time points.
3. *Inducing mice:* For adult inductions, doxycycline is administered via both drinking water and IP injection (*see* **Note 4**). Drinking water containing 2 mg/mL doxycycline and 10 mg/mL sucrose is administered for 1 week. IP injections are performed using a 20 mg/mL solution of doxycycline/PBS at a final concentration of 50 μg/g mouse. IP injections are administered every other day (total 3 injections delivered on days 1, 3, 5 of drinking water treatment).

   For embryo inductions, doxycycline is administered via retro-orbital (RO) injection [8]. RO injections are performed using a 20 mg/mL solution of doxycycline/PBS at a final concentration of 50 μg/g mouse. For most studies, a single injection is used to perform static barcoding. However, multiple inductions can be administered to gain multilevel phylogenetic lineage information, as previously demonstrated [6].

3.2 NGS Library Preparation

Library preparation is performed using nested PCRs (i.e., two separate PCR amplification steps using nested primers), which we have found increases the efficiency of CARLIN transcript/barcode capture and reduces amplification bias compared with non-nested PCRs. All reaction preparation steps should be carried out on ice. Furthermore, great care should be taken to avoid PCR contamination between samples. When generating libraries, a water-only control reaction should be included alongside template reactions to allow detection of non-specific PCR amplification.

3.3 Bulk DNA Library Preparation and Sequencing (See Note 5)

1. Extract DNA from tissues using the preferred method.
2. UMI-tagging PCR step: Load up to 250 ng DNA into UMI-tagging reaction:

Component	Volume (μL)
Q5 buffer	5
dNTP (10 mM)	0.5
RT bulk CARLIN primer (10 μM)	2.5
Q5 polymerase	0.25
DNA	10
H_2O	6.75
Total	**25**

3. Load into a thermocycler machine using the following program (lid temperature: 105 °C)

Step	1	2	3	4	5	6	7
Temperature (°C)	98 °C	98 °C	67 °C	72 °C	To step 3 5×	72 °C	4 °C
Time	30"	10"	30"	30"		2'	∞

4. To perform PCR purification using 1× AMPure beads (25 μL), resuspend the aliquot of AMPure beads by vortexing for 30 s.
5. Add required volume of AMPure beads to product (in PCR tube) and pipette up and down 15×.
6. Incubate 5 min at room temperature.
7. Place tube strip on magnet.
8. Remove and discard supernatant without disturbing beads.
9. Add 200 μL 80% ethanol (prepared same day), wait 30 s.
10. Remove the ethanol using p200 pipette.

11. Repeat **steps 9** and **10** for a total of two washes.
12. Remove remaining ethanol carefully with p20 pipette. Leave a couple of minutes to air dry.
13. Remove from magnet and add 20 μL buffer EB to resuspend beads. Pipette mix 15× and incubate 5 min at room temperature.
14. Place back on magnet and transfer 20 μL supernatant to new tube and discard beads. If there is some bead carryover, place back on magnet and transfer supernatant to a new tube.
15. Amplification round #1: Load product into PCR amplification reaction:

Component	Volume (μL)
Q5 buffer	10
dNTP (10 mM)	1
NGS2_F primer (10 μM)	2.5
CARLIN_1 primer (10 μM)	2.5
Q5 polymerase	0.5
DNA from **step 4**	20
H_2O	13.5
Total	**50**

16. Load into a thermocycler machine using the following program (lid temperature: 105 °C)

Step	1	2	3	4	5	6	7
Temperature (°C)	98 °C	98 °C	69 °C	72 °C	To **step 2** 14×	72 °C	4 °C
Time	30"	10"	30"	30"		2'	∞

17. Perform PCR purification using 1× AMPure beads (50 μL), as described above.
18. Amplification round #2: Load product into PCR amplification reaction:

Component	Volume (μL)
Q5 buffer	5
dNTP (10 mM)	0.5
NGS2_F primer (10 μM)	1.25

(continued)

Component	Volume (μL)
NGS1_Bulk_R primer (10 μM)	1.25
Q5 polymerase	0.25
DNA from **step 7**	10
H_2O	6.75
Total	**25**

19. Load into a thermocycler machine using the following program (lid temperature: 105 °C)

Step	1	2	3	4	5	6	7
Temperature (°C)	98 °C	98 °C	68 °C	72 °C	To **step 2** 14×	72 °C	4 °C
Time	30"	10"	30"	30"		2'	∞

20. Perform PCR purification using 1× AMPure beads (25 μL), as described above (**steps 4–14**).
21. Indexing PCR: Load product into amplification/indexing reaction, making sure to take note of which index primer sequence is used for each library:

Component	Volume (μL)
Q5 buffer	5
dNTP (10 mM)	0.5
P5 primer (10 μM)	1.25
NG2R_I# primer (10 μM)	1.25
Q5 polymerase	0.25
DNA from **step 10**	2.5
H_2O	14.25
Total	**25**

22. Load into a thermocycler machine using the following program (lid temperature: 105 °C)

Step	1	2	3	4	5	6	7
Temperature (°C)	98 °C	98 °C	62 °C	72 °C	To **step 2** 7×	72 °C	4 °C
Time	30"	10"	30"	30"		2'	∞

23. Run 5 μL of each sample on a gel to check for amplification and library size. In general, libraries show a strong band at ~500 bp and also a smear of smaller DNA fragments representing edited CARLIN barcodes. Well-edited (>60% editing) samples may only show a smear.
24. Quantify libraries using Qubit. If sequencing multiple libraries, libraries can be pooled together; representations can be equalized by combining equal nanogram quantities of each library. Alternatively, libraries can be equalized according to the original number of cells, so that libraries containing a higher number of cells can be overrepresented in the final library pool compared with libraries with low numbers of cells. This can help ensure similar sequencing depths are obtained between different libraries.
25. Bring total volume of libraries to be sequenced to 50 μL with H_2O and purify with 0.8× AMPure (40 μL) beads.
26. Quantify pooled library concentration using Qubit and/or Tapestation.
27. Perform Kapa library quantification to determine sequencing-standard concentration.
28. Load sequencing libraries into MiSeq at 10 pM concentration with 5% PhiX spike-in using the following cycling parameters: read 1: 250 cycles, index: 6 cycles, read 2: 250 cycles (*see* **Note 6**).

3.4 Bulk RNA Library Preparation and Sequencing

1. Resuspend whole tissue/cells in 1 mL Trizol and follow manufacturer's guidelines for extracting RNA (*see* **Note 7**). We recommend adding 1.5 μL GlycoBlue to each tube before isopropanol addition to aid visualization of RNA pellet.
2. Reverse transcription: Load up to 500 ng RNA (or whole sample if <150,000 cells) into RT reaction:

Component	Volume (μL)
RNA (500 ng or whole sample if <150,000 cells)	10
RT bulk CARLIN (2 μM)	2
dNTP (10 mM)	1
H_2O	To 13
Total	**13**

Load reaction into thermocycler (lid temperature: 105 °C) for the first PCR cycle using the following program:

Step	1	2	3	4	5
Temperature (°C)	65 °C	4 °C	55 °C	70 °C	4 °C
Time	5'	∞	60'	15'	∞

At **step 2**, ensure the reaction has been incubated at 4 °C for at least 1 min. Then, remove the tubes and add the following master mix to each tube on ice:

Component	Volume (μL)
First-strand buffer (in Superscript III kit)	4
0.1 M DTT (in Superscript III kit)	1
RNaseOUT	1
Superscript III	1
Total	7

Pipette mix and load the PCR tubes back into the thermocycler, which is still on **step 2** (4 °C, ∞). Press "skip step" to progress to the next stage of the PCR.

3. Perform PCR purification using 1× AMPure beads (20 μL), as described above (**steps 4–14** in Subheading 3.3).
4. Amplification round #1: Load product into PCR amplification reaction:

Component	Volume (μL)
Q5 buffer	10
dNTP (10 mM)	1
NGS2_F primer (10 μM)	2.5
CARLIN_1 primer (10 μM)	2.5
Q5 polymerase	0.5
DNA from **step 3**	20
H_2O	13.5
Total	**50**

5. Load into a thermocycler machine using the following program (lid temperature: 105 °C)

Step	1	2	3	4	5	6	7
Temperature (°C)	98 °C	98 °C	69 °C	72 °C	To **step 2** 14×	72 °C	4 °C
Time	30"	10"	30"	30"		2'	∞

6. Perform DNA purification using 1× AMPure beads (50 μL), as described above (**steps 4–14** in Subheading 3.3).
7. Amplification round #2: Load product into PCR amplification reaction:

Component	Volume (μL)
Q5 buffer	5
dNTP (10 mM)	0.5
NGS2_F primer (10 μM)	1.25
NGS1_Bulk_R primer (10 μM)	1.25
Q5 polymerase	0.25
DNA from **step 6**	10
H_2O	6.75
Total	25

8. Load into a thermocycler machine using the following program (lid temperature: 105 °C)

Step	1	2	3	4	5	6	7
Temperature (°C)	98 °C	98 °C	68 °C	72 °C	To **step 2** 14×	72 °C	4 °C
Time	30"	10"	30"	30"		2'	∞

9. Perform DNA purification using 1× AMPure beads (25 μL), as described above.
10. Indexing PCR: Load DNA into amplification/indexing reaction, making sure to take note of which index primer sequence is used for each library:

Component	Volume (μL)
Q5 buffer	5
dNTP (10 mM)	0.5

(continued)

Component	Volume (μL)
P5 primer (10 μM)	1.25
NG2R_I# primer (10 μM)	1.25
Q5 polymerase	0.25
DNA from **step 9**	2.5
H_2O	14.25
Total	**25**

11. Load into a thermocycler machine using the following program (lid temperature: 105 °C)

Step	1	2	3	4	5	6	7
Temperature (°C)	98 °C	98 °C	62 °C	72 °C	To **step 2** 7×	72 °C	4 °C
Time	30"	10"	30"	30"		2'	∞

12. Run 5 μL of samples on a gel to check for amplification and library size. In general, libraries show a strong band at ~500 bp and also a smear of smaller DNA fragments representing edited CARLIN barcodes. Well-edited (>60% editing) samples may only show a smear.
13. Quantify libraries using Qubit. If sequencing multiple libraries, libraries can be pooled together; representations can be equalized by combining equal nanogram quantities of each library. Alternatively, libraries can be equalized according to the original number of cells, so that libraries containing a higher number of cells can be overrepresented in the final library pool compared with libraries with low numbers of cells. This can help ensure similar sequencing depths are obtained between different libraries.
14. Bring total volume of libraries to be sequenced to 50 μL with H_2O then purify with 0.8× AMPure (40 μL) beads.
15. Quantify pooled library concentration using Qubit and/or Tapestation.
16. Perform Kapa library quantification for sequencing-standard concentration quantification.
17. Load sequencing libraries into MiSeq at 10 pM concentration with 5% PhiX spike-in using the following cycling parameters: read 1: 250 cycles, index: 6 cycles, read 2: 250 cycles (*see* **Note 8**).

3.5 Single-Cell RNA-Sequencing Library Preparation and Sequencing

1. Prepare 5 ng/μL dilution of the library from step 2.3 of Chromium Single Cell 3′ Reagent Kits protocol (or equivalent step)—*See* **Note 9**.
2. Amplification round #1: Load product into PCR amplification reaction (*see* **Note 10**):

Component	Volume (μL)
Kapa HiFi HotStart ReadyMix	12.5
CARLIN_1-bio primer (10 μM)	0.75
P5-PR1 primer (20 μM)	0.75
Library from **step 1**	1
H_2O	10
Total	**25**

3. Load into a thermocycler machine using the following program (lid temperature: 105 °C)

Step	1	2	3	4	5	6	7
Temperature (°C)	95 °C	98 °C	65 °C	72 °C	To **step 2** 9×	72 °C	4 °C
Time	3'	20"	15"	15"		1'	∞

4. Perform DNA purification using 1× AMPure beads (25 μL), as described above.
5. If performing Amplification round #1 without biotin-labeled primers (*see* **Note 10**), skip **steps 6–15** and continue with protocol at **step 16**.
6. For Dynabead purification, resuspend Dynabeads in a vial by vortexing for 30 s.
7. For each library required, transfer 5 μL Dynabeads to PCR tube and place on magnet for 2 min.
8. Remove and discard supernatant without touching bead pellet.
9. Take tube off magnet and resuspend bead pellet in 20 μL binding solution. Resuspend by pipetting carefully to not cause bubbles/foaming.
10. Place tube on magnet for 2 min and remove supernatant.
11. Take tube off magnet and resuspend bead pellet in 20 μL binding solution carefully.

12. Add 20 μL of PCR product from **step 3** and pipette up and down.
13. Leave on roller at RT for 3 h. After 1 h, pipette up and down to resuspend beads if they have settled.
14. Place tube on magnet and remove supernatant: this is a library that does not contain biotin so you can set aside or discard.
15. Wash bead-DNA complex twice with 40 μL washing solution: once with 40 μL water and once with 40 μL EB buffer. Finally, resuspend in 20 μL EB buffer. Beads do not affect PCR amplification.
16. Amplification round #2: Load product into PCR amplification reaction:

Component	Volume (μL)
Kapa HiFi HotStart ReadyMix	25
CARLIN_2 primer (10 μM)	1.5
P5-PR1 primer (10 μM)	1.5
Library from **step 5**	10
H_2O	12
Total	**50**

17. Load into a thermocycler machine using the following program (lid temperature: 105 °C)

Step	1	2	3	4	5	6	7
Temperature (°C)	95 °C	98 °C	65 °C	72 °C	To **step 2** 9×	72 °C	4 °C
Time	3'	20"	15"	15"		1'	∞

18. Perform PCR purification using 1× AMPure beads (50 μL), as described above.
19. Indexing PCR: Load product into amplification/indexing reaction, making sure to take note of which index primer sequence is used for each library:

Component		Volume (μL)	
Kapa HiFi HotStart ReadyMix		12.5	
SI primer	Dual index if using Chromium Dual Index Kit	0.75	1.5
Single Index from Chromium Single Index Kit		0.75	

(continued)

Component	Volume (μL)
Library from **step 8**	10
H_2O	1
Total	**25**

20. Load into a thermocycler machine using the following program (lid temperature: 105 °C)

Step	1	2	3	4	5	6	7
Temperature (°C)	95 °C	98 °C	55 °C	72 °C	To **step 2** 8×	72 °C	4 °C
Time	3'	20"	15"	15"		1'	∞

21. Run 5 μL of samples on a gel to check for amplification and library size. In general, libraries show a strong band at ~600 bp and also a smear of smaller DNA fragments representing edited CARLIN barcodes. Well-edited (>60% editing) samples may only show a smear.
22. Quantify libraries using Qubit. If sequencing multiple libraries, libraries can be pooled together; representations can be equalized by combining equal nanogram quantities of each library. Alternatively, libraries can be equalized according to the original number of cells, so that libraries containing a higher number of cells can be overrepresented in the final library pool compared with libraries with low numbers of cells. This can help ensure similar sequencing depths are obtained between different libraries.
23. Bring total volume of libraries to be sequenced to 50 μL with H_2O then purify with 0.8× AMPure (40 μL) beads.
24. Quantify pooled library concentration using Qubit and/or Tapestation.
25. Perform Kapa library quantification for sequencing-standard concentration quantification.
26. Load sequencing libraries into MiSeq at 10 pM concentration with 5% PhiX spike-in using the following cycling parameters: read 1: 28 cycles, index (i7): 8 cycles, read 2: 350 cycles. Check 10× protocol for required read 1 and index cycling parameters. Depending on the protocol and indexes used, different cycling parameters may be required (*see* **Note 11**).

4 Notes

1. CARLIN library preparation has been optimized with specific molecular biology reagents listed (reverse transcriptase, polymerase). Use of alternative reagents may require additional optimization of annealing temperature or PCR cycle number, for example.
2. In the published CARLIN paper, the CARLIN amplification from 10× libraries used the polymerase provided by 10× genomics. Since only limited volumes of this are supplied with each kit, and the polymerase is not sold separately, we have switched to using Kapa polymerase with no adverse effects on the efficiency of barcode or transcript capture.
3. Note that the DARLIN line [7] has higher editing efficiency at embryonic induction time points. When available, it may be a preferred model for embryonic barcoding.
4. We have observed that editing efficiency (the proportion of non-template CARLIN DNA) is variable from mouse to mouse, even in mice induced alongside each other in the same litter and cage. Because of this, prior to performing large or expensive experiments, we recommend testing the editing efficiency of CARLIN in each mouse by collecting tissue (e.g., tail/peripheral blood) and performing a PCR amplification of the CARLIN array (primer sequences listed in Table 1). The PCR product can then be run on a 2% agarose gel alongside unedited controls and checked for a DNA "smear" indicating edited DNA. Alternatively, for a more quantitative method, fragment analysis can be performed on CARLIN PCR products amplified using fluorescent-labeled primers. Editing efficiency is roughly correlated to the proportion of DNA fragments that are not full-length CARLIN.
5. Bulk DNA CARLIN sequencing may be preferable over bulk RNA CARLIN sequencing when amplification from a single copy of CARLIN is required, when cell number is not limiting, or when clonal complexity (i.e., number of clones present) is low. However, due to the low efficiency of amplifying the barcodes from DNA, we recommend performing bulk RNA sequencing in all cases where higher efficiency amplification, and therefore higher barcode detection sensitivity, is desired.
6. Steps should be taken to ensure that sequencing depth is sufficient to cover the majority of barcodes/transcripts present in the sample. The CARLIN pipeline uses a cut-off of minimum 10 reads for every unique molecular identifier (UMI, equivalent to transcript) present in the library. Since each tran-

script can be over-/underrepresented in the final library due to PCR amplification bias, we recommend sequencing bulk DNA libraries at a minimum depth of 30 reads per cell. The following equation can be used to determine sequencing depths:

$$\text{Sequencing reads per library} \\ = \text{estimated number of cells in sample} \times 30$$

An indication of sequencing depth can be determined from the Read Breakdown found in the "Results.txt" file, an output of the CARLIN pipeline. The "Common_UMIs" entry shows the number and proportion of reads that have been retained following the filtering based on minimum read thresholds; the sequencing depth of the library will roughly correlate with this percentage.

7. We have successfully generated libraries from as few as ~250 cells (sorted by FACS) using this protocol. It is especially important to use GlycoBlue when low numbers of cells are used, so that the RNA pellet can be visualized during the Trizol extraction.

8. Steps should be taken to ensure that sequencing depth is sufficient to cover the majority of barcodes/transcripts present in the sample. The CARLIN pipeline uses a cut-off of minimum 10 reads for every unique molecular identifier (UMI, equivalent to transcript) present in the library. Since each cell may contain multiple transcripts, and each transcript can be over-/underrepresented in the final library due to PCR amplification bias, we recommend sequencing bulk RNA libraries at a minimum depth of 50 reads per cell. This is higher than the bulk DNA sequencing protocols due to the higher number of transcript copies per cell. The following equation can be used to determine sequencing depths:

$$\text{Sequencing reads per library} \\ = \text{estimated number of cells in sample} \times 50$$

An indication of sequencing depth can be determined from the Read Breakdown found in the "Results.txt" file, an output of the CARLIN pipeline. The "Common_UMIs" entry shows the number and proportion of reads that have been retained following the filtering based on minimum read thresholds; the sequencing depth of the library will roughly correlate with the percentage.

9. Technical replicates of the library prep can be performed to reduce the effect of PCR errors on final data. Each technical replicate will need to be assigned a separate index, and following analysis, barcodes or UMIs that are detected only in one replicate library can be excluded.

10. We have observed modest but variable increases in barcode capture when using the biotin-labeled primers compared with non-biotin-labeled primers. If time and/or money is limited, the biotin enrichment steps can be omitted without major effects on the efficiency of barcode capture.

11. Steps should be taken to ensure that sequencing depth is sufficient to cover the majority of barcodes/transcripts present in the sample. The CARLIN pipeline uses a cut-off of minimum 10 reads for every unique molecular identifier (UMI, equivalent to transcript) present in the library. As each cell may contain multiple transcripts, and each transcript can be over-/underrepresented in the final library due to PCR amplification bias, we recommend sequencing single-cell libraries at a minimum depth of 100 reads per cell. This is higher than the bulk RNA-sequencing protocols due to the slightly lower quality of sequencing performance due to the long read 2 cycle number. The following equation can be used to determine recommended sequencing depths:

$$\text{Sequencing reads per library} = \text{estimated number of cells in sample} \times 100$$

An indication of sequencing depth can be determined from the Read Breakdown found in the "Results.txt" file, an output of the CARLIN pipeline. The "Common_CBs" entry shows the number and proportion of reads that have been retained following the filtering based on minimum read thresholds; the sequencing depth of the library will roughly correlate with the percentage.

Acknowledgments

We are grateful to current and past members of the Camargo lab, especially Fernando Garcia-Osorio who developed early protocols for preparing CARLIN next-generation sequencing libraries. SB acknowledges support from EMBO (ALTF 798-2018), the Wellcome Trust (grant number 215920/Z/19/Z), and the NIH (grant number 1K99HL164969). FDC is funded by grants R01HL128850, RC2DK131963, and R01HL158192, the Edward P Evans Foundation, and an Alex Lemonade Crazy 8 award.

References

1. Kretzschmar K, Watt FM (2012) Lineage tracing. Cell 148:33–45
2. Baron CS, van Oudenaarden A (2019) Unravelling cellular relationships during development and regeneration using genetic lineage tracing. Nat Rev Mol Cell Biol 20:753–765
3. VanHorn S, Morris SA (2021) Next-generation lineage tracing and fate mapping to interrogate development. Dev Cell 56:7–21
4. McKenna A, Findlay GM, Gagnon JA et al (2016) Whole-organism lineage tracing by combinatorial and cumulative genome editing. Science 353:aaf7907
5. Raj B, Wagner DE, McKenna A et al (2018) Simultaneous single-cell profiling of lineages and cell types in the vertebrate brain. Nat Biotechnol 36:442–450
6. Bowling S, Sritharan D, Osorio FG et al (2020) An engineered CRISPR-Cas9 mouse line for simultaneous readout of lineage histories and gene expression profiles in single cells. Cell 181:1410–1422.e27
7. Li L, Bowling S, McGeary SE et al (2023) A mouse model with high clonal barcode diversity for joint lineage, transcriptomic, and epigenomic profiling in single cells. Cell 186:5183–5199
8. Yardeni T, Eckhaus M, Douglas Morris H et al (2011) Retro-orbital injections in mice. Lab Anim (NY) 40:155–160

Chapter 15

Single-Cell Profiling of Lineages and Cell Types in the Vertebrate Brain

Bushra Raj

Abstract

CRISPR-Cas tools have recently been adapted for cell lineage tracing during development. Combined with single-cell RNA sequencing, these methods enable scalable lineage tracing with single-cell resolution. Here, I describe, scGESTALTv2, which combines cumulative CRISPR-Cas9 editing of a lineage barcode array with transcriptional profiling via droplet-based single-cell RNA sequencing (scRNA-seq). The technique is applied in developing zebrafish brains to generate mutations in the barcode array during development. The recorded lineages along with cellular transcriptomes are then extracted via scRNA-seq to define cell relationships among thousands of profiled brain cells and dozens of cell types.

Key words CRISPR-Cas9, Barcoding, Single cell, Lineage tracing, Zebrafish, scRNA-seq, Neurodevelopment, Brain dissociation, Genetics

1 Introduction

Lineage tracing is a method used to track cell fate decisions during development, thereby revealing where cells originated and how cells are related to one another. In the past 5 years, the CRISPR-Cas9 nuclease system has been extensively adapted for lineage tracing purposes in many model systems [1–8]. The major advantage of such technology is that Cas9-induced stochastic mutations are cumulative, combinatorial, and heritable [9, 10]. As cells divide, Cas9-induced indels are passed onto daughter cells where additional mutations can accumulate, and so forth until all target sites are mutated or the Cas9-sgRNA reagents are exhausted. Target sites are predefined and localized to compact arrays known as "barcode arrays." Such arrays comprise a series (e.g., 5–10) of specially selected or designed unique sgRNA target sites. This system is dynamic, enabling tracking over several cell divisions. Furthermore, the partially stochastic nature of Cas9-induced

Jorge García-Marqués and Tzumin Lee (eds.), *Lineage Tracing: Methods and Protocols*, Methods in Molecular Biology, vol. 2886, https://doi.org/10.1007/978-1-0716-4310-5_15,

double-strand break repair generates sufficient unique indels to enable unambiguous labeling of thousands of ancestral cells.

Pioneering work in zebrafish using genome editing of synthetic target arrays for lineage tracing (GESTALT) demonstrated the power of Cas9-based lineage recorders for clonal tracing and lineage tree reconstruction [2]. More recently, GESTALT has been paired with single-cell RNA sequencing to result in scGESTALT, allowing simultaneous cell-type identification and lineage recording [1, 3]. Here, I describe an optimized version, scGESTALTv2 lineage recorder, which results in higher barcode recovery from single cells, and its application in profiling cell types and lineage relationships in the zebrafish brain [3]. The protocol involves the following: (1) performing lineage barcode editing using Cas9 protein, synthesized sgRNAs, and transgenic zebrafish embryos; (2) identifying animals with successful barcode editing; (3) brain dissociation steps; (4) performing single-cell RNA sequencing using the 10× Genomics platform; (5) generating the scGESTALTv2 amplicon library; (6) sequencing the transcriptome and amplicon libraries.

2 Materials

2.1 scGESTALTv2 Lineage Barcode Editing

1. EnGen sgRNA Synthesis Kit, *Streptococcus pyogenes* (New England Biolabs (NEB)).
2. sgRNA oligonucleotides:
   dsR-1 TTCTAATACGACTCACTATAGGCGTGAACTTCCCCTCCGAGTTTTAGAGCTAGA,
   dsR-2 TTCTAATACGACTCACTATAGTGATGCAGAAGAAGACCATGTTTTAGAGCTAGA,
   dsR-3 TTCTAATACGACTCACTATAGAGCGCCTGTACCCCCGCGAGTTTTAGAGCTAGA,
   dsR-4 TTCTAATACGACTCACTATAGCCCTGAAGCTGAAGGACGGGTTTTAGAGCTAGA,
   dsR-5 TTCTAATACGACTCACTATAGCTGCCCGGCTACTACTACGGTTTTAGAGCTAGA,
   dsR-6 TTCTAATACGACTCACTATAGGACATCACCTCCCACAACGGTTTTAGAGCTAGA,
   dsR-7 TTCTAATACGACTCACTATAGGAGCAGTACGAGCGCACCGGTTTTAGAGCTAGA,
   dsR-lowX TTCTAATACGACTCACTATAGGAGTTCAAGTCCATCTACAGTTTTAGAGCTAGA.
3. RNA Clean & Concentrator kit.
4. Nuclease-free water.
5. Cas9 NLS protein, *S. pyogenes*.

6. Phenol red: Dilute 0.5% (wt/vol) phenol red stock 1:5 with nuclease-free water. Use this diluted stock for injections, and store at room temperature (23–26 °C) for at least 1 year.
7. NanoDrop.
8. Microinjection glass capillaries.
9. Microloader tips for glass capillaries.
10. Injection molds.
11. Petri dishes, 10 cm.
12. Microinjection setup: Treadlite II Foot Switch, Pneumatic PicoPump, Piconozzle Kit v2, manipulator, magnetic stand, and iron plate (or equivalent setup).
13. Micrometer for needle calibration.
14. Blue water: Combine 20 L of water in a carboy with 20 mL of 0.1% (wt/vol) Methylene blue and 5 g of Instant Ocean sea salt, and adjust pH to 7.0 with sodium bicarbonate buffer. Store at room temperature for up to 1 month (*see* **Note 1**).
15. scGESTALTv2 barcode transgenic: available by request from Raj lab.
16. Wild-type strains (e.g., TL/AB, TL, or AB; available from ZIRC, http://zebrafish.org).

2.2 Identification of Successful scGESTALTv2 Barcode Editing

1. PCR machine.
2. Q5 High-Fidelity 2× Master Mix.
3. DNA electrophoresis system.
4. Agarose.
5. 100 bp DNA ladder.
6. Oligonucleotides:
   scGESTALT-F AGCGCGTGATGAACTTCGA
   scGESTALT-R TGTGGTATGGCTGATTATGATCTAG.

2.3 Brain Dissociation

1. Papain dissociation system.
2. Neurobasal media.
3. B27 supplement, serum-free, 50×: Upon receiving the product, thaw the solution completely at room temperature. Make 500 μL aliquots and store at −20 °C for up to 1 year.
4. DPBS with no calcium, no magnesium.
5. 1% Bovine serum albumin (BSA)/Dulbecco's phosphate-buffered saline (DPBS) solution: Weigh 1.25 g of BSA powder and dilute with 25 mL of DPBS (no calcium, no magnesium) in a 50-mL tube to make a 5% BSA/DPBS stock solution. Add 4 mL of 5% BSA/DPBS solution to 16 mL of DPBS in a 50-mL tube to make a 1% BSA/DPBS solution. Store the 5%

and 1% BSA/DPBS solutions at 4 °C for up to 1 week or at − 20 °C for longer storage.

6. 20-μm cell strainer.
7. 0.4% (wt/vol) Trypan blue.
8. Hemocytometer.
9. Benchtop centrifuges for 1.5- and 15-mL tubes.
10. Refrigerated centrifuge for 1.5-mL tubes.
11. Minifuge.
12. Water bath set to 37 °C.
13. Sylgard 10-cm Petri dish: Prepare 10-cm Petri dishes with Sylgard silicone as described previously [10].
14. Dissection microscope with light source.
15. Fluorescent microscope with DsRed filter.
16. Insect pins.
17. Fine forceps.
18. 95%O_2: 5%CO_2 gas tank fitted with a gas regulator and small diameter Tygon E-3603 tubing for bubbling gas.

2.4 10× Genomics scRNA-seq

1. Chromium controller.
2. Chromium Next GEM Single Cell 3′ Kit v3.1.

2.5 scGESTALTv2 Barcode Library Amplification

1. PCR machine.
2. Q5 High-Fidelity DNA polymerase.
3. Oligonucleotides:

   GP10×1 AGTTCATGCGCTTCAAGGTGC

   GP10×2 GTGACTGGAGTTCAGACGTGTGCTCTTCCGATCT GCTGCTTCATCTACAAGGTGAAG

   10×PCR1F CTACACGACGCTCTTCCGATCT

   10×P7Dual_Pi CAAGCAGAAGACGGCATACGAGAT XXXXXXXXXX GTGACTGGAGTTCAGACGTGT

   10×P5Dual_Pi AATGATACGGCGACCACCGAGATCTACAC XXXXXXXXXX ACACTCTTTCCCTACACGACGCTCTTCCGATCT

   XXXXXXXXXX represents index sequences from the 10× Genomics Dual Index Kit TT, Set A.

   Pi represents unique index primer pairs. For example, 10×P7Dual_PH1 and 10×P5Dual_PH1 represent primer pairs corresponding to index H1 of the Dual Index Kit TT, Set A primer list.
4. SPRIselect beads.
5. Nuclease-free water.

6. Qubit.
7. Qubit dsDNA HS Assay Kit and Assay Tubes.

2.6 Sequencing

1. NovaSeq S4 300 cycle kit (or other kits depending on sequencing preference).
2. 2100 Bioanalyzer or 4200 TapeStation system.

3 Methods

3.1 scGESTALTv2 Lineage Barcode Editing

1. Generate sgRNAs targeting sites 1–8 of the scGESTALTv2 barcode using an EnGen sgRNA synthesis kit according to the manufacturer's guidelines.
2. Purify sgRNAs with RNA Clean & Concentrator kit according to the manufacturer's guidelines and elute with 50 μL nuclease-free water.
3. Quantify sgRNAs using NanoDrop and dilute each sgRNA to ~350 ng/μL with nuclease-free water. Combine equal volumes (e.g., 5 μL) of sgRNAs 1–8 to make an equimolar pool.
4. Store the sgRNA pool in 1.5 μL aliquots at −80 °C for at least 1 year. Avoid freeze–thaw of sgRNA pools. Each aliquot is sufficient for one injection mix.
5. The day before performing microinjections, set up 4–6 overnight tanks with one scGESTALTv2 F_1 male and one wild-type female per tank for mating (*see* **Note 2**). Additionally, if necessary, prepare glass needles, embryo microinjection dish, and microinjector as previously described [10].
6. On the morning of microinjections, assemble the following reagent mix (3.3 μL total): 1 μL nuclease-free water, 0.3 μL diluted phenol red, 1 μL sgRNA pool (350 ng/ μL), and 1 μL SpCas9 NLS protein. Vortex to mix and centrifuge briefly.
7. Incubate the injection mix at room temperature for 5 min to enable Cas9-sgRNA complexes to assemble. Hold the mix on ice until ready to start injections.
8. Start mating of scGESTALTv2 and wild-type fish.
9. While fish are mating, prepare and calibrate needles for microinjection [10]. Load 1.5 μL of the Cas9-sgRNA injection mix into the needle.
10. Collect fertilized eggs into Petri dishes with blue water and arrange one-cell stage embryos for injections [10]. Inject 1.5 nL of Cas9-sgRNA mix through the chorion.
11. Transfer injected embryos with blue water into a new Petri dish and incubate at 28.5 °C. Keep 20–50 uninjected control embryos to assess the health of the clutch. At the end of the

day, remove any unfertilized or damaged embryos from the injected batches.

12. At 24 hpf, screen injected embryos and discard those that are DsRed positive as these represent embryos where lineage barcode editing was unsuccessful (*see* **Note 3**). Incubate injected embryos that are DsRed-negative at 28.5 °C until 5 dpf and then transfer to appropriate tanks in a zebrafish facility to raise to juveniles (21–23 dpf).

3.2 Identification of Juveniles with Successful scGESTALTv2 Barcode Editing

1. The day prior to performing brain dissociations for scRNA-seq, identify individual juveniles with successfully edited scGESTALTv2 barcodes (*see* **Note 3**). Fin clip 3–5 juveniles and prepare genomic DNA as described previously [10]. Transfer each fin-clipped fish to single tanks to recover and number them uniquely.
2. Set up PCR reactions (10 μL): 5 μL Q5 2× Master Mix, 0.5 μL of 10 μM scGESTALT-F primer, 0.5 μL of 10 μM scGESTALT-R primer, 1 μL genomic DNA, and 3 μL nuclease-free water.
3. Run the following program on a thermocycler: 98 °C, 30 s; [98 °C, 10 s; 64 °C, 20 s; 72 °C, 15 s] × 30 cycles; 72 °C, 2 min; 8 °C, hold.
4. Load PCR products on 1.5% agarose gel with 100 bp DNA ladder according to standard procedures. Run the gel and visualize bands. Successfully edited scGESTALTv2 barcodes will appear as smears ranging from ~120 to 250 bp. Wild-type animals will not show any PCR product.
5. Select 1–3 juveniles with the highest degree of PCR band smearing for scRNA-seq experiments the next day, and keep each selected juvenile in separate single tanks overnight.

3.3 Brain Dissociation

1. The day prior to performing brain dissociations for scRNA-seq, prepare 1% BSA precoated tubes. Add 1 mL of 1% BSA/DPBS solution to 4–6 low-adhesion 1.5-mL microcentrifuge tubes and rotate overnight at 4 °C. Additionally, add 4 mL of 1% BSA/DPBS solution to one 15-mL tube and incubate at 4 °C overnight (*see* **Note 4**). This precoating step prevents cells from sticking to the walls of the tube during the single-cell suspension preparation.
2. On the morning of performing brain dissociations, discard 1% BSA/DPBS solutions used to coat the 1.5- and 15-mL tubes. Spin tubes and remove any residual solution. The tubes are now ready to be used in the dissociation steps described below.
3. Bring one of the successfully edited juvenile fish (identified in Subheading 3.2) to be used in the experiment to the bench.

The steps below assume that one brain is being processed for dissociation and scRNA-seq.

4. Start setting up for brain dissociation. Transfer 25 mL Neurobasal media to a 50-mL tube. Add 500 μL of 50× B27 supplement. Oxygenate the solution by passing 95%O_2:5%CO_2 gas through a tubing attached to a sterile 5-mL serological pipette for 2 min, as previously described [10]. Hold Neurobasal/B27 media on ice.
5. Prepare papain solution for dissociation as described by the manufacturer. Oxygenate the papain/DNase mix by passing 95%O_2:5%CO_2 gas through a tubing attached to a sterile 5-mL serological pipette for 2 min (*see* **Note 5**).
6. Incubate papain/DNase mix in a 37 °C water bath while prepping the single-cell brain suspension as described below (**steps 7–13**).
7. Anesthetize the selected juvenile fish with Tricaine according to standard procedures.
8. Add 1 mL DPBS to a 1% BSA/DPBS precoated 1.5-mL tube and hold on ice.
9. Add 10 mL ice-cold Neurobasal/B27 media to a 10-cm Sylgard dish.
10. Transfer the anesthetized fish to the Sylgard dish. Add additional Tricaine to maintain anesthesia. Pin the fish just posterior of the head, in the middle of the trunk, and near the tail using three insect pins.
11. Using a pair of fine forceps, remove the jaw, eyes, heart, and gut tissues. Pierce the skin on top of the head and gently peel it back to expose the brain. Using fine forceps, gently scoop the brain out taking care not to lose part of the hindbrain in the process. Typically, dissection takes 3–5 min per brain.
12. Transfer the brain tissue to the 1.5-mL tube with DPBS from **step 8** using fine forceps or with a P1000 pipette + tip. Minimize the amount of Neurobasal media transferred (*see* **Note 6**).
13. Briefly spin down the tube using a benchtop minifuge (2-s spin at room temperature). Discard DPBS and add 1 mL of ice-cold DPBS. Repeat the spin and hold the tube at room temperature.
14. Remove the papain/DNase mix from the water bath (**step 6**). Discard DPBS in the 1.5-mL tube from **step 13** and transfer 1 mL of the papain/DNase solution to this tube.
15. Transfer the entire contents of the 1.5-mL tube from **step 14** to a 1% BSA/DPBS precoated 15-mL tube (from **steps 1** to **2**).
16. Oxygenate the solution as described in **steps 4** and **5**, taking care not to cause the brain tissue to stick to the walls of the tube.

17. Incubate the brain tissue + papain solution in a 37 °C water bath for 20 min.
18. Gently triturate the tissue 20 times with a p1000 pipette set at 800 μL. Do not triturate too vigorously and avoid introducing bubbles. Generally, pipetting at a rate of 1 s per the up and down motion of the pipette is desirable (*see* **Note 7**).
19. Transfer the entire contents of the 15-mL tube to a new 1% BSA/DPBS precoated 1.5-mL microcentrifuge tube. Add 100 μL of resuspended ovumucoid protease inhibitor and 50 μL of resuspended DNase (Vial 3) from the Papain Dissociation Kit. Gently pipette 5× to mix.
20. Spin the tube at 450 g for 5 min at 4 °C.
21. Discard supernatant. A small cell pellet is usually visible. Gently resuspend the pellet with 1.3 mL of ice-cold DPBS.
22. Filter the resuspension through a 20-μm cell strainer placed on top of a 1% BSA/DPBS precoated 1.5-mL tube. Perform all steps on ice with pre-chilled tubes.
23. Spin at 450 g for 5 min at 4 °C.
24. Discard the supernatant and resuspend the pellet with 150 μL of ice-cold 1% BSA/DPBS. Hold on ice. This is a suggested resuspension volume. The objective is to obtain cells at a concentration of ~1000–1500 cells/μL. It may be necessary to adjust the volume depending on cell recovery, and this can be optimized during trial runs.
25. Determine cell concentration and viability. Add 10 μL of the cell suspension to 10 μL of Trypan blue in a 1% BSA/DPBS precoated 1.5-mL tube. Mix by gentle pipetting and load 10 μL onto a hemocytometer. Count cells and determine the concentration according to the hemocytometer's manufacturer. Cell viability should be >85% (ideally >90%; *see* **Note 8**).
26. If concentration of cells is 1000–1500 cells/μL, proceed immediately with loading the scRNA-seq microfluidics chip (*see* Subheading 3.4). If concentration is higher (especially, if higher than 2000 cells/μL), dilute with appropriate volume of ice-cold 1% BSA/DPBS and repeat the cell concentration count as detailed in **step 25**.

3.4 10× Genomics scRNA-seq

1. Perform scRNA-seq as described in the Chromium Next GEM Single Cell 3′ v3.1 manufacturer's manual. Generally, we aim to recover 10,000 cells per loaded channel of the microfluidics chip. Up to eight channels can be loaded simultaneously. For each individual brain, we aim to sequence at least 15,000–30,000 cells, which correspond to 2–3 channels of the chip.

2. Generate transcriptome libraries as described in the Chromium Next GEM Single Cell 3′ v3.1 manufacturer's manual. 10 μL of the cDNA is used for generating transcriptome libraries. The remaining cDNA can be used to amplify and generate the scGESTALTv2 amplicon library as described in the Subheading 3.5.

3.5 scGESTALTv2 Barcode Library Amplification

1. PCR amplify the GESTALTv2 barcode library using a 3-step PCR protocol. Set up the first PCR (30 μL): 15 μL Q5 2× Master Mix, 1.5 μL of 10 μM 10×PCR1F primer, 1.5 μL of 10 μM GP10×1 primer, 4 μL cDNA (from **step 2** in Subheading 3.4), 8 μL nuclease-free water (*see* **Note 9**).
2. Run the following program on a thermocycler: 98 °C, 30 s; [98 °C, 10 s; 68 °C, 25 s; 72 °C, 15 s] × 14 cycles; 72 °C, 2 min; 8 °C, hold.
3. Add 20 μL nuclease-free water to PCR reaction to bring the volume to 50 μL. Purify the PCR product with 0.8× volume SPRIselect beads (40 μL) and elute with 20 μL DNA elution buffer.
4. Measure the concentration of 1 μL of eluate using Qubit, according to the manufacturer's instructions. The concentration is typically 2–6 ng/ μL.
5. Set up the second PCR amplification (30 μL): 15 μL Q5 2× Master Mix, 1.5 μL of 10×PCR1F primer, 1.5 μL of 10 μM GP10×2 primer, 3 μL PCR1 product (from **step 3**), 9 μL nuclease-free water.
6. Run the following program on a thermocycler: 98 °C, 30 s; [98 °C, 10 s; 68 °C, 25 s; 72 °C, 15 s] × 12–13 cycles; 72 °C, 2 min; 8 °C, hold.
7. Add 20 μL nuclease-free water to PCR reaction to bring the volume to 50 μL. Purify the PCR product with 0.8× volume SPRIselect beads (40 μL) and elute with 20 μL DNA elution buffer.
8. Measure the concentration of 1 μL of eluate using Qubit according to the manufacturer's instructions. The concentration is typically 5–8 ng/ μL.
9. Dilute the PCR2 product (from **step 7**) 1:10 with nuclease-free water. For example, add 1.5 μL of PCR2 product to 13.5 μL of nuclease-free water.
10. Set up the third PCR amplification (30 μL): 15 μL Q5 2× Master Mix, 1.5 μL of 10 μM 10×P7Dual_Pi primer, 1.5 μL of 10 μM 10×P5Dual_Pi primer, 5 μL PCR2 product (from **step 9**), 7 μL nuclease-free water (*see* **Note 10**).

11. Run the following program on a thermocycler: 98 °C, 30 s; [98 °C, 10 s; 68 °C, 25 s; 72 °C, 15 s] × 6–8 cycles; 72 °C, 2 min; 8 °C, hold.
12. Add 20 μL nuclease-free water to PCR reaction to bring the volume to 50 μL. Purify the PCR product with 0.8× volume SPRIselect beads (40 μL) and elute with 20 μL DNA elution buffer.
13. Measure the concentration of 1 μL of eluate using Qubit according to the manufacturer's instructions. The concentration is typically 2–4 ng/μL.

3.6 Sequencing of Transcriptome and scGESTALTv2 Barcode Libraries

1. Analyze the transcriptome (from **step 2** in Subheading 3.4) and scGESTALTv2 amplicon libraries with Tapestation or Bioanalyzer according to the manufacturer's guidelines.
2. Determine the molar concentration of each library.
3. Make a pooled 10-nM library containing a mix of the transcriptome and GESTALTv2 amplicon libraries. The ratios of the libraries in the pool can be adjusted based on the desired sequencing depth for each. For example, 60% transcriptome and 40% amplicon. When pooling multiple libraries ensure there is no overlap in indices.
4. Sequence the pool on the NovaSeq S4 300 cycle kit (partial lanes are acceptable) with standard Illumina primers. Read 1 is 150 cycles, Index 1 is 10 cycles, Index 2 is 10 cycles, Read 2 is 150 cycles. It is also possible to adjust Read 1 and Read 2 such that their combination adds up to 300 cycles. This will depend on the sequencing core's limitations especially if sharing the flow cell with other "standard" samples. It is recommended that Read 2 is at least 150 cycles to ensure coverage of the DsRed barcode cassette and Read 1 is at least 28 cycles to sequence the 10× cell barcode and unique molecular identifier (UMI). Alternatively, the transcriptome and scGESTALTv2 amplicon libraries can be sequenced separately. For example, sequence the transcriptome on the NextSeq 75 cycle kit and scGESTALTv2 on the MiSeq 300 cycle kit (add 20% PhiX spike-in).

4 Notes

1. Can substitute E3 media for blue water throughout the protocol.
2. The use of male scGESTALTv2 F1s is recommended to avoid maternal contribution of DsRed expression that can confound monitoring of editing success.

3. Since sgRNAs used in the lineage tracing experiments target the DsRed cassette, editing will result in loss of DsRed signal. To distinguish transgenic animals with successful editing of DsRed barcode cassette from wild-type animals that lack the DsRed cassette, PCR amplification of the scGESTALTv2 barcode is performed using genomic DNA from the tail fin of juvenile animals.
4. During the protocol, four to six 1.5-mL microcentrifuge tubes and one 15-mL tube are required per brain dissociation. If dissociating more than one brain, adjust the number of tubes to be precoated accordingly.
5. Do not bubble gas directly into the solution as it will cause frothing and protein denaturation and reduce papain activity. Instead, slowly move the pipette in a circular motion above the liquid's surface being careful not to cause the liquid to spill over (adjust the gas pressure if necessary).
6. If processing more than one brain, perform the first dissection and transfer the brain tissue to DPBS in a 1.5-mL tube as described in **step 12** in Subheading 3.3. Hold the tube on ice. Perform the next dissection and collect the second brain in a separate 1% BSA/DPBS precoated 1.5-mL tube on ice. Label the tubes individually. Proceed with dissociation **steps 13–26** while keeping the two brains in separate tubes throughout the protocol. Processing of more than two brains at once is not recommended.
7. Typically, triturating 20 times with p1000 is sufficient for efficient tissue dissociation. However, depending on the strength of the papain, if large pieces of tissues are still visible, it may be necessary to perform additional pipetting or increase incubation time by 3–5 min. If longer incubation time at 37 °C is necessary, repeat **step 16** in Subheading 3.3 just before re-incubation if the solution is turning pink (indicating high pH) to restore the pH to neutral (should be an orange-red color).
8. High cell viability is required to proceed with scRNA-seq experiments. Poor viability will result in fewer cells passing data quality filtering steps and can introduce more ambient RNA (i.e., non-cell-associated RNA, such as free-floating RNA from dead or dying cells) into cell suspension, compromising single-cell measurements.
9. This PCR amplifies a longer scGESTALT fragment using a forward primer that binds in the DsRed transgene upstream of the scGESTALT barcode. We have found that using a nested PCR strategy improves specificity of amplification.
10. If multiplexing libraries, use different 10×P7Dual_Pi and 10×P5Dual_Pi primer variants for each library. Generally, we

ensure that these primers are from a different part of the index plate than that used for the transcriptome libraries. For example, if multiplexing two samples (Sample X and Sample Y) in the same sequencing run, we use index A1 (Sample X) and A2 (Sample Y) for the transcriptome libraries and index H1 (Sample X) and H2 (Sample Y) for the amplicon library. Thus, when demultiplexing, amplicon reads from index H1 will correspond to transcriptome reads from index A1 for Sample X.

References

1. Raj B, Wagner DE, McKenna A et al (2018) Simultaneous single-cell profiling of lineages and cell types in the vertebrate brain. Nat Biotechnol 36:442–450
2. McKenna A, Findlay GM, Gagnon JA et al (2016) Whole-organism lineage tracing by combinatorial and cumulative genome editing. Science (1979) 353:aaf7907
3. Raj B, Farrell JA, Liu J et al (2020) Emergence of neuronal diversity during vertebrate brain development. Neuron 108:1058–1074.e6
4. Alemany A, Florescu M, Baron CS et al (2018) Whole-organism clone tracing using single-cell sequencing. Nature 556:108–112
5. Spanjaard B, Hu B, Mitic N et al (2018) Simultaneous lineage tracing and cell-type identification using CRISPR–Cas9-induced genetic scars. Nat Biotechnol 36:469–473
6. Simeonov KP, Byrns CN, Clark ML et al (2021) Single-cell lineage tracing of metastatic cancer reveals selection of hybrid EMT states. Cancer Cell 39:1150–1162.e9
7. Bowling S, Sritharan D, Osorio FG et al (2020) An engineered CRISPR-Cas9 mouse line for simultaneous readout of lineage histories and gene expression profiles in single cells. Cell 181:1410–1422.e27
8. Xie L, Liu H, You Z et al (2023) Comprehensive spatiotemporal mapping of single-cell lineages in developing mouse brain by CRISPR-based barcoding. Nat Methods 20: 1244–1255
9. Yao M, Ren T, Pan Y et al (2022) A new generation of lineage tracing dynamically records cell fate choices. Int J Mol Sci 23(9): 5021
10. Raj B, Gagnon JA, Schier AF (2018) Large-scale reconstruction of cell lineages using single-cell readout of transcriptomes and CRISPR–Cas9 barcodes by scGESTALT. Nat Protoc 13(11):2685–2713

Chapter 16

StarTrack: Mapping Cellular Fates with Inheritable Color Codes

M. Figueres-Oñate, Jorge García-Marqués, A. C. Ojalvo-Sanz, and Laura López-Mascaraque

Abstract

StarTrack is a powerful multicolor genetic tool designed to unravel cellular lineages arising from neural progenitor cells (NPCs). This innovative technique, based on retrospective clonal analysis and built upon the PiggyBac system, creates a unique and inheritable "color code" within NPCs. Through the stochastic integration of 12 distinct plasmids encoding six fluorescent proteins, StarTrack enables precise and comprehensive tracking of cellular fates and progenitor potentials. The versatility of this tool is further enhanced by the potential of combining multiple promoters. Whether through the use of fluorescent integrable constructs or driving the expression of the PiggyBac transposase, StarTrack broadens the horizons for lineage tracing from progenitors of multiple origins.

StarTrack revolutionized our understanding of cellular origins and lineages, offering an invaluable resource for researchers in the field of neural development and lineage tracing. This protocol provides a comprehensive overview of the technique's capabilities and applications, shedding light on its significance within the scientific community.

Key words Lineage tracing, Clonal analysis, Progenitor potential, Cell fate, In utero electroporation, StarTrack, Genetic tagging

1 Introduction

StarTrack is a multicolor genetic tool that generates a unique and inheritable "color code" in neural progenitor cells (NPCs) and their complete cell progeny [1]. This allows the analysis of the fate and cell potential of single progenitors [1–14].

To assign cell lineages, this clonal method relies on the retrospective analysis of targeted cells. To create stable clonal marks, this tool involves the random integration of 12 distinct Piggybac plasmids, which are inserted at genomic TTAA sites [15]. Six of these plasmids encode a different cytoplasmic fluorescent protein (XFP) from a selection of six: mT-Sapphire (a UV-excitable monomeric GFP mutant XFP with a significant Stokes shift), mCerulean (the

Jorge García-Marqués and Tzumin Lee (eds.), *Lineage Tracing: Methods and Protocols*, Methods in Molecular Biology, vol. 2886, https://doi.org/10.1007/978-1-0716-4310-5_16,

brightest monomeric fluorescent cyan protein), EGFP (a weak dimer-enhanced green fluorescent protein), YFP (a commonly used weak dimer yellow fluorescent protein), Kusabira Orange (mKO, the brightest orange monomer), and mCherry (a photostable red monomer). The other six plasmids encode the same XFPs fused to the H2B histone, which directs the protein to the nucleus. The stochastic integration of these 12 constructs by the PiggyBac transposase can lead to a theoretical number of 4096 potential combinations. Furthermore, variations in the number of copies for a single fluorophore may introduce variability in the fluorescent signal. By categorizing these combinations into two groups based on intensity variation, the potential permutations exponentially increase to more than 16 million [1]. Consequently, this extensive range of combinatorial possibilities within NPCs ensures the assignment of clones to cells that exhibit identical color coding, localization patterns, and intensity levels, affirming that these cells are indeed sibling cells originating from the same progenitor and forming a clone.

StarTrack is an exceptionally versatile tool that offers the flexibility to modify the promoters driving the expression of the XFPs according to the specific research objectives (Fig. 1). If the aim is to trace the complete offspring from a single progenitor, regardless of their cell type, the UbC-StarTrack method is a suitable choice published by Figueres-Oñate in 2016. In this StarTrack version, both the plasmids carrying the XFPs and the PiggyBac transposase are regulated by the ubiquitous promoter Ubiquitin C [1]. Moreover, these plasmids incorporate two loxP sites, enabling the disruption of nonintegrated copies after tamoxifen (Tx) administration [16].

Alternatively, for studying glial lineages, two distinct versions of the StarTrack tool have been developed. For astrocytes, GFAP-StarTrack would be the ideal option. In this system, the XFP reporters are expressed under the control of the GFAP promoter, as elucidated by García-Marqués and López-Mascaraque in 2013 [3]. Additionally, to track the NG2-glia cell lineage, the NG2-StarTrack system employs plasmids encoding an NG2 promoter, as outlined by Sanchez-Gonzalez in 2020 [17].

An overview of the general workflow for this methodology involves several key steps (Fig. 2). The desired StarTrack variant is selected first depending on the research needs. Subsequently, a DNA mixture is injected into the brain ventricles, and electroporation is conducted by directing the DNA to the specific region where the progenitors of interest are situated. Here, we will focus on the embryonic targeting of neural progenitors; however, postnatal electroporation can also be performed. Afterward, the tissue is collected at the designated study time point and processed through sectioning and serial collection. Subsequently, a thorough imaging acquisition process is carried out to capture all the images of the

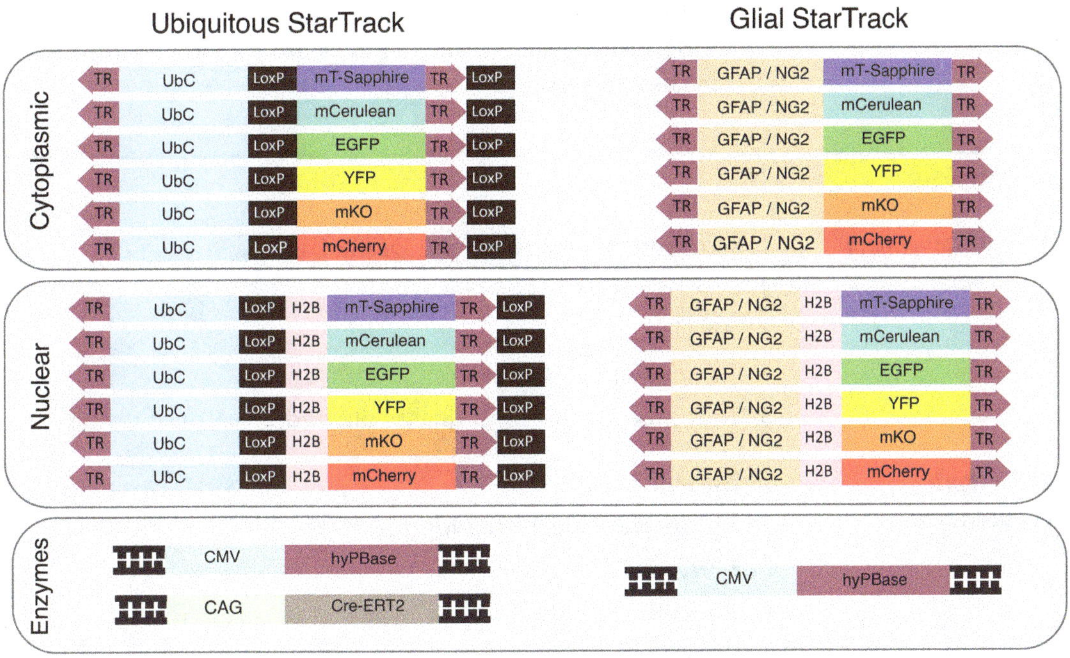

Fig. 1 Versatile StarTrack methods for lineage tracing. On the left, UbC-StarTrack provides comprehensive lineage tracing, independent of cell type, by utilizing ubiquitous promoters for XFPs and PiggyBac transposase. The plasmids are equipped with loxP sites to facilitate the removal of nonintegrated copies after tamoxifen administration [1]. On the right, GFAP-StarTrack is employed to monitor astroglial lineages, incorporating XFP reporters under the control of the GFAP promoter [3], while NG2-StarTrack is used for tracking NG2-glia lineages, employing plasmids encoded with the NG2 promoter [11]

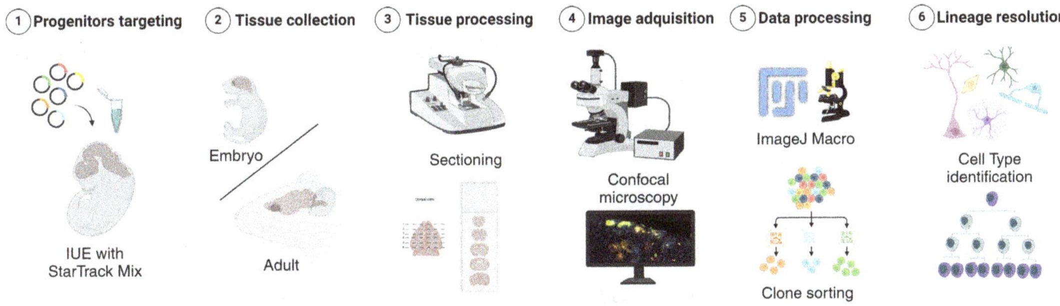

Fig. 2 Clonal analysis workflow and timeline. (1) To conduct clonal analysis, it is essential to inject the StarTrack mixture into the ventricle of embryos via in utero electroporation (IUE) and allow the embryos to develop until the specified stage. (2) Animals will be euthanized at the chosen analysis time point, which may encompass both embryonic and adult stages. (3) The tissue should then be sectioned, typically using a vibratome, and sequentially collected in a rostro-caudal manner. (4) Subsequently, the acquisition of images, covering all six fluorescent channels, is carried out using confocal microscopy. (5) Once the images are prepared, an ImageJ macro is employed to categorize cells based on their color codes. (6) Finally, cell types are determined, and lineages are revealed

StarTrack-targeted cells. Color codes are assigned using a self-developed ImageJ macro, and lineages are established through a retrospective analysis.

2 Materials

2.1 DNA Constructs Amplification

1. Competent *Escherichia coli* DH5α bacteria.
2. StarTrack plasmids.
3. Ampicillin.
4. LB medium and plates.
5. Maxiprep kits for high-quality plasmid DNA purification.

2.2 In Utero Electroporation

1. Microinjector.
2. Square wave electroporation system.
3. Ultrasound machine.
4. Radiant heating plate.
5. Anaesthetic equipment.
6. Fiber optic dual gooseneck illuminator.
7. Glass micropipette puller.
8. Saline 0.9%.
9. Ethanol 70%.
10. Isoflurane: inhalatory anesthesia.
11. Fluoroquinolone: antibiotic used to treat bacterial infections.
12. Nonsteroidal anti-inflammatory analgesics (NSAIDs) diluted in saline.
13. Timed-pregnant mice at the required stage.
14. Gauze sponge.
15. Glass capillary.
16. Cotton swabs.
17. Forceps with a tip width of 0.8 mm.
18. Extra Fine Bonn Scissors, cutting edge 13 mm.
19. Clamp.
20. Absorbable suture.
21. Nonabsorbable suture.
22. Electroconductive ultrasound gel.
23. Electrode tweezer with platinum disk electrodes.

2.3 Tissue Processing

1. Tamoxifen.
2. Surgery instrumentation (forceps, scissors, etc.).
3. Paraformaldehyde 4%.
4. Slicing equipment (vibratome or cryostat).
5. Sucrose (for cryostat sectioning).
6. Agar (for vibratome sectioning).
7. Mounting media.
8. Brushes.
9. Slides and cover slides.
10. Phosphate buffer saline.

2.4 Data Compilation

1. Confocal microscope.
2. Fiji software.
3. Computer with image processing capabilities.

3 Methods

3.1 Construct Design

The StarTrack methodology involves the electroporation of individual DNA constructs, each encoding different fluorescent proteins under the control of various promoters, into target cells. To fully understand the potential of this technology, it is essential to comprehend the dynamics of the system under study. In our case, the neural development process occurs sequentially, with neurons being generated first, followed by glial cells [18]. Upon examining fluorescent reporter expression from integrable and nonintegrable constructs within the adult cell progeny after progenitor targeting, we revealed distinct expression patterns depending on the electroporation stage and cell type. We observed that neuronal lineages emerged from progenitors after minimal divisions in proliferative zones [19]. By contrast, glial lineages originated from progenitors that had undergone a larger number of divisions resulting in the loss of the nonintegrable fluorescent mark (Fig. 3). This demonstrates the critical role of construct integration in labeling glial cells derived from early NPCs. Nonetheless, it is important to note that the expression pattern of episomal copies is contingent on the number of divisions of the targeted progenitor, rather than on the specific cell type (*see* **Note 1**). Collectively, these findings provide valuable insights into the design of strategies for tracking the fate of specific cell populations related to lineage differentiation.

As a result, all StarTrack plasmids are designed based on the PiggyBac transposon system and require the co-electroporation of a transposase enzyme. The transposase precisely identifies terminal repeats within these constructs and subsequently integrates the

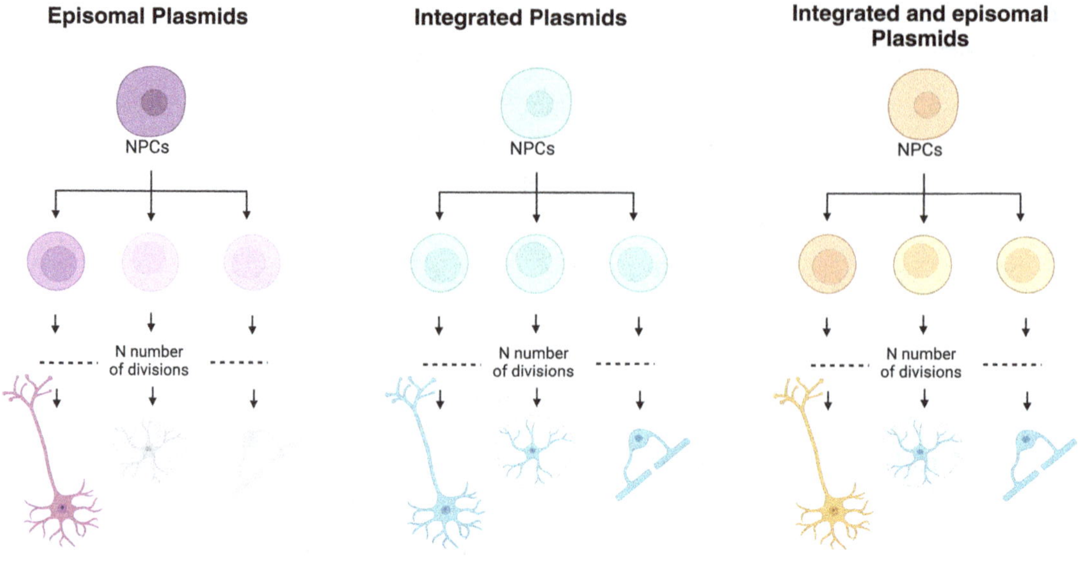

Fig. 3 Importance of the expression of integrated vectors vs. episomal copies. Cells produced by progenitors that underwent few divisions (e.g., neurons produced across development), will retain the expression of episomal copies (in purple). Thus, to target cells that arose from progenitors with a high cell division rate (e.g., glia), it is necessary to use integrable plasmids (blue). When performing co-electroporation of integrated (blue) and nonintegrated (purple) plasmids expressing different colors (yellow cells), the progeny will display different color codes, giving rise to splitting errors when performing clonal analysis

flanking regions of these constructs into the genome of electroporated cells (Fig. 4a, b). This integration process results in the creation of unique color codes through the stochastic integration of fluorescent constructs.

Furthermore, to establish a stable color code for sibling cells of different types, such as neurons and glia, it is crucial to inhibit the persistent copies of nonintegrated or episomal plasmids. To address this challenge, the UbC-StarTrack incorporates the Cre-LoxP genetic recombination system. This strategy enables us to selectively remove all electroporated constructs that have not been integrated into the genome via the transposase. This elimination of nonintegrated constructs occurs when Cre-recombinase is activated, a process induced by the administration of tamoxifen (Fig. 4c). Typically, tamoxifen is administered 2 days after electroporation, facilitating the removal of episomal plasmids and ensuring that only integrated constructs contribute to the long-term color coding of cells. However, in the case of the StarTrack versions that specifically target glial lineages under the GFAP or NG2 promoters, there is no imperative need for inhibiting episomal plasmids. This is because these glial lineages develop over multiple cell division cycles following electroporation, naturally eliminating any nonintegrated constructs.

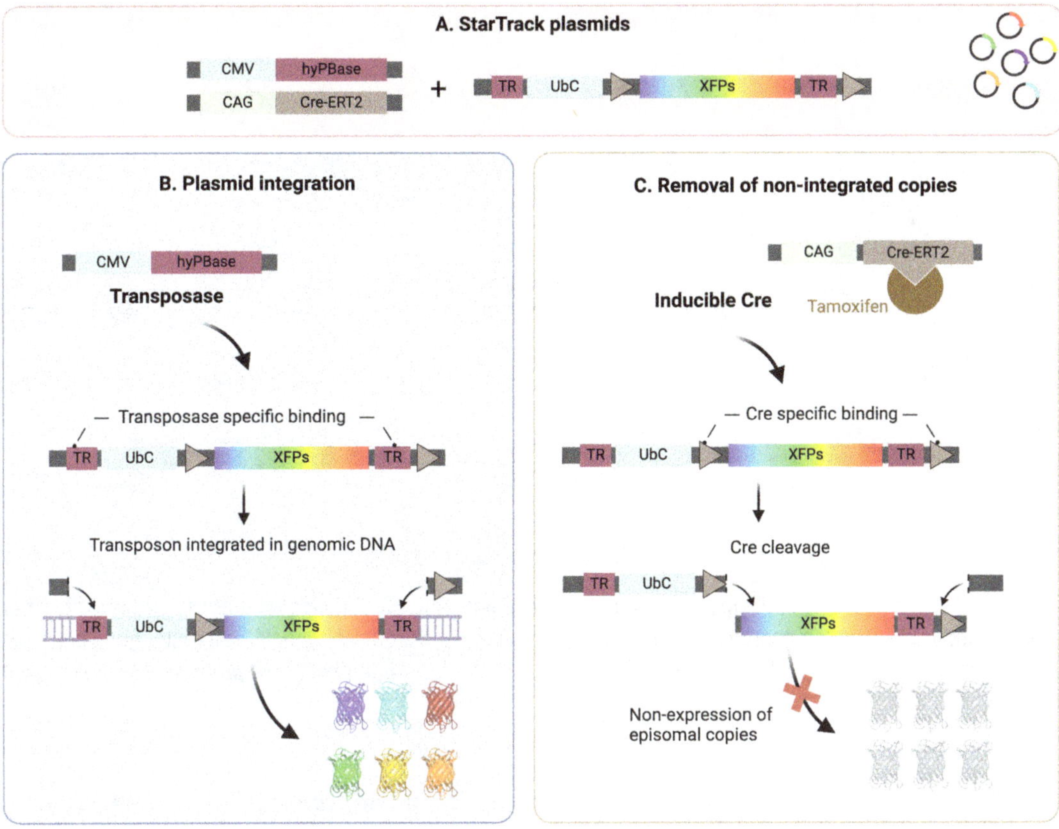

Fig. 4 Integration and inhibition of StarTrack plasmids. (**a**) UbC-StarTrack plasmids are co-electroporated along with the transposase and an inducible Cre plasmid. (**b**) The transposase will recognize the terminal repeats (TRs) of the StarTrack constructs, integrating the flanked region into the cell genome, resulting in stable and heritable targeting. (**c**) The removal of nonintegrated copies is achieved by activating the Cre enzyme through tamoxifen administration. Nonintegrated copies contain two loxP sites, whereas integrated copies retain only one. Thus, Cre will specifically bind to the two loxP sites of episomal copies and cleave them, preventing their expression

3.2 Cloning

StarTrack vectors feature multiple cloning sites (MCS) to facilitate the easy interchange of promoters or fluorescent proteins through straightforward cloning strategies. Unique restriction enzyme sites allow for the interchange of promoters, as well as the addition or removal of custom sequences.

Figure 5 illustrates the various possibilities for designing new versions of the StarTrack plasmids. All employed enzymes produce sticky ends to ensure the correct maintenance of the sequence reading frame. In some cases, cloning strategies may involve different restriction enzymes, such as the interchange of the Ubiquitin C promoter in UbC-StarTrack plasmids using NheI and BamHI. Alternatively, a single enzyme can be used for cloning purposes, such as excising the fluorescent reporter in UbC-StarTrack plasmids with EcoRI.

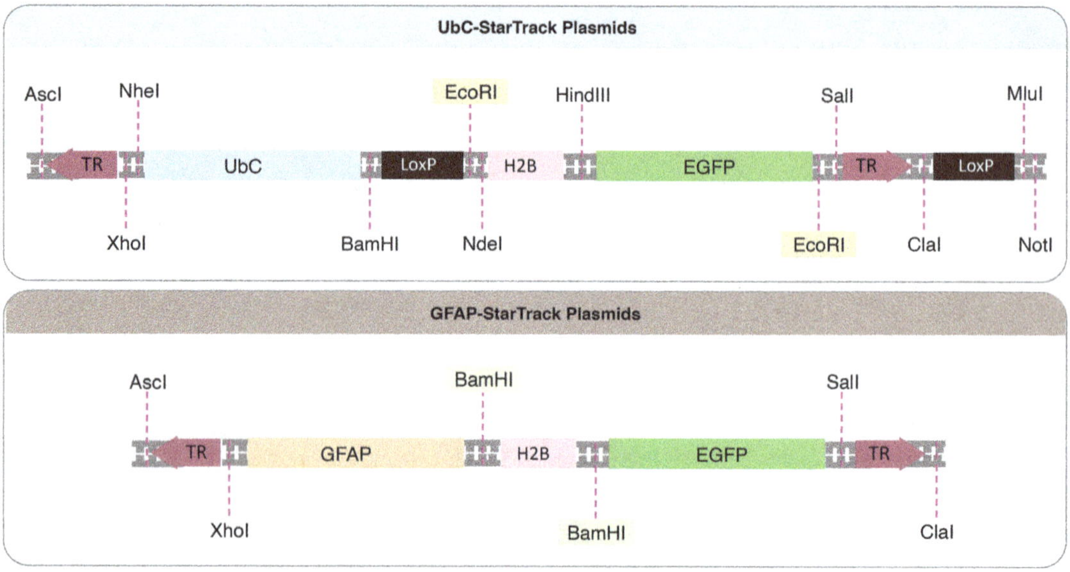

Fig. 5 Clonal strategies for modifying UbC-StarTrack plasmids. UbC-StarTrack plasmids are designed to offer flexibility and customization according to specific study needs. Consequently, single-end restriction sites have been incorporated into the UbC-StarTrack plasmids to facilitate the interchange of promoters or fluorescent reporters with ease

3.3 Plasmid Amplification and Mixing

1. StarTrack plasmids can be introduced into *E. coli* DH5α bacteria through a heat shock procedure.
2. Bacteria containing the StarTrack plasmids are selected by adding ampicillin to the LB medium, where the bacteria will grow overnight.
3. Plasmids are amplified and purified from the bacterial culture using commercial Maxiprep kits for high-quality plasmid DNA purification.
4. Transformed bacteria can be stored at −80 °C mixed with 20% sterile glycerol.
5. To prepare the StarTrack plasmid mixture, it is necessary to follow the criteria below:
   - Add the necessary micrograms (μg) of each XFP plasmid to have an equal final μg amount for any of them.
   - Calculate the total amount of XFP plasmids (μg) and add 1/3 μg of transposase and 1/3 μg of Cre-recombinase if necessary (*see* **Note 2**).
   - Ensure the total final concentration of all plasmids in the mixture falls within the range of 1.8–2.2 μg/μL. This concentration range is critical for the successful execution of the experiment.

- Add Fast Green dye (0.05%) to improve visualization during plasmid injection.

6. The StarTrack plasmid mixture can be stored at −20 °C for years. Keep one stock solution without Fast Green to avoid dye precipitation.

3.4 In Utero Electroporations (IUE)

1. Before the IUE procedure, it is necessary to prepare the glass capillaries by narrowing them using a puller (*see* **Note 3**).
2. Thoroughly sterilize the surgical area and all instruments using a 70% ethanol solution. Proceed to anesthetize pregnant mice by administering a 2% isoflurane/O_2 mixture through inhalation. Once anesthetized, place the pregnant mice on a radiant heating plate set at 37 °C and sustain anesthesia by maintaining a 1.5% isoflurane/O_2 mixture.
3. Administer the fluoroquinolone antibiotic and the nonsteroidal anti-inflammatory analgesics subcutaneously.
4. Clean the abdominal region with 70% ethanol followed by saline 0.9% and perform a precise skin incision, uncovering the uterine horns while placing them on sterile gauze (*see* **Note 4**).
5. Identify the target ventricle by transilluminating the embryo's head. For embryos younger than E12, use ultrasound assistance if possible. For older embryos, cold light transillumination is appropriate.
6. Using a glass capillary, inject 1–2 μL of the StarTrack plasmid mixture. Ensure that the mixture is prepared at a concentration of 1.8–2 μg/μL in distilled water and delivered into the identified ventricle (*see* **Note 5**).
7. After the injection, place the positive electrode tweezer at the region of interest and apply five square electrical pulses with a current duration of 50 ms, delivered at intervals of 950 ms. The size of the tweezer electrode can be adjusted based on the target region. For most IUE applications, 3-mm diameter paddle electrodes are ideal. The voltages should be modified according to the embryonic stage (*see* **Note 6**).
8. Carefully return the embryos to the abdominal cavity. You can use moistened cotton swabs with saline to assist with this step.
9. Close the incision site with a 5/0 resorbable suture for the inner layers and a 3/0 non-resorbable suture for the skin layer.
10. Place the pregnant dams in a sterile environment and maintain them at 37 °C to facilitate the recovery process and reduce the risk of postsurgery infection. For postnatal electroporation, *see* **Note** 7.

3.5 Removal of Episomal Copies

1. Prepare a solution of 20 mg/mL tamoxifen in corn oil. Carefully measure out the required amount of tamoxifen and dissolve it in prewarmed corn oil. Place the mixture in a suitable tube and ensure it is protected from light with tin foil. Place the tube in a roller, and set the temperature to 45 °C. Allow the mixture to dissolve for 1–2 h under these conditions. After dissolution, store your stock solution at +4 °C for up to a week to maintain its stability. Before injecting, ensure the tamoxifen solution is thoroughly mixed by warming it at 45 °C for 30 min. This protocol ensures the proper preparation and storage of tamoxifen solution for experimental use, maintaining its integrity and efficacy.
2. A single intraperitoneal dose of tamoxifen (7.5 mg/kg body weight) is injected to remove nonepisomal copies (*see* **Note 8**).

3.6 Tissue Processing

1. When reaching the desired stage, animals are subjected to perfusion with a 4% PFA solution, after which their brains are retrieved.
2. Following perfusion, brains are then embedded in agarose to facilitate subsequent serial sectioning. Depending on the specific requirements of the clonal analysis, brains can be sectioned in various orientations, including coronal, sagittal, or horizontal. The optimal thickness for these slices, as recommended, is set at 50 μm, using the vibratome settings (*see* **Note 9**). Samples are gathered sequentially to facilitate the reconstruction for accurate identification of the distribution of sibling cells.
3. Cell types are defined by different immunohistochemistry experiments. Depending on the confocal microscopy available for image acquisition either one or two antibodies could be detected (*see* **Note 10**).
4. After the immunohistochemistry process is completed, the samples are mounted using Mowiol mounting media, carried out to maintain the fluorescent labeling for extended durations, guaranteeing the persistence of the fluorescent signals (*see* **Note 11**).

3.7 Image Acquisition

1. After analyzing the labeling in the sections through epifluorescence microscopy, employing suitable filter cubes for GFP (FF01-473/10), mCherry (FF01-590/20), and Cy5 (FF01-628/40-25), sections will be chosen for confocal imaging. Each fluorophore is captured using an independent channel on the confocal microscope. It is crucial to establish specific capture conditions for each microscope to prevent spectral crosstalk. The conditions outlined in Table 1 have been validated on the Stellaris confocal microscope (Leica), specifying the excitation and emission wavelengths in nanometers for each fluorophore (*see* **Note 12**).

Table 1
Confocal conditions for StarTrack with 10× Stelaris-8

Table 1. Confocal conditions for StarTrack with 10x Stelaris-8						
	mT-Sapphire	mCerulean	EGFP	YFP	mKO	mCherry
Excitation	405	440	489	522	548	587
Emission	525-540	455-490	497-510	529-544	556-580	597-625
Intensity	6.5%	8.01%	2,1%	4%	2%	8.01%

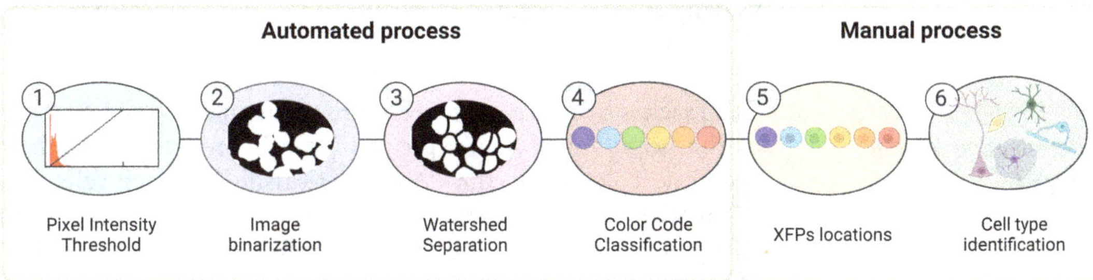

Fig. 6 Workflow of data processing. (1) Pixel intensity threshold: separation of cells from the background. (2) Image binarization where cells are depicted in white against a black background. (3) Watershed separation is applied to isolate individual cell particles. (4) Color coding of cells based on the presence or absence of XFPs. (5) Identification of the nuclear or cytoplasmic localization of XFPs to group cells within a clone. (6) Cell type identification: sibling cells resolution

3.8 Data Processing

We developed an ImageJ macro to assist in the categorization of labeled cells, primarily relying on their segregation according to their expression of various fluorophores (*see* **Note 13**). Here, we outline the fundamental steps incorporated into the script for classifying cells based on their color coding (Fig. 6).

1. *Background Separation*: Through pixel intensity thresholding, targeted cells will be selected and differentiated from the background. This threshold will be determined for each fluorescent protein that has been acquired in a different confocal channel. In addition, values of the minimum fluorescence intensity have to be determined for each fluorescent channel as well as the minimal cell size to later remove artifactual signals (*see* **Note 14**).
2. *Binary Image Generation*: A binary image is then generated, where labeled cells are represented in white, while the background is set to black. This binary image will help isolate the cells from the surrounding background.
3. *Watershed Separation*: Watershed separation on the binary image will be later performed. This process will singularize each cell particle, representing them as individual points. This

step helps prevent overlapping regions from being counted as a single cell. Segmented objects will be then categorized and filtered based on different sizes. This step helps eliminate selections that may correspond to background artifacts.

4. *Color-Code Classification*. Cells are finally categorized based on their color codes, which indicate the presence or absence of each fluorophore (e.g., YFP, mKO, mCerulean, mCherry, mT-Sapphire, and EGFP).
5. *Localization Identification*: This step will not be automated, and the nuclear or cytoplasmic localization of the XFPs will be visually assessed, ultimately resulting in the classification of the final clones.

3.9 Lineage Determination (Clonal Analysis)

1. StarTrack is a retrospective technique that enables the inference of the potential of neural progenitor cells by examining their progeny. To establish the clonal relationship between cells, it is essential to identify distinct cell types (Fig. 7). Cell types are categorized based on both their morphology and immunohistochemical markers, which are visualized in far-red channels. By utilizing a Stellaris confocal microscope equipped with a white light laser, it becomes feasible to identify cells using two different antibodies at the same slice, thereby enhancing the accuracy of cell identification.
2. Based on our experience, certain combinations of color codes tend to be more prevalent than others. Consequently, color codes consisting of just one fluorophore, for example, may be less reliable for identifying sibling cells. The combinations that offer the most accurate identification of sibling cells are those found in the less frequent combinations [11].

4 Notes

1. The expression pattern of episomal copies in neuronal cells might differ when conducting postnatal electroporations. This is particularly relevant in the context of interneurons migrating to the olfactory bulb after postnatal progenitor targeting, as they are generated after fewer progenitor cell divisions. As a result, some of these interneurons might retain the expression of episomal plasmids over time [19].
2. As an example, in a StarTrack mix with 10 μg of each of the 12 fluorescent reporters (totaling 120 μg), you will need to add 40 μg of hyPBase and 40 μg of Cre-recombinase.
3. The capillary should have a thin enough profile for precision but must not be so delicate that it is prone to bending or breaking during the injection. When using the Narishige Puller

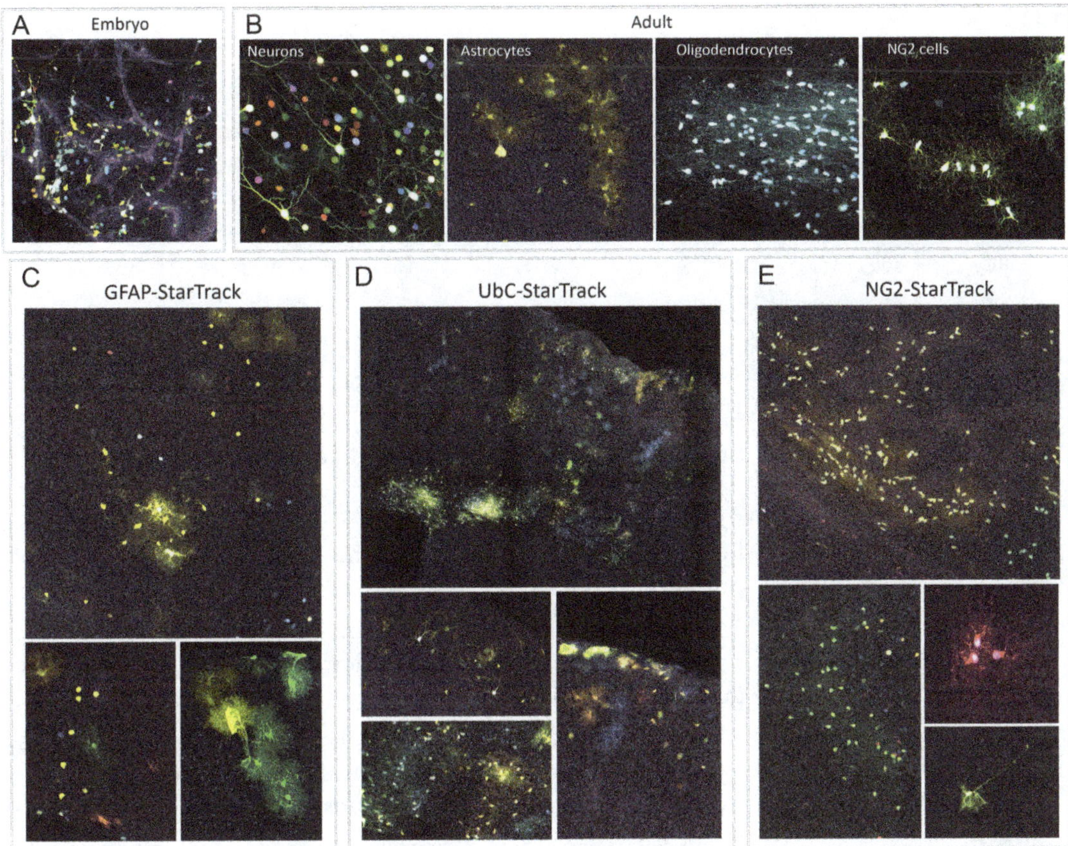

Fig. 7 StarTrack labeling. In vivo cell targeting using different StarTrack approaches. (**a**) In vivo targeting with UbC-StarTrack can be employed to assess sibling relations among immature cells at embryonic stages. (**b**) Different cell types labeled with StarTrack presented recognizable neural morphologies in the adult brain, such as neurons, astrocytes, oligodendrocytes, and NG2 cells. (**c**) GFAP-StarTrack can be used to study clonal relationships within adult astrocytes widespread throughout the cortex. The clonally related cells exhibited different color codes. (**d**) UbC-StarTrack labels neural cells, including neurons and glial cells, to further study their clonal relationships. (E) NG2-StarTrack enables the targeting of NG2 cells, displaying the largest clones found in the mouse cerebral cortex

PC-100 model, the following recommended conditions should be applied: Step 1: Maintain a temperature of 70 °C with two heavyweights and one lightweight. Capillary tip could be trimmed to perform smoother injections.

4. It is of utmost importance to maintain the embryos' moisture by consistently applying saline 0.9% throughout the procedure to ensure their welfare.
5. To confirm successful ventricle filling, ensure that the StarTrack plasmid mixture is labeled with 0.05% Fast Green and is well mixed.
6. Adjust the voltage according to developmental stages and target regions. For example, when targeting pallial-NPCs,

Table 2
Confocal conditions for StarTrack with 20× objective TCS-SP5

Table 2. Confocal conditions for StarTrack with 20x objective TCS-SP5.

	mT-Sapphire	mCerulean	EGFP	YFP	mKO	mCherry
Excitation	405	458	488	514		561
Emission	520-540	468-480	498-509	520-535	560-580	601-620
Intensity	30%	40%	20%	15%		35%

recommend 33 mV for E12 mice, 35 mV for E14, 37 mV for E16, and 100 mV for P0.

7. StarTrack plasmids can be used in postnatal mice (P0–P2). After anesthetizing the pups by hypothermia, inject plasmids following steps 5–6. Before step 7, it is important to cover the head of the pup with electroconductive gel. In step 7, apply pulses of 100 mV.
8. In some cases, the administration of tamoxifen during pregnancy may lead to delivery problems, being necessary the exogen administration of oxytocin to facilitate the labor.
9. A thickness of 50 μm has proven to be the ideal choice, striking a perfect balance between immunohistochemistry requirements and facilitating clonal reconstruction.
10. The Stellaris confocal system is equipped with a white laser for effectively distinguishing between fluorophores. It facilitates the concurrent use of two extra fluorophores in the far-red channel, which are suitable for immunohistochemistry. In our experience, two immunohistochemistry procedures can be performed alongside StarTrack electroporation by using the secondary antibodies Alexa Fluor 647 and 800 or Alexa Fluor 633 and 750 combinations.
11. While Mowiol is the mounting medium we are most experienced with and is particularly effective at preserving the fluorescent proteins in StarTrack, any alternative antifade mounting medium would be appropriate.
12. Conditions for either inverted or direct confocal SP5 (Leica) have been described in detail in Figueres-Oñate et al. [1] (see Table 2).
13. Please note that while we used ImageJ for creating our script based on Java language, any alternative interface and programming language can be used.
14. The minimum intensity level may differ for each channel and should be individually determined for each fluorophore.

References

1. Figueres-Oñate M, García-Marqués J, López-Mascaraque L et al (2016) UbC-StarTrack, a clonal method to target the entire progeny of individual progenitors. Sci Rep 6:33896. https://doi.org/10.1038/srep33896
2. Martín-López E, García-Marques J, Núñez-Llaves R, López-Mascaraque L (2013) Clonal astrocytic response to cortical injury. PLoS One 8:e74039. https://doi.org/10.1371/journal.pone.0074039
3. García-Marqués J, López-Mascaraque L (2013) Clonal identity determines astrocyte cortical heterogeneity. Cereb Cortex 23: 1463–1472. https://doi.org/10.1093/cercor/bhs134
4. Garcia-Marques J, Nunez-Llaves R, Lopez-Mascaraque L (2014) NG2-glia from pallial progenitors produce the largest clonal clusters of the brain: time frame of clonal generation in cortex and olfactory bulb. J Neurosci 34: 2305–2313. https://doi.org/10.1523/JNEUROSCI.3060-13.2014
5. Parmigiani E, Leto K, Rolando C et al (2015) Heterogeneity and bipotency of astroglial-like cerebellar progenitors along the interneuron and glial lineages. J Neurosci 35:7388–7402. https://doi.org/10.1523/JNEUROSCI.5255-14.2015
6. Cerrato V, Parmigiani E, Betizeau M et al (2018) Multiple origins and modularity in the spatiotemporal emergence of cerebellar astrocyte heterogeneity. PLoS Biol 16(9):e2005513
7. Figueres-Oñate M, Sánchez-González R, López-Mascaraque L (2020) Deciphering neural heterogeneity through cell lineage tracing. Cell Mol Life Sci 78:1971. https://doi.org/10.1007/s00018-020-03689-3
8. Bribián A, Figueres-Oñate M, Martín-López E, López-Mascaraque L (2016) Decoding astrocyte heterogeneity: new tools for clonal analysis. Neuroscience 323:10. https://doi.org/10.1016/j.neuroscience.2015.04.036
9. Olmos-Carreño CL, Figueres-Oñate M, Scicolone GE, López-Mascaraque L (2022) Cell fate of retinal progenitor cells: in ovo {UbC}-{StarTrack} analysis. Int J Mol Sci 23:12388. https://doi.org/10.3390/ijms232012388
10. Figueres-Oñate M, Sánchez-Villalón M, Sánchez-González R, López-Mascaraque L (2019) Lineage tracing and cell potential of postnatal single progenitor cells in vivo. Stem Cell Rep 13:700–712. https://doi.org/10.1016/j.stemcr.2019.08.010
11. Sánchez-González R, Figueres-Oñate M, Ojalvo-Sanz AC, López-Mascaraque L (2020) Cell progeny in the olfactory bulb after targeting specific progenitors with different UbC-startrack approaches. Genes (Basel) 11. https://doi.org/10.3390/genes11030305
12. Redmond SA, Figueres-Oñate M, Obernier K et al (2019) Development of ependymal and postnatal neural stem cells and their origin from a common embryonic progenitor. Cell Rep 27:429–441.e3. https://doi.org/10.1016/j.celrep.2019.01.088
13. Herrero-navarro Á, Puche-aroca L, Moreno-juan V, et al (2020) Astrocytes and neurons share brain region-specific transcriptional signatures. bioRxiv preprint 1–23
14. García-Marqués J, López-Mascaraque L (2016) Clonal mapping of astrocytes in the olfactory bulb and rostral migratory stream. Cereb Cortex bhw071:2195. https://doi.org/10.1093/cercor/bhw071
15. Ivics Z, Li MA, Mátés L et al (2009) Transposon-mediated genome manipulation in vertebrates. Nat Methods 6:415. https://doi.org/10.1038/NMEtH.1332
16. Hoess RH, Abremski K (1985) Mechanism of strand cleavage and exchange in the Cre-lox site-specific recombination system. J Mol Biol 181:351–362. https://doi.org/10.1016/0022-2836(85)90224-4
17. Sánchez-González R, Bribián A, López-Mascaraque L (2020) Cell fate potential of NG2 progenitors. Sci Rep 10:1–12. https://doi.org/10.1038/s41598-020-66753-9
18. Cadwell CR, Bhaduri A, Mostajo-Radji MA et al (2019) Development and arealization of the cerebral cortex. Neuron 103:980–1004. https://doi.org/10.1016/j.neuron.2019.07.009
19. Figueres-Oñate M, García-Marqués J, Pedraza M et al (2015) Spatiotemporal analyses of neural lineages after embryonic and postnatal progenitor targeting combining different reporters. Front Neurosci 9:1–11. https://doi.org/10.3389/fnins.2015.00087

Chapter 17

Design and Generation of TEMPO Reagents for Sequential Labeling and Manipulation of Vertebrate Cell Lineages

Isabel Espinosa-Medina

Abstract

During development, cells undergo a sequence of specification events to form functional tissues and organs. To investigate complex tissue development, it is crucial to visualize how cell lineages emerge and to be able to manipulate regulatory factors with temporal control. We recently developed TEMPO (Temporal Encoding and Manipulation in a Predefined Order), a genetic tool to label with different colors and genetically manipulate consecutive cell generations in vertebrates. TEMPO relies on CRISPR to activate a cascade of fluorescent proteins which can be imaged in vivo. Here, we explain the steps to design, generate, and express TEMPO constructs in zebrafish and mice.

Key words CRISPR, Single-strand annealing, Spatiotemporal labeling, In vivo, Vertebrates, Lineage reconstruction

1 Introduction

The formation of complex tissues and organs depends on spatial and temporal patterning as well as cell lineage specification [1, 2]. Tissues undergo cell proliferation and specification events during development, influenced by molecular factors that appear in a precise order to determine cell fate decisions over time [3–5]. To determine the mechanisms of cell specification in the context of developing tissues and organs, it is crucial to be able to visualize and manipulate cell lineages with temporal control.

Advances in single-cell sequencing have unveiled the transcriptional timing of many cell specification factors during development [6, 7]. However, these technologies require tissue dissociation, resulting in loss of spatial information. Ideally, tracking and manipulating cell lineages should be done by live imaging within the tissue context. However, limitations such as tissue impenetrability, lack of highly multiplexed labeling, and phototoxicity often impair the visualization of cell interactions and lineage progression live

Jorge García-Marqués and Tzumin Lee (eds.), *Lineage Tracing: Methods and Protocols*, Methods in Molecular Biology, vol. 2886, https://doi.org/10.1007/978-1-0716-4310-5_17,

[8]. Alternative strategies have emerged, which allow reconstruction of cell histories via retrospective analysis of sequentially incorporated cell labels. Some of these strategies rely on sequential cell barcoding and in situ sequencing of multiple DNA barcodes, requiring tissue fixation [9–16]. Other strategies allow multiclonal labeling of cell lineages in vivo but lack temporal resolution to distinguish subsequent cell generations [17]. The technology CLADES was developed in Drosophila to solve these limitations [18]. This strategy allows differential labeling of consecutively produced neurons, revealing both spatial and temporal information of defined cell lineages in the same sample.

Recently, we developed a genetic tool called TEMPO, which allows sequential labeling and manipulation of consecutive cell generations in vertebrates, including zebrafish and mice [19]. TEMPO relies on CRISPR/Cas9 to sequentially activate a cascade of fluorescent proteins in a precise order (Fig. 1). The TEMPO system consists of two main components: a TEMPO reporter encoding three conditional fluorescent reporters in different coding frames, and a conditional cascade of gRNAs which, along with Cas9, allow consecutive activation and inactivation of the TEMPO fluorescent reporters. Importantly, the compact design of TEMPO allows alternative expression of three (or more) fluorescent reporters from a single transgene. The orderly activation of each fluorescent reporter relies on frame shifts driven by a programmed cascade of Cas9/gRNA cuts and repair events. This system is based on a homology-based dsDNA repair mechanism called single-strand annealing (SSA) [20], which happens in dividing cells and allows predictable and scarless activation of the fluorescent proteins. This way, only one fluorescent protein can be expressed at a time, and TEMPO color transitions are irreversible. Finally, TEMPO not only allows sequential labeling of consecutive cell generations, but it can also be used to deploy genetic perturbations in a temporal manner. For this, we coupled the expression of an effector gene (e.g., a constitutively active or a dominant negative protein) to that of one of the TEMPO fluorescent reporters. This allows generation of defined temporal windows of genetic perturbations and simultaneous visualization of control and perturbed cells in the same sample.

Here, we outline methods to generate TEMPO reagents that can be expressed in vertebrate models. We first explain the steps to design and clone the TEMPO reporter and Cas9/gRNA cascade transgenes. Then, we outline protocols to design and clone genetic perturbation motifs along with the TEMPO labeling cascade. Finally, we provide guidelines to express and analyze TEMPO transgenes in zebrafish and in mice.

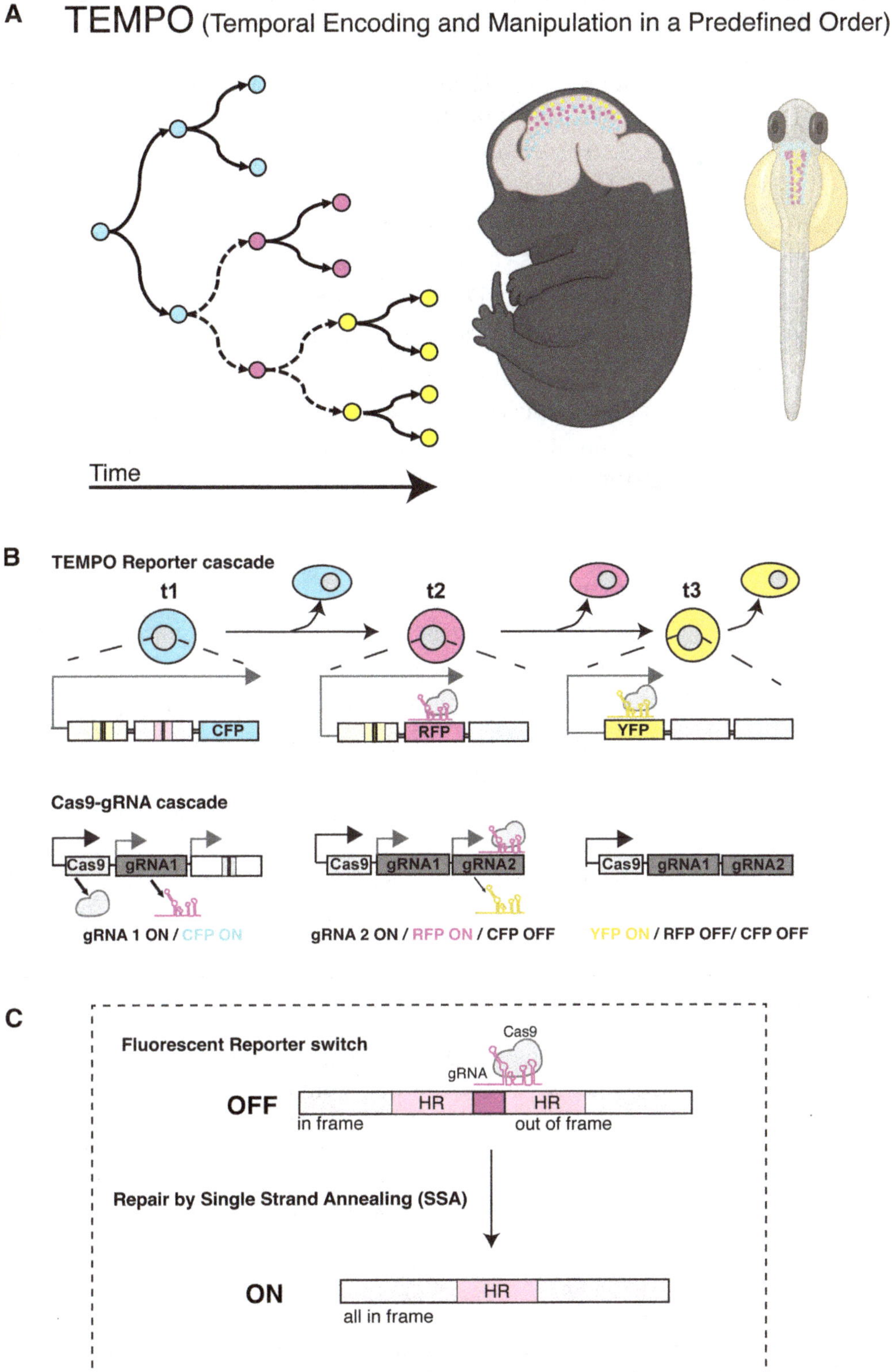

Fig. 1 (**a**) TEMPO allows sequential labeling of cell generations in their tissue context in vertebrates. Dashed lines in the lineage schematic represent an undefined number of cell divisions. On the right, two schematic representations of TEMPO-labeled neurons in the developing brain of a mouse or zebrafish embryo (**b**) TEMPO

2 Materials

1. Restriction enzymes (KpnI, SnaBI, BsiWI, AatII) and associated digestion buffers.
2. Entry vectors for multisite Gateway cloning: pME-MCS, p3E-MCS, and pDest (*see* Subheading 3.4).
3. DNA Gel purification kit.
4. Gibson Assembly 2× Master Mix.
5. High-fidelity DNA polymerase.
6. LR Clonase II for Multisite Gateway recombination.
7. Plasmid miniprep and midiprep kits.
8. Chemically competent bacteria (*Escherichia coli*).
9. Ampicillin and Kanamycin containing agar plates and LB culture media.
10. Liquid culture tubes.
11. 37 °C shaking incubator.
12. Table-top centrifuge.
13. Programmable nanoinjector for zebrafish injections.
14. Fast green and phenol red solutions.
15. Glass-pulled pipettes for zebrafish and mouse microinjections.
16. ECM 830 Electroporator (with electroporator tweezertrodes 3 mm or 5 mm).
17. Nanoject III for mouse in utero electroporation.
18. Antibodies to amplify TEMPO proteins: rat anti-mCherry Invitrogen M11217; RRID: AB_25366; mouse anti-V5 Invitrogen R96025; RRID: AB_2556564.

3 Methods

3.1 Initial Considerations: Choosing TEMPO Fluorescent Reporters and Number of Labeling Steps

Before designing TEMPO genetic constructs, users need to decide what fluorescent proteins they want their system to express, and the ideal number of sequential labeling steps for their application. The original TEMPO reporter sequence was designed using mCerulean3.0, mCherry, and mCitrine as fluorescent protein (FP) reporters (from now on named CFP, RFP, and YFP) (Fig. 1a) [19], but other reporters can be used following the

Fig. 1 (continued) relies on the consecutive activation of a cascade of fluorescent reporters, driven by a Cas9-guided gRNA cascade. t = time (**c**) Schematic depicting a Fluorescent Reporter Switch and its repair mechanism via single-strand annealing (SSA)

same design principles described in the next section. First, users should consider what microscope and excitation wavelengths are available to image their samples. Second, for a given excitation and emission spectrum, what fluorophore exhibits the most appropriate characteristics for a given application (e.g., brightness, photostability, and folding speed for live-imaging experiments, or most compatible excitation/emission spectra for combinatorial labeling with transgenic lines). Finally, among the available options, which fluorescent reporters differ the most in their DNA sequences (*see* **Note 1**).

Regarding the number of sequential labeling steps, this will depend on the length of the lineage tracked. TEMPO cascade progression relies on cell division and our observations suggest that each color transition requires more than one cell division to happen (e.g., ~2.5 cell divisions on average) in asymmetrically dividing neural progenitor cells [19]. Longer lineages will thus require more labels to distinguish the multiple cell generations, while shorter lineages might not reach the last TEMPO color. Thus, prior knowledge on the number of cell divisions in a lineage would be beneficial to better decide on the number of steps to add to the TEMPO fluorescent reporter cascade. However, in most cases, this number will be unknown to the user. In that case, a compromise between number of steps added, length of the transgene, and number of channels available to image will have to be established. We have demonstrated the functionality of up to four labeling steps in mice and zebrafish. However, users will have to test additional fluorescent labeling steps in their specific experimental conditions. Moreover, it is possible that the speed of reporter cascade progression differs across lineages. Thus, when using TEMPO in a different model system or cell population for the first time, we recommend establishing the timing of cascade progression first (*see* **Note 2**).

3.2 TEMPO Reporter Cascade: Sequence Design

Start by creating a computer file with the required sequences for a given model system: fluorescent reporter sequences and peptide linker sequences (e.g., T2A, P2A [21]) (*see* **Note 3**). To design the constructs, use an online gene design tool that allows detailed sequence annotation and visualization of both the nucleic and amino acid sequences. The goal is to design a polycistronic construct comprising fluorescent protein sequences translatable from different open reading frames. As an example, we will demonstrate the design of a three-step TEMPO sequence, but the same logic applies to the four-step TEMPO construct [19]. The main steps for the TEMPO reporter construct design are the following (Fig. 2):

1. Create a tri-cistron by adding the FP sequences one after the other and separating them by peptide sequences like T2A or P2A. The order is inverse to that in which the cascade will

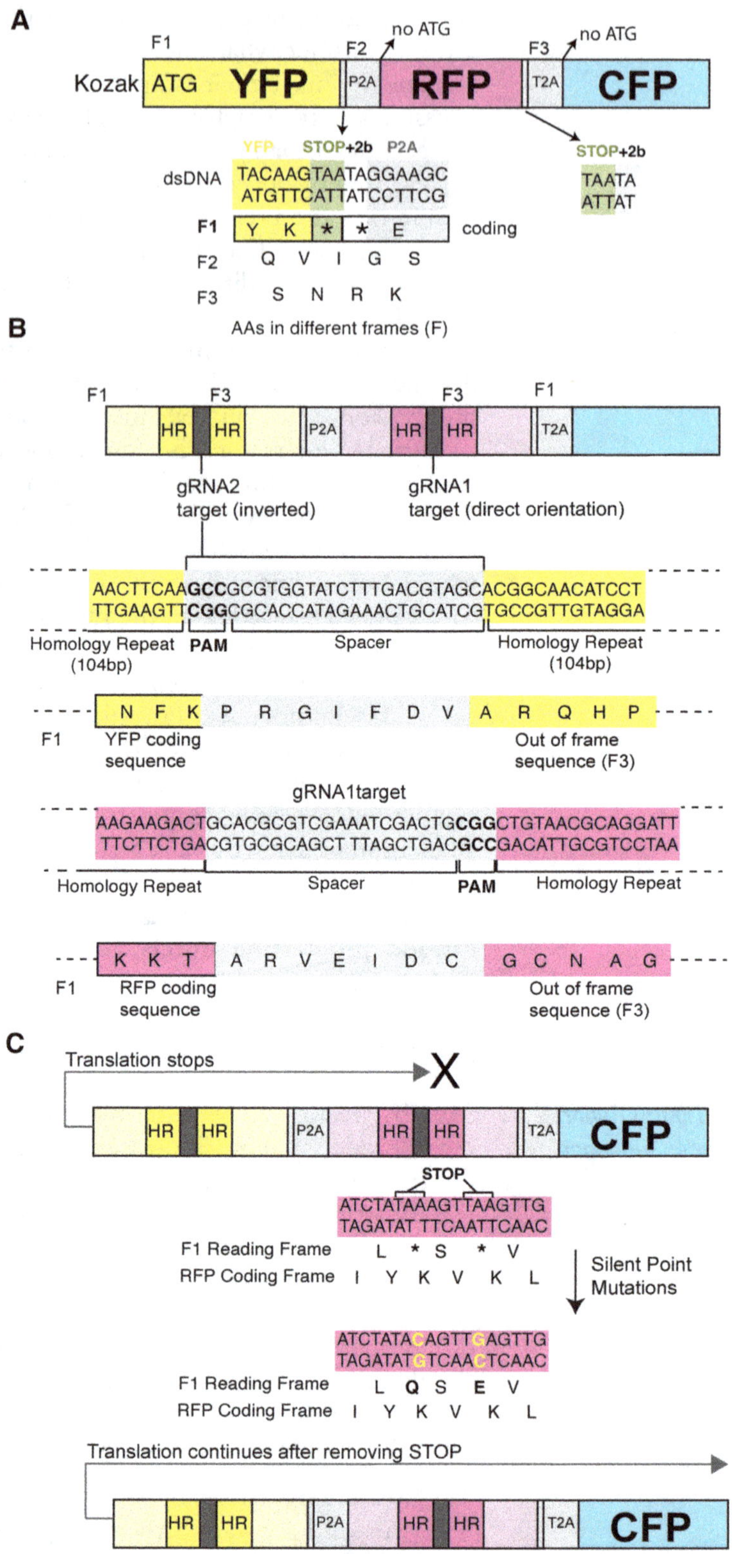

Fig. 2 Design of the TEMPO fluorescent reporter cascade (**a**) Polycistronic construct formed by three fluorescent proteins (YFP, RFP, CFP) separated by peptide (2A) sequences. STOP codons and extra base pairs were added at the

transition (the first fluorophore to be translated will be the last in the tri-cistronic sequence, while the last to be translated is the first in the sequence).

2. Remove the start codons (ATG) from the second and third FPs, immediately after the 2A sequences. This modification will allow translation to start only in the first ATG of the sequence.
3. Add a STOP codon after each fluorescent reporter sequence. We recommend using the preferred STOP codons depending on the species.
4. By visualizing both the nucleic and amino acid sequences, add two random base pairs (bp) after each STOP codon right before the 2A peptide sequences, so that each FP sequence can be translated from a different reading frame. Make sure the 2 bp added does not generate a START codon (ATG) in any of the frames (Fig. 2a).
5. To generate activatable TEMPO reporter switches, we use the concept of CaSSA switches. These are reporter sequences that are disrupted by a gRNA target flanked by two direct repeats. CaSSA switches get activated after Cas9/gRNA-driven DNA break and repair via single-strand annealing (SSA) [22]) (Figs. 1 and 2b).
6. To convert FP sequences into CaSSA switches, users should first select the position in which to introduce a gRNA target that will disrupt translation of a functional FP. We recommend consulting protein database (PDB) to find amino acids important for chromophore binding, to disrupt their translation and thus impair fluorescence. Once selected, introduce the gRNA target sequence (20 bp + PAM) for the first active gRNA of the cascade (see (c) gRNA cascade design below). The orientation of the target can be direct or inverted (*see* **Note 4**).
7. Select a sequence of 104 bp from the FP sequence upstream of the gRNA target and repeat it right after. This will result in two direct repeats, also called homology repeats (HRs), flanking the gRNA target. This is required for the FP to get activated via Cas9/gRNA-guided cut and DNA repair by SSA (*see* **Note 5**).
8. Verify that the reading frame downstream of the gRNA target has now switched to a different one and will not produce a functional protein (Fig. 2b).

Fig. 2 (continued) indicated locations to allow each FP to be expressed from different coding frames (F1, F2, F3) (**b**) Inactivation of YFP or RFP reporters by introduction of gRNA target sequences and duplication of homology repeats (HRs) (**c**) Silent point mutations (base pairs highlighted in yellow) were introduced to allow translation to proceed after disruption of upstream FP coding frames

9. Repeat this same process for the second FP sequence downstream of the first modified one.
10. The new noncoding reading frames after each gRNA target will likely contain STOP codons. To allow translation to proceed beyond the disrupted FP sequences up to the third, intact FP sequence at the end of the tri-cistron, one will need to mutate these STOP codons in the noncoding sequences so the translation can read through. However, the mutations that eliminate the STOP codons need to be silent point mutations in the coding FP sequences. For this, it is recommended to always visualize all three reading frames simultaneously to eliminate the STOP codons in the noncoding frames, using silent point mutations that preserve the amino acid sequence in the coding frames. Annotate any bp changes made in the original FP sequences (e.g., G to T for stop removal) (Fig. 2c).
11. Look for any cryptic START codon (Kozak + ATG in the species of study (*see* **Note 6**) that is present along the sequence and, if possible, introduce silent point mutations to remove it. This will prevent translation of the downstream FP from alternative start sites, which could result in translation of FPs in the incorrect order.
12. Look for any unwanted repeated sequences (>30 bp) other than the FP direct repeats, which could lead to undesired recombination (*see* **Note 1**).
13. Finally, perform an in silico experiment to verify that the construct will work as expected if targeted by Cas9/gRNA and repaired by SSA (Fig. 3):
14. Open the sequence of the final construct which comprises two disrupted FPs upstream of an intact FP sequence. Always visualize the DNA and amino acid sequences of all three reading frames simultaneously.
15. Confirm that the translation, which starts in the first ATG of the first FP of the sequence, does not stop until the STOP codon of the third and intact FP sequence. This open reading frame should only show one functional amino acid sequence, for the last 2A peptide and FP. The other amino acid sequences upstream of that, in that open reading frame, will not encode fully functional FPs.
16. Next, "Repair" the second FP in the construct sequence by eliminating the gRNA target and one of the HR repeats. Verify that now, the open reading frame shows a functional amino acid sequence only for the second FP of the construct. The last FP sequence of the construct should now appear out of frame and cannot be translated from the open reading frame even if the STOP was skipped.

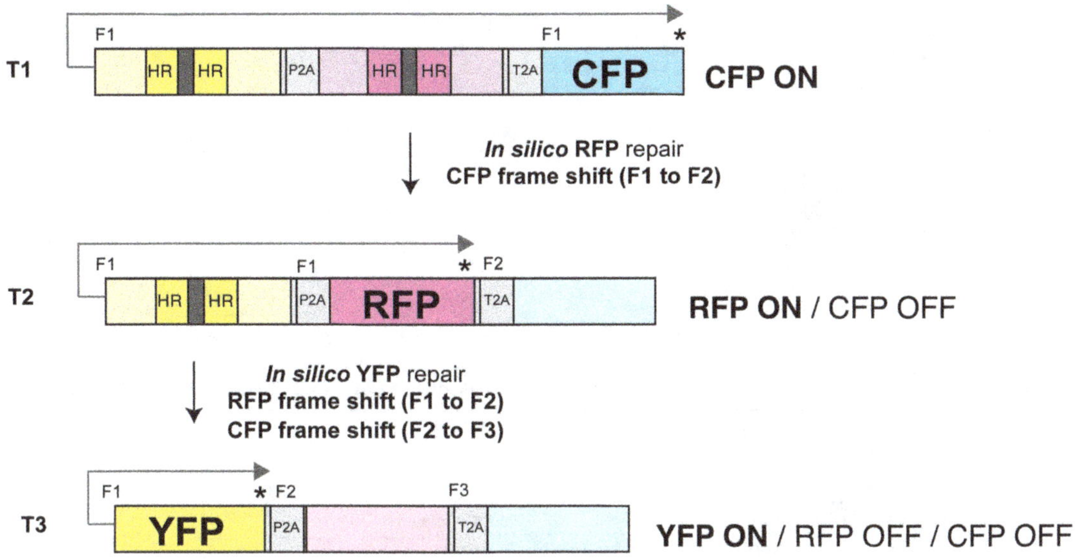

Fig. 3 In silico repair scheme for the TEMPO reporter cascade. T1: CFP is in-frame and translation can read through the disrupted YFP and RFP and can only produce an active CFP. T2: after removal of the gRNA target and one of the homology repeats (HRs) of the disrupted RFP, only RFP is in frame (F1) and now CFP is out of frame (F2). T3: after repair of the YFP switch by removing the gRNA target and one of the HR, only YFP is in-frame (F1) while both RFP and CFP are placed out of frame (F2 and F3, respectively)

17. Finally, "Repair" the first FP in the construct sequence by following the same steps as above. Verify that now the open reading frame would only translate the first FP until its STOP codon, while the other two FP downstream will remain out of frame.

3.3 gRNA Cascade: Sequence Design

The activation of the TEMPO reporters in a predefined order requires a cascade of Cas9/gRNA-driven double-stranded breaks (DSB) followed by SSA repairs. To this end, in addition to the TEMPO reporter construct, users will need to deploy constructs containing Cas9 and gRNAs which are activated in order. In a three-step TEMPO cascade in which one FP is preactivated, and two FP need to be sequentially activated, users will need to provide two gRNAs, each of which will activate one FP in order (Fig. 1b). The first gRNA will be active (gRNA1), while the second gRNA will need to be inactive, and only activated by the first one (gRNA2-switch). The concept of gRNA switch is like that of a CaSSA switch: a gRNA switch is disrupted by the presence of two direct repeats flaking a gRNA1 target. gRNA2 can only be activated after Cas9/gRNA1 driven cut and SSA repair (Fig. 4). Generation of a gRNAswitch which does not function in the inactive form, but with high repair efficiency, required a long optimization process [Supplementary Material 19] and we recommend to reuse this gRNAswitch design for optimal results (*see* **Note 7**). The steps followed for the design of the gRNAs used in TEMPO are detailed below:

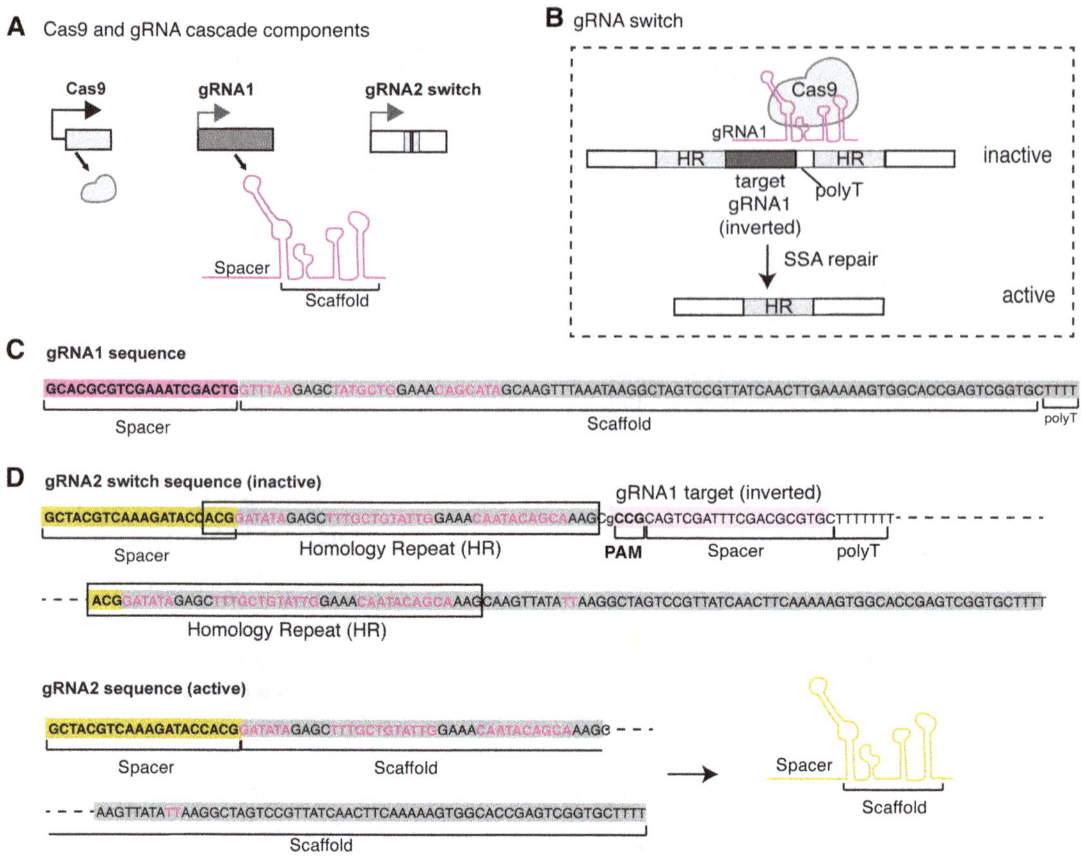

Fig. 4 (**a**) Components of a gRNA cascade (**b**) Scheme of a gRNAswitch, including two homology repeats (HRs) flanking an inverted gRNA target and a polyT transcriptional termination signal. The gRNA2switch is activated via Cas9/gRNA1 guided cut and single-strand annealing (SSA) repair (**c**) gRNA1 sequence. The spacer and scaffold domains are underlined. (**d**) Sequences of the inactive gRNA2switch and active gRNA2 after repair by SSA. The gRNAswitch contains two HRs which include 3 bp from the spacer, and 40 bp from the scaffold region. Base pairs that accept some level of modification. (Based on Briner et al. [5]) are labeled in pink

1. Choose two gRNA sequences (or more for longer cascades), including the spacer and scaffold motifs (Fig. 4). The gRNA sequences used in our original design were chosen using Benchling based on their low off-target [23] and their high on-target activity [24] with the lowest score being 75 (on-target activity) and 96 (off-target activity). These gRNAs did not target any of the genomes in our species of study (zebrafish or mice) (*see* **Note 8**). Each of these gRNAs will recognize its corresponding target (spacer+PAM) in the TEMPO-disrupted fluorescent reporter sequences.
2. gRNA1: This gRNA targets the first disrupted FP (RFP in our original TEMPO design).
3. gRNA2switch: The design of gRNA2switch includes the target sequence of gRNA1 in reverse orientation, between two 43 bp

direct repeats of the scaffold region. After Cas9/gRNA1 guided cut and SSA repair, a functional gRNA2 will be produced and will target the second disrupted FP (YFP in our design).

4. The scaffold sequences of the different gRNAs should not be identical, to prevent recombination between them after Cas9/gRNA-guided double-strand break. For this, we introduced nucleotide changes that should not affect gRNA structure or function [25] (Fig. 4, pink labeled nucleotides) (*see* **Note 9**).

3.4 Cloning TEMPO and Cas9/gRNA Expression Plasmids

TEMPO relies on the expression of two main components: a fluorescent Reporter Cascade and Cas9/gRNA Cascade. For this system to work efficiently, the Reporter Cascade and the gRNA2switch should be cloned in different plasmids (*see* **Note 10**).

Cloning the TEMPO Reporter Plasmid (Fig. 5)

We recommend using multisite Gateway cloning to combine the TEMPO reporter transgene with species-specific promoters and other elements of interest contained in entry vectors (p5E-Promoter, pME-TEMPO, p3E-Promoter-gRNA1, pDest). To clone the three-step TEMPO transgene into the multiple cloning site (MCS) of a pME-MCS vector [26]:

1. Obtain dsDNA of four fragments either by gene synthesis or PCR of readily available plasmids. Fragment 1 contains YFP (or the last FP expressed in the TEMPO cascade) up to the gRNA2 target sequence. Fragment 2 contains the gRNA2 target and the downstream YFP sequence flanked by P2A followed by the first part of the disrupted RFP (the second FP of any TEMPO cascade) up to the gRNA1 target. Fragment 3 contains the gRNA1 target and the downstream RFP sequence flanked by T2A. Fragment 4 contains CFP (or the first active FP expressed in the TEMPO cascade). dsDNA fragments 1 and 4 contain homology arms (25–30 bp) that match the digested pME-MCS vector arms, while the other fragments contain homology regions between them for Gibson assembly (*see* **Note 11**).
2. Digest and gel purify pME-MCS.
3. Gibson assembly the five components into a single reaction:
   - X μL digested pME-MCS (100 ng).
   - X μL dsDNA fragment 1 (0.05 pmol).
   - X μL dsDNA fragment 2 (0.05 pmol).
   - X μL dsDNA fragment 3 (0.05 pmol).
   - X μL dsDNA fragment 4 (0.05 pmol).

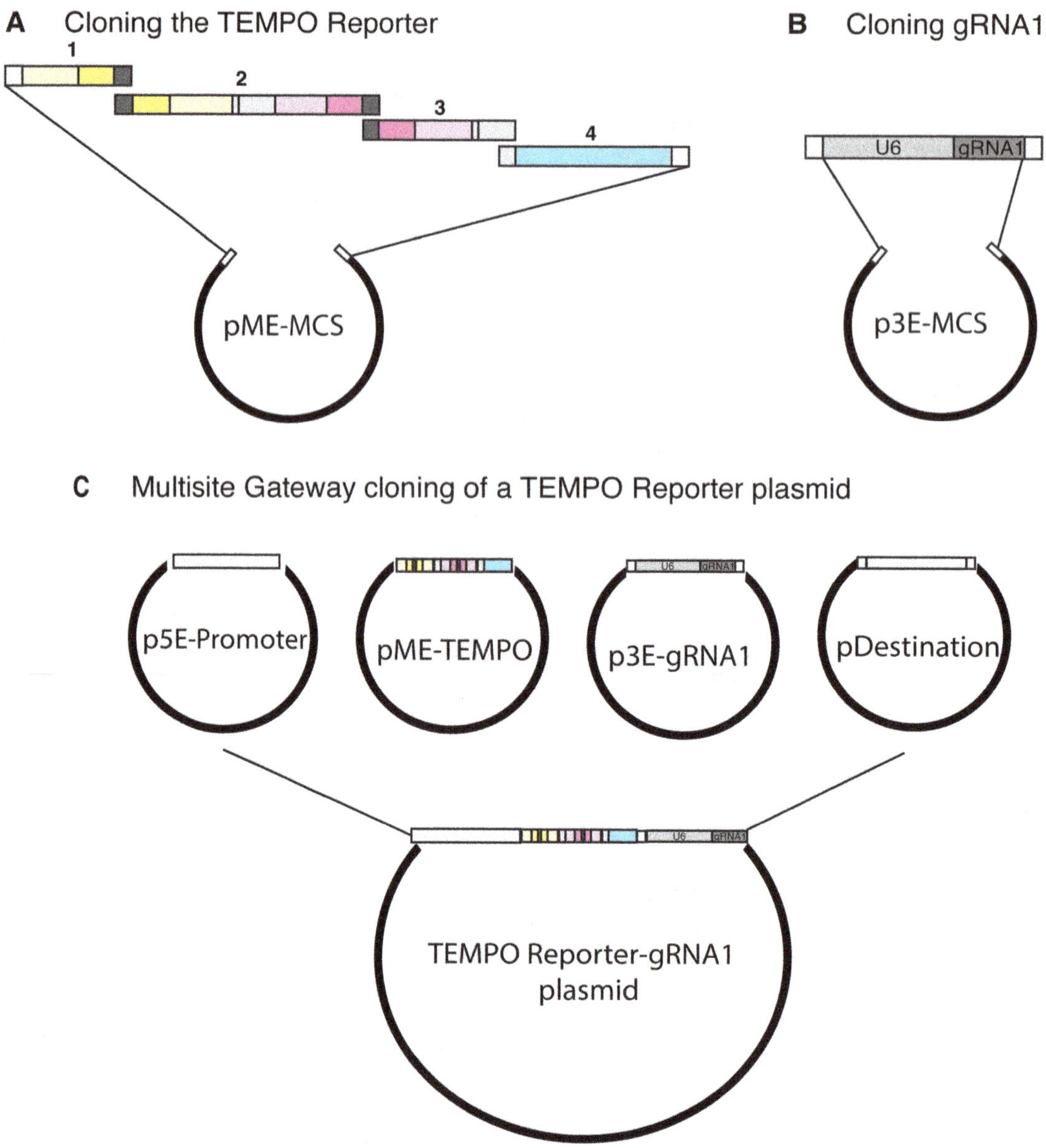

Fig. 5 Cloning strategy for the TEMPO reporter cascade plasmid. Gibson Assembly of (**a**) TEMPO fluorescent reporter fragments into a digested pME-MCS and (**b**) U6 driven gRNA1 into a digested p3E-MCS vector (**c**) Multisite Gateway cloning of the final TEMPO reporter cascade plasmid using entry vectors (p5E contains a promoter chosen by the user). Note that the polyA termination signal can be included at the end of the pME-TEMPO vector, at the beginning of the p3E-U6gRNA1 vector, or in the pDestination vector

- 5 μL Gibson master mix.
- X μL H_2O.
- 10 μL total.
- Incubate reaction at 50 °C for 60 min.

4. Transform Gibson assembly mix into competent cells and grow colonies on LB agar Kanamycin plates.
5. Culture colonies with LB + Kanamycin and sequence plasmid (pME-TEMPO).

In parallel, clone the gRNA1 sequence into the multiple cloning site (MCS) of the p3E-MCS vector [26] (Fig. 4b).

1. Obtain dsDNA of a species-specific type-III U6 promoter followed by gRNA1 including homology arms matching a digested p3E-MCS plasmid (*see* **Note 12**).
2. Digest and gel purify p3E-MCS.
3. Gibson assembly the two components into a single reaction:
   - X μL digested pME-MCS (100 ng).
   - X μL dsDNA fragment 1 (0.1 pmol).
   - 5 μL Gibson master mix.
   - X μL H_2O.
   - 10 μL total.
   - Incubate reaction at 50 °C for 30 min.
4. Transform Gibson assembly mix into competent cells and grow colonies on LB agar Kanamycin plates.
5. Culture colonies with LB + Kanamycin and sequence plasmid (p3E-gRNA1).

Use multisite Gateway cloning to generate the final expression vector by combining the following elements: p5E-Promoter, pME-TEMPO, p3E-U6gRNA1, pDestination vector with genetic elements of interest (Fig. 4c) (*see* **Notes 13** and **14**).

1. Gateway reaction (*see* **Note 15**)
   - X μL pDest (20 pmol).
   - X μL p5E-Promoter (10 pmol).
   - X μL pME-TEMPO (10 pmol).
   - X μL p3E- U6gRNA1 (10 pmol).
   - 1 μL LR Clonase-II.
   - X μL H20.
   - 5 μL total.
   - Incubate reaction at 25 °C overnight.
2. Transform Gateway mix into Top10 competent cells, grow colonies on LB agar Ampicillin plates.
3. Culture colonies with LB + Ampicillin and sequence plasmid.

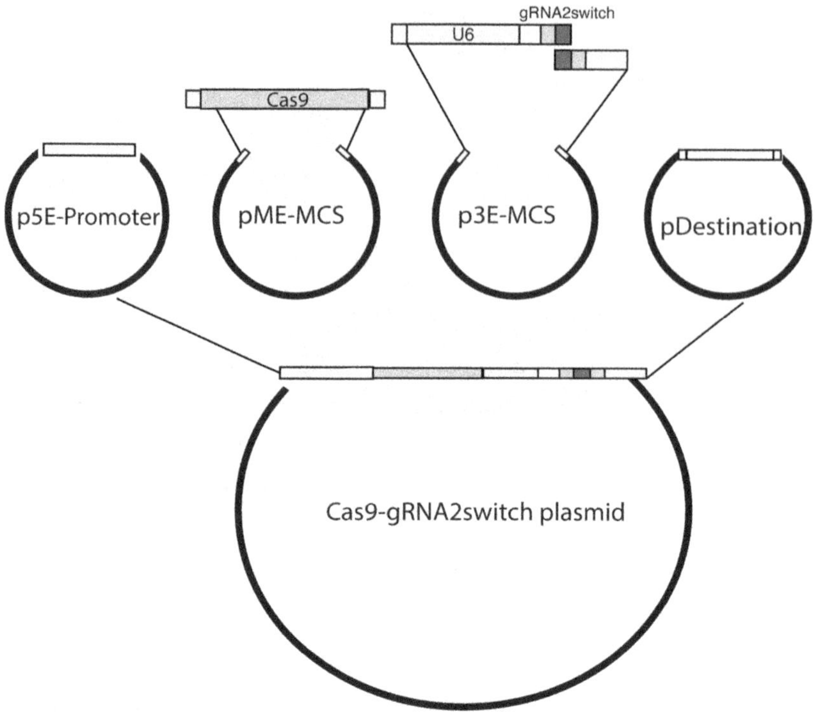

Fig. 6 Cloning strategy for the Cas9 and gRNA cascade plasmid. Gibson assembly of (**a**) a Cas9 fragment into a digested pME-MCS and (**b**) two fragments spanning the U6-driven gRNA2switch into a digested p3E-MCS vector. (**c**) Multisite Gateway cloning of the final Cas9 and gRNA2switch plasmid using entry vectors (p5E contains a promoter of interest)

Cloning the Cas9 and gRNA2switch Plasmid (Fig. 6)

We also recommend using multisite Gateway cloning to combine the Cas9 transgene with species-specific promoters, the gRNA2switch expression vector, and other elements of interest.

To clone the Cas9 transgene into the multiple cloning site (MCS) of the pME-MCS vector:

1. Obtain a codon-optimized Cas9 fragment either by gene synthesis or PCR of readily available plasmids, including homology arms matching the digested pME-MCS vector arms.
2. Digest and gel purify pME-MCS.
3. Gibson assembly the two components into a single reaction:
   - X μL digested pME-MCS (100 ng).
   - X μL dsDNA fragment 1 (0.1 pmol).
   - 5 μL Gibson master mix.
   - X μL H_2O.
   - 10 μL total.
   - Incubate reaction at 50 °C for 30 min.

4. Transform Gibson assembly mix into competent cells and grow colonies on LB agar Kanamycin plates.
5. Culture colonies with LB + Kanamycin and sequence plasmid (pME-Cas9).

In parallel, clone the gRNA2switch into the multiple cloning site (MCS) of a p3E-MCS vector:

1. Obtain dsDNA of two gene fragments (Fig. 6): fragment 1 contains a species-specific U6 promoter followed by the first half of the gRNA2 up to the inverted gRNA1 target sequence, and fragment 2 contains the remaining sequence of the gRNA2switch including the gRNA1 target as the overlapping region.
2. Digest and gel purify p3E-MCS.
3. Gibson assembly the three components into a single reaction:
   - X μL digested pME-MCS (100 ng).
   - X μL dsDNA fragment 1 (0.05 pmol).
   - X μL dsDNA fragment 2 (0.05 pmol).
   - 5 μL Gibson master mix.
   - X μL H20.
   - 10 μL total.
   - Incubate reaction at 50 °C for 30 min.
4. Transform Gibson assembly mix into competent cells and grow colonies on LB agar Kanamycin plates.
5. Culture colonies with LB + Kanamycin and sequence plasmid (pME-gRNA2switch).

Use multisite Gateway cloning to generate the final expression vector by combining the following elements: p5E-Promoter, pME-Cas9, p3E-U6gRNA2switch, and pDestination vector with elements required for genetic integration (*see* **Note 16**).

1. Gateway reaction.
   - X μL pDest (20 pmol).
   - X μL p5E-Promoter (10 pmol).
   - X μL pME-Cas9 (10 pmol).
   - X μL p3E-U6gRNA2switch (10 pmol).
   - 1 μL LR Clonase-II.
   - X μL H_2O.
   - 5 μL total.
   - Incubate reaction at 25 °C overnight.

2. Transform Gateway mix into Top10 competent cells and grow colonies on LB agar Ampicillin plates.
3. Culture colonies with LB + Ampicillin and sequence plasmid.

3.5 Introducing Temporal Perturbations

A powerful feature of the TEMPO technology is the possibility of introducing temporal perturbations coupled with sequential fluorescent labeling [19]. This allows, for instance, to determine how different molecular factors are involved in the cell specification cascade, by activating or inactivating effector molecules along time. To this end, we generated TEMPO-effector constructs, which couple activation and inactivation of an effector gene together with a TEMPO reporter, leaving the other reporter frames intact and allowing sequential labeling of manipulated and control cell populations (Fig. 7) (*see* **Note 17**).

3.6 Designing and Cloning Effector Genes to Generate TEMPO-Effector Reporter Cascades

To express an effector gene simultaneously with a TEMPO reporter gene, users should clone the effector sequence in frame with that of the TEMPO reporter, separated by a 2A peptide (Fig. 7) (*see* **Note 18**).

1. First, find and copy the sequence of the effector of interest.
2. Open the file containing the TEMPO reporter cascade gene design and paste the effector sequence in frame, immediately downstream of the FP sequence to be co-expressed with it. Alternatively, if an effector is to be expressed with the first FP (CFP) of the TEMPO cascade, the effector can be placed immediately upstream of CFP, followed by a 2A-CFP sequence.
3. Introduce a 2A peptide sequence between the end of the FP sequence and the beginning of the effector gene. If the 2A sequence used is the same as one already used in the original TEMPO reporter design, we recommend introducing four silent point mutations to that sequence to avoid unwanted repeat recombination.
4. The STOP codon and 2 bp present at the end of the FP in the original TEMPO design should now be at the end of the effector gene.
5. To allow translation to proceed beyond the disrupted FP sequence and its associated effector gene, users will need to introduce silent point mutations to eliminate STOP codons in the noncoding sequences so the translation can read through. As explained in section (b), the mutations need to be silent point mutations in the coding effector sequence.

To express the effector gene in a different temporal window, follow the same steps above, but place the effector sequence in the corresponding coding frame, immediately after the associated FP of

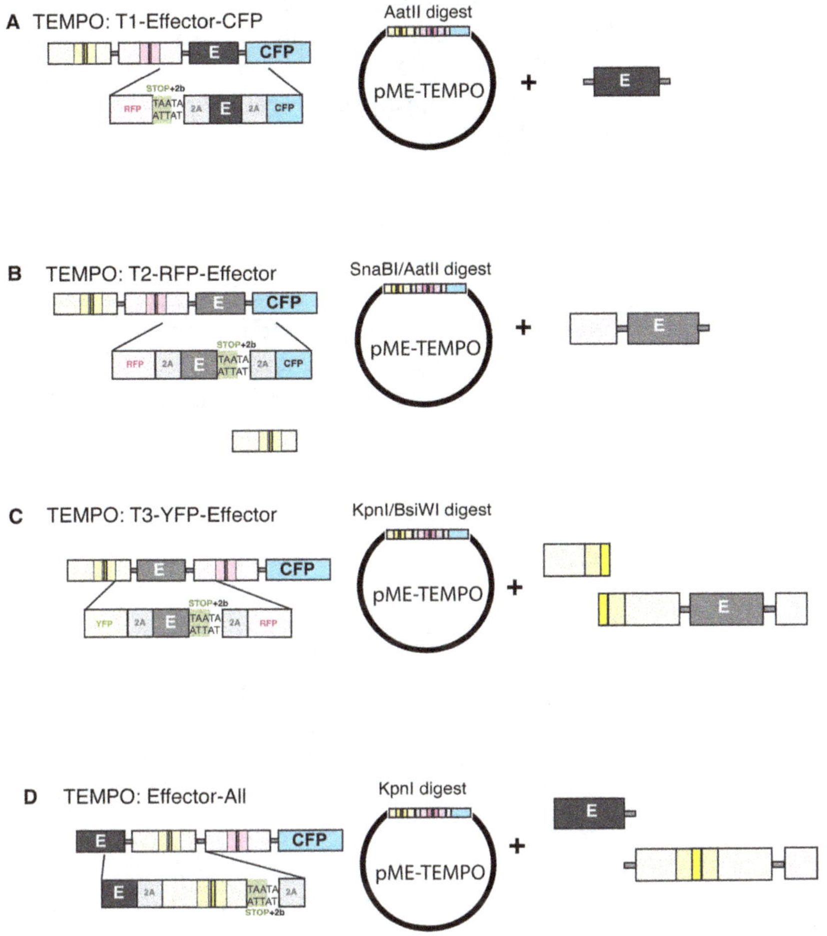

Fig. 7 Design and cloning strategy for TEMPO-effector plasmids. Expression of an Effector [E] protein along with the reporter in the first T1 (**a**), second T2 (**b**), third T3 (**c**) temporal windows, or in all temporal windows (All) (**d**). Details of the junction where the Effector sequence is introduced, highlighting the position of the STOP +2 bp for each of the designs. On the right, cloning strategies to generate entry vectors by Gibson assembly of different fragments and digested vectors

the TEMPO reporter cascade. To clone the TEMPO-effector plasmid, we recommend introducing the effector gene in the previously generated pME-TEMPO plasmid as follows:

Effector in the First Temporal Window (TEMPO: T1-Effector-CFP) (Fig. 7a)

In this case, we introduced the effector gene upstream of the first FP of the TEMPO reporter cascade (CFP) but placing it downstream should also work.

1. Digest pME-TEMPO with AatII, dephosphorylate, and gel purify the 5.6 kb band.

2. Obtain a dsDNA fragment containing the effector gene flanked by homology arms that match the digested ends of the pME--TEMPO plasmid.

Effector in the Second Temporal Window (TEMPO: T2-RFP-Effector) (Fig. 7b)

1. Digest pME-TEMPO with SnaBI/AatII and gel purify the 5.4 kb band.
2. Obtain a dsDNA fragment containing the sequence from SnaBI-RFP-T2A-effector-AatII, which has homology arms matching the digested plasmid.

Effector in the Third Temporal Window (TEMPO: T3-YFP-Effector) (Fig. 7c)

1. Digest pME-TEMPO with KpnI/BsiWI and gel purify the 4.6 kb band.
2. Obtain two fragments: (1) a PCR amplified fragment containing KpnI-YFP-gRNA2target and (2) a fragment containing gRNA2target-YFP-P2A-Effector-P2A'-RFP-BsiWI. Obtaining two fragments that overlap at the gRNA target sequence for subsequent Gibson cloning is recommended instead of amplifying all as a single fragment (*see* **Note 11**).

Effector Expressed Constitutively in All Temporal Windows (TEMPO: Effector-All) (Fig. 7d)

1. Digest pME-TEMPO with KpnI and gel purify the 4.6 kb band.
2. Obtain two dsDNA fragments: (1) a KpnI-Effector-T2A fragment and (2) a T2A-YFP- P2A'-RFP-KpnI fragment, which both overlap at the new T2A peptide sequence and contain homology arms matching the digested plasmid.

   The subsequent steps are common for all cloning strategies above:

3. Gibson assemble the digested pME-TEMPO plasmid with the dsDNA fragments obtained for each cloning strategy above (follow the same Gibson protocol as in section (d)).
4. Transform Gibson assembly mix into competent cells, grow colonies in LB with Kanamycin, and sequence plasmids.
5. Use multisite Gateway cloning to generate the final TEMPO-effector vectors by combining the following elements: p5E-Promoter, pME-TEMPO-effector, p3E-U6gRNA1, pDestination vector with elements required for genetic integration.

While performing an experiment using TEMPO-effector vectors, users will need to co-express both, the TEMPO-effector vector, and the Cas9/gRNA2switch vector for the temporal cascade to progress (*see* Fig. 1b and Subheadings 2 and 3). In addition, users should always perform a control experiment in parallel, in which only the original TEMPO and Cas9/gRNA2switch plasmids are expressed and compare cells expressing the different reporters as well as their spatial distribution in the region of interest.

3.7 TEMPO Expression in Zebrafish

This protocol describes the injection and genomic integration of TEMPO plasmids into 1-cell-stage zebrafish embryos using Tol-based genomic transposition (*see* **Note 13**). Subsequent analysis can be performed in mosaic F0 larvae, or alternatively, users can grow the obtained fish to adulthood for stable line generation.

TEMPO Injection into Zebrafish Embryos

1. First, obtain Tol mRNA by in vitro transcription from a linearized vector (Tol1: pCS2-Tol1.zf1 vector [27] (Addgene # 61388), and Tol2: pCS2FAtransposase vector [28] (Tolkit plasmid #396). We use mMessage mMachine SP6 Transcription kit and purify the mRNA using an RNAeasyMini Kit following the manual guidelines.
2. Purify TEMPO and Cas9/gRNA plasmid DNA by column purification and elute in elution buffer at least at 300 ng/μL.
3. Set up the nanoinjector station and injection needles in advance.
4. Prepare the injection mix and keep it at 4 °C:
   - 5 μL mRNA (final concentration: 25 ng/μL).
   - X μL TEMPO reporter cascade plasmid (0.025 pmol).
   - AND/OR.
   - X μL Cas9/gRNA2switch plasmid (0.025 pmol).
   - X μL Molecular Biology grade H_2O.
   - 0.75 μL Phenol Red dye
   - 10 μL total.

   When several plasmids are co-injected, the total amount of DNA should not exceed 40 ng/μL.
5. Cross zebrafish adults (3–18 months old) to generate embryos. Inject 1-cell-state embryos with 1–2 nL of the injection mix. We recommend injecting at least 100 embryos for each experiment.
6. Grow injected embryos at 28.5 °C and use a fluorescence stereoscope to select positive larvae.

7. To perform analysis of TEMPO cascade transitions in F0 fish, users should co-inject both the TEMPO and Cas9 plasmids (*see* **Note 19**). To generate stable transgenic lines, users should inject each of the plasmids separately, and raise the TEMPO reporter line, and the Cas9/gRNAswitch line independently, so that recombination does not happen in the parent fish.

TEMPO Analysis and Zebrafish Line Propagation

The expression patterns of the TEMPO and Cas9 plasmids will be determined by the promoters chosen by the user. It is possible to include a secondary marker (e.g., heart or eye fluorescent markers; or Halotag co-expressed with Cas9) to facilitate transgenic zebrafish screening and line propagation (*see* **Note 20**). When using TEMPO for the first time, users should analyze the pattern of expression at different developmental times to determine the transition timing in their lineage of interest. While a preliminary analysis can be done in F0 larvae injected with the two plasmids, the number of cells co-expressing the plasmids will be sparse. Thus, we recommend raising each transgenic line to adulthood and analyzing the cell populations of interest in stable F1 fish obtained from the cross of TEMPO-reporter fish with Cas9/gRNA2switch fish. Furthermore, imaging TEMPO reporters in zebrafish can be done directly in live samples, which allows tracking fluorescent reporter transitions in the same animal over time. In addition, if immunostaining for other markers of interest is required, the signal for the TEMPO reporters can be amplified with antibodies specific to the reporters or tags. In the original TEMPO design, CFP was tagged with V5 tag. We then used anti-V5 tag antibody to amplify CFP and also anti-mCherry antibody, but we did not have to amplify the YFP signal as it was very bright even after fixation of the samples.

3.8 TEMPO Expression in Mouse Cortical Neurons

This protocol describes the electroporation of TEMPO plasmids into the mouse brain during development, to label cortical progenitors. The TEMPO and Cas9 plasmids used here carry PiggyBac transposition elements to get inserted into the mouse genome.

Preparation of the Plasmid Mix for Mouse Cortical Electroporation

1. First, obtain a nonintegrative plasmid expressing PiggyBac transposase from the ubiquitous promoter CAG (e.g., CAG-PB [19]). Co-electroporation of this plasmid is required if PiggyBac transposition for the other plasmids is needed.
2. Purify all plasmids by column purification and elute in Molecular Biology H2O at least at 2.5 μg/μL.

3. Prepare the injection mix and keep it at 4 °C:
   - X μL PB-CAG (0.7 μg/μL).
   - X μL TEMPO reporter cascade plasmid (0.7 μg/μL).
   - X μL Cas9/gRNA2switch plasmid (0.7 μg/μL).
   - X μL sterile filtered PBS 1×.
   - 50 μL total.

   The three plasmids should be mixed at a 1:1:1 molar ratio, with a maximum concentration of 2.1 μg/μL and the solution will be supplemented with fast green dye prior to the electroporation (*see* **Note 21**).

In Utero Electroporation of TEMPO Plasmids in the Developing Mouse Cortex

1. In utero electroporation in our TEMPO study [19] was performed on embryonic day E12.5, E14.5, or E16.5 timed-pregnant C57BL/6 J mice (*see* **Note 22**). TEMPO plasmids can also be electroporated at postnatal stages and will undergo color transitions in dividing cells.
2. In utero electroporation was performed as previously described [29, 30]. Specific parameters used for the injection and electroporation of TEMPO plasmids:
   - Mice were anesthetized by using an isoflurane–oxygen mixture [2% (vol/vol) isoflurane in O_2].
   - The right lateral brain ventricle of each embryo was injected with ~280–350 nL of plasmid mix.
   - Immediately after ventricle injection, the head of each embryo was placed between tweezer electrodes, and up to eight pulses of 50 ms and 45 V were applied at 950 ms intervals.

Analysis of TEMPO-Labeled Cortical Neurons.

After in utero electroporation, pregnant mice can be either euthanized to recover the samples at embryonic stages or allowed to give birth for analysis of TEMPO-positive cells at postnatal stages. Embryonic brains can be screened for TEMPO reporter fluorescence directly under a stereoscope before immersing the brains in fixative for downstream processing. Live imaging of organotypic slices from brain cortices electroporated with TEMPO is also possible following previously published protocols [31]. After sample fixation, we recommend amplifying the signal for the TEMPO reporters using immunostaining on brain slices following standard histology protocols (*see* **Note 23**). The imaging filters used to image the original TEMPO reporters are as follows: 458 for CFP, 514 for YFP, and 568 for RFP. If other fluorophores are used as alternative TEMPO reporters, we recommend using a software for spectra visualization to define the imaging parameters. For optimal analysis and quantification of TEMPO-expressing neurons, the following aspects should be considered. First, while most neurons will be labeled by only one FP, some will express two or even three FP (very rare events <1%). We typically leave these out of the downstream analysis, as they might have incorporated more than one copy of the TEMPO reporter plasmid, which could confound result interpretation. Second, TEMPO requires co-expression of two plasmids to function, and users should consider in their analysis those cells that incorporated both plasmids. For this, it is convenient to label Cas9 or use a secondary marker to label the Cas9/gRNA2switch plasmid. In the original TEMPO version, we co-expressed a Halotag protein along with Cas9 and performed immunostaining for TEMPO reporters and Halotag, which allowed us to identify and quantify those cells that expressed both markers. In the future, one could obtain plasmid variants that allow TEMPO and Cas9 expression plasmids to be interdependent so that only cells expressing both would show fluorescence.

4 Notes

1. Avoid the use of repeated reporter sequences. For TEMPO cascade transitions to occur, there will be Cas9/gRNA-guided double-stranded breaks that could result in unwanted recombination and deletion of sequences between repeated elements. To find alternative fluorescent reporter sequences that do not derive from the same family, we recommend consulting available websites such as FPbase (https://www.fpbase.org/) or equivalent.
2. Ideally, this should be done by performing live imaging of the same sample along time. Alternatively, if live imaging over long periods of time is not available, users could take snapshots from different samples at consecutive developmental stages to obtain an estimate of timing for each TEMPO transition. However, the latter cannot be used to extract conclusions about single-cell lineage progression as each sample analyzed will be independent from the other.
3. In the original TEMPO design [19], we did not perform codon optimization of the fluorescent reporter sequences and used the same ones for both mice and zebrafish expression, with good results. However, users should first verify that reporters and peptide linkers have been tested in their model of study.

4. We did not find noticeable differences in TEMPO reporter transition efficiency between reporters disrupted with either a direct or an inverted gRNA target sequence.
5. The length of the homology repeats directly impacts the probability of repair by SSA [32]. We found that repeats between 101 bp and 104 bp in the reporter sequences have high repair efficiency via SSA (>99%). However, others have found increases in SSA repair efficiency when extending the repeats up to 400 bp [32].
6. Look for consensus Kozak sequences of your model organism of study [33]. Introduce silent point mutations to remove those present within the TEMPO reporter sequence and only leave the first one present upstream of the first Start codon intact.
7. Our original gRNA2switch design [18] used 22 bp homology repeats flanking the gRNA1 target, and its efficiency was ~50%. We tested eight different gRNAswitch versions to improve the repair efficiency and reduce the leakiness of the switch. The modifications tested include extension of the homology repeats in the gRNA scaffold region, addition of pol-III transcriptional termination signals, addition of hairpin motifs, or reverse orientation of the target [19]. The most efficient gRNA version generated (~70% of repair efficiency) contains 43 bp homology repeats in the scaffold region flanking a reverse gRNA target, and a 7-polyT termination signal immediately after the target to prevent transcription of the downstream scaffold which could cause leakiness of the disrupted gRNA.
8. Verify the optimal gRNA targeting and lack of off-target activity if using a different species of study. This can be done by using the CRISPR analysis tool in Benchling or equivalent software and mapping the genome of the species of study.
9. If a longer TEMPO cascade is used, additional gRNAswitches will be needed. We have optimized two gRNAswitches in our previous study [19]. To generate additional ones, users would just need to substitute the gRNA spacer and target sequences of the ones shown here. We recommend introducing a couple of nucleotide changes in the scaffold region as well, to prevent unwanted recombination between nearby gRNAs. Changes in the scaffold sequence should be done avoiding modification of the secondary structure as shown in [25].
10. We previously attempted at expressing both the Reporter Cascade and the gRNAswitch in the same plasmid. However, this led to long intertarget deletions because of simultaneous cuts at the gRNA1 target in both the disrupted FP and the gRNA2switch. In practice, this means that we need two

different integration loci for the TEMPO reporter sequence and the gRNAswitches to obtain functional repairs.

11. To subclone any of the repeated sequences within TEMPO or gRNAswitches, we recommend obtaining (either via PCR or gene synthesis) separated fragments with overlaps at the gRNA target which will then be assembled by Gibson cloning. This strategy helps prevent PCR fusion of the repeated elements which often happens if trying to amplify one single fragment containing the homology repeats in these plasmids.
12. We used polymerase type-III U6 promoters to transcribe the gRNAs used in our study, allowing ubiquitous and high levels of expression. For spatiotemporal control of gRNA expression, it could be possible to express gRNAs from Pol-II promoters as shown before [22]. In all cases, species-specific promoters will need to be used to allow transcription of the gRNAs.
13. We integrate TEMPO plasmids by transposition into the zebrafish or mouse genomes. If transposition is the method to integrate the TEMPO transgenes into the genome, users should choose pDestination vectors containing the elements recognized by the transposase, which will excise the transgene from the plasmid and drive transposon integration into various genomic positions containing a transposition footprint [34]. In addition to integration elements, pDestination vectors should also contain polyA transcriptional termination sequences.
14. The choice of promoter should be determined by the user, based on the specific spatiotemporal expression desired. In our experience, using binary amplification systems (e.g., tTA/tetOFF [35]) helped boost TEMPO expression and increased the obtained fluorescent signal. An example of this is the use of the TRE promoter to express TEMPO, combined with expression of tTA from a specific promoter [19].
15. We recommend diluting the entry vectors first, to a concentration of 10 pmol/μL for p5E, pME, p3E, or 20 pmol/μL for pDest, and then add 1 μL of each to the final reaction mix.
16. The use of a strong promoter, or an amplification system to boost the levels of Cas9 expression is desired, as in our experience this is the limiting factor for efficient TEMPO cascade progression. As an example, we originally expressed Cas9 in zebrafish neuronal progenitors under the *her4.1* promoter [36]. However, a higher number of neurons are labeled and undergo TEMPO transitions when using her4.1:tTA coupled with TRE:TEMPO as a binary tetOFF amplification system [19, 31]. Further, spatiotemporal control of cascade progression could be a very helpful feature to study lineage progression. This was implemented in the system CLADES in

Drosophila [18] and we envision that a similar strategy could be deployed in vertebrates by using inducible promoters (e.g., heat shock inducible promoter or tetON/tetOFF system).

17. To date, we have tested temporal overexpression of effector genes along with specific TEMPO reporters for gain-of-function studies (e.g., CyclinB1, CyclinD1) [19]. However, the same strategy could be applied to express dominant negative proteins for temporal loss-of-function studies. Future implementations could include sequential activation of RNAi or sequential gene knock-out.
18. Any effector placed immediately before a 2A peptide will incorporate extra amino acids in its C-terminal, while those effectors immediately after a 2A peptide signal will incorporate a Proline in its N-terminal [21]. Users should confirm first whether any of these modifications could affect the function of the effector of interest. For example, a myristoylation signal for membrane targeting of a protein will not function if expressed after a 2A peptide [37].
19. It is recommended to add a secondary marker co-expressed with Cas9, such as Halotag, to be able to analyze only the cells that co-express both TEMPO plasmid and Cas9 plasmid.
20. If using a secondary marker, select a fluorescent protein that does not have sequence similarities with the TEMPO reporters, to avoid unwanted deletions caused by recombination between these.
21. We typically use 5 μL per electroporated mouse but always have at least double that amount in case the needle needs to be replaced.
22. After crossing the mice, a vaginal plug is indicative of successful mating, and this is considered as 0.5 days of pregnancy and recorded as E0.5.
23. In postnatal animals that undergo transcardial perfusion, in our experience, TEMPO fluorescence will not be apparent by just looking at the whole dissected brain under a fluorescence scope. We recommend obtaining sections or performing immunostaining to amplify the signal.

References

1. Holguera I, Desplan C (2018) Neuronal specification in space and time. Science 362:176–180
2. Kohwi M, Doe CQ (2013) Temporal fate specification and neural progenitor competence during development. Nat Rev Neurosci 14:823–838
3. Oberst P, Agirman G, Jabaudon D (2019) Principles of progenitor temporal patterning in the developing invertebrate and vertebrate nervous system. Curr Opin Neurobiol 56:185–193
4. Telley LAG, Prados J, Amberg N et al (2019) Temporal patterning of apical progenitors and

their daughter neurons in the developing neocortex. Science 364(6440):eaav2522

5. Sagner A, Zhang I, Watson T et al (2021) A shared transcriptional code orchestrates temporal patterning of the central nervous system. PLoS Biol 19:e3001450
6. Cao J, Spielmann M, Qiu X et al (2019) The single-cell transcriptional landscape of mammalian organogenesis. Nature 566(7745): 496–502
7. Ma Z, Zhang X, Zhong W et al (2023) Deciphering early human pancreas development at the single-cell level. Nat Commun 14:5354
8. Garcia-Marques J, Espinosa-Medina I, Lee T (2021) The art of lineage tracing: from worm to human. Prog Neurobiol 199:101966
9. Alemany A, Florescu M, Baron CS et al (2018) Whole-organism clone tracing using single-cell sequencing. Nature 556:108–112
10. Chan MM, Smith ZD, Grosswendt S et al (2019) Molecular recording of mammalian embryogenesis. Nature 570:77–82
11. Choi J, Chen W, Minkina A et al (2022) A time-resolved, multi-symbol molecular recorder via sequential genome editing. Nature 608:98
12. Cotterell J, Vila-Cejudo M, Batlle-Morera L et al (2020) Endogenous CRISPR/Cas9 arrays for scalable whole-organism lineage tracing. Development 147:dev184481
13. McKenna A, Findlay GM, Gagnon JA et al (2016) Whole-organism lineage tracing by combinatorial and cumulative genome editing. Science 353:aaf7907
14. Raj B, Gagnon JA, Schier AF (2018) Large-scale reconstruction of cell lineages using single-cell readout of transcriptomes and CRISPR-Cas9 barcodes by scGESTALT. Nat Protoc 13:2685–2713
15. Spanjaard B, Hu B, Mitic N et al (2018) Simultaneous lineage tracing and cell-type identification using CRISPR-Cas9-induced genetic scars. Nat Biotechnol 36:469–473
16. Espinosa-Medina I, Garcia-Marques J, Cepko C et al (2019) High-throughput dense reconstruction of cell lineages. Open Biol 9:190229
17. Richier B, Salecker I (2015) Versatile genetic paintbrushes: Brainbow technologies. Wiley Interdiscip Rev Dev Biol 4:161–180
18. Garcia-Marques J, Espinosa-Medina I, Ku KY et al (2020) A programmable sequence of reporters for lineage analysis. Nat Neurosci 23:1618–1628
19. Espinosa-Medina I, Feliciano D, Belmonte-Mateos C et al (2023) TEMPO enables sequential genetic labeling and manipulation of vertebrate cell lineages. Neuron 111(3): 345–361
20. Bhargava R, Onyango DO, Stark JM (2016) Regulation of single-strand annealing and its role in genome maintenance. Trends Genet 32:566–575
21. Liu Z, Chen O, Wall JBJ et al (2017) Systematic comparison of 2A peptides for cloning multi-genes in a polycistronic vector. Sci Rep 7(1):2193
22. Garcia-Marques J, Yang CP, Espinosa-Medina I et al (2019) Unlimited genetic switches for cell-type-specific manipulation. Neuron 104: 227–238. e227
23. Hsu PD, Scott DA, Weinstein JA et al (2013) DNA targeting specificity of RNA-guided Cas9 nucleases. Nat Biotechnol 31:827–832
24. Doench JG, Hartenian E, Graham DB et al (2014) Rational design of highly active sgRNAs for CRISPR-Cas9-mediated gene inactivation. Nat Biotechnol 32:1262–1267
25. Briner AE, Donohoue PD, Gomaa AA et al (2014) Guide RNA functional modules direct Cas9 activity and orthogonality. Mol Cell 56: 333–339
26. Kwan KM, Fujimoto E, Grabher C et al (2007) The Tol2kit: a multisite gateway-based construction kit for Tol2 transposon transgenesis constructs. Dev Dyn 236:3088–3099
27. Horstick EJ, Jordan DC, Bergeron SA, Tabor KM, Serpe M, Feldman B, Burgess HA (2015) Increased functional protein expression using nucleotide sequence features enriched in highly expressed genes in zebrafish. Nucleic Acids Res 43(7):e48. https://doi.org/10.1093/nar/gkv035. Epub 2015 Jan 27. PMID: 25628360; PMCID: PMC4402511
28. Clark KJ, Urban MD, Skuster KJ, Ekker SC (2011) Transgenic zebrafish using transposable elements. Methods Cell Biol 104:137–49. https://doi.org/10.1016/B978-0-12-374814-0.00008-2. PMID: 21924161; PMCID: PMC3454445
29. Loulier K, Lathia JD, Marthiens V et al (2009) beta1 integrin maintains integrity of the embryonic neocortical stem cell niche. PLoS Biol 7:e1000176
30. Wang C, Mei L (2013) In utero electroporation in mice. Methods Mol Biol 1018:151–163
31. Wiegreffe C, Feldmann S, Gaessler S et al (2017) Time-lapse confocal imaging of migrating neurons in organotypic slice culture of embryonic mouse brain using in utero electroporation. J Vis Exp 125:55886
32. Sugawara N, Ira G, Haber JE (2000) DNA length dependence of the single-strand annealing pathway and the role of Saccharomyces

cerevisiae RAD59 in double-strand break repair. Mol Cell Biol 20(14):5300–5309

33. Grzegorski SJ, Chiari EF, Robbins A et al (2014) Natural variability of Kozak sequences correlates with function in a zebrafish model. PLoS One 9(9):e108475. https://doi.org/10.1371/journal.pone.0108475
34. Ivics Z, Li MA, Mates L et al (2009) Transposon-mediated genome manipulation in vertebrates. Nat Methods 6:415–422
35. Knopf F, Schnabel K, Haase C et al (2010) Dually inducible TetON systems for tissue-specific conditional gene expression in zebrafish. Proc Natl Acad Sci USA 107: 19933–19938
36. Yeo SY, Kim M, Kim HS et al (2007) Fluorescent protein expression driven by her4 regulatory elements reveals the spatiotemporal pattern of Notch signaling in the nervous system of zebrafish embryos. Dev Biol 301:555–567
37. Hadpech S, Jinathep W, Saoin S et al (2018) Impairment of a membrane-targeting protein translated from a downstream gene of a "self-cleaving" T2A peptide conjunction. Protein Expr Purif 150:17–25

Chapter 18

Computational Methods for Lineage Reconstruction

Irepan Salvador-Martínez

Abstract

The recent development of genetic lineage recorders, designed to register the genealogical history of cells using induced somatic mutations, has opened the possibility of reconstructing complete animal cell lineages. To reconstruct a cell lineage tree from a molecular recorder, it is crucial to use an appropriate reconstruction algorithm. Current approaches include algorithms specifically designed for cell lineage reconstruction and the repurposing of phylogenetic algorithms. These methods have, however, the same objective: to uncover the hierarchical relationships between cells and the sequence of cell divisions that have occurred during development. In this chapter, I will use the phylogenetic software FastTree to reconstruct a lineage tree, in a step-by-step manner, using data from a simulated CRISPR-Cas9 recorder. To ensure reproducibility, the code is presented as a Jupyter Notebook, available (together with the necessary input files) at https://github.com/irepansalvador/lineage_reconstruction_chapter.

Key words Computational methods, Lineage reconstruction, FastTree

1 Introduction

In recent years, the development and utilization of genetic lineage recorders have offered a new perspective on studying cell lineages. One of the game-changing tools in this field is CRISPR-based molecular technology. By harnessing the power of CRISPR-Cas9, scientists can introduce targeted genetic modifications into cells, allowing the recording of their lineage relationships through mutations accumulated over time. The process involves the insertion of a synthetic construct into the genetic material of cells, which can accumulate stochastic mutations upon the induction of CRISPR-Cas9 activity during cell differentiation. These mutations serve as a molecular archive, akin to DNA barcodes, which can be read and deciphered through sequencing techniques. By analyzing the patterns of mutations, researchers can reconstruct the cell lineage, shedding light on the developmental paths taken by individual cells.

Jorge García-Marqués and Tzumin Lee (eds.), *Lineage Tracing: Methods and Protocols*, Methods in Molecular Biology, vol. 2886, https://doi.org/10.1007/978-1-0716-4310-5_18,

The explosion of lineage tracing technologies has also highlighted the importance of using appropriate lineage reconstruction methods. In some cases, researchers have tailored algorithms specifically designed for cell lineage reconstruction [1, 2], while others have applied phylogenetic reconstruction algorithms to reconstruct cell lineages [3, 4]. A recent benchmarking effort conducted via a crowdsourcing initiative by the DREAM challenge [5] encouraged participants from diverse fields to submit lineage reconstruction algorithms to reconstruct lineages of different natures and sizes. The best performers from this benchmarking study included a distance-based method "DCLEAR" [1] and a Machine learning algorithm "AMberRland" [2]. Interestingly, the widely used phylogenetic reconstruction algorithm FastTree [6] (used as a reference algorithm in the study) performed very well, suggesting it is a reliable and effective tool for cell lineage reconstruction.

In this chapter, I will use FastTree to reconstruct a cell lineage as I consider it has several advantages that make it a suitable choice for cell lineage reconstruction. First, its speed and efficiency are noteworthy. With the increasing amount of data generated by lineage tracing technologies, it is crucial to have a reconstruction method that can handle large datasets in a reasonable amount of time. In this sense, FastTree's ability to reconstruct phylogenetic trees from thousands or even millions of sequences quickly and accurately makes it a good candidate. Additionally, FastTree uses an approximately maximum-likelihood method, a robust and reliable approach for reconstructing phylogenetic trees. Last, an important practical advantage is its ease of installation and use, making it accessible to a wide range of researchers with varying levels of computational expertise.

2 Requirements

This chapter, including the example code for lineage reconstruction, is written in a Jupyter Notebook with Python as a programming language. Jupyter Notebooks are documents that combine computer code, plain text, data, rich visualizations like 3D models, charts, graphs and figures, and interactive controls. They provide a fast interactive environment for exploring and visualizing data that can be shared with others and therefore facilitate code reproducibility. The Jupyter default programming language (called "kernel" in the Jupyter ecosystem) is Python, although many other languages (e.g., Julia, R, Matlab, Octave) are also supported and can be used by installing their specific kernels. For detailed information on Jupyter Notebooks, visit https://docs.jupyter.org/en/latest/.

The specific requirements to run this notebook are as follows:

1. Jupyter Notebook package.
2. Python packages: Biopython and DendroPy.
3. FastTree software.
4. Input files (FASTA file and transition matrix file).

2.1 Jupyter Notebook

To run this notebook, it is necessary to install the Jupyter Notebook package. Here, we will use pip, the recommended installation tool for Python. For additional information on the installation of Jupyter, please visit the official Jupyter webpage at https://jupyter.org/install.

To install Jupyter, run the following command in the terminal:

```
pip install notebook
```

Once installed we can open Jupyter Notebook with the following command:

```
jupyter notebook
```

This will open a dashboard in your default browser where you can see the files located in the folder from where you executed the command. If the command is run from the same folder where this notebook is located, the dashboard should look similar to the image shown in Fig. 1.

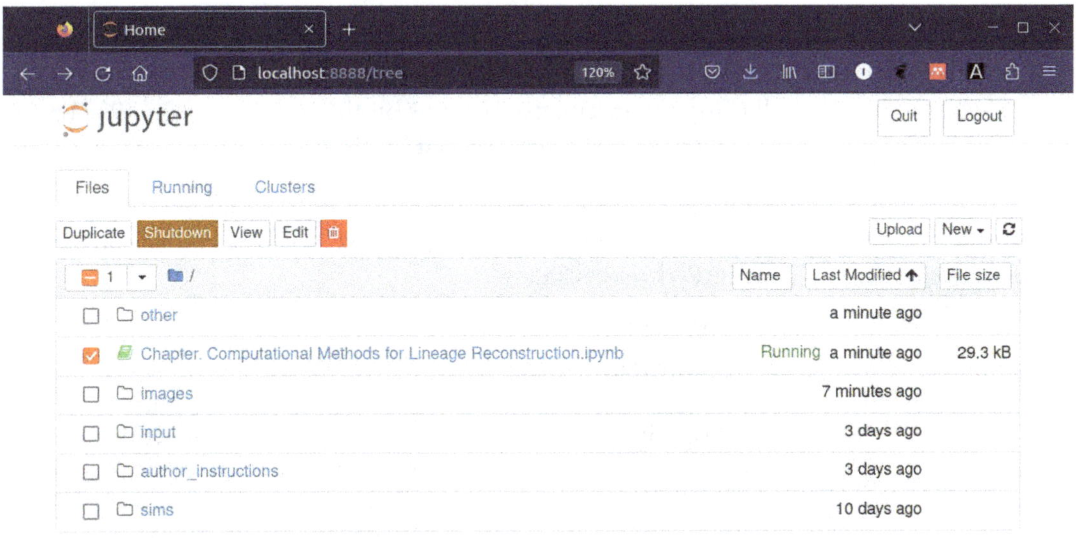

Fig. 1 Jupyter Notebook dashboard. The dashboard user interface shows a list of the notebooks, files, and subdirectories in the directory where the notebook server is started

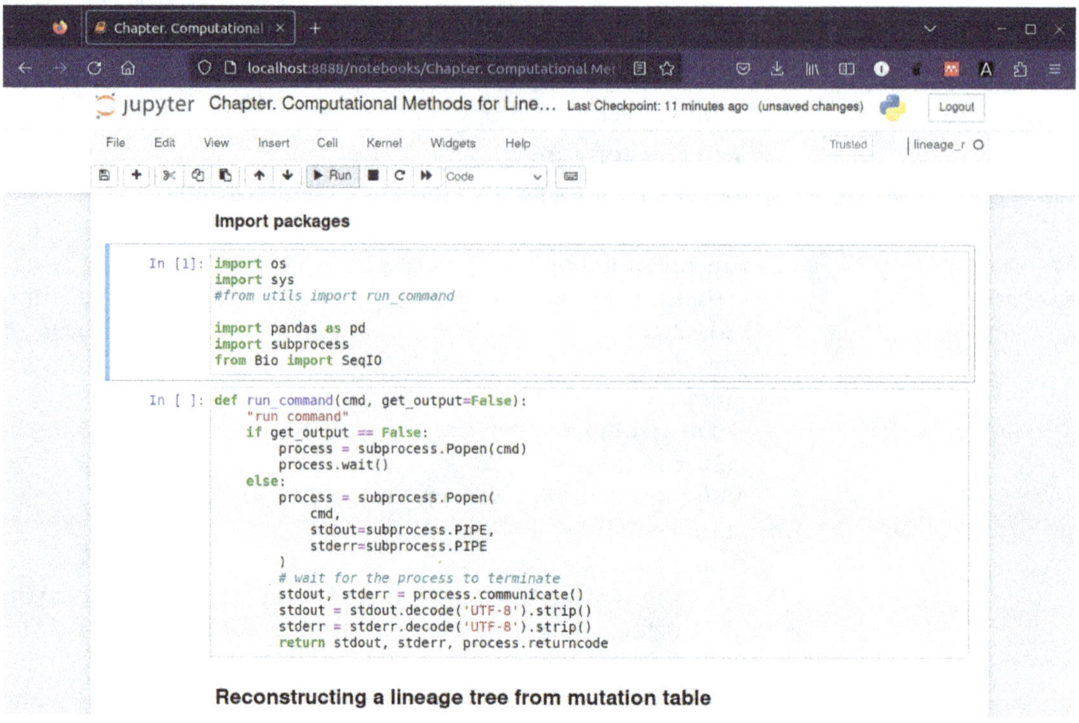

Fig. 2 Notebook editor. The Notebook used in this chapter is shown as opened in the Jupyter Notebook editor, from where the code can be run interactively

Clicking on the Jupyter Notebook file (".ipynb") will open the notebook on a new browser tab. The notebook can be run cell-by-cell using the interactive buttons on the top bar (*see* Fig. 2). To run a specific cell, first select the cell by clicking on it and then click the "Run" button.

2.2 Python Packages: Biopython and Dendropy

After installing the Jupyter Notebook, we need to install the following Python packages:

1. Biopython
2. DendroPy

2.2.1 Biopython

The Biopython Project [7] is an open-source collection of non-commercial Python tools for computational biology and bioinformatics, created by an international association of developers. With Byopython, we can parse the most common bioinformatics file formats into Python-utilizable data structures. Here, we will use Biopython to parse a FASTA file into a list with sequences and into another list with cell IDs. Visit the Biopython Tutorial and Cookbook webpage (http://biopython.org/DIST/docs/tutorial/Tutorial.html) to get detailed documentation, including information to get started with Biopython, and specific documentation on a number of modules.

To install Biopython run, this command in the terminal:

```
pip install biopython
```

2.2.2 DendroPy

DendroPy is a Python library for phylogenetic computing. It provides functions for the simulation, processing, and manipulation of phylogenetic trees and character matrices. It also supports the reading and writing of phylogenetic data in a range of formats, such as NEXUS, NEWICK, and Phylip.

In our example, we will use DendroPy to visualize the reconstructed lineage tree. It can also be used to compare and summarize trees, to prune tree nodes, or to extract a subtree of interest. A detailed tutorial on how to use the DendroPy library, with annotated practical examples can be found at https://dendropy.org/primer/.

To install DendroPy, run this command in the terminal:

```
pip install DendroPy
```

2.2.3 FastTree Software

FastTree is a phylogenetic software that infers approximately maximum-likelihood phylogenetic trees from alignments of nucleotide or protein sequences. A great advantage of FastTree is that it can handle alignments with up to a million sequences in a reasonable amount of time and memory. The easiest way to use FastTree is to download a binary executable file. Ready-to-execute binary files for the different OS are provided at www.microbesonline.org/fasttree/#Install.

In this example, I am using the Linux 64-bit executable (+SSE) double-precision executable binary file ("FastTreeDbl" binary file). The double-precision version is better to resolve very short branch lengths (i.e., nearly identical sequences) accurately. For more details on FastTree, visit the webpage at http://www.microbesonline.org/fasttree/ and see the original paper of Price et al. [6].

2.3 Input Files

2.3.1 FASTA File

For this example, I assume sequencing data from a CRISPR-Cas9 lineage recorder is already available and has already passed the relevant quality controls. To use a phylogenetic reconstruction algorithm like FastTree, the mutations observed in the sequencing reads need to be coded using a character state matrix and formatted in a file that can be recognized by the software. In this example, we will use the FASTA format, a text-based format for representing either nucleotide sequences or amino acid (protein) sequences, in which nucleotides or amino acids are represented using single-letter codes. In general, a FASTA file usually contains multiple sequences, where each sequence is preceded by a header line that starts with the ">" character. This header usually includes the gene ID name, the sequencing read ID, or any other identifier. The next line after the header contains the actual sequence.

Using the following command, we can see the first 20 lines of the FASTA file we will use as an example.

```
!head -20 INPUT/05div_30_targets_20_states.fasta
```

[1]:

```
>unmutated
000000000000000000000000000000
>cell_001
0000000000-------H0100G0000000
>cell_002
0000000000-------H0100G0000000
>cell_003
000000---------------AG0030000
>cell_004
000000---------------AG0030000
>cell_005
-H0_0-----B4009000D00000000000
>cell_006
-H0_0-----B4009000D0000000D000
>cell_007
-H0_00000000009000D00005020000
>cell_008
-H0_00000000-----------------D
>cell_009
002-----------D000000000-I000D
```

2.3.2 Transition Table

A transition (or substitution) matrix describes the frequency at which a character in a nucleotide sequence or a protein sequence changes to other character states over evolutionary time. In the case of protein sequences, the scores correspond to a substitution model that includes also amino acid stationary frequencies. According to the FastTree webpage, the file to be used as a transition matrix needs to fulfill the following requirements (for further details visit the FastTree webpage):

The transition matrix file must be tab-delimited, with the columns A R N D C Q E G H I L K M F P S T W Y V * (in that order). Each entry gives the rate at which one amino acid changes to another. The last column ("*") should have the stationary distribution. Each row must have the row's name as its first field (in order, A R N ... V). The transition matrix Q and the stationary distribution π should satisfy the following:

1. Each amino acid is reachable: each element of π must be positive.
2. The frequencies of the amino acids sum to 1: $\Sigma\pi_i = 1$.
3. Each amino acid can mutate away to other amino acids: each element on the diagonal of Q is negative.
4. Each element off the diagonal of Q is nonnegative.
5. When at the stationary distribution, the frequencies of each amino acid do not change: $Q \cdot \pi = 0$.
6. The total frequency of all amino acids does not change as the distribution evolves: the sum of each column M is zero.
7. The average rate of mutation is 1: $\Sigma Q_{ii}\pi_i = -1$.
8. Evolution is time reversible: $Q_{ij}/Q_{ji} = \pi_i/s_j$.

Here, we will use a simplified transition matrix, where any state can reach another state with the same probability. Following the requirements of FastTree on the transition table, we will use a transition table that contains a 20 × 20 matrix, with column and row names corresponding to the 20 amino acid (AA) symbols and values of 0.052631579 for all the cells, except for the diagonal that has values of −1. An additional column, named *, has values of 0.05 for all rows. We can look at the transition table file with the following command (for readability, only the first 50 characters of the first ten lines are shown):

```
!head -10 INPUT/transition_table.txt | cut -c-50
```

[2]:

```
A    R    N    D    C    Q    E    G    H    I    L    K
     M    F    P    S    T    W    Y    V    *

A    -1          0.052631579 0.052631579 0.052631579 0.0526315

R    0.052631579 -1          0.052631579 0.052631579 0.0526315

N    0.052631579 0.052631579 -1          0.052631579 0.0526315

D    0.052631579 0.052631579 0.052631579 -1          0.0526315

C    0.052631579 0.052631579 0.052631579 0.052631579

Q    0.052631579 0.052631579 0.052631579 0.052631579

E    0.052631579 0.052631579 0.052631579 0.052631579

G    0.052631579 0.052631579 0.052631579 0.052631579

H    0.052631579 0.052631579 0.052631579 0.052631579
```

3 Methods

In this section, I will describe how to reconstruct a lineage tree from sequences coming from a CRISPR-Cas9 recorder. The code used for this example is based on the FastTree reconstruction example provided by Gong et al. [5], available at https://github.com/ofirr/dream_examples. The code blocks can be executed if this notebook is downloaded, and the software requirements are met (see above). To help with code interpretability, before each code block, I describe its objective with a brief text. Additional comments are also included within the code blocks (for example describing the meaning of a value or a parameter). These comments are always preceded by the # character.

3.1 Import Packages

As a first step, we will import all the necessary packages to run this example.

```
import os
import sys
import pandas as pd
import subprocess
import dendropy      # See Requirements section to install this
from Bio import SeqIO  # See Requirements section to install this
```

[3]:

The following code block will print the version of packages used in this notebook.

```
import pkg_resources
from IPython.display import display
root_packages = ['biopython', 'dendropy','ipython',
        'subprocess','jupyterlab', 'pandas']
root_packages_list = []
for m in pkg_resources.working_set:
 if m.project_name.lower() in root_packages:
  root_packages_list.append([m.project_name,m.version])
display(pd.DataFrame(root_packages_list,columns=["package",
"version"]).set_index("package").transpose())
```

[4]:

package	jupyterlab	ipython	DendroPy	biopython	pandas
version	3.0.3	7.19.0	4.6.1	1.78	1.1.3

3.2 Functions

This is a general function to execute a shell command and return its output as a string. We will use it to run FastTree and retrieve the reconstructed tree as a string (in a NEWICK format).

```
def run_command(cmd, get_output=False):
    "run command"
    if get_output == False:
        process = subprocess.Popen(cmd)
        process.wait()
    else:
        process = subprocess.Popen(
            cmd,
            stdout=subprocess.PIPE,
            stderr=subprocess.PIPE
        )
        # wait for the process to terminate
        stdout, stderr = process.communicate()
        stdout = stdout.decode('UTF-8').strip()
        stderr = stderr.decode('UTF-8').strip()
        return stdout, stderr, process.returncode
```

[5]:

3.3 Input and Output

Before running FastTree, we will define the INPUT and OUTPUT folders. In this example, the input folder should contain two files (*see* Subheading 2 for a description of these files):

1. FASTA file with CRISPR-Cas9 recorder sequences
2. Transition table (necessary for FastTree)

```
# Define folder for input and output
input_dir = 'INPUT/'
output_dir = 'OUTPUT/'

# Input file (without the ".fasta")
name = '05div_30_targets_20_states'
# Define name for output file with the modified sequences using AA symbols
fasta_mod_file = os.path.join(input_dir, (name + '_for_fasttree.fasta'))
# Define name for output file with the reconstructed tree (in NEWICK format)
newick_output = os.path.join(output_dir, (name + '.newick'))
```

[6]:

3.4 Read the Input Sequences

We will open the FASTA file with the encoded sequences coming from the CRISPR-Cas9 recorder and store the IDs and sequences in the ids and seqs lists, respectively.

```
ids = []
seqs = []

for seq_record in SeqIO.parse(input_dir + name + ".fasta", "fasta"):
  ids.append(seq_record.id)
  seqs.append(str(seq_record.seq))
```

[7]:

We can print the first ten values of ids and seqs lists we just created to make sure they are correct.

```
for i in range(10):
   print('>{}\t{}\r'.format(ids[i],seqs[i]))
```

[8]:

```
>unmutated    000000000000000000000000000000
>cell_001     0000000000-------H0100G0000000
>cell_002     0000000000-------H0100G0000000
>cell_003     000000---------------AG0030000
>cell_004     000000---------------AG0030000
>cell_005     -H0_0-----B4009000D00000000000
>cell_006     -H0_0-----B4009000D0000000D000
>cell_007     -H0_00000000009000D00005020000
>cell_008     -H0_00000000-----------------D
>cell_009     002-----------D000000000-I000D
```

Note This example has been generated with a computer simulation implemented as in Salvador-Martínez et al. 2019 [3]. Briefly, in this simulation, a cell is implemented as a vector of $m = 30$ targets. The simulation starts with one cell, with all its targets in an unmutated state. The initial cell then undergoes $d = 5$ cell divisions, growing into a population of $N = 32$ cells ($N = 2^d$). Following each cell division, each unmutated target can mutate (with a probability $\mu_d = 0.05$) to one of several possible mutated states. Once a target is mutated, it can no longer change, either to revert to the unmutated state or to transit to a new state. In this example, I have assigned a "0" to an umutated CRISPR-Cas9 target, and the numbers "1–9" and letters "A–J," to the most frequent mutations. If a target is missing due to an intertarget deletion, we assign it the character "–". The code for the simulations is available in the Github repository at https://github.com/irepansalvador/CRISPR_recorders_sims.

3.5 Translate Characters into Amino Acid Symbols

Then, we will translate the encoded mutations into amino acid characters, so the file can be read by FastTree. We will use a Python "dictionary" to assign an AA character to the encoded mutations in our FASTA file according to an associative array. First, we create the symbol_map dictionary:

```
symbol_map = {
   '0':'A', '1':'R',
   '2':'N', '3':'D',
   '4':'C', '5':'Q',
   '6':'E', '7':'G',
   '8':'H', '9':'I',
   'A':'L', 'B':'K',
   'C':'M', 'D':'F',
   'E':'P', 'F':'S',
   'G':'T', 'H':'W',
   'I':'Y', 'J':'V', # Here I "merge" two rare characters as
   '_':'V',          # we can only have 20 AA characters
   '-':'-', # Intertarget deletion considered as missing information
}
```
[9]:

Then, we use the dictionary to modify each sequence in the `seqs` list. The new AA-encoded sequences are stored in the `seqs_trans` list. Then, we print out the first ten sequences of the `seqs_trans` list to visually check that they are correct.

```
seqs_trans = []
tmp = []
# getting length of list
Nseqs = len(seqs)

# Iterating the sequences
for i in range(Nseqs):
  df = pd.DataFrame(list(seqs[i]))
  df = df[0].map(symbol_map)
  seqs_trans.append(''.join(df.values))

for i in range(10):
   print('>{}\t{}\r'.format(ids[i],seqs_trans[i]))
```

[10]:

```
>unmutated    AAAAAAAAAAAAAAAAAAAAAAAAAAAAAA
>cell_001     AAAAAAAAAA-------WARAATAAAAAAA
>cell_002     AAAAAAAAAA-------WARAATAAAAAAA
>cell_003     AAAAAA---------------LTAADAAAA
>cell_004     AAAAAA---------------LTAADAAAA
>cell_005     -WAVA-----KCAAIAAAFAAAAAAAAAAA
>cell_006     -WAVA-----KCAAIAAAFAAAAAAAFAAA
>cell_007     -WAVAAAAAAAAAAIAAAFAAAAQANAAAA
>cell_008     -WAVAAAAAAAAA-----------------F
>cell_009     AAN-----------FAAAAAAAAA-YAAAF
```

Finally, we create a new FASTA file with the new AA-encoded sequences. This is the file that will be used as input for FastTree later on.

```
with open(fasta_mod_file, 'w') as f:
  print("writing " + fasta_mod_file)
  for i in range(Nseqs):
    f.write('>{}\r\n{}\r\n'.format(ids[i],seqs_trans[i]))
```

[11]:

```
writing INPUT/05div_30_targets_20_states_for_fasttree.fasta
```

3.6 Reconstruct the Lineage Tree and Visualize It

First, we will create a variable with the path to the executable FastTree file and a function to run FastTree as a shell command and retrieve its output in a string variable. The output should be the reconstructed tree in NEWICK format.

```
fasttree_path = "/home/irepan/software/FastTreeDbl" # Indicate path to executable file
transition_table = input_dir + "transition_table.txt"

def run_fasttree(newick_output, fasta_mod_file):
  run_command([fasttree_path,
          '-pseudo',
          '-trans', transition_table,
          '-out', newick_output,
          fasta_mod_file])

  reconstructed_tree = dendropy.Tree.get_from_path(
    newick_output,"newick")
  return reconstructed_tree
```

[12]:

Then, we use the `run_fasttree` function we just created to reconstruct the lineage tree using the FASTA file as the input. The output, a tree in the NEWICK format, will be stored in the `rec_tree` variable.

```
rec_tree = run_fasttree(newick_output, fasta_mod_file)
```

[13]:

To have a proper representation of the tree, it is important that we define the "root" of the tree using the "unmutated" cell included in the FASTA file.

```
# root the output of FastTree at the leaf node representing the unmutated construct
rec_tree.reroot_at_node(rec_tree.find_node_with_taxon_label('unmutated'),
update_bipartitions=False)
```

[14]:

```
[14]: <Node object at 0x7f1ac7a55040: 'None' (<Taxon 0x7f1ac7a550a0  'unmutated'>)>
```

Finally, we can display the reconstructed tree using the `as_ascii_plot()` function of DendroPy. We can define the width of the displayed tree as number of characters. Here, we define 80 as the width.

```
# Alternatively we can also display the branch lengths by uncommenting the following command
#print(rec_tree.as_ascii_plot(width=80, plot_metric='length'))

print(rec_tree.as_ascii_plot(width=80))
```

[15]:

```
                                       /------- cell 015
                               /-------------+
                               |       \------- cell 016
                               |
/----------------------------------------------+        /------- cell 001
|                              |    /------+
|                              |    |   \------- cell 002
|                              \------+
|                                  |    /------- cell 003
|                                  \------+
|                                       \------- cell 004
|
|                                       /------- cell 007
|                                  /------+
|                                  |    \------- cell 008
|                     /--------------------+
+                     |             |    /------- cell 005
|                     |             \------+
|                     |                  \------- cell 006
|    /--------------------------+
|    |                |                 /------- cell 013
|    |                |    /--------------------+
|    |                |    |             \------- cell 014
|    |                \------+
|    |                     |   /------------------- cell 012
|    |                     \------+
|    |                         |    /------------- cell 011
|    |                         \------+
\------+                               |   /------- cell 009
     |                             \------+
     |                                  \------- cell 010
     |
     |                                  /------- cell 017
     |                             /------+
     |   /----------------------------------------+    \------- cell 018
     |   |                         |
     |   |                         \------------- cell 020
     |   |
     \------+                              /------------- cell 026
         |   /----------------------------------+
         |   |                      |   /------- cell 019
         |   |                     \------+
         |   |                          \------- cell 025
         \------+
             |    /------------------------------------------ cell 023
             |    |
             |    |                /------------- cell 024
             \------+    /-------------------+
                  |   |             |   /------- cell 021
                  |   |            \------+
                  \------+                  \------- cell 022
                      |
                      |    /--------------------------- cell 028
                      |    |
                      \------+                /------- cell 027
                          |    /-------------+
                          |    |        \------- cell 032
                          \------+
                               |        /------- cell 029
                               |   /------+
                               \------+    \------- cell 030
                                   |
                                   \------------- cell 031
```

3.6.1 Pruning a Tree

In case you are interested in looking at the reconstructed tree using only a subset of cells, you can "prune" the reconstructed tree using the cell IDs using the function `extract_tree_with_taxa_labels()` of DendroPy. This option is especially useful when handling trees with hundreds or thousands of cells. In the following example, we show the topology of a subset tree containing only the first 12 cells.

```
# Cell IDs for pruning the tree
cells_for_subset_tree = ["cell 001","cell 002","cell 003","cell 004",
                "cell 005","cell 006","cell 007","cell 008",
                "cell 009","cell 010","cell 011","cell 012",
               ]
# tree1 is a subset of the reconstructed tree containing only the first 12 cells
tree1 = rec_tree.extract_tree_with_taxa_labels(labels=cells_for_subset_tree)

# print the subtree topology
print(tree1.as_ascii_plot(width=60))
```

[16]:

```
                                        /---------- cell 001
                              /---------+
                              |         \---------- cell 002
/---------------------------------------+
|                             |         /---------- cell 003
|                             \---------+
|                                       \---------- cell 004
|
+                                       /---------- cell 007
|                             /---------+
|                             |         \---------- cell 008
|         /-------------------+
|         |                   |         /---------- cell 005
|         |                   \---------+
\---------+                             \---------- cell 006
          |
          |         /------------------------------ cell 012
          \---------+
                    |         /-------------------- cell 011
                    \---------+
                              |         /---------- cell 009
                              \---------+
                                        \---------- cell 010
```

3.6.2 Compare with Reference Tree

Although this is not always possible, in the case of having a ground truth tree (or "reference tree"), we could quantify the accuracy of our lineage reconstruction using DendroPy. As in our example we are using a simulated tree, and we have this possibility. A simple metric that can be used to compare two phylogenetic trees is the Robinson–Foulds (RF) distance. Here, we will use a scaled implementation of the RF distance, so the RF distance can have a minimum value of 0 (of the reconstructed tree is identical to the reference tree) and a maximum value of 1 if none of the bifurcations of the reference tree is found in the reconstructed tree). This provides a measure of the global accuracy of the inferred lineage tree.

The first step is to read the reference tree file, available as a newick file, creating a DendroPy tree object.

```
REF_tree = dendropy.Tree.get(
  path=os.path.join(input_dir,'{}_REF.nw'.format(name)),
  schema="newick",
  )
```

[17]:

A DendroPy requirement to compare trees is that the trees share a common "TaxonNamespace."

```
# establish common taxon namespace
tns = dendropy.TaxonNamespace()

# ensure all trees loaded use common namespace
tree1 = dendropy.Tree.get(
    data=REF_tree.as_string(schema="newick"),
    schema='newick', taxon_namespace=tns)
tree2 = dendropy.Tree.get(
    data=rec_tree.as_string(schema="newick"),
    schema='newick', taxon_namespace=tns)
```

[18]:

Then, we calculate the number of splits (or bipartitions) that differ between the reference and the reconstructed tree using the `dendropy.calculate.treecompare.find_missing_bipartitions()` function.

```
# splits in REF tree missing in reconstructed tree
len(dendropy.calculate.treecompare.find_missing_bipartitions(tree1, tree2))
```

[19]:

[19]: 19

We finally divide the number of wrong splits between the total number of splits of the reference tree.

```
from dendropy.calculate import treecompare

#dendropy.calculate.treecompare.find_missing_bipartitions(tree1, tree2)
split_diff = len(dendropy.calculate.treecompare.find_missing_bipartitions(tree1, tree2))
all_splits = len(tree1.encode_bipartitions())
split_diff / all_splits
```

[20]:

[20]: 0.3064516129032258

Our result is 0.306, from which we can conclude that 30% of the bifurcations in the reference tree were not found in the reconstructed tree.

4 Conclusion

In conclusion, the use of FastTree as the reconstruction algorithm in this study is justified by its speed, accuracy, and ease of use. Its ability to handle large datasets, estimate evolutionary relationships, and produce reliable results make it a good choice for analyzing the genetic lineage recorders generated through CRISPR-based molecular technology.

However, it is important to highlight that currently there is no consensus on a reconstruction method, and further benchmarking of existing and new reconstruction methods is still necessary. As was shown in the DREAM challenge, the performance of the reconstruction algorithms depended on the nature and size of the lineage trees to be reconstructed. It is therefore recommended that researchers consider choosing an appropriate reconstruction method based on the features (size, depth, number of targets) of the lineage tree they want to reconstruct.

4.1 Note About the OS

This notebook has been created with a computer with Ubuntu OS (version 20.04.6) and using Python version 3.8.5, as it can be seen from the output of the following commands:

```
!lsb_release -a
```

[21]:

```
No LSB modules are available.
Distributor ID: Ubuntu
Description:    Ubuntu 20.04.6 LTS
Release:        20.04
Codename:       focal
```

```
!python --version
```

[22]:

```
Python 3.8.5
```

In Ubuntu, as in many other Linux distributions, Python comes preinstalled. Instructions to install Python in different OS (Windows, Mac, Linux), as well as detailed documentation, tutorials, and guides, can be found on the official webpage at https://www.python.org/.

References

1. Gong W, Kim HJ, Garry DJ et al (2022) Single cell lineage reconstruction using distance-based algorithms and the R package, DCLEAR. BMC Bioinformatics 23:103
2. Prusokiene A, Prusokas A, Retkute R (2023) Machine learning based lineage tree reconstruction improved with knowledge of higher level relationships between cells and genomic barcodes. NAR Genom Bioinform 5:lqad077
3. Salvador-Martínez I, Grillo M, Averof M et al (2019) Is it possible to reconstruct an accurate cell lineage using CRISPR recorders? elife 8: e40292
4. McKenna A, Findlay GM, Gagnon JA et al (2016) Whole-organism lineage tracing by combinatorial and cumulative genome editing. Science 353:aaf7907
5. Gong W, Granados AA, Hu J et al (2021) Benchmarked approaches for reconstruction of in vitro cell lineages and in silico models of C. elegans and M. musculus developmental trees. Cell Syst 12:810–826.e4
6. Price MN, Dehal PS, Arkin AP (2010) FastTree 2 – approximately maximum-likelihood trees for large alignments. PLoS One 5:e9490
7. Cock PJA, Antao T, Chang JT et al (2009) Biopython: freely available Python tools for computational molecular biology and bioinformatics. Bioinformatics 25:1422–1423

Chapter 19

GEMLI: Gene Expression Memory-Based Lineage Inference from Single-Cell RNA-Sequencing Datasets

A. S. Eisele and D. M. Suter

Abstract

Gene expression memory-based lineage inference (GEMLI) is a computational tool allowing to predict cell lineages solely from single-cell RNA-sequencing (scRNA-seq) datasets and is publicly available as an R package on GitHub. GEMLI is based on the occurrence of gene expression memory, i.e., the gene-specific maintenance of expression levels through cell divisions. This represents a shift away from experimental lineage tracing techniques based on genetic marks or physical cell lineage separation and greatly eases and expands lineage annotation. GEMLI allows to study cell lineages during differentiation in development, homeostasis, and regeneration, as well as disease onset and progression in various physiological and pathological contexts. This makes it possible to dissect cell type-specific gene expression memory, to discriminate symmetric and asymmetric cell fate decisions, and to reconstruct individual multicellular structures from pooled scRNA-seq datasets. GEMLI is particularly promising for its ability to identify small lineages in human samples, a context in which no other lineage tracing methods are applicable. In this chapter, we provide a detailed protocol of the GEMLI R package usage on gene expression matrices derived from standard scRNA-seq on various platforms. We cover the use of the main function to predict cell lineages and how to adjust its parameters to different tasks. We also show how lineage information is extracted, visualized, and fine-tuned. Finally, we describe the use of the package's functions for the detailed analysis of the predicted cell lineages. This includes the analysis of gene expression memory, cell type composition of individual large lineages, and identification of lineages at the transition point between two cell types.

Key words Lineage tracing, Cell family, Lineage inference, Gene expression memory, Memory genes, Monotonically expressed genes, Multicellular structure, Clonality, Gene expression stability, Fate decisions

1 Introduction

GEMLI is a computational tool allowing to annotate cell lineages, i.e., cells with a common ancestor, using only scRNA-seq data as an input [1]. It is available as an R package on GitHub under https://github.com/UPSUTER/GEMLI. GEMLI relies on the fact that differences in gene expression between cell lineages can be maintained transiently over several cell divisions [2–8]. This gene

Jorge García-Marqués and Tzumin Lee (eds.), *Lineage Tracing: Methods and Protocols*, Methods in Molecular Biology, vol. 2886, https://doi.org/10.1007/978-1-0716-4310-5_19,

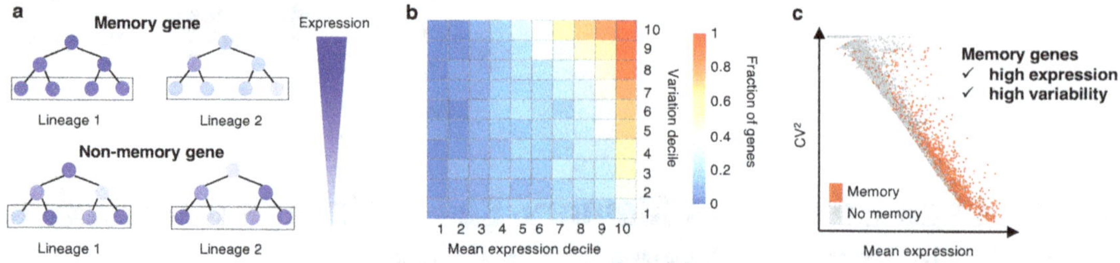

Fig. 1 Genes with high gene expression memory are highly expressed and variable. (**a**) Scheme of memory gene and nonmemory gene expression in different lineages over several cell generations. Color intensity represents gene expression levels. (**b**) Representative scheme of the abundance of memory genes as a function of mean and variation deciles in a scRNA-seq dataset. (**c**) Representative scheme of the relation of the coefficient of variation squared (CV^2) and mean gene expression for memory genes and nonmemory genes in a scRNA-seq dataset

expression memory is gene-specific, with some genes (that we call memory genes) showing strong stability in their expression level over cell division (Fig. 1a). More in detail, memory genes either show very conserved expression levels within individual lineages or are only expressed in a small number of cell lineages [1]. Memory genes can be identified in various cell types, including self-renewing and differentiating stem- and immune cells [1–4, 9–13]. Interestingly, memory genes often correspond to housekeeping genes and show characteristic expression distributions across cell types. Memory genes are mostly highly expressed and highly variable (Fig. 1b, c) [1].

GEMLI uses the characteristic expression distributions of memory genes across cell types to predict cell lineages de novo from scRNA-seq datasets in three steps (Fig. 2). First, genes with a high expression mean and high variability are selected, which enriches for memory genes in the 30–70% range. More in detail, the genes selected are the 2% highest expressed genes, the 60% most variable among the 10% highest expressed genes, and the 10% most variable among the 40% highest expressed genes ((mean expression quantile >=98) or (mean expression quantile >=90 and expression variability quantile >=40) or (mean expression quantile >=60 and expression variability quantile >=90); variability being defined as mean-corrected CV^2 calculated as the residual of the CV^2 to a linear fit of CV^2 and mean expression; Fig. 2b). Second, the selected genes are taken as input for the GEMLI repetitive iterative clustering algorithm. During each iterative clustering, the input gene set is randomly subsampled to 75% (*sample_size* parameter; see GEMLI parameters below). This gene set subsample is then used to cluster the cells of the dataset iteratively as follows. During the first iteration, two clusters are generated. In every subsequent iteration round, cells of each previously defined cluster are clustered again into two clusters using the same procedure. The iteration is ended

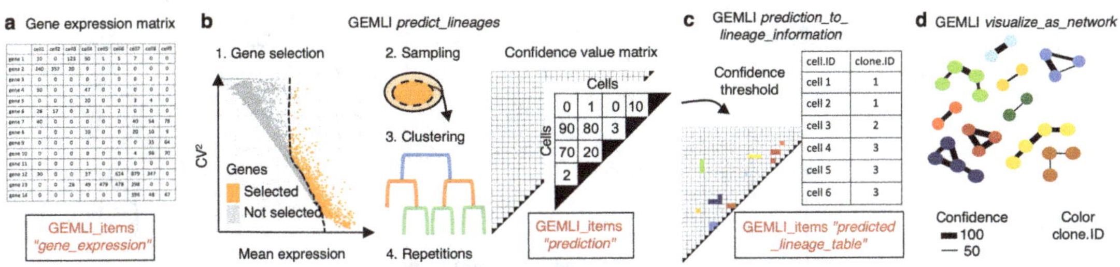

Fig. 2 Overview of the GEMLI lineage prediction algorithm and main functions. (**a**) The input for GEMLI is a scRNA-seq gene expression matrix stored as "gene_expression" element in the "GEMLI_items" list. (**b**) The GEMLI *predict_lineages* function selects genes based on variability and mean expression. These are subsampled and used as input for an iterative hierarchical clustering. Sampling and clustering are repeated to obtain a cell-by-cell confidence value matrix indicating for each cell pair how often it clustered together across clustering repetitions, and thus how likely it is to be part of the same lineage. This matrix is stored as "prediction" element in the "GEMLI_items" list. (**c**) Using the GEMLI function *prediction_to_lineage_information*, a confidence value threshold is set to transform the cell-by-cell confidence value matrix into a lineage information table. This table is stored as a "predicted_lineage_table" element in the "GEMLI_items" list. (**d**) The *visualize_as_network* GEMLI function creates a network representation of lineages predicted above a specific confidence value threshold

for clusters meeting a predetermined size of 2–3 cells (*desired_cluster_size* parameter, see below). This results in a cell-by-cell matrix indicating how each cell pair clusters together in the final cluster of 2–3 cells. This iterative hierarchical clustering represents one cell lineage prediction (Fig. 2b). It is repeated 100 times (*repetitions* parameter, see below) to generate a confidence level for the lineage predictions. More in detail, the cell-by-cell matrices generated in the 100 repetitions of the iterative hierarchical clustering are summed into a confidence level cell-by-cell matrix (Fig. 2b). For example, a cell pair ending up in the same final cluster in 30 of the 100 predictions will receive a value of 30. Finally, setting a threshold on the confidence level can then be used for lineage assignment (Fig. 2c). Cell lineages are assigned as cell groups of which members are clustering together above a confidence level threshold. The GEMLI R package encompasses functions to generate the cell-by-cell lineage prediction matrix as well as to extract lineage annotations and make a lineage network visualization (Fig. 2d). Furthermore, it contains functions to extract memory genes, analyze the cell type composition of predicted lineages, and find differentially expressed genes (DEG) specific to lineages with specific cell type composition.

GEMLI lineage prediction performance has been benchmarked on >100 scRNA-seq datasets of various cell types with experimentally derived ground truth lineage annotation [1]. In particular, both in vitro and in vivo-derived samples have been analyzed as well as lineages being composed of a single cell type (symmetric) and lineages with members in several cell types (asymmetric). Scoring of GEMLI performance was performed based on precision

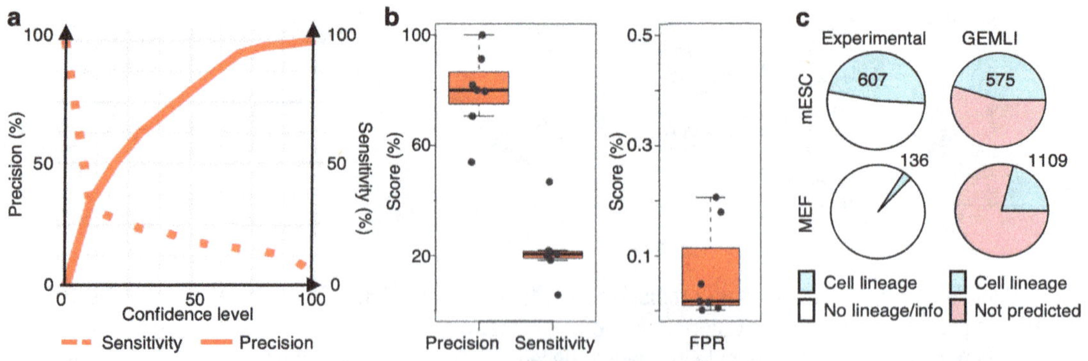

Fig. 3 Scoring of GEMLI performance across cell types. (**a**) Curve of GEMLI lineage prediction precision and sensitivity in relation to confidence level in a mESC dataset. (**b**) Precision, sensitivity, and false-positive rate (FPR) of GEMLI predictions across datasets of seven cell types (mESC, mouse embryonic fibroblasts (MEF), CD8+ T cells, L1210 leukemia cells, hematopoietic stem, progenitor cells, and hematopoietic stem cells), at confidence level 50. (**c**) The number of cells that are lineage-annotated experimentally and using GEMLI predictions shown using the example of a mESC and MEF dataset

(true positives (TP)/(TP + false positives (FP))), sensitivity (TP/(TP+ false negatives (FN))), and false-positive rate (FPR; FP/(FP+ true negatives (TN))) at different confidence levels. All scores were calculated based on cell pairs within lineages by comparing predictions at a confidence level above a given threshold in the cell-by-cell prediction matrix output by GEMLI to the experimentally derived lineage annotation. With increasing confidence levels of GEMLI lineage predictions, precision increases while sensitivity and FPR decrease. At confidence levels of 50%, an average precision and sensitivity of 80% ± SD 15%, and 22% ± SD 12%, respectively, have been reached on benchmarking datasets (Fig. 3). The FPR at a confidence level of 50% remained always very low and negligible with an average of 0.07% ± SD 0.08%. Setting the confidence level threshold (*cutoff* parameter, see GEMLI parameters below) during GEMLI usage allows thereby to optimize precision or sensitivity of lineage predictions depending on the needs for further analysis. Crucially, as GEMLI allows to predict cell lineages for all cells of a scRNA-seq dataset, which is not possible for many experimental lineage assignment strategies, the final number of lineage-annotated cells using GEMLI is either comparable or substantially increased as compared to experimentally derived ground truth lineage annotation.

We present here the detailed protocol of the GEMLI R package usage on gene expression matrices derived from standard scRNA-seq on various platforms. First, we cover the installation of GEMLI, the use of the main function to predict cell lineages (*predict_lineages*) and how to adjust its parameters to different tasks or to increase precision or sensitivity, respectively. In particular, we cover the influence of the GEMLI *predict_lineages* parameters *sample_size*, *desired_cluster_size*, and *repetitions* on results. We describe how the *predict_lineages_multiple_sizes* and *cluster_stability_plot* functions can help to estimate lineage size distributions and decide on a suitable *desired_cluster_size* parameter. Second, we show how to extract lineage annotation with the *prediction_to_lineage_information* function and how to generate a prediction network visualization using the *visualize_as_network* function. We describe how in both instances the *cutoff* parameter sets a confidence level threshold and allows to gain predictions with a particularly high precision or sensitivity depending on user needs. We cover how predicted lineages can be trimmed to specific sizes using the *suggest_network_trimming_to_size* and *trim_network_to_size* functions. Finally, we describe the use of the package functions for (i) the analysis of gene expression memory using the *memory_gene_calling* function; (ii) the analysis of the cell type composition of individual lineages using the *cell_type_composition_plot* function; (iii) the analysis of gene expression specific to cell lineage being composed of certain cell types using the *extract_cell_fate_lineages cell_fate_DEG_calling* and *DEG_volcano_plot* functions. In the last part, we cover GEMLI usage when a scRNA-seq dataset is partly lineage-annotated through other means. In particular, we explain the analysis of ground truth lineages and how the *test_lineage_prediction* function allows to score GEMLI predictions against ground truth lineage annotations to optimize parameters for lineage predictions in the unannotated part of a scRNA-seq dataset

2 Materials

For the prediction of lineages in scRNA-seq datasets using GEMLI, the GEMLI R package and a scRNA-seq dataset in the form of a filtered and normalized gene expression matrix are required. For some additional features of GEMLI, cell type annotation of the scRNA-seq dataset is needed. Finally, some GEMLI functions allow to input a ground truth lineage annotation to improve lineage predictions, extend lineage annotation, or perform further analysis of lineages. More details on the GEMLI R package, scRNA-seq datasets, cell type annotation, and ground truth lineage information are given in the following paragraphs.

GEMLI R package The GEMLI R package is available for installation at GitHub under https://github.com/UPSUTER/GEMLI. It has been developed with R (version 4.0.2) and relies on several other R packages as dependencies, namely: igraph (>=1.4.3) HiClimR (>=2.2.1), dplyr (>=1.1.2), Seurat (>=4.3.0), ggplot2 (>=3.4.2), ggrepel (>=0.9.3), clustree (>=0.5.0), UpSetR (>=1.4.0), reshape (>=0.8.9), and tidyr (>=1.3.0). These packages have to be installed and loaded before starting to use the GEMLI R package.

ScRNA-seq dataset The scRNA-seq dataset should encompass cells that are related over a small number of generations (*see* **Note 1**). Datasets that represent samples of large cell populations are not suited. The dataset should be present as a quality-controlled read-count normalized gene expression matrix in which row names correspond to genes and column names to cells ("cell.IDs"). GEMLI can be used on datasets produced using various sequencing platforms, such as the 10× Genomic Chromium platform, InDrop scRNA-seq, SmartSeq sequencing, SCRB sequencing, and Disco sequencing [14] (*see* **Note 2**). Each of these platforms has dedicated pipelines to process the raw sequencing data, which we will not cover here. Also, quality control may depend on the platform. GEMLI has been benchmarked on datasets filtered on cell read number and mitochondrial read fraction (*see* **Note 3**). The gene expression matrix should be normalized by read count (*see* **Note 4**). Sequencing depth should lie within the recommended ranges for scRNA-seq, i.e., above 8000 reads/cell (*see* **Note 5**). GEMLI performs better for small datasets, but good performance has been achieved for datasets up to 11,000 cells.

Cell type annotation For some functions of GEMLI allowing to analyze predicted lineages, a cell type annotation of the scRNA-seq dataset is required. Different scRNA-seq analysis software allow to generate such annotation, often using curated cell type marker lists or machine learning, which we will not cover here. The cell type annotation should be present in the form of a dataframe with columns "cell.ID" and "cell.type."

Ground truth lineage annotation Importantly, ground truth lineage annotation is not required for basic GEMLI usage. If ground truth lineage annotation is available for (part of) a scRNA-seq dataset, it can be used to improve lineage predictions with GEMLI and to analyze these ground truth lineages with respect to gene expression memory, cell type composition, or gene expression. Ground truth lineage information should be present as a named vector with values corresponding to lineage number ("clone.ID") and names corresponding to cells ("cell.ID").

Example datasets To describe GEMLIs usage in this chapter, we use three example datasets available at https://github.com/UPSUTER/GEMLI/tree/main/Example. All figures are made using this data. The datasets are as follows: (i) A scRNA-seq dataset of mouse embryonic stem cells (mESCs; GEMLI_example_data_matrix.RData) [1]. (ii) A scRNA-seq of intestinal crypts

(GEMLI_crypts_example_data_matrix.RData) and its cell type annotation (GEMLI_crypts_example_cell_type_annotation.RData) [14]. (iii) A breast cancer scRNA-seq dataset (GEMLI_cancer_example_norm_count.RData) [15] and its cell type annotation (GEMLI_cancer_example_cell_type_annotation.RData) [1].

3 Methods

3.1 Installation of the GEMLI R Package

To install the GEMLI R package, the R library devtools is used. The current GEMLI R package version is v0. The package has several dependencies (listed in the Methods section) which have to be installed independently (*see* **Note 6**).

```
# Installing the GEMLI R package
> install.packages("devtools")
> library("devtools")
>  install_github("UPSUTER/GEMLI", subdir="GEMLI_package_v0")
```

3.2 GEMLI Lineage Predictions

3.2.1 Creating a GEMLI_items List

GEMLI operates on a "GEMLI_items" list, which is a list of elements in which both GEMLI inputs and outputs are stored under predefined names. To start using GEMLI for basic lineage annotations, the "GEMLI_items" list has to be created. The scRNA-seq datasets in the form of a quality-controlled and normalized gene expression matrix of read counts are loaded into this list under the name "gene expression." The gene expression matrix should have genes as row names (*see* **Note 7**), and cells ("cell.IDs") as column names. For additional features of GEMLI, cell type annotation of cells in the gene expression matrix has to be uploaded under the name "cell_type." Cell type annotation should be a dataframe with column names "cell.ID" and "cell.type." The cell IDs should match the ones present in the gene expression matrix stored as "gene expression" element in the "GEMLI_items" list.

```
# Making a GEMLI_items list
> GEMLI_items = list()
> GEMLI_items[['gene_expression']] = gene_expression_matrix
> GEMLI_items[['cell_type']] = cell_type_annotation

# Example of a stored gene expression matrix
> GEMLI_items[['gene_expression']][1:6,1:3]
                              AAACGAACAGGTGTGA-1  AAAGGTAGTTGCTTGA-1
AACCACAAGTTTGTCG-1
ENSMUSG00000033845                 14.761746           12.9570026
13.240645
ENSMUSG00000025903                  3.163231            1.2340002
```

```
4.878132
ENSMUSG00000033813          6.326463          5.5530011
4.181256
ENSMUSG00000002459          0.000000          0.0000000
0.000000
ENSMUSG00000085623          0.000000          0.0000000
0.000000
ENSMUSG00000033793          3.163231          0.6170001
1.393752

# Example of a stored cell type annotation
> GEMLI_items[['cell_type_annotation']][1:6,]
     cell.ID cell.type
1 AAACGAACAGGTGTGA-1      Stemcell
2 AAAGGTAGTTGCTTGA-1      Stemcell
3 AACCACAAGTTTGTCG-1      Endothelial
4 GAGTGAGTCCAGTACA-1      Gobletcell
5 AGAGAATAGGTCATAA-1      Endothelial
6 CACAGATAGTGATGGC-1      Gobletcell
```

3.2.2 Basic GEMLI Lineage Predictions

GEMLI lineage predictions are performed on the scRNA-seq gene expression matrix stored as "GEMLI_items" element "gene expression" using the *predict_lineages* function. The function performs all steps of the GEMLI prediction algorithm, from calculation of the gene-wise variability and mean expression in the dataset, selection of the GEMLI input gene set, to the repetitive iterative hierarchical clustering itself. It does so by calling on the two other functions *potential_markers* and *calculate_correlations,* which the user does not interact with. It outputs a cell-by-cell matrix containing the number of times each cell pair clustered together across the clustering repetitions. This number represents the confidence level for each cell pair to be part of the same lineage. The matrix is stored in the "GEMLI_items" list as "prediction" element.

```
# Running the predict_lineages function
> GEMLI_items = predict_lineages(GEMLI_items)

# Example of a stored prediction result
> GEMLI_items[['prediction']][1:3,1:5]
                    AGAGAATAGGTCATAA-1 AGAGCAGCAAGTGATA-1
AGATGCTTCAAAGACA-1
AAACGAACAGGTGTGA-1                   0                  0
0
AAAGGTAGTTGCTTGA-1                  87                  0
0
AACCACAAGTTTGTCG-1                  76                  0
0
```

```
AAGCCATGTTCCACGG-1                 0                  0
0
AAGCGAGGTACGGCAA-1                 0                 40
0
```

3.2.3 Adjusting GEMLI Prediction Parameters

The *predict_lineages* function has several parameters which can be adjusted (*see* **Note 8**). These are the parameters *repetitions*, *sample_size*, *desired_cluster_size*, and *fast*. The *repetitions* parameter corresponds to the number of repetitions of the iterative hierarchical clustering, which sets the maximal confidence level a cell pair can reach. Its default value is 100. The *repetitions* parameter has only a small influence on prediction performance, even though higher values tend to increase precision. The *sample_size* parameter corresponds to the fraction of genes of the GEMLI input gene set, which is used in each clustering repetition. Its default value is 75% (*sample_size* = c(2/3)). A lower *sample_size* value can increase precision but decreases sensitivity. The *desired_cluster_size* parameter is the size of clusters at which clustering iterations are stopped. Its default value is *desired_cluster_size* = c (2,3). The *desired_cluster_size* parameter influences both precision and sensitivity. Precision will be high for predictions with a parameter of 2–3 until reaching the average size of lineages present (*see* also Subheading 3.2.4). The *desired_cluster_size* parameter does not correspond to the final lineage size predicted (*see* **Note 9** and Subheading 3.3.3). If the *fast* parameter is set to true (default *fast* = F), the correlation for clustering is calculated as Pearson correlation using the fastCor function of the HiClimR R package. This speeds up predictions but leads to lower prediction performance.

3.2.4 GEMLI Prediction Stability to Estimate the Size of Large Lineages

The stability of GEMLI predictions using the *predict_lineages* function with varying *desired_cluster_size* parameters can give an indication of the ideal parameter value used and the true lineage sizes present. In particular, it reliably hints at the size of multicellular structures such as intestinal crypts (*see* **Note 10**). To study the stability of predictions, two GEMLI functions are used: *predict_lineages_multiple_sizes* and *cluster_stability_plot*. The *predict_lineages_multiple_sizes* function will perform GEMLI predictions (with default parameters *repetitions* = 10, and *sample_size* = (2/3)) with all *desired_cluster_size* parameters in between the values given in the *minimal_maximal_cluster_size* parameter (default value *minimal_maximal_cluster_size* = c(2–50)). It outputs a lineage prediction annotation using the *cutoff* parameter value (*see* also Subheading 3.3.1) as the confidence level threshold (default *cutoff* = 5 leading to a confidence level threshold of 5/10 (50%)). The output is stored in the GEMLI_items list "prediction_multiple_sizes" element in the form of a dataframe with cell IDs as row names and the maximal cluster size used during the predictions as column names.

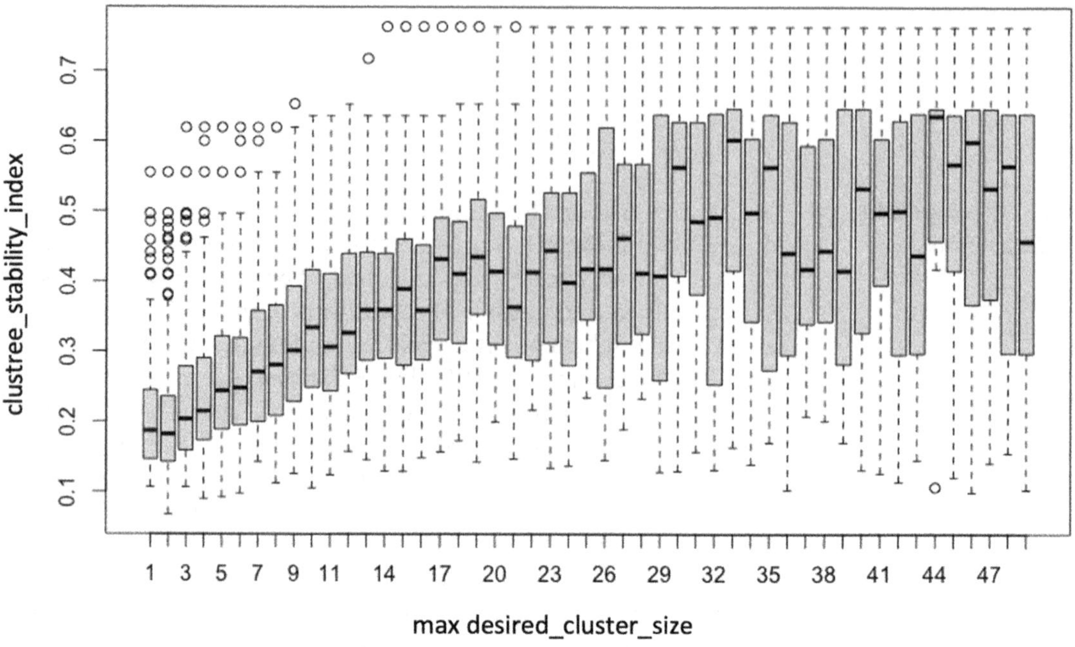

Fig. 4 Output of a *cluster_stability_plot* function call. The cluster_stability_plot function outputs a boxplot of the cluster stability index vs. the maximal value of the *desired_cluster_size* parameter used. The stability index plateaus when the maximal lineage size in the dataset is reached. This is the *desired_cluster_size* value where sensitivity will be maximal at high precision. Based on the plot a *desired_cluster_size* = *c(2–20)* should be chosen to optimize sensitivity of lineage predictions in the represented dataset

Using the *cluster_stability_plot* function, a cluster stability index (R package clustree) is calculated for the GEMLI_items "prediction_multiple_sizes" element. It outputs a plot of the cluster stability index vs. the maximal value of the *desired_cluster_size* parameter used (Fig. 4). The cluster stability index values plateau when the maximal lineage size in the dataset is reached (*see* **Note 11**). This is the value where sensitivity will be maximal at high precision.

```
# Cell_type_composition_plot results with different type para-
meter
> GEMLI_items = predict_lineages_multiple_sizes(GEMLI_items)

# Example of a stored prediction with multiple sizes result
> GEMLI_items[['prediction_multiple_sizes']][1:5,1:5]
        cell.ID              K1      K2      K3      K4
1 AAACGAACAGGTGTGA-1    1       1       1       1
2 AAAGGTAGTTGCTTGA-1    1       1       1       1
3 AACCACAAGTTTGTCG-1    10      10      9       8
4 GAGTGAGTCCAGTACA-1    10      10      9       8
5 AGAGAATAGGTCATAA-1    11      11      3       9

# Cell_type_composition_plot results with different type para-
```

```
meter
> GEMLI_items = cluster_stability_plot(GEMLI_items)(Fig 4)
```

3.3 GEMLI Lineage Prediction Extraction and Visualization

3.3.1 Extracting GEMLI Lineage Information

The *predict_lineages* function outputs a cell-by-cell matrix with confidence levels for every cell pair to be related. Setting a threshold on the confidence level can then be used to extract lineage information. This is performed using the *prediction_to_lineage_information* function. The confidence level threshold is set using the *cutoff* parameter in the function call. Cell lineages are defined by default as cell groups for which any members are clustering together above the confidence level threshold. The *cutoff* parameter should be set with respect to the repetition number used during the GEMLI predictions. We recommend to use values corresponding to at least 50% of the repetition number (default is a threshold of 50 (*cutoff*= 50) for predictions with 100 repetitions). Higher *cutoff* values will result in increased precision and decreased sensitivity. The *predictions_to_lineage_information* function outputs a table with columns "cell.ID" and the lineage ("clone.ID") as "GEMLI_items" element "predicted_lineage_table" and a named character vector (characters are lineages (clone IDs) and names are cell IDs) in the "GEMLI_items" element "predicted_lineages."

```
# Extracting lineages using the_prediction_to_lineage_informa-
tion function
> GEMLI_items = prediction_to_lineage_information(GEMLI_i-
tems, cutoff=50)

# Example of a stored prediction lineage table
> GEMLI_items$predicted_lineage_table[1:5,]
     cell.ID               clone.ID
[1,] "AAACGAACAGGTGTGA-1" "1"
[2,] "AAAGGTAGTTGCTTGA-1" "2"
[3,] "AACCACAAGTTTGTCG-1" "3"
[4,] "AAGCCATGTTCCACGG-1" "4"
[5,] "AAGCGAGGTACGGCAA-1" "5"

# Example of a stored prediction lineage vector
> GEMLI_items$predicted_lineages[1:4]
AAACGAACAGGTGTGA-1 AAAGGTAGTTGCTTGA-1 AACCACAAGTTTGTCG-1
AAGCCATGTTCCACGG-1
                 1                  2                  3
4
```

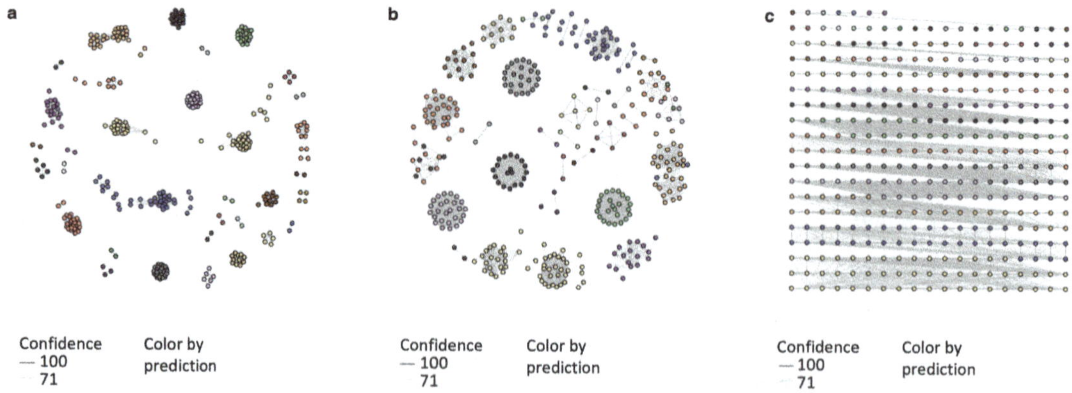

Fig. 5 Output of a *visualize_as_network* function call with different parameters. (**a**) The *visualize_as_network* function with parameter *layout_style* = "fr" outputs a network representation in which individual lineages are well separated from each other. The strength and number of connections between the cells of individual lineages are however hidden. (**b**) The *visualize_as_network* function with parameter *layout_style* = "kk" outputs a network representation in which the strength and number of connections between the cells of individual lineages are visible. Lineages can however be overlaying each other. (**c**) The *visualize_as_network* function with parameter *layout_style* = "grid" outputs a grid representation of lineage predictions

3.3.2 Visualizing GEMLI Predictions

To visualize the predicted GEMLI lineages in the form of a network, GEMLI encompasses the function *visualize_as_network*, which is based on the igraph R package. The function uses the "GEMLI_items" element "prediction." This means that as for the extraction of lineages, a *cutoff* parameter exists, which is the confidence level threshold used to define lineages (*see* Subheading 3.3.1) (*see* **Note 12**). In the resulting network, cells (vertexes) related with a confidence level above the threshold are connected by a line. The thickness of the line corresponds to the confidence that these two cells belong to the same lineage. The *visualize_as_network* function has a number of additional parameters. By adjusting the *layout_style* parameter, three different network layout algorithms can be used: Fruchterman–Reingold ("fr"), Kamada–Kawai ("kk"), or Grid ("grid"). The default parameter is *layout_style* = "fr". The three layout styles are differently suited to show predictions with different lineage number and size (Fig. 5). The *max_edge_width* parameter (default value is *max_edge_width* = 5) gives the range of connection line thickness. The *display_orphan* parameter indicates whether or not single cells are included in the representation. The default value is *display_orphan* = F is recommended for readability in most circumstances. *Include_labels* specifies whether or not the lineage number (clone IDs) is included as number in the representation (default value is *include_labels* = F). The parameters *ground_truth* and *highlight_FPs* relate to the optional presence of a ground truth lineage annotation (*see* Subheading 3.5). The parameter *cell_type_colors* (default value is *cell_type_colors* = F) allows to color the network by cell type annotation (*see* Subheading 3.4.2).

```
# Example visualization of with different layout styles (Fig 5)
> visualize_as_network(GEMLI_items, cutoff=70, layout_sty-
le="fr")
> visualize_as_network(GEMLI_items, cutoff=70, layout_sty-
le="kk")
> visualize_as_network(GEMLI_items, cutoff=70, layout_sty-
le="grid")
```

3.3.3 Trimming GEMLI Lineage Predictions

Trimming lineages predicted by GEMLI can further increase prediction precision. By default, lineages predicted by GEMLI are defined as cell groups of which any members are clustering together above a chosen confidence level threshold. It can therefore happen that within a predicted lineage, some cells are more likely to be related than others. This is visible in particularly high cell pair confidence values, and in multiple high confidence cell pairs that strengthen the lineage relationships within a group of cells. By contrast, it can happen that one cell is connected to an otherwise well-connected group of cells by a single high confidence value, which implies that this cell might be wrongly assigned to the lineage. Furthermore, in some instances, the exact lineage sizes within a dataset might be known beforehand, e.g., if only sister cells or two-cell lineages have been sequenced. Finally, the user might be specifically interested in lineages of a certain size. Two functions in the GEMLI package allow to adjust the lineage size in these cases: The *suggest_network_trimming_to_size* function generates a network visualization (*see* Subheading 3.3.2) of the predictions with a suggested trimming to the size highlighted in red (Fig. 6a). In the function call, the size to which lineages should be trimmed is given in the *max_size* function parameter. Its default value is *max_size* = 2. Other *suggest_network_trimming_to_size* parameters are as for the *visualize_as_network* function (*see* Subheading 3.3.2). To implement the suggested trimming, the *trim_network_to_size* function is used with the same *max_size* and *cutoff* parameter. This generates a new "GEMLI_items" list in which both the "prediction" (confidence value cell-by-cell matrix) and the "predicted_lineages" (lineage information) elements are altered. After trimming lineages, the result can again be visualized with the *visualize_as_network* function (Fig. 6b).

```
# Example visualization of suggested network trimming (Fig 6a)
> suggest_network_trimming_to_size(GEMLI_items, max_size=2,
cutoff=50)
```

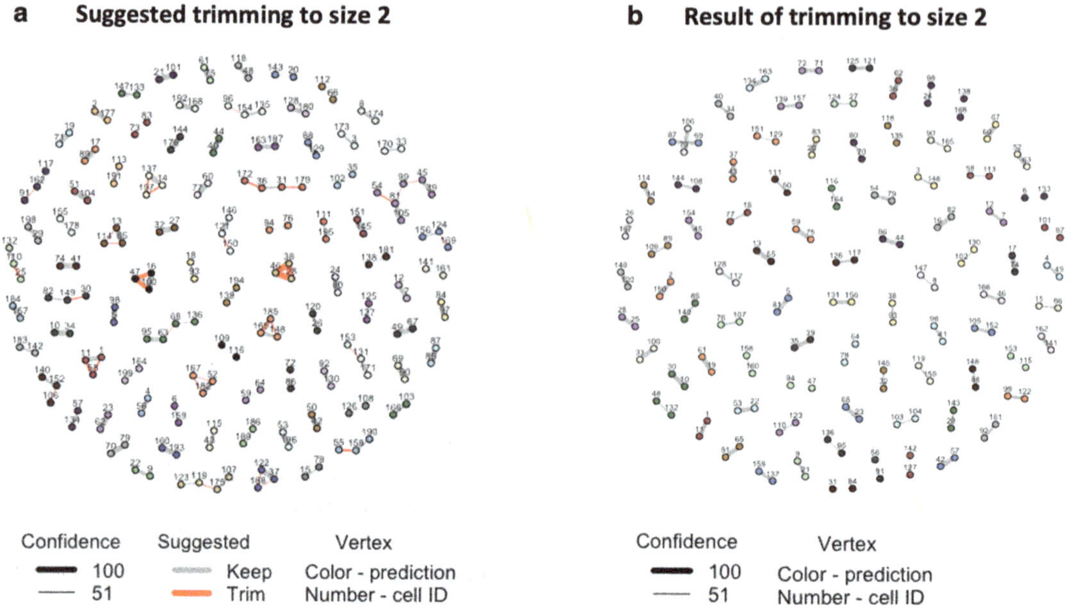

Fig. 6 Representation of lineage network trimming. (**a**) The *suggest_network_trimming_to_size* function outputs a lineage network representation in which lines connecting cells are red when trimming is suggested. It is visible, e.g., in the upper right part that the three-cell lineage encompassing cells 137, 14, and 197 is suggested to be trimmed to the 137 and 14 cell pair as the confidence value for this connection is much higher than for the connections of both cells to cell 197. (**b**) Output of the *visualize_as_network* function after performing trimming suggested in a using the *trim_network_to_size* function with parameter *max_size* = 2. The cell connections highlighted in red in panel (a) have been removed. It is visible that all lineages now have the same size

```
# Running the trim_network_to_size function
> GEMLI_items = trim_network_to_size(GEMLI_items, max_size=2,
cutoff=50)

# Example visualization of trimmed lineages (Fig 6b)
> visualize_as_network(GEMLI_items, cutoff=50)
```

3.4 GEMLI Lineage Analysis

GEMLI encompasses functions to analyze the predicted lineages with respect to (1) gene expression memory, (2) cell type composition, and (3) DEG for cells in lineages composed of specific cell types. An overview of these is given in Fig. 7. For all of these analyses, the "predicted_lineages" element in the "GEMLI_items" list is used. This means that after the *predict_lineages* function, the *prediction_to_lineage_information* function has to be run (*see* Subheading 3.3.1) before starting the analysis. To analyze predicted lineages with other tools or integrate the lineage information in other analyses, the "GEMLI_items" element "predicted_lineages" can be extracted and saved. Correspondingly, the GEMLI functions

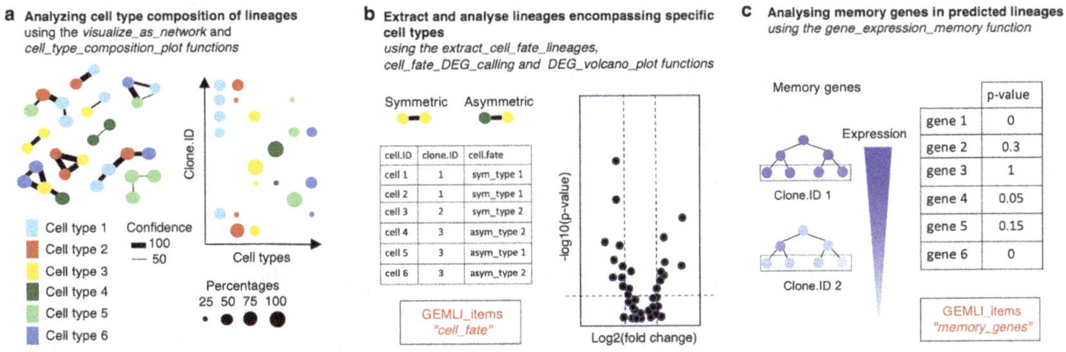

Fig. 7 Output of a *visualize_as_network* function call with coloring by cell type. (**a**) If no colors have been specified for cell types, the *visualize_as_network* function with parameter *cell_type_colors* = T outputs a lineage network representation with a random cell type coloring. (**b**) If colors for cell types have been specified in a "GEMLI_items" list element "cell_type_color," the *visualize_as_network* function with parameter *cell_type_colors* = T outputs a lineage network representation with the corresponding custom cell type coloring. In the example, the composition of individual lineages corresponding to intestinal crypts is represented. Individual cell types in lineages can easily be distinguished as colors are diverse

for the analysis of lineages can also be used for lineage annotations gained through other means (*see* Subheading 3.5).

3.4.1 Call Memory Genes on Predicted Lineages

GEMLI allows to call memory genes on predicted cell lineages using the function *memory_gene_calling*. We define memory genes as genes with high variability (coefficient of variability squared CV^2) in mean gene expression in different cell lineages. During memory gene calling, the lineage value is compared to the distribution of values from repeated (20×) random size-matched samples of cells to calculate a *p*-value. A threshold can then be set based on *p*-value to define memory genes. The *valid_lineage_sizes* parameter indicates the size of the lineages on which memory genes should be called (default *valid_lineage_sizes* = (2:5). The *cell_fate* parameter allows to restrict memory gene calling to a specific type of lineages with respect to their cell type composition (*see* Subheading 3.4.3). The parameters *ground_truth* relate to the optional analysis of a ground truth lineage annotation (*see* Subheading 3.5). The result of the *memory_gene_calling* run is stored in the "memory_genes" element of the "GEMLI_items" list. It consists of a table with gene names as row names, and the columns "*p*-value" that indicate *p*-values of how likely each gene is to be a memory gene.

```
# Running the memory_gene_calling function
> GEMLI_items <- memory_gene_calling(GEMLI_items)

# Example of a resulting memory genes table
> GEMLI_items$memory_genes[1:5,]
```

```
                     p-value
ENSMUSG00000022055 0
ENSMUSG00000095180 0
ENSMUSG00000023905 0
ENSMUSG00000098178 0
ENSMUSG00000068220 0
```

3.4.2 Analyze the Cell Type Composition of Predicted Lineages

Lineages predicted by GEMLI can be analyzed for their cell type composition. This can for example help in resolving the composition of large lineages as individual intestinal crypts or organoids, but also in identifying small lineages switching from one cell type to another, such as, for example, during the progression of an in situ to an invasive carcinoma. To analyze the cell type annotation of predicted lineages, the gene expression matrix has to be present in the "GEMLI_items" list as element "cell_type" (*see* Subheading 3.2.1). To identify the cell type composition of predicted lineages, the *visualize_as_network* function (*see* Subheading 3.3.2) can be used with parameter *cell_type_colors* = T. In the resulting network, cell vertexes will be colored by cell type (Fig. 8). The coloring is random by default (Fig. 8a). Specific colors can be added to specific cell types by adding a dataframe specifying these to the "GEMLI_items" list as element "cell_type_color." The dataframe should have the columns "cell.type" and "color." If the "cell_type_color" element is present in the "GEMLI_items" list, it will automatically be

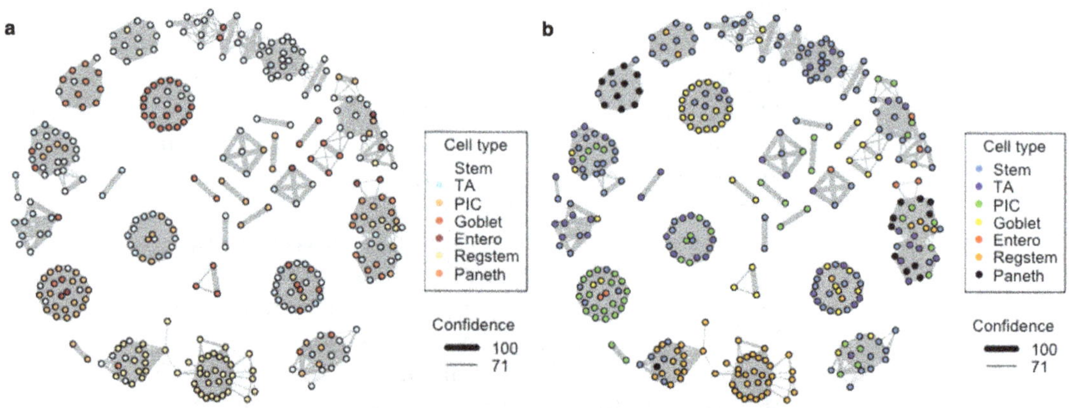

Fig. 8 Overview of GEMLI functions for the analysis of predicted lineages. (**a**) The cell type composition of predicted GEMLI lineages can be visualized in the form of a network using the *visualize_as_network* function or as table, bubble, or upsetR plot using the *cell_type_composition_plot* function. (**b**) Predicted GEMLI lineages encompassing specific cell types can be extracted using the *extract_cell_fate_lineages* function and stored in the "GEMLI_items" list element "cell_fate." Gene expression profiles of specific types of lineages can be compared using the *cell_fate_DEG_calling*. Results are stored as "GEMLI_items" list element "cell_fate_analysis" and can be visualized using the *DEG_volcano_plot* functions. (**c**) Memory genes for predicted GEMLI lineages can be called using the *gene_expression_memory* function. Results are stored in the "memory_genes" element of the "GEMLI_items" list

used in the *visualize_at_network* call with parameter *cell_type_colors* = T (Fig. 8b). Importantly, adding custom color specifications can also allow to highlight just one or two selected cell types.

```
# Example visualization of predictions with cell type coloring
> visualize_as_network(GEMLI_items, cutoff=70, cell_type_co-
lors=T)(Fig 8)

# Specifying colors per cell type
> cell.type <- unique(GEMLI_items[['cell_type']]$cell.type)
> color <- c("#5386BD", "blue", "green", "gold", "red",
"white", "black")
> cell_type_color <- data.frame(cell.type, color)
> GEMLI_items[['cell_type_color']] = cell_type_color

# Example visualization with custom coloring of cell types
> visualize_as_network(GEMLI_items, cutoff=70, cell_type_co-
lors=T)(Fig 9)
```

To obtain a more quantitative overview of the cell type composition of individual lineages, the function *cell_type_composition_plot* can be used. The most important parameter of this function is *type*, which can have the values c("bubble"), c("upsetR"), or c("plain"). This outputs either a bubble or an upsetR plot or a table of the frequency of different cell type combinations (Fig. 9). Depending on the number of cell types and lineages, and the size of lineages, different representations will be most appropriate. As for the *visualize_as_network_function*, the *cell_type_composition_plot* parameter *cell_type_colors* indicates if custom colors should be used when using *type* = c("bubble"). The parameter *ground_truth* relates to the optional analysis of a ground truth lineage annotation (*see* Subheading 3.5).

```
# Cell_type_composition_plot results with different type para-
meter
> cell_type_composition_plot(GEMLI_items, type=c("bubble"))
(Fig 9a)
> cell_type_composition_plot(GEMLI_items, type=c("plain"))
(Fig 9b)
> cell_type_composition_plot(GEMLI_items, type=c("upsetR"))
(Fig 9c)
```

3.4.3 Extract-Specific Lineage Types Based on Cell Type Composition

For further analysis of gene expression profiles of specific lineages, it can be useful to select lineages based on their cell type composition. In particular, one can compare gene expression profiles of cells in

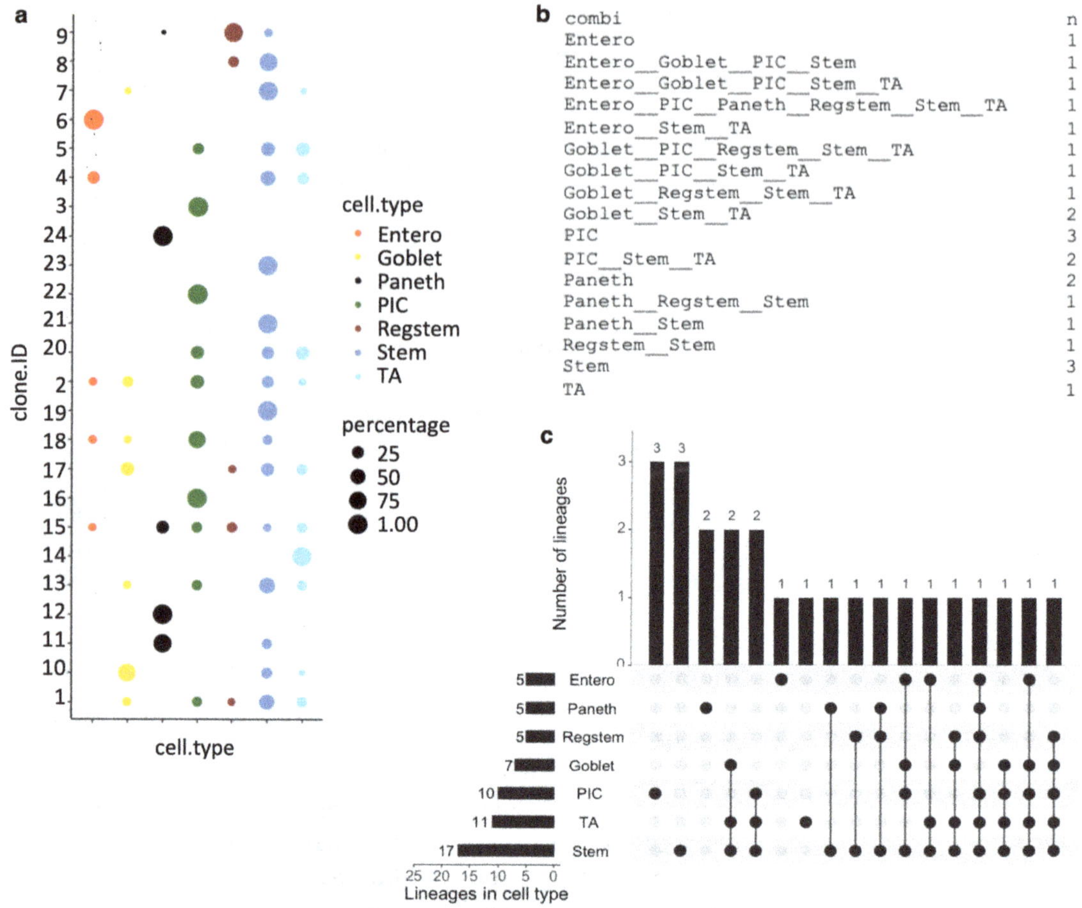

Fig. 9 Output of a *cell_type_composition_plot* function call with different *type* parameters. (**a**) A bubble plot of lineage cell type composition is generated with *cell_type_composition_plot* function parameter *type* = ("bubble"). In the example shown, the composition of individual lineages corresponding to intestinal crypts is represented. It is visible that most lineages encompass several cell types. (**b**) The *cell_type_composition_plot* function with parameter *type* = ("plain") outputs a table with columns "combi" and "n" indicating cell type combinations and the number of lineages in them, respectively. It is visible, e.g., that lineages composed of only cell types "Stem" and "PIC" are the most prevalent. (**c**) The *cell_type_composition_plot* function with parameter *type* = ("upsetR") outputs a upsetR plot of lineage cell type composition. It is visible, e.g., that the "Stem" cell type occurs in the highest number of lineages

lineages in which all members are in one cell type (symmetric; sym) to lineages in which cells are in two or more cell types (asymmetric; asym). This is possible with the function *extract_cell_fate_lineages*. The function has three parameters. *Selection* is a character vector that specifies the cell types based on which lineages (two or more) are chosen. The *unique* parameter specifies whether selected lineages can encompass other cell types than the ones indicated in the *selection* parameter (default is unique = F). Finally, the *threshold* parameter is a numeric vector indicating the minimal percentage the cell types in the *selection* should represent within the chosen lineages. Giving high values in the *threshold* parameter guarantees to select only lineages with a relatively even cell type composition, e.g., choosing lineages with a 50/50 cell type composition and disregarding lineages in which only 10% (1/10) of cells are of a specific cell type. The parameter *ground_truth* relates to the optional analysis of a ground truth lineage annotation (*see* Subheading 3.5). The function output is stored in "GEMLI_items" list element "cell_fate_analysis." It is a data frame with a column "cell.fate" with label sym or asym and cell types separated by an underscore. This "cell.fate" designation allows to subsequently analyze only a specific cell type in asymmetric or symmetric cell lineages.

```
# Running the extract_cell_fate_lineages function
> GEMLI_items<-extract_cell_fate_lineages(GEMLI_items, selec-
tion=c("inv_tumor", "DCIS"), unique=FALSE, threshold=c
(10,10))

# Example of extracted_cell_fate_lineages result
> GEMLI_items[['cell_fate_analysis']][1:10,]
  cell.ID              clone.ID cell.type cell.fate
 AAACCCACATCCGTGG-3       2826 DCIS       NA_DCIS
 AAACCCATCCTTATAC-4         54 DCIS       asym_DCIS
 AAACGAACAACACGTT-2       1466 inv_tumor sym_inv_tumor
 AAAGAACCAACAGCTT-3        726 DCIS       sym_DCIS
 AAAGAACGTCGAATGG-2       1467 inv_tumor sym_inv_tumor
 AAAGGATCAGAGTTCT-4       4383 inv_tumor sym_inv_tumor
 AAAGGATTCGCCAATA-4       4385 inv_tumor sym_inv_tumor
 AAAGGGCCAGTCGGAA-3        754 inv_tumor sym_inv_tumor
 AAAGGGCGTAAGAACT-1         10 inv_tumor sym_inv_tumor
 AAAGGGCGTAGTTCCA-1         11 inv_tumor sym_inv_tumor
```

3.4.4 Calling Differentially Expressed Genes (DEG) Between Cells in Specific Lineage Types

To call DEG specific for cells of specific cell types in specific lineage types, GEMLI encompasses the function *cell_fate_DEG_calling*. It is based on the Seurat R package findMarker function and uses the lineage type assignments in the "GEMLI_items" list element "cell_fate_analysis," created using the function *extract_cell_fate_lineages* (*see* Subheading 3.4.3). The *cell_fate_DEG_calling* function has four parameters. Two identity parameters (*ident1*, and *ident2*) specify the two cell categories to be compared. These must correspond to the lineage type ("cell.fate") designations in the "GEMLI_items" list "cell_fate_analysis" element. The parameters *min.pct* and *logfc.threshold* are the corresponding parameters of the Seurat FindMarker function. *Min.pct* indicates the minimum fraction of cells in either of the compared cell populations in which a gene should be expressed to be tested (default value is *min.pct* = 0.05). *Logfc.threshold* limits the DEG calling to genes which show on

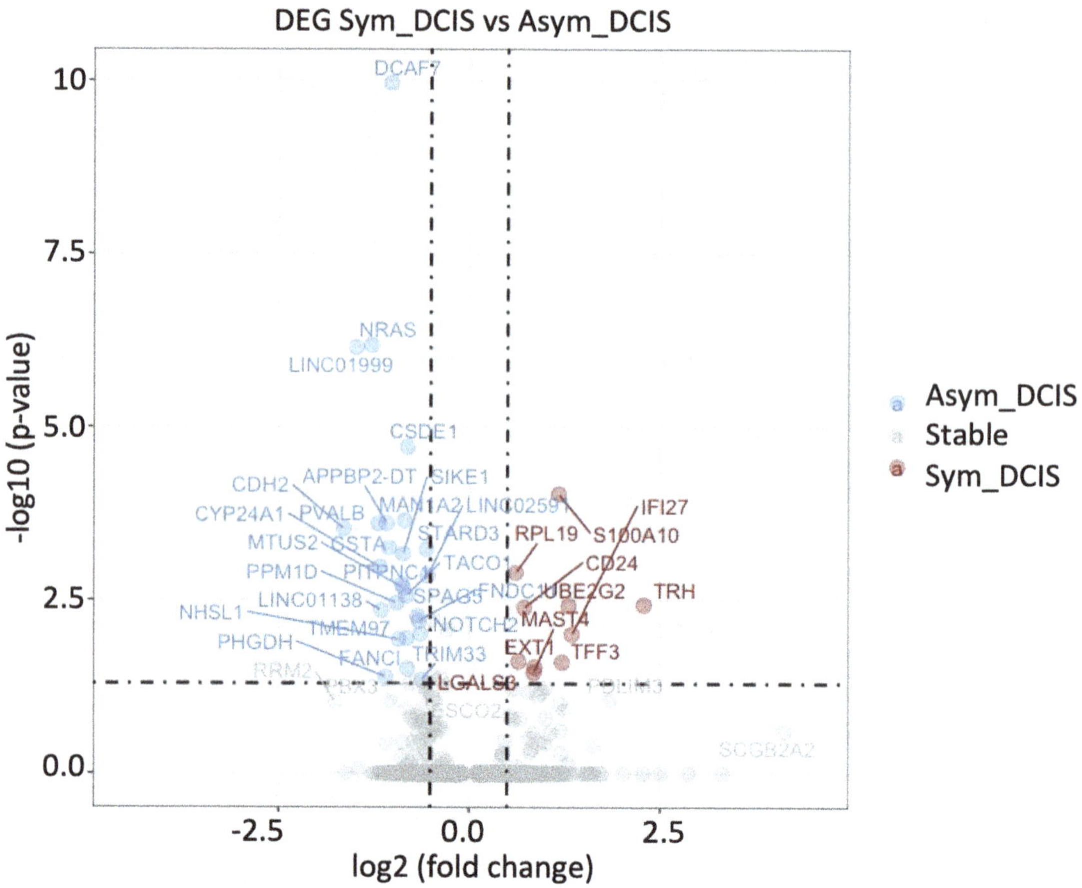

Fig. 10 Output of a *DEG_volcano_plot* function call. In the volcano plot generated by the *DEG_volcano_plot* function, DEG called between the identities specified in the *idents* parameters of the *cell_fate_DEG_calling* function call are shown. Represented are DEG called between breast DCIS cells in lineages with (asymmetric; asym) and without (symmetric; sym) invasive carcinoma cell members. The top DEGs are named and DEGs with a *p*-value under 0.05 and fold change above 0.5 are colored according to the *ident* in which they are highest expressed. It is visible that the gene DCAF7 is highly expressed in DCIS cells that are part of asymmetric lineages, i.e., also encompassing invasive tumor cells

average at least *x*-fold differences (log-scale) between the two compared cell populations (default value is *logfc.threshold* = 0.1). The output of the *cell_fate_DEG_calling* function is a dataframe stored as "GEMLI_items" element "DEG." It corresponds to Seurats findMarker output and contains the columns "p_val," "avg_log2FC," "pct.1," "pct.2," and "p_val_adj." The DEG calling results can then be visualized in the form of a volcano plot using the *DEG_volcano_plot* function (Fig. 10). Its *name1* and *name2* parameters should specify the *ident1* and *ident2* parameters used during DEG calling but do not have to be identical.

```
# Running the cell_fate_DEG_calling function
> GEMLI_items<-cell_fate_DEG_calling(GEMLI_items, ident1="-
sym_DCIS", ident2="asym_DCIS", min.pct=0.05, logfc.thresh-
old=0.1)

# Example of the cell_fate_DEG_calling result
> GEMLI_items[['DEG']][1:10,]
                 p_val avg_log2FC pct.1 pct.2    p_val_adj
DCAF7     4.308339e-15 -1.0275658 0.985 1.000 1.086520e-10
NRAS      2.630887e-11 -1.2821469 0.516 0.766 6.634834e-07
LINC01999 2.824883e-11 -1.4817245 0.736 0.961 7.124072e-07
CSDE1     7.762498e-10 -0.8055031 0.952 0.974 1.957624e-05
S100A10   3.676826e-09  1.1729996 0.839 0.610 9.272588e-05
MAN1A2    8.900499e-09 -0.8487466 0.868 0.961 2.244617e-04
APPBP2-DT 9.767327e-09 -1.0797417 0.495 0.792 2.463222e-04
PVALB     9.882215e-09 -1.1913536 0.304 0.623 2.492196e-04
CDH2      1.169570e-08 -1.6410588 0.264 0.558 2.949538e-04
CSTA      2.204180e-08 -1.0506112 0.198 0.506 5.558721e-04

# Running the DEG_volcano_plot function
> DEG_volcano_plot(GEMLI_items, name1="Sym_DCIS", name2="A-
sym_DCIS")(Fig 10)
```

3.5 Testing GEMLI Lineage Predictions Against Ground Truth

GEMLI does encompass several functions that allow scoring of predictions against a ground truth lineage annotation. These functions have been especially important in the development of the method, but can also be helpful for users. They allow, for example, to expand lineage annotation to a full scRNA-seq dataset when only a fraction is annotated. Prediction performance can be tested on the lineage-annotated part of the dataset with different parameters and then be applied with optimized parameters to the unannotated part of the dataset. Furthermore, GEMLI allows to analyze ground truth lineages, e.g., to call memory genes, analyze the cell type composition, or call DEG for cells in specific ground truth lineages.

3.5.1 Adding Ground Truth Lineage Information to GEMLI_items List

Ground truth lineage information can be added to the "GEMLI_items" list as the element "ground_truth." The ground truth lineage information should be present as a named vector with values corresponding to "clone.IDs" and names corresponding to "cell.IDs." The "cell.IDs" should match the ones present in the gene expression matrix stored as the "gene expression" element in the "GEMLI_items" list.

```
# Adding ground truth lineage information to GEMLI_items list
> GEMLI_items[['ground_truth']] = ground_truth_lineage_infor-
mation
```

```
# Example of a stored ground truth lineage information
> GEMLI_items[['ground_truth']][1:5]
CACAGATAGTGATGGC-1 TATCTTGGTACGGGAT-1 AAACGAACAGGTGTGA-1 AGA-
GAATAGGTCATAA-1
                   2                  2                 2
7
```

3.5.2 Testing GEMLI Lineage Predictions Against Ground Truth

When a "GEMLI_items" list contains both a lineage prediction in the form of a cell-by-cell confidence matrix stored as "prediction" element and a ground truth lineage information in the form of a "ground_truth" element, the *test_lineage_prediction* function can be used to test the predicted lineages against ground truth. The function outputs the number of true-positive (TP) predictions and false-positive (FP) predictions, as well as precision and sensitivity for various confidence levels (Conf). This is stored in the form of a table in the "testing_results" element of the "GEMLI_items" list. The output can be further visualized in the form of a curve showing precision and sensitivity with respect to the confidence value threshold by setting the parameter *plot_results* = T (Fig. 11). Additional parameters in the function call are *valid_fam_sizes*, which indicates an interval of predicted lineage sizes to consider (default is *valid_fam_sizes* = c(2:5), and *max_interval*, which indicates the

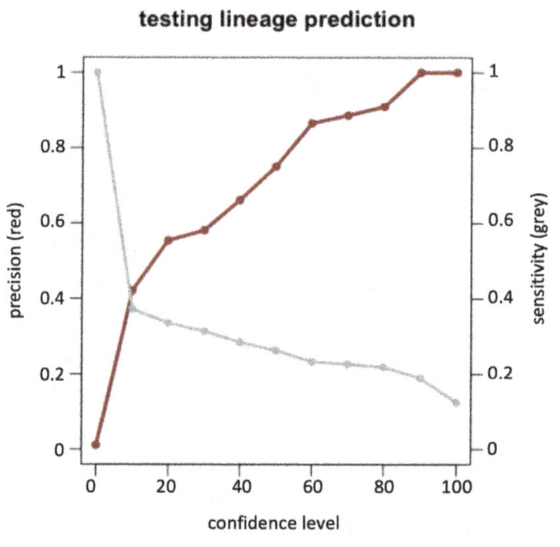

Fig. 11 Output of a *test_lineages* function call. When using the *test_lineages* function on a "GEMLI_items" list with both a "prediction" and "ground truth" element, a plot is generated showing precision (red) and sensitivity (gray) with respect to the confidence level of lineage predictions. The confidence level is binned into ten intervals and indicated as percentage. Precision and sensitivity are calculated based on cell pairs within lineages. Both are indicated as fractions. In the plot, it is visible that, e.g., a confidence level of 50% results in a precision of 70% and a sensitivity of 30%

number of repetitions used during lineage predictions (default is *max_interval* = 100).

```
# Running the test_lineages function
> GEMLI_items = test_lineages(GEMLI_items)

# Example of the output of the test_lineages function
> GEMLI_items$testing_results
Conf TP    FP  precision sensitivity
0    274 24062 0.01125904   1.0000000
10   104   128 0.44827586   0.3795620
20    90    82 0.52325581   0.3284672
30    84    56 0.60000000   0.3065693
40    80    36 0.68965517   0.2919708
50    68    22 0.75555556   0.2481752
60    64    14 0.82051282   0.2335766
70    62    10 0.86111111   0.2262774
80    58     2 0.96666667   0.2116788
90    46     0 1.00000000   0.1678832
100   34     0 1.00000000   0.1240876

# Running the test_lineages function with parameter plot_re-
sults=T
> GEMLI_items = test_lineages(GEMLI_items, plot_results=T)
(Fig 11)
```

The ground truth lineage information can also be highlighted in the network representation of predictions generated with the *visualize_as_network* function (*see* Subheading 3.3.2). When setting the parameter *ground_truth* = T, cells will be colored in the representation according to their ground truth lineage information (Fig. 12a). When setting the parameter *highlight_FPs* = T, false-positive connections between cells are colored in red (Fig. 12b). The highlight_FPs = T setting can also be used when coloring cells by cell type (*see* Subheading 3.4.2).

```
# Running the visualize_as_network function highlighting
ground truth
> visualize_as_network(GEMLI_items, cutoff=70, ground_-
truth=T) (Fig 12a)
> visualize_as_network(GEMLI_items, cutoff=70, high-
light_FPs=T) (Fig 12b)
```

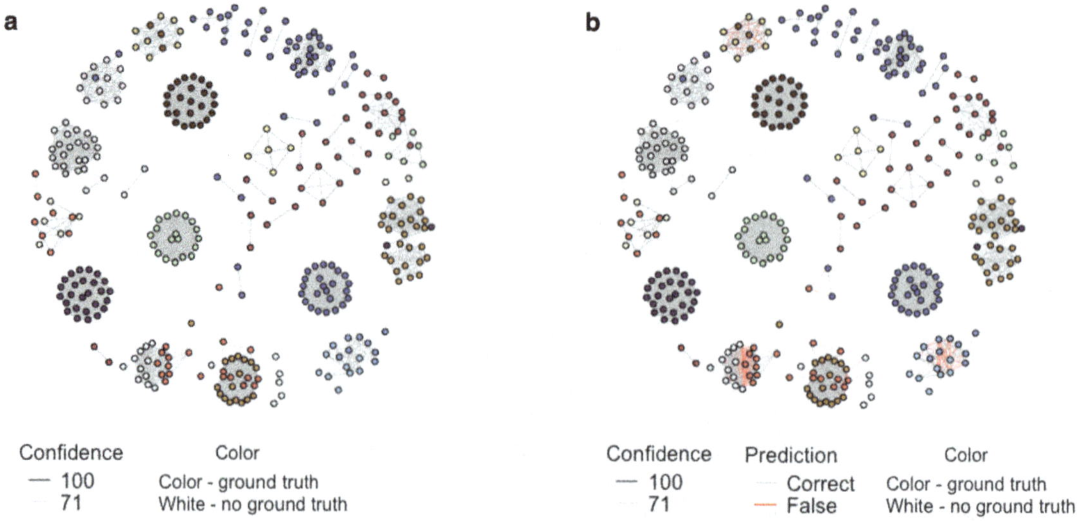

Fig. 12 Output of a *visualize_as_network* function call highlighting ground truth lineage information. (**a**) If the "GEMLI_items" list comprises ground truth lineage information as element "ground truth," the *visualize_as_-network* function with parameter *ground_truth* = T outputs a lineage network representation in which cells are colored according to ground truth lineage information. (**b**) If the "GEMLI_items" list comprises ground truth lineage information as element "ground truth," the *visualize_as_network* function with parameter *high-light_FPs* = T outputs a lineage network representation in which predicted cell connections not present in the ground truth are colored red. In the plot, it is visible that most predictions are correct but that two ground truth lineages are merged in the lower left corner

3.5.3 Analyzing Ground Truth Lineages

The GEMLI functions for the analysis of lineages with respect to gene expression memory, cell type composition, and DEG can also be applied to ground truth lineages. This is possible by adjusting the parameter *ground_truth* = T. This parameter can be set for the function *memory_gene_calling* (*see* Subheading 3.4.1), *cell_type_-composition_plot* (*see* Subheading 3.4.2), and *extract_cell_fate_li-neages* (*see* Subheading 3.4.3). If *extract_cell_fate_lineages* has been run with parameter *ground_truth* = T, the functions *cell_fa-te_DEG_calling* and *DEG_volcano_plot* (*see* Subheading 3.4.4) will automatically yield results specific to ground truth lineages.

4 Notes

1. The dataset should encompass cells related over a small number of divisions. Datasets that represent samples of large cell populations are not suited as the chance that they encompass cells related over a small number of divisions is low. This means, for example, that scRNA-seq data of blood samples is rather unsuitable. Importantly, however, GEMLI precision stays high also for the predictions of more distantly related cell lineages. This means that in datasets in which cells are more distantly related rather no lineages than false lineages are predicted.

2. GEMLI has been benchmarked on datasets produced using 10× Genomic Chromium platform, InDrop scRNA-seq, SmartSeq sequencing, SCRB sequencing, and Disco sequencing [14]. As the gene expression variability GEMLI is using for predictions has biological sources, we expect GEMLI to perform well on quantitative scRNA-seq datasets from other sequencing platforms too.
3. We recommend filtering on cell read number and mitochondrial read fraction. Other filtering methods can be used but have not yet been benchmarked.
4. We recommend filtering by read count. Other normalization methods can be used but have not yet been benchmarked.
5. Sequencing depth should ideally lie above 8000 reads/cell. This is above the recommended ranges for most commercially available scRNA-seq platforms. In particular, the widely used 10× Genomics Chromium assay has a recommended sequencing depth of around 20,000 reads/cell.
6. Most GEMLI R package dependencies are available on CRAN, Bioconductor, or GitHub. They can be installed with the commands install.packages("library"), BiocManager::install(c (library)), or as GEMLI using install_github, respectively. Loading is performed using library ("library_name").
7. Gene names can be provided in any format. For some downstream analysis, as the calling of DEG between different lineage types, it is recommended to use gene symbols. This makes plots such as the volcano plot generated by *DEG_volcano_plot* function easily readable.
8. The parameters *repetitions*, *sample_size*, *desired_cluster_size*, and *fast* in the *lineage_prediction* function call should be adjusted as a function of the final aim of the lineage analysis. In some instances, a maximal precision is essential. For highly precise predictions in a cell population, *repetitions* should be set high (e.g., 200), *sample_size* and *desired_cluster_size* should have small values (e.g., ½ and 2,3), and the default value of *fast* = F should be used. For highly precise predictions of larger lineages, the *desired_cluster_size* can be adjusted as described in Subheading 3.3.3 to increase sensitivity while keeping a high precision.
9. The *desired_cluster_size* parameter in the *predict_lineages* function call does correspond to the cluster size at which clustering iterations are stopped. This does not necessarily correspond to the size of predicted lineages as cells can cluster differently during the clustering repetitions. Predicted lineages can be larger than the *desired_cluster_size*.

10. For predictions in scRNA-seq datasets of cell populations, the recommended *desired_cluster_size parameter* is the default value of 2,3. Such datasets will most often contain a majority of two-cell lineages (even if more distantly related than sister cells) due to the loss of cells during scRNA-seq using most sequencing platforms.
11. For predictions in scRNA-seq datasets of cell populations, no plateau is reached, but the index is continuously increasing (*see* also **Note 10**).
12. When many lineages are predicted, using a *cutoff* parameter of 50 yields crowded network visualizations. To ease the representation, higher *cutoff* values can be used, or the dataset restricted to a random subset.

References

1. Eisele AS, Tarbier M, Dormann AA et al (2022) Barcode-free prediction of cell lineages from scRNA-seq datasets. bioRxiv. PMID: 38553478. https://doi.org/10.1101/2022.09.20.508646
2. Shaffer SM, Emert BL, Reyes Hueros RA et al (2020) Memory sequencing reveals heritable single-cell gene expression programs associated with distinct cellular behaviors. Cell 182:947–959
3. Meir Z, Mukamel Z, Chomsky E et al (2020) Single-cell analysis of clonal maintenance of transcriptional and epigenetic states in cancer cells. Nat Genet 52:709–718
4. Kimmerling RJ, Lee Szeto G, Li JW et al (2016) A microfluidic platform enabling single-cell RNA-seq of multigenerational lineages. Nat Commun 7:10220
5. Kumar RM, Cahan P, Shalek AK et al (2014) Deconstructing transcriptional heterogeneity in pluripotent stem cells. Nature 516:56–61
6. Hormoz S, Singer ZS, Linton JM et al (2016) Inferring cell-state transition dynamics from lineage trees and endpoint single-cell measurements. Cell Syst 3:419–433
7. Frieda KL, Linton JM, Hormoz S et al (2017) Synthetic recording and in situ readout of lineage information in single cells. Nature 541: 107–111
8. Phillips NE, Mandic A, Omidi S et al (2019) Memory and relatedness of transcriptional activity in mammalian cell lineages. Nat Commun 10:1208
9. Mold JE, Weissman MH, Ratz M et al (2022) Clonally heritable gene expression imparts a layer of diversity within cell types.bioRxiv. PMID: 38340731. https://doi.org/10.1016/j.cels.2024.01.004
10. Harmange G, Hueros RAR, Schaff D et al (2022) Disrupting cellular memory to overcome drug resistance. bioRxiv. PMID: 37932277. https://doi.org/10.1038/s41467-023-41811-8
11. Rückert T, Lareau CA, Mashreghi MF et al (2022) Clonal expansion and epigenetic inheritance of long-lasting NK cell memory. Nat Immunol 23:1551–1563
12. Wehling A, Loeffler D, Zhang Y et al (2022) Combining single-cell tracking and omics improves blood stem cell fate regulator identification. Blood 140:1482–1495
13. Wang K, Hou L, Wang X et al (2023) PhyloVelo enhances transcriptomic velocity field mapping using monotonically expressed genes. Nat Biotechnol 42:778–789
14. Bues J, Biočanin M, Pezoldt J et al (2022) Deterministic scRNA-seq captures variation in intestinal crypt and organoid composition. Nat Methods 19:323–330
15. Janesick A, Shelansky R, Gottscho AD et al (2022) High resolution mapping of the breast cancer tumor microenvironment using integrated single cell, spatial and in situ analysis of FFPE tissue. Nat Comm 14:8353

Chapter 20

Lineage Analysis at Single-Cell Resolution by Twin-Spot MARCM with Lineage-Restricted Drivers

Angelica Previero, Ching-Po Yang, and Tzumin Lee

Abstract

Cell lineage analysis is primarily undertaken to understand cell fate specification and diversification along a cell lineage tree. Built with dual repressible markers, twin-spot mosaic analysis with repressible cell markers (MARCM) labels the two daughter cells made by a common precursor in distinct colors. The power of twin-spot MARCM to systematically subdivide complex lineages is exemplified in studies of *Drosophila* neural stem-cell lineages. To achieve targeted lineage analysis, lineage-restricted drivers can be created by restricting and immortalizing gene expression to a lineage of interest. Here, we describe detailed methods to perform lineage analysis in *Drosophila* neural stem-cell lineages utilizing lineage-specific drivers. Our innovative lineage tracing method can help resolve complex tissue development.

Key words Twin-spot MARCM, Neuronal birth order, Single-cell lineage reconstruction, Mosaic labeling, Lineage specificity, Binary sister fate decision, *Drosophila* brain development

1 Introduction

A zygote will give rise to all the cells of an organism through a series of cell divisions. Lineage analysis involves trailing a cell's history of cell divisions from its postmitotic state to the original progenitor, for all cells deriving from a common progenitor. Complete lineage analysis should therefore record all cell divisions. The mosaic analysis with repressible cell markers (MARCM) method creates a genetically labeled homozygous mutant cell in an otherwise heterozygous background [1]. MARCM relies on mitotic recombination in a progenitor cell that leads to loss of heterozygosity in the two daughter cells and results in de-repression of a reporter gene in the mutant daughter and its progeny. Of note is MARCM's unique ability to label homozygous mutant clones in an otherwise wild-type background, enabling mosaic analysis of gene functions in the complex nervous system [2, 3]. However, MARCM leaves the progeny of the other daughter cell unlabeled. Addressing this

Jorge García-Marqués and Tzumin Lee (eds.), *Lineage Tracing: Methods and Protocols*, Methods in Molecular Biology, vol. 2886, https://doi.org/10.1007/978-1-0716-4310-5_20,

limitation, twin-spot MARCM (ts-MARCM) employs two reporter genes repressed by corresponding RNAi transgenes prior to mitotic recombination (Fig. 1) [4]. This achieves labeling of sister clones arising from the paired homozygous daughter cells (following mitotic recombination) in distinct colors as twin spots. However, each individual experiment, generating one single twin-spot clone, is only a snapshot of the development of the lineage. As they record the outcomes of individual cell divisions, the full collection of twin-spot clones needs to be used for comprehensive cell lineage analysis.

However, since complex tissues arise from diverse cell lineages, it is not possible to perform lineage analysis using twin-spot clones if the parent lineage is unknown. It is therefore logical to first determine the entire progeny born from a founding progenitor by inducing ts-MARCM at the beginning of tissue development, i.e., making full-size clones. Comparing full-size clones born from different progenitors across samples gives light to the number and diversity of lineages comprising a tissue. The presence of identifiable, characteristic clones would indicate tissue development through stereotyped cell lineages, indicated by neurons born in fixed patterns and timings along the axis of development [5–8].

In tissues with minimal cell migration, the differentially labeled sister clones arising from a progenitor typically exist next to each other as twin spots. However, the sister clones may vary greatly in size and morphology due to differences in patterns of cell division and programmed cell death. The relationship between the cells within a clone or between sister clones with multiple cells cannot be ascertained without additional knowledge. Thus, reconstructing an entire stereotyped lineage entails mapping sublineage patterns within the greater lineage [9]. This involves analyzing related twin-spot clones generated at serial time points [6–8]. Essential to reconstructing cell division patterns are sublineage clones with single, identifiable cells. The existence of more than one twin-spot clone in the same region can complicate analysis, making it difficult to ascertain sister relationships or map detailed clone morphology. Therefore, the generation of isolated clones by limited clone induction may be required.

Also challenging is the identification of clones belonging to the same lineage across samples. The task of identifying clonally related sister clones can be greatly aided by restricting twin-spot labeling only to the lineage(s) of interest rather than sampling all dividing cells regardless of lineage origin [10]. Built with binary transgene induction systems (e.g., GAL4/UAS), ts-MARCM selectively labels the postmitotic cells positive for a tissue-specific driver (e.g., a GAL4 enhancer trap) (Fig. 1). Knowing the expression pattern of the employed driver is thus crucial to correctly interpret ts-MARCM results. Ideal drivers for lineage analysis should selectively label all the progeny generated by the lineage(s) of interest, without labeling nearby cells. Such lineage-restricted drivers can be

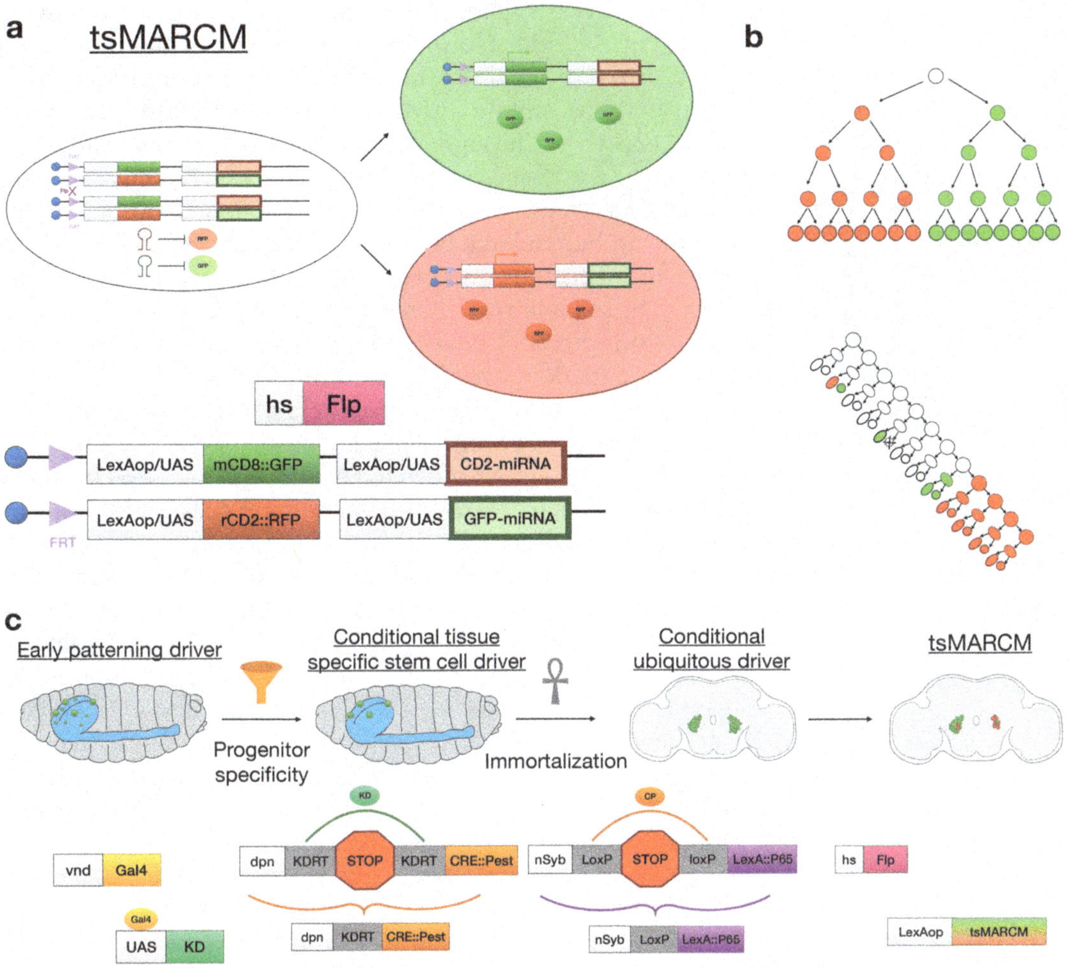

Fig. 1 Genetic strategy of lineage-restricted ts-MARCM. (**a**) ts-MARCM reporter genes. In ts-MARCM, a homologous chromosome harbors a transgene expressing a GFP-RNAi and the transgene expressing rCD2::RFP, both downstream of an FRT sequence. The other homologous chromosome harbors a transgene expressing a CD2 -RNAi and the transgene expressing mCD8::GFP, both downstream of the same FRT. Without recombination, the transcripts of both fluorescent proteins would be repressed by their respective RNAi. However, heat-shocked induced mitotic recombination in a progenitor cell leads to one daughter cell inheriting two copies of the GFP-RNAi and rCD2::RFP genes, and the other daughter cell to inherit two copies of the CD2-RNAi and mCD8::GFP genes, leading to the expression of both fluorescent proteins in the respective daughter cells and their progeny. (**b**) ts-MARCM patterns of expression in the labeled clones change in the case of symmetric and asymmetric cell division. (Top) Symmetric cell division would yield two labeled clones of equal size. (Bottom) The *Drosophila* type I NB division pattern. NBs divide asymmetrically, and GMC progeny produces two neurons. Thus, different scenarios are possible, and programmed cell death can also be identified in the case of the NB+1 and 1+0 scenarios. From left to right: GMC division (1+1), GMC division with apoptosis (1+0), and NB division (NB+2) scenarios are presented. (**c**) Driver-related and immortalization transgenes. The first required transgene for lineage restriction is the early patterning gene driver. Such drivers can either lead to direct expression of the recombinase KD, or they can drive KD expression via binary induction by the addition of a UAS-KD transgene, as the driver used to label our lineage of interest, vnd-Gal4. To exclusively label stem-cell lineages, expression needs to be restricted to progenitors, through a conditional tissue-specific, stem-cell driver. In the case of *Drosophila* NBs, dpn is used. Only progenitor cells positive for the tissue-specific KD would be able to activate the expression of the

engineered from early spatial patterning genes expressed in progenitor cells (Fig. 1) [11]. However, such genes are often transiently expressed in the progenitors, begging the inclusion of an immortalization step to sustain driver activity throughout the entire lineages [12, 13]. Through recombinases expressed under the control of a tissue-specific stem-cell promoter, activating the expression of a driver under the control of a ubiquitous promoter, expression can be rendered active exclusively in the progenitors of select lineages and thus immortalized in all their postmitotic descendants [10, 14]. One can then iteratively induce twin-spot clones at select developmental time points to map the targeted lineage trees with fine spatiotemporal resolution (Fig. 1).

In this protocol chapter, we describe the materials and methods to track a specific neuronal lineage in the adult fly brain by ts-MARCM with a lineage-restricted driver (Fig. 2). The *Drosophila* central brain originates from ~100 neural stem cells per hemisphere [5]. One neural stem cell (neuroblast, NB) can undergo ~100 serial asymmetric cell divisions to produce a series of ~100 intermediate precursors (ganglion mother cell, GMC) that each divide once to make two postmitotic neurons following each stem-cell asymmetric cell division [15]. A complete fly neuronal lineage thus consists of ~100 serially derived sister-neuron pairs. Using lineage-restricted drivers, we can induce clones in specific neuronal lineages at a specific developmental time via transient FLP expression. Mitotic recombination can occur in dividing neural stem cells (neuroblast, NB) or intermediate precursors (ganglion mother cell, GMC), which yield either asymmetric sister clones (a multicellular NB clone paired with a two-cell GMC clone) or paired single-cell clones (GMC division). To accomplish single-cell reconstruction of a protracted neuronal lineage, each cell division needs to be captured. It takes ~1.5 h for a NB division, while it takes ~4.2 h for GMCs to mature and divide [16, 17]. Thus, dividing a full-size NB clone into a series of single-cell clones can be accomplished by generating twin-spot clones in serial 2-h windows. Notably, each fly neural stem cell produces a unique sequence of distinct sister-neuron pairs, indicating lineage-guided neuron fate specification and diversification.

The superb trackability and striking stereotypy of fly neuronal lineages greatly aid single-cell lineage analysis of the central brain by ts-MARCM. ts-MARCM is remarkable in studying stem-cell biology in other fly tissues [18–20]. In larger animals, MADM (*see* Chapter 13 by Shi) with a mice-specific genetic design has

Fig. 1 (continued) recombinase CRE::PEST. Finally, the last set of driver transgenes needed is to immortalize expression in all postmitotic cells born from that specific progenitor, by taking advantage of a conditional ubiquitous driver, including drivers ubiquitously expressed within a tissue. The activity of such driver is conditional as it is dependent on CRE::PEST activity in the respective progenitor

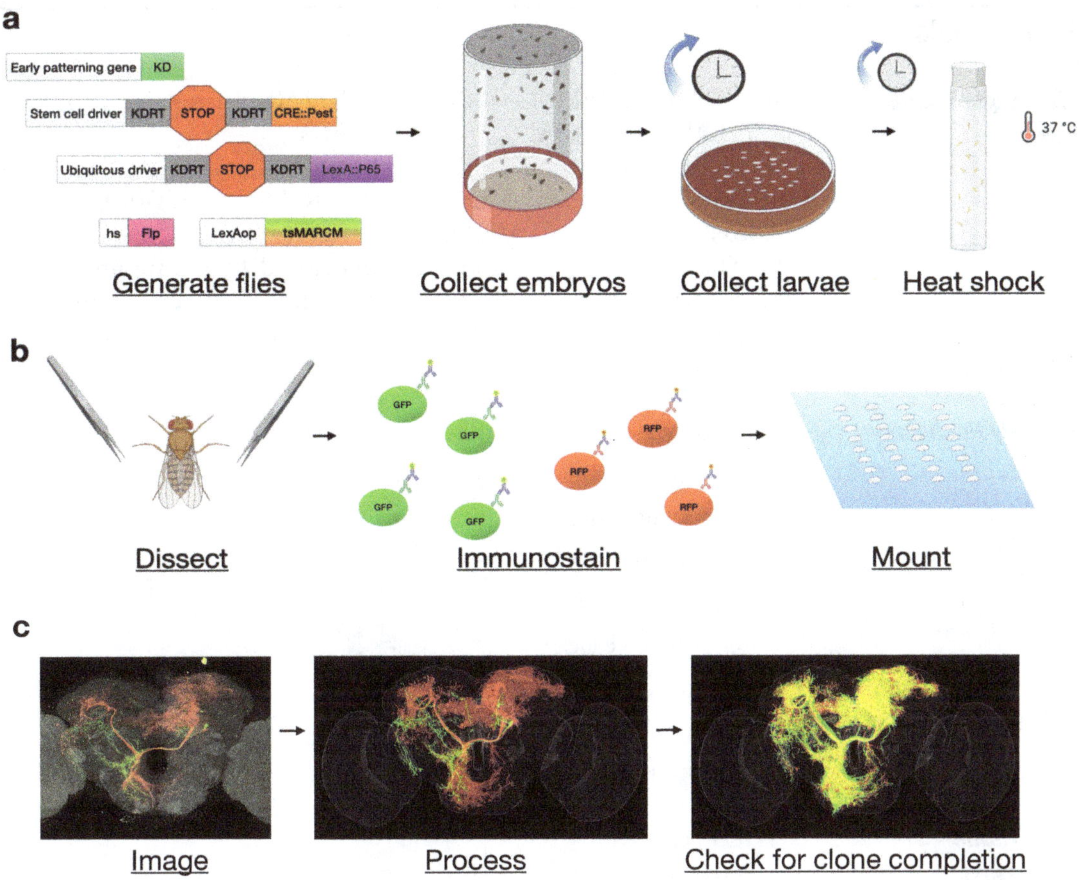

Fig. 2 Workflow from fly generation to clone analysis. (**a**) Animal husbandry. In the case of NB lineage development, to resolve the entire lineage of interest, we recommend heat-shocking every 2 h from 18 h ALH to 16 h APF. (**b**) Tissue preparation. The tissue of interest in young adult flies is dissected, stained, and mounted to prepare for imaging. (**c**) Image processing. After raw images are taken, they are processed for clarity purposes and to segment out various clones. To check for complete lineage coverage, all the identified sister-neuron pairs (in this case in the CREa1 lineage) need to be superimposed on a standard brain to reconstitute the full-size NB+2 clone generated at the beginning of larval development

analogously achieved lineage analysis with single-cell-cycle resolution in the complex mouse brain. However, the limited throughput of recording one cell cycle at one time in each animal by twin-spot labeling calls for further sophisticated systems to resolve more dynamic lineages with fine temporal resolution in higher model organisms (*see* Chapters 17 and 21 by Espinosa-Medina and García-Marqués).

2 Materials

2.1 Experimental Flies

vnd-T2A-Gal4/FM6C; lexAop2-rCD2::RFP-insulator-lexAop2-GFP-RNAi, FRT40A.

hs-FLP,dpn > KDRT-stop-KDRT>Cre PEST; lexAop2-mCD8::GFP-insulator-lexAop2-rCD2-RNAi, FRT40A; nSyb>loxP-stop-loxP>LexA::P65,UAS-KD1.

The flies described above are generated by combining driver transgenes, conditional tissue-specific stem-cell drivers, conditional ubiquitous drivers, and ts-MARCM transgenes (Table 1) (*see* **Note 1**).

2.2 Fly Husbandry

1. Embryo-collection cages.
2. 60-mm Petri dishes.
3. Grape agar premix.
4. Grape juice.
5. *Para*-hydroxybenzoate.
6. Granular yeast.
7. 50-mL syringes without needle.
8. Fly food.
9. Empty narrow *Drosophila* vials.

2.3 Tissue Processing

1. Forceps.
2. Dissecting dish.
3. 2-mL Eppendorf tube.
4. Phosphate-buffered saline.
5. Normal goat serum.
6. Anti-mCD8 antibody, rat monoclonal.
7. Anti-RFP antibody, rabbit polyclonal.
8. Anti-nc82, mouse monoclonal.
9. Anti-rat, Alexa488, goat.
10. Anti-rabbit, Alexa568, goat.
11. Anti-mouse, Alexa647, donkey.
12. Coverslips.
13. Poly-L-lysine solution.
14. Slides.
15. Paraformaldehyde 20% solution.
16. DPX mountant, a mixture of Distyrene, a Plasticizer, and Xylene.
17. Xylene.

Table 1
List of available transgenic fly stocks for conducting lineage analysis using ts-MARCM

Transgene	Source and reference	BDSC stock	Labeled lineages
Postmitotic drivers / early patterning genes drivers			
GH146-Gal4	Yu et al. [6]	BDSC:30026	Partial mPN of Alad1
acj6-Gal4	Yu et al. [6]	BDSC:30025	ALad1
13C01-Gal4	Awasaki et al. [10]		FLAa2, EBa1, ALl1, VLPa1, VLPa2, VLPl1, VLPl2, VLPl4, LHl1 and LHl4.
44F03-Gal4	Awasaki et al. [10]		ALad1, ALl1, ALv1, ALv2, EBa1 and LALv1.
44F03-KD	Awasaki et al. [10]		ALad1, ALl1, ALv1, ALv2, EBa1 and LALv1.
vnd-T2A-Gal4	Lee et al. [8]		SMPad1, SMPp&v1, SLPpm3, CREa1, CREa2, WEDd1, AOTUv1, AOTUv3, AOTUv4, VLPa2, VESa1, VESa2, ALv1, LALv1, FLAa1, FLAa2, FLAa3, and WEDa1
msh-T2A-Gal4	Chen et al. [26]		VNC lineages 1A, 9A, 14A and 21A
stg-KD	Wang et al. [27]		Type II NB
UAS-KD	Awasaki et al. [10]	BDRC:55790	
Conditional tissue specific stem cell driver			
dpn > FRT-stop-FRT>Cre::PEST	Awasaki et al. [10]	BDSC:56165	
dpn > KDRT-stop-KDRT>Cre::PEST	Awasaki et al. [10]	BDSC:56164	

(continued)

Table 1 (continued)

Transgene	Source and reference	BDSC stock	Labeled lineages
Conditional ubiquitous driver			
nSyb > FRT-stop-FRT>LexA::P65	Awasaki et al. [10]	BDSC:56163	
act5C > loxP-stop-loxP>LexA::P65	Lee et al. [8]		
nSyb > loxP-stop-loxP>LexA::P65	Awasaki et al. [10]	BDSC:56166	

Transgene	Source and reference	BDSC stock	Chromosome
Twin-spot MARCM reagents	UAS and lexAop of tsMARCM transgenes on chromosome arm 2L,2R,3L and 3R are availabel on Bloomington stock center (https://bdsc.indiana.edu/stocks/misc/twinspot_marcm.html)		
hs-FLP	Awasaki et al. [10]		
lexAop2-rCD2::RFP-insulator-lexAop2-GFP-RNAi, FRT40A	Awasaki et al. [10]		2L
lexAop2-mCD8::GFP-insulator-lexAop2-rCD2-RNAi, FRT40A	Awasaki et al. [10]		2L
lexAop2-mCD8::GFP-insulator-lexAop2-rCD2-RNAi (attP2)	Awasaki et al. [10]	BDSC:56173	3L
lexAop2-mCD8::GFP-insulator-lexAop2-rCD2-RNAi (attP40)	Awasaki et al. [10]	BDSC:56174	2L
lexAop2-mCD8::GFP-insulator-lexAop2-rCD2-RNAi (VK00027)	Awasaki et al. [10]	BDSC:56182	3R

lexAop2-mCD8::GFP-insulator-lexAop2-rCD2-RNAi (su(Hw)attP5)	Awasaki et al. [10]	BDSC:56750	2R
lexAop2-rCD2::RFP-insulator-lexAop2-GFP-RNAi (attP2)	Awasaki et al. [10]	BDSC:56175	3L
lexAop2-rCD2::RFP-insulator-lexAop2-GFP-RNAi (attP40)	Awasaki et al. [10]	BDSC:56176	2L
lexAop2-rCD2::RFP-insulator-lexAop2-GFP-RNAi (su(Hw)attP5)	Awasaki et al. [10]	BDSC:56177	2R
lexAop2-rCD2::RFP-insulator-lexAop2-GFP-RNAi (VK00027)	Awasaki et al. [10]	BDSC:56183	3R
lexAop-mCD8::GFP-insu-lexAop-rCD2i (attP18),FRT19A/FM6C; Pin/CyO	Awasaki et al. [10]		X
lexAop-rCD2::RFP-insu-lexAop-GFPi (attP18),FRT19A/FM6C; Pin/CyO	Awasaki et al. [10]		X
UAS-mCD8::GFP-insulator-UAS-rCD2-RNAi (VK00027)	Awasaki et al. [10]	BDSC:56169	3R
UAS-mCD8::GFP-insulator-UAS-rCD2-RNAi (attP2)	Awasaki et al. [10]	BDSC:56170	3L
UAS-mCD8::GFP-insulator-UAS-rCD2-RNAi (su(Hw)attP5)	Awasaki et al. [10]	BDSC:56171	2R
UAS-mCD8::GFP-insulator-UAS-rCD2-RNAi (attP40)	Awasaki et al. [10]	BDSC:56172	2L
	Awasaki et al. [10]	BDSC:56185	2L

(continued)

Table 1 (continued)

Transgene	Source and reference	BDSC stock	Labeled lineages
UAS-mCD8::GFP-insulator-UAS-rCD2-RNAi (attP40),FRT40A; Tm3,Sb/TM6B			
UAS-rCD2::RFP-insulator-UAS-GFP-RNAi (VK00027)	Awasaki et al. [10]	BDSC:56178	3R
UAS-rCD2::RFP-insulator-UAS-GFP-RNAi (attP2)	Awasaki et al. [10]	BDSC:56179	3L
UAS-rCD2::RFP-insulator-UAS-GFP-RNAi (su(Hw)attP5)	Awasaki et al. [10]	BDSC:56180	2R
UAS-rCD2::RFP-insulator-UAS-GFP-RNAi (attP40)	Awasaki et al. [10]	BDSC:56181	2L
UAS-rCD2::RFP-insulator-UAS-GFP-RNAi (attP40),FRT40A; Tm3,Sb/TM6B	Awasaki et al. [10]	BDSC:56184	2L
UAS-mCD8::GFP-insulator-UAS-rCD2-RNAi (su(Hw)attP8), FRT19A; Pin/CyO	Awasaki et al. [10]		X
UAS-rCD2::RFP-insulator-UAS-GFP-RNAi (su(Hw)attP8), FRT19A; Pin/CyO	Awasaki et al. [10]		X

18. Ethyl alcohol, pure.
19. Nail polish.
20. Mounting media.
21. Electrical tape.

2.4 General Equipment

1. Water bath.
2. Dissecting microscope.
3. Confocal fluorescent microscope.
4. Microwave.
5. Paintbrush.
6. Coverslip staining jar.
7. Nutating (3D) mixer.
8. Fly incubator kept at 25 °C.

2.5 Data Analysis Tools

1. Fiji.
2. Adobe Photoshop.
3. Python.
4. Flybase 2.0.
5. VVDviewer (https://github.com/takashi310/VVD_Viewer/releases).
6. ImageJ skeletonize3D macro (https://imagej.net/Skeletonize3D).
7. *Drosophila* standard brain (JFRC2010, https://github.com/VirtualFlyBrain/DrosAdultBRAINdomains).

3 Methods

3.1 Fly Food Preparation

1. Prepare grape juice agar plates. To make grape juice agar plates, add a packet of grape agar premix to 300 mL of H_2O in a 1-L beaker while continuously mixing gently with a spatula.
2. Microwave until boiling point and until the agar is fully dissolved (*see* **Note 2**). Mix periodically.
3. Add 150 mL of grape juice.
4. Fill with H_2O up to 500 mL and mix the solution well.
5. Let the solution cool to ~50 °C and add *para*-hydroxybenzoate to a final concentration of 0.03%. Mix the solution well.
6. Pour the mix into 60-mm Petri dishes (~5–10 mL) and let the dishes set for 2 h.
7. Cap the plates, seal each plate with parafilm, and move to 4 °C.

3.2 Fly Husbandry

1. On day 0, make the following cross of ~300 virgin females and ~100 males:

   Virgin females: vnd-T2A-Gal4/FM6C; lexAop2-rCD2::RFP-insulator-lexAop2-GFP-RNAi, FRT40A.

   Males: hs-FLP,dpn > KDRT-stop-KDRT>Cre PEST; lex-Aop2-mCD8::GFP-insulator-lexAop2-rCD2-RNAi, FRT40A; nSyb>loxP-stop-loxP>LexA::P65,UAS-KD1.

2. On day 3, flip the cross into a cage with a grape juice plate with some granular yeast sprinkled on (*see* **Note 3**).
3. Let the flies lay embryos overnight at 25 °C.
4. The following day, 22 h after the plate was placed in the cage, start removing the already hatched larvae (*see* **Note 4**).
5. At 24 h, start collecting the newly hatched larvae.
6. Prepare multiple food vials by adding a layer of soft food (food collected from other vials and/or bottles and then passed through a 50-mL syringe) to a vial already containing food.
7. Collect newly hatched larvae with a wet paintbrush. Add ~80 larvae at a time to each prepared food vial.
8. Collect larvae in 2-h batches for 8 h and mark their hatching time.
9. Repeat **steps 2–8** until the entire NB postembryonic development is covered. Heat-shock timings and intervals may be adjusted depending on the question being asked, the lineages of interest, and the developmental windows when the cells of interest are actively dividing.
10. Heat shock the larvae for 15–40 min at 37 °C in a water bath (*see* **Note 5**). The timings for such heat shocks should follow the logic below:
    (a) For early and mid-stage larvae, the timings of the heat shocks should cover the whole developmental span of the larvae from 18 h after larval hatching (ALH) to 84 h ALH in 2-h windows.
    (b) For late-stage larvae, once larvae start coming up from the food, the age is determined by counting how many hours pass between the heat shock and pupae formation. Heat shock to cover all developmental windows between 22 h before pupae formation (BPF) and 0 h BPF in 2-h increments.
    (c) For pupae, synchronize all animals at 0 h after pupae formation (APF) and perform heat shocks every 2 h to cover all developmental windows between 0 after pupae formation (APF) and 16 APF.

11. Keep animals at 25 °C until eclosion.
12. Dissect 4–7-day-old *Drosophila* females according to the tissue preparation protocol below.

3.3 Tissue Preparation

1. Dissect brains from 4- to 7-day-old *Drosophila* females on ice in 1× phosphate-buffered saline (PBS) in a dissection plate (*see* **Note 6**).
2. Transfer the brains to a 2-mL Eppendorf (*see* **Note 7**).
3. Add 2 mL of 2% paraformaldehyde (PFA) in PBS (*see* **Note 8**).
4. Fix brains for 50 min at room temperature on a nutating mixer. Do not fix it for more than 1 h.
5. Carefully discard PFA and wash the brains three times in 1x PBS with 0.5% Triton-X100 (PBT) for 10 min per wash on the nutating mixer.
6. Prepare the primary antibody mix by adding the below indicated antibodies to PBT supplemented with 5% of normal goat serum (NGS) (*see* **Note 9**). Prepare 200 μL per sample.
   (a) Rat monoclonal anti-mCD8 antibody (1:100).
   (b) Rabbit anti-RFP antibody (1:1000).
   (c) Mouse monoclonal anti-nc82 antibody (1:50).
7. Discard the final PBT wash volume (*see* **Note 10**). Add the primary antibody mix to the brains and incubate overnight on a nutating mixer at 4 °C.
8. Discard the primary antibody mix and wash the brains three times in PBT for 10 min each at room temperature on the nutating mixer.
9. Prepare the secondary antibody mix by adding the below indicated antibodies to PBT supplemented with 5% of normal goat serum (NGS). Prepare 200 μL per sample:
   (a) Anti-rat, Alexa488, Goat (1:500).
   (b) Anti-rabbit, Alexa568, Goat (1:500).
   (c) Anti-mouse, Alexa647, Donkey (1:500).
10. Discard the final PBT wash volume (*see* **Note 10**). Add the secondary antibody mix and incubate overnight on a nutating mixer at 4 °C in the dark.
11. Coat the needed coverslips in 500 μL of poly-L-lysine and dry at room temperature (*see* **Note 11**).
12. Discard the secondary antibody mix and wash the brains three times in PBT for 10 min each at room temperature on the nutating mixer.
13. Postfix the samples in 2 mL of 4% PFA in PBS for 4 h at room temperature.

14. Discard PFA and wash the brains three times in PBT for 10 min per wash on the nutating mixer.
15. Discard the final PBT wash. Rinse the sample in 2 mL of PBS.
16. With respect to mounting, it is optional to dehydrate the sample before mounting. For conventional mounting with a glycerol-based mounting media, follow **steps 17–23**. For dehydration instructions, follow **steps 24–28**.
17. *For conventional mounting*: Add 1:1 volume mounting media to remaining PBS in the tube.
18. Cut four ~5 mm × 5 mm squares of electrical tape and place them so that they are positioned at the four corners of the coverslip you are going to use (*see* **Note 12**).
19. Transfer the brains on a slide and align them in columns using forceps (*see* **Note 13**).
20. Place a coverslip on top of the slide, with its four corners resting on the electrical tape.
21. Pipette the mounting media from the edge of the coverslip to fill the entire space in between the slide and the coverslip, and gently push out any bubble (*see* **Note 14**).
22. Seal the slide and coverslip with nail polish.
23. Let the slide dry in the dark for 1 h.
24. *To dehydrate and mount*: Transfer the brains to the poly-L-lysine coated coverslip, orient them correctly, and align them in columns. After alignment, allow the brains to stick to the coverslip.
25. To dehydrate your samples, place the coverslip in Coplin staining jars with the following solutions for 10 min each: 30% EtOH in 1× PBS, 50% EtOH in 1× PBS, 75% EtOH in 1× PBS, 95% EtOH in 1× PBS, 100% EtOH (repeat twice), 100% xylene (repeat twice). Xylene should be handled in the hood.
26. In the hood, add a few drops of DPX on a slide, before placing the coverslip face down on the slide.
27. Seal the slide and coverslip with DPX using a paintbrush.
28. Let the slide dry in the dark overnight in the hood.

3.4 Imaging

1. Image the brains of interest using confocal// microscopy with the following parameters:
   (a) Z-stack in 1-μm intervals.
   (b) 0.7 zoom factor.
   (c) 1024 × 1024 pixel resolution.
   (d) Tile scan to independently image each fly brain hemisphere to later merge together.

3.5 Analysis

1. Using the nc82 image for each sample, align the images to a standard brain (JFRC2010, https://github.com/VirtualFlyBrain/DrosAdultBRAINdomains).
2. The vnd enhancer labels a total of 18 lineages. After stochastically labeling only a few lineages at a time through light heat shock, manually identify the clones belonging to the lineage of interest, here CREa1, based on cell body location and neuropil projection (*see* **Note 5**). To do so, use the predefined neuropil masks on the fly brain template.
3. Once clones belonging to the lineage of interest are identified, divide the clones by general morphological types with shared projection and targeting areas (Fig. 3).
4. Furthermore, characterize each specific morphological neuron type (Fig. 3).
5. For presentation purposes, perform 3D interactive segmentation and align each segmented neuron to the same standard whole brain [21–23]. Then, use VVDviewer for 3D rendering for 2D presentation.

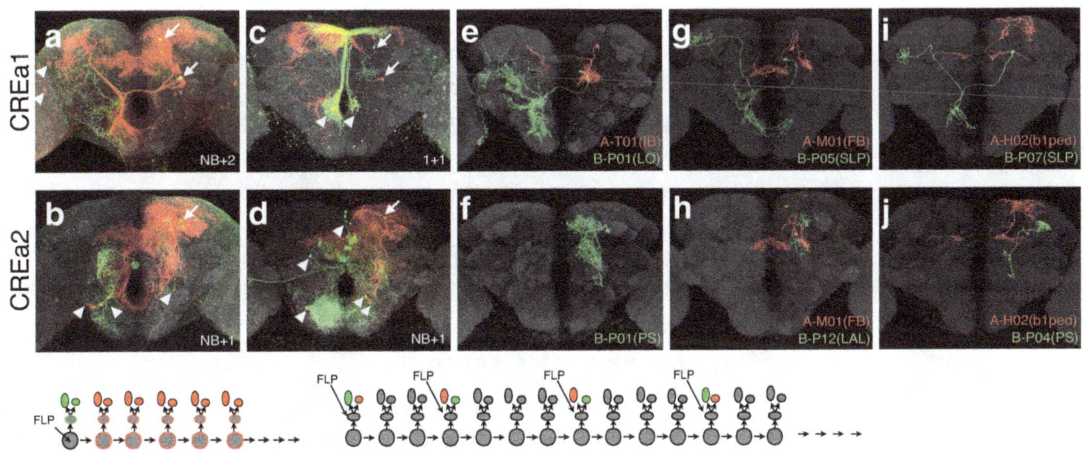

Fig. 3 ts-MARCM clone patterning in two lineages of the *Drosophila* central brain. (**a**–**d**) Raw images presenting different patterns of labeling in the CREa1 and CREa2 lineages, with arrowheads pointing at background clones, belonging to other lineages also labeled by vnd, and arrows at clones of interest. When inducing ts-MARCM clones, recombination can happen in any dividing cell within the lineage. In (**a**), recombination occurred in the NB and both sister neurons produced by the GMC (a case with no cell death) that originated from that specific asymmetric division were labeled with GFP. The rest of the clone was labeled with RFP (NB+2). In (**c**, **e**, and **g**–**j**), recombination occurred in a GMC (cases with no cell death), thus the two fluorescent proteins differentially label the two neurons born by that specific GMC (1+1). Cell death can easily be highlighted with ts-MARCM. In (**b**, **d**, **f**), only one neuron from a GMC division is labeled (NB+1), suggesting the other underwent apoptosis. (**e**, **f**) Processed images portraying respectively 1+1 and 1+0 clones of the neurons born at the beginning of the lineage. (**g**, **j**) The CREa1 and CREa2 lineages are related lineages, with some of their progeny being of the same type. Such neurons can be born at similar stages during development (**g**, **h**) or at more distant times (**i**, **j**). In either case, it is important to rely on the respective sister cell to correctly assign the clone to the correct lineage

6. For each time point of heat-shock induction, resolve birth order by calculating the percentage of individually labeled sister neurons recovered that belong to a specific morphological neuronal type (*see* **Note 5**). Furthermore, also consider neurons belonging to morphological neuronal types appearing in the previous and consecutive time points of heat-shock induction. Moreover, the NB clone can also be informative. In certain cases, neurons born within a lineage innervate distinct compartments in a stereotypical manner, such as the ALad1 neurons dendrites innervating the antennal lobe glomeruli [6]. However, in most lineages, morphological types are recurrent, meaning that the projections of the neurons in the NB clone are not distinct enough to establish the identity of the uniquely labeled sister neurons. NB clone size can also be used by counting the cells being labeled, inferring then the cell count of each morphological neuron type. However, NB clones are more commonly obtained at the beginning and at the end of lineage development. During third instar larvae, NB clones are rarely being generated.
7. Check for complete lineage coverage. Superimpose all the identified unique sister-neuron pairs artificially labeled in green on a standard brain. On the same brain, superimpose the full-size NB+2 clone generated at the beginning of larval development artificially labeled in red.

4 Notes

1. Successful ts-MARCM involves generating flies harboring the following transgenes.
   (a) *Driver-related transgenes*:
      (i) Gal4 drivers expressed in specific postmitotic neuron types: Many of the current drivers label specific sets of postmitotic neurons. For example, GH146-Gal4 labels most, and acj6-Gal4 labels all the monoglomerular projection neurons (mPNs) born from the anterodorsal antennal lobe (ALad1) NB [6]. These Gal4 drivers were used to drive expression of ts-MARCM clones to resolve the ALad1 lineage [6].
      (ii) Gal4 drivers expressed in specific NBs (early patterning drivers): In this case, the *vnd-T2A-Gal4* and *UAS-KD* constructs are used to specifically drive the expression of the KD recombinase in vnd-expressing cells. The *vnd-T2A-Gal4* line was generated through a method called Golic+ [26]. As mentioned in Awaski et al. [10], any Gal4 line driving expression in a

specific NB can be exploited for neuronal lineage tracing once immortalized. Other drivers expressed in the desired progenitor cells may be chosen depending on the question being asked.

(b) *Immortalization transgenes*: The *conditional tissue-specific stem-cell driver dpn>KDRT-stop-KDRT>Cre::PEST* is necessary to drive the expression of CRE only in KD-expressing NBs, which are the only NBs able to remove the stop cassette through KD-mediated recombination. Moreover, the *conditional ubiquitous driver nSyb>loxP-stop-loxP>LexA::p65* is needed to ensure sustained expression of LexA::p65 in the whole lineage, by nSyb ubiquitously driving expression in fly neurons. It is important to notice that this genetic strategy leads to the expression of the LexA::p65 transcription factor in all 18 lineages made by vnd+ NBs in the fly central brain. Conditional tissue-specific stem-cell drivers may be chosen depending on the question being asked. An alternative conditional ubiquitous driver to lead expression in tissues that are not neuronal is *act5C>loxP-stop-loxP>LexA::p65*.

(c) *ts-MARCM transgenes*: the experimental flies need to be the cross from two transgenic *Drosophila* lines, each homozygous for one of the two homologous chromosomes needed for ts-MARCM (Fig. 1). One parental line for the 2L chromosome arm carries *FRT40A, lexAop-mCD8::GFP-insulated spacer-lexAop-rCD2i*. The other 2L parental line carries *FRT40, lexAop-rCD2::RFP-insulated spacer-lexAop-GFPi*. One parental line also carries *hs-Flp*. Mild heat-shock conditions will stochastically induce recombination in a limited number of dividing cells. Only the LexA::p65 expressing lineages in which recombination occurred will be fluorescently labeled. If using a different Gal4 transgene for other lineages, the other transgenes locations may be affected, especially the transgenes needed for ts-MARCM expression. Fly lines with ts-MARCM transgenes on all three major chromosomes are available.

2. To be able to tell if the agar is fully melted, look out for an increase in the size of the bubbles when the mixture is boiling.
3. Yeast is going to help increase the number of embryos being laid. However, too much yeast added to the plate is going to make it more difficult to find the hatched larvae both while clearing the plate as well as during larval collection.
4. If several larvae have already hatched, pour a 20% sucrose solution onto the plate and wait for the larvae to float. Pour the sucrose solution out of the plate to remove the larvae.

5. Heat-shock duration might need to be optimized for different genetics. If no clones are being labeled, increase the heat-shock duration. If too many clones are being labeled, decrease the heat-shock duration.
6. Do not dissect for more than 30 min to preserve tissue morphology and the quality of the staining.
7. Using a 20-μL pipette tip coated with PBT to transfer the brain is going to lower the chances of the brains sticking on the side of the pipette and therefore minimize tissue loss.
8. PFA is a hazardous substance, and appropriate precautions should be taken while handling it. All PFA-related steps should be carried under the hood, and waste that has come in contact with PFA should be appropriately disposed.
9. If the staining level is too low, potential causes might include over-fixation or too much light exposure. On the other hand, if there is too much background staining, potential causes might include too high antibody concentration or incomplete tissue washing.
10. Remove as much PBT as possible while leaving just enough so that the brains will remain covered during the buffer exchange. Removing all the PBT buffer should be avoided because it will result in the brains drying out, while leaving too much leftover PBT buffer should be avoided because it will affect the antibody dilution.
11. To prevent dust accumulation on the coverslips, place them in a covered chamber (e.g., empty pipette tip box).
12. Placing a coverslip underneath the slide can be used as reference while placing the four pieces of tape so that they are in the right location at the respective corners.
13. If the brains are floating and not adhering to the slide, remove some of the mounting media. If it is difficult to flip the brains so that the ventral side is in contact with the slide, add some mounting media.
14. Be very gentle while adding mounting media to minimize any brain moving from its position.

Acknowledgments

We thank Janelia Workstation, Janelia FlyLight, and Janelia Fly Core for technical support. We thank Rosa Miyares for editing the chapter. This work was supported by Howard Hughes Medical Institute. Some of the icons used in the figures were from BioRender.com.

References

1. Lee T, Luo L (1999) Mosaic analysis with a repressible cell marker for studies of gene function in neuronal morphogenesis. Neuron 22(3):451–461
2. Lee T, Luo L (2001) Mosaic analysis with a repressible cell marker (MARCM) for Drosophila neural development. Trends Neurosci 24(5):251–254
3. Lin S, Lai SL, Yu HH et al (2010) Lineage-specific effects of Notch/Numb signaling in post-embryonic development of the Drosophila brain. Development 137(1):43–51
4. Yu HH, Chen CH, Shi L et al (2009) Twin-spot MARCM to reveal the developmental origin and identity of neurons. Nat Neurosci 12(7):947–953
5. Das A, Gupta T, Davla S et al (2013) Neuroblast lineage-specific origin of the neurons of the Drosophila larval olfactory system. Dev Biol 373(2):322–337
6. Yu HH, Kao CF, He Y et al (2010) A complete developmental sequence of a Drosophila neuronal lineage as revealed by twin-spot MARCM. PLoS Biol 8(8):e1000461
7. Lin S, Kao CF, Yu HH et al (2012) Lineage analysis of Drosophila lateral antennal lobe neurons reveals notch-dependent binary temporal fate decisions. PLoS Biol 10(11): e1001425
8. Lee YJ, Yang CP, Miyares R et al (2020) Conservation and divergence of related neuronal lineages in the *Drosophila* central brain. elife 9:e53518
9. Ren Q, Awasaki T, Huang YF et al (2016) Cell class-lineage analysis reveals sexually dimorphic lineage compositions in the Drosophila brain. Curr Biol 26(19):2583–2593
10. Awasaki T, Kao CF, Lee YJ et al (2014) Making Drosophila lineage-restricted drivers via patterned recombination in neuroblasts. Nat Neurosci 17(4):631–637
11. Urbach R, Technau GM (2003) Molecular markers for identified neuroblasts in the developing brain of Drosophila. Development 130(16):3621–3637
12. Lacin H, Chen HM, Long X et al (2019) Neurotransmitter identity is acquired in a lineage-restricted manner in the *Drosophila* CNS. elife 8:e43701
13. Nern A, Pfeiffer BD, Rubin GM (2015) Optimized tools for multicolor stochastic labeling reveal diverse stereotyped cell arrangements in the fly visual system. Proc Natl Acad Sci USA 112(22):E2967–E2976
14. Nern A, Pfeiffer BD, Svoboda K et al (2011) Multiple new site-specific recombinases for use in manipulating animal genomes. Proc Natl Acad Sci USA 108(34):14198–14203
15. Wodarz A, Huttner WB (2003) Asymmetric cell division during neurogenesis in Drosophila and vertebrates. Mech Dev 120(11): 1297–1309
16. Homem CC, Reichardt I, Berger C et al (2013) Long-term live cell imaging and automated 4D analysis of drosophila neuroblast lineages. PLoS One 8(11):e79588
17. Bowman SK, Rolland V, Betschinger J et al (2008) The tumor suppressors Brat and Numb regulate transit-amplifying neuroblast lineages in Drosophila. Dev Cell 14(4): 535–546
18. de Navascués J, Perdigoto CN, Bian Y et al (2012) Drosophila midgut homeostasis involves neutral competition between symmetrically dividing intestinal stem cells. EMBO J 31(11):2473–2485
19. Sahai-Hernandez P, Nystul TG (2013) A dynamic population of stromal cells contributes to the follicle stem cell niche in the Drosophila ovary. Development 140(22): 4490–4498
20. Gunage RD, Reichert H, VijayRaghavan K (2014) Identification of a new stem cell population that generates Drosophila flight muscles. elife 3:e03126
21. Wan Y, Otsuna H, Chien CB et al (2012) Interactive extraction of neural structures with user-guided morphological diffusion. Proc IEEE Symp Biol Data Vis 2012:1–8
22. Rohlfing T, Maurer CR (2003) Nonrigid image registration in shared-memory multiprocessor environments with application to brains, breasts, and bees. IEEE Trans Inf Technol Biomed 27(1):16–25
23. Wan Y, Otsuna H, Chien CB et al (2009) An interactive visualization tool for multi-channel confocal microscopy data in neurobiology research. IEEE Trans Vis Comput Graph 15(6):1489–1496
24. Stocker RF, Heimbeck G, Gendre N (1997) Neuroblast ablation in Drosophila P[GAL4] lines reveals origins of olfactory interneurons. J Neurobiol 32(5):443–456

25. Jefferis GS, Marin EC, Stocker RF et al (2001) Target neuron prespecification in the olfactory map of Drosophila. Nature 414(6860): 204–208
26. Chen HM, Huang Y, Pfeiffer BD et al (2015) An enhanced gene targeting toolkit for Drosophila: Golic+. Genetics 199(3):683–694
27. Wang YC, Yang JS, Johnston R et al (2014) Drosophila intermediate neural progenitors produce lineage-dependent related series of diverse neurons. Development 141(2) 253–258

Chapter 21

CLADES: A Programmable Cascade of Genes for Cell Lineage Analysis and Manipulation

Inés González-Aspe, Beatriz Madariaga Puchol, Blanca López Arrabal, Brooke Bosonac, and Jorge García-Marqués

Abstract

In the *Drosophila* brain, neuronal diversity originates from approximately 100 neural stem cells, each dividing asymmetrically. Precise mapping of cell lineages at the single-cell resolution is crucial for understanding the mechanisms that direct neuronal specification. However, existing methods for high-resolution lineage tracing are notably time-consuming and labor-intensive. Here, we outline the best practices for lineage tracing using CLADES (cell lineage access driven by an edition sequence), a revolutionary approach to neuronal lineage tracing that addresses the limitations of previous methods. CLADES effectively traces the birth order of neurons using approximately 100 samples. The technique relies on a genetic cascade of reporter activations and deactivations that delineate lineage progression through color-coded markers. This system not only facilitates the detailed mapping of neuronal lineages but also holds the potential to be applied to tracking biological events and producing cell types for therapeutic purposes.

Key words Lineage tracing, *Drosophila*, CRISPR, CaSSA, Neuronal specification, Genetic cascades, Genetic tools

1 Introduction

In the *Drosophila melanogaster* brain, the vast array of neuronal diversity originates from about 100 neural stem cells (neuroblasts) for each hemisphere, each neuroblast undergoing approximately 100 proliferation cycles [1, 2]. The dominant lineage model observed in the brain is asymmetrical (Type I), where each neuroblast divides asymmetrically to produce another neuroblast and one transient intermediate progenitor, the ganglion mother cell (GMC). The GMC then divides once more, giving rise to two sibling neurons (Fig. 1). This lineage type, due to its simplicity, is particularly suited for lineage tracing and analysis using imaging-based methodologies. Following this pattern, each neuroblast that undergoes about 100 mitoses will produce only 200 neurons. In

Jorge García-Marqués and Tzumin Lee (eds.), *Lineage Tracing: Methods and Protocols*, Methods in Molecular Biology, vol. 2886, https://doi.org/10.1007/978-1-0716-4310-5_21,

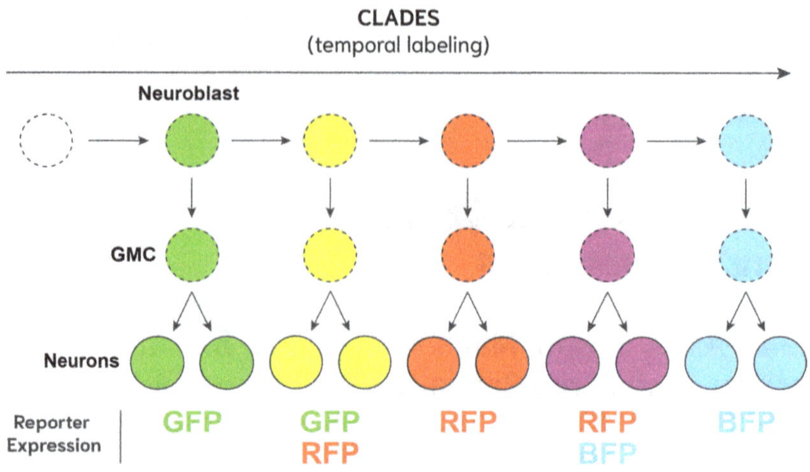

Fig. 1 CLADES: a system for tracing asymmetrical cell lineages in the *Drosophila* brain. Scheme outlining the fundamentals of CLADES. This system triggers a sequence of activation and inactivation of reporter genes, designed to follow a predetermined order. As a result, the color-coded markers inherited by subsequent generations of cells provide a chronological record of birth order, enabling the determination of each cell's position within the lineage hierarchy based on its unique color signature. Since Cas9 is exclusively expressed in the stem cell (neuroblast), the cascade progression results in progeny labeled in different colors. The sequence of activation and deactivation in the scheme is as follows: GFP (activation), RFP (activation), GFP (deactivation), BFP (activation), and RFP (deactivation)

contrast, cell production in symmetrical lineages grows exponentially, with each progenitor dividing to produce two progenitors that will continue dividing.

The gold standard for lineage tracing at single-cell resolution in *Drosophila* is the twin-spot mosaic analysis with a repressible cell marker (twin-spot MARCM, [3]), which has been crucial to mapping parts of some neuronal lineages. Twin-spot MARCM utilizes induced interchromosomal recombination to create permanent DNA alterations, thus marking the two sibling cells arising from the same mitosis in different colors. By inducing this recombination at different developmental times, researchers can label neurons resulting from each stem cell division, thus mapping the birth order of neuronal generation for specific cell lineages. Despite its unrivaled resolution, twin-spot MARCM requires the processing of thousands of brains to trace a single neuronal lineage. This constraint significantly hinders its utility for functional studies, such as those designed to investigate the contributions of specific genes to the neuronal lineage framework.

To overcome these challenges, CLADES (cell lineage access driven by an edition sequence) was created as a more scalable approach for lineage tracing ([4], Fig. 1). This method initiates a cascade of gene reporter activation and deactivation, marking cells with unique reporter combinations. The progression of this cascade, characterized by a specific sequence of reporter combinations, is limited to the neuroblast. Neurons inherit the color expressed by the neuroblast at the moment of their birth, enabling the determination of progeny birth order based on the cascade's color sequence. Although CLADES does not provide single-cell resolution lineage mapping within individual brains, it enables the reconstruction of neuronal lineages at single-cell resolution by analyzing approximately 100 processed brains. Given that CLADES can effortlessly provide an overview of lineage structure from a single brain specimen, it is optimally suited for examining the impact of genetic manipulations on neuronal lineage structure.

CLADES was developed as a tool to efficiently trace the birth order of neurons within an individual brain. This approach is unique among retrospective imaging-based tools due to its ability to provide information on the internal structure of cell lineage trees. While previous retrospective approaches are only able to label clones of cells arising from the same progenitor [3, 5, 6], CLADES can further divide these clones into subgroups based on birth order. This potential is crucial for unraveling the mechanisms of cell differentiation, especially concerning the influence of temporal specification factors. With CLADES, an asymmetrical cell lineage can be divided into five distinct segments, each representing the neuronal progeny derived from the neural stem cell at a specific temporal window (*see* **Note 1**).

The utilization of CLADES for *Drosophila* lineage tracing involves a single genetic cross, exemplifying its utility in tracing asymmetrical cell lineages with minimal effort. While tailored for *Drosophila*, the principles underlying CLADES provide a framework for the creation of more versatile and effective lineage-tracing methodologies, applicable to a wider array of organisms. Beyond its usefulness for lineage tracing, CLADES holds promise for future applications that involve the activation of endogenous gene cascades. This advanced functionality could offer deeper insights into the developmental trajectories of therapeutic cell types, unveiling the intricate processes that guide their formation and function.

2 Key Concepts

CLADES operates through a designed sequence of reporter gene activation and deactivation, with the progression of this cascade being confined to the neuroblast (Fig. 1). Consequently, the neuronal descendants inherit the specific marker expressed by the neuroblast at their inception. The color sequence of the reporters is predetermined, enabling a straightforward inference of neuronal birth order based on the specific color marker each neuron inherits. Neurons that emerge earlier in development will exhibit the initial colors in the reporter sequence, whereas neurons that arise later will

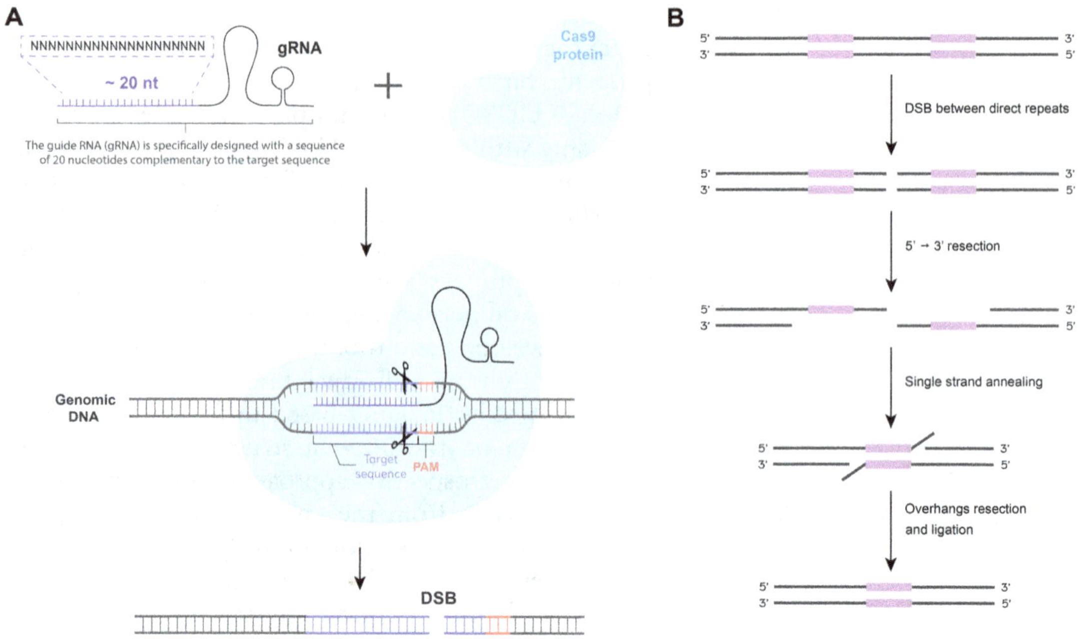

Fig. 2 CLADES is based on CRISPR/Cas9 and SSA. (**a**) Cartoon depicting the functioning of Cas9, the CRISPR-derived endonuclease that cleaves dsDNA in a region defined by a 20-bp sequence, which is encoded in a gRNA molecule. (**b**) Scheme to illustrate the mechanism of SSA, a highly conserved eukaryotic DNA repair pathway specialized in repairing cuts in repetitive DNA. SSA repairs DSB that occurs between repeated DNA sequences by excising one repeat along with the intervening sequence. This multistep repair process is initiated by resection of the 5′ ends flanking the DSB, facilitating the annealing of the remaining repeats, and concluding with the ligation of the final nicks

display combinations of reporters that are activated in the terminal phases of the sequence.

To achieve such precise control of reporter activation and inactivation, CLADES relies on the foundational concept of CaSSA, another approach we developed for precise genetic manipulations [7]. CaSSA relies on two critical elements: (1) Cas9, a CRISPR nuclease capable of inducing targeted double-strand breaks (DSB) in DNA at locations defined by a 20-bp sequence, which is encoded in the guide RNA (gRNA, [8, 9]), and (2) the single-strand annealing (SSA) process, a DNA repair mechanism that is evolutionary conserved across eukaryotes ([10], Fig. 2). When a double-strand break occurs in a DNA region between two direct repeats, SSA repairs the DNA by removing one of the repeats, as well as the intervening sequence between both repeats. This predictable pattern of DNA repair allows for the design of genetic switches in which a gRNA target site is flanked by two DNA repeats. In CaSSA, these DNA repeats are part of a reporter gene. Prior to repair, the genetic switch brings the downstream part of the reporter out of frame, thus encoding an inactive reporter gene, which is then switched on following the predictable SSA-mediated modification. In the presence of Cas9 and the corresponding gRNA, the endonuclease cleaves the DNA at the target site in the genetic switch. The subsequent repair via SSA removes one of the repeats along with the intervening region, which results in an active form of the reporter gene.

Instead of using CaSSA genetic switches with reporter genes, CLADES applies these genetic switches to engineer gRNAs that are activated by distinct gRNAs. The main challenge in this strategy was that, in contrast to CaSSA where reporter genes are long and can accommodate larger 400 bp DNA repeats, gRNAs are small and can only contain repeats limited to approximately 20 bp in length. Although this limitation reduces the efficiency of SSA, these small repeats nonetheless achieve a reasonably high level of efficiency. Another concern was the development of a genetic switch that ensured the inactivity of the gRNA until it was specifically activated. Despite various studies suggesting that the structure of the gRNA is very sensitive to minor modifications [11], we needed extensive optimization to engineer a gRNA configuration in which the integration of a switch cassette completely abolished the gRNA's function in its nonactivated state. The capability to activate a gRNA with a different gRNA introduces the possibility of constructing gRNA cascades. Within a cascade, an initial gRNA (trigger gRNA) can initiate a series of subsequent gRNA activations. Each gRNA within this cascade is also designed to trigger a distinct reporter gene activation or inactivation, facilitating a coordinated sequence of expression of reporter gene combinations.

Due to Cas9-induced DSB and SSA being highly stochastic events, there is no guaranteed sequential congruence between the activation of gRNAs and the subsequent reporter gene activation or inactivation. For instance, we can imagine a cascade in which gRNA#1 is responsible for the activation of gRNA#2 and a green fluorescent protein (GFP) reporter, and gRNA#2 is responsible for activating gRNA#3 and the red fluorescent protein (RFP) reporter. Attributable to its stochasticity, it is plausible that the RFP reporter will be activated by gRNA#2 prior to the activation of the GFP reporter by gRNA#1, despite that gRNA#2's activation is downstream of gRNA#1's activation in the cascade. This emphasizes the necessity of establishing a robust coupling mechanism between gRNA activation and the corresponding reporter gene's modulation to ensure precise and predictable reporter outcomes.

3 The CLADES Construct

To ensure synchronization between the activation of gRNAs and reporter genes in the CLADES system, we devised a system where conditional gRNAs are intricately integrated within the open reading frame (ORF) of the reporter gene, which is transcribed

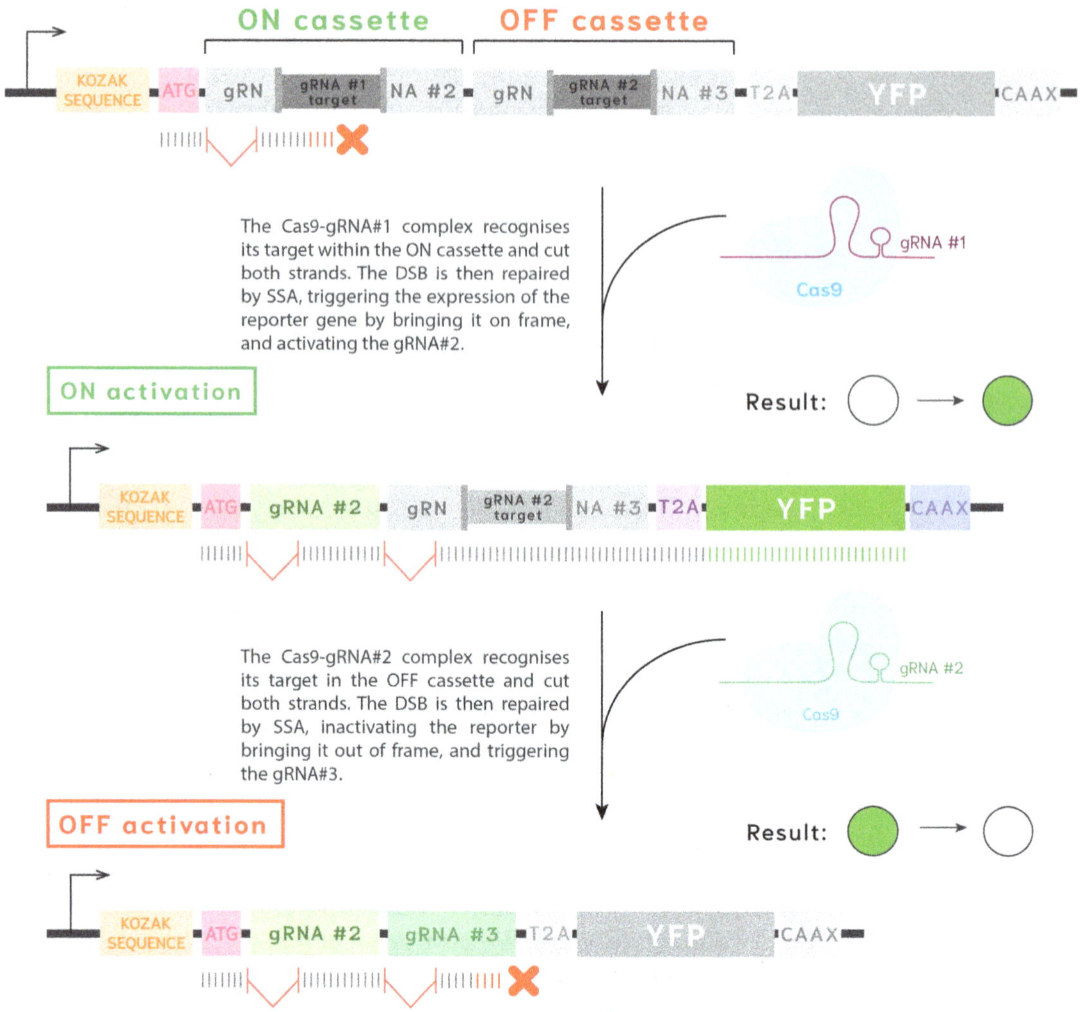

Fig. 3 Structure of the CLADES construct. A CLADES construct consists of a reporter gene whose expression is controlled by two regulatory cassettes, termed the ON and OFF cassettes. Each cassette contains a conditional gRNA that is activated by another gRNA. The cartoon illustrates that upon co-expression of Cas9 and a matching gRNA#1, the resulting Cas9-gRNA#1 complex cleaves the DNA at the gRNA#2 target site. This DSB is subsequently repaired by SSA, thereby generating a functional gRNA#2. This event brings the reporter in frame, leading to its expression. Concurrently, the presence of Cas9 and a different gRNA#2 triggers the activation of gRNA#3 within the OFF cassette, which shifts the reporter gene out of frame, deactivating it

from a type II promoter (Fig. 3). In this configuration, each CLADES construct is composed of two distinct genetic cassettes, each with a conditional gRNA under the control of a U6 promoter. The first cassette, termed the "ON" cassette, controls reporter gene activation. Conversely, the "OFF" cassette is designed to initiate the reporter gene's inactivation. These cassettes are strategically positioned upstream within the same ORF as the reporter gene to ensure their concerted function. Furthermore, to enhance the system's applicability, particularly in neural lineage tracing, the reporter is conjugated to a CAAX peptide sequence ([12], *see* **Note 2**). This refinement strategically anchors the reporter to the membrane, enabling a precise delineation of intricate axonal projections and facilitating the comprehensive mapping of neuronal axonal connections. This enhancement is pivotal for detailed neurobiological analyses, providing a granular view of neuronal architecture and connectivity essential for understanding complex neural networks.

In the original configuration, the reporter gene is intentionally positioned out of frame, rendering it inactive. Upon the co-expression of Cas9 and the corresponding gRNA, the ON cassette undergoes repair through SSA, which not only activates the gRNA within the cassette but also realigns the reporter gene into the correct ORF, thereby initiating its expression. This active gRNA then has the potential to catalyze subsequent steps in the cascade. Conversely, the reporter gene's expression can be selectively switched off by activating the gRNA within the OFF cassette, a process that alters the ORF of the reporter gene, rendering it inactive once more (*see* **Note 3**). This dual cassette system allows for dynamic control over the reporter gene's state, enabling precise temporal and spatial regulation of gene expression in response to the activation of specific gRNAs (*see* **Note 4**).

Based on this design, we tested a first construct in which the reporter was inactive in the initial status since the reporter was out of frame. This first construct exhibited fluorescence, suggesting that splicing sites could be acting to bring the reporter into the frame. For this reason, we created a second design in which all predicted splicing sites were removed. However, this construct showed no fluorescence even after generating a construct in which the ON cassette was preactivated. It is unknown why there was a lack of reporter expression in the construct depleted of splicing sites, although we hypothesize that the splicing might be removing some integral sequences blocking translation. Finally, we designed a construct to which we added strong splicing sites to produce a controlled pattern of splicing compatible with our design.

In summary, a CLADES construct includes a reporter gene, which is under the transcriptional control of a type II promoter. This reporter's ORF is strategically disrupted by two key elements, the ON and OFF cassettes, positioned upstream. Each cassette harbors a gRNA that is transcribed from a U6 promoter. Initially, these cassettes place the reporter gene out of frame, with the ORF terminating within the ON cassette. The ON cassette's activation triggers a targeted deletion within the gRNA region through SSA, thereby activating the gRNA and realigning the reporter gene into the correct reading frame. This realignment enables the concurrent expression of the reporter and the initial protein segments encoded by the ON and OFF cassettes, separated from the reporter by a T2A peptide sequence. Activation of the gRNA in the OFF shifts the reporter gene out of frame once again, and terminates the ORF within the OFF cassette.

4 Two Versions of CLADES

The original CLADES architecture (version 1.0) as depicted in Fig. 4, uses the Actin5C promoter to induce ubiquitous reporter expression. This version relies on two CLADES constructs: the first initiates the expression of yellow fluorescent protein (YFP), which is additionally tagged with the V5 epitope to augment its detection via immunostaining; the second construct is responsible for RFP expression.

The cascade begins with a trigger gRNA that activates the ON cassette within the YFP construct, leading to the expression of green fluorescence in the cells. The activation of the ON cassette is coupled with the activation of a second gRNA that, consequently, activates a third gRNA integrated into the ON cassette of the RFP construct. In this step of the cascade, cells express YFP and RFP, resulting in a yellow color. The final phase of the cascade involves the activation of the YFP construct's OFF cassette, ceasing YFP expression and turning cells into red. The sequential color sequence expected is as follows: (1) absence of color, (2) green (YFP expression), (3) yellow (concurrent YFP and RFP expression), and (4) red (exclusive RFP expression).

One significant limitation of CLADES 1.0 is the temporal delay inherent in the activation of the first reporter by the trigger gRNA, which precludes labeling entire cell lineages from the first mitosis.

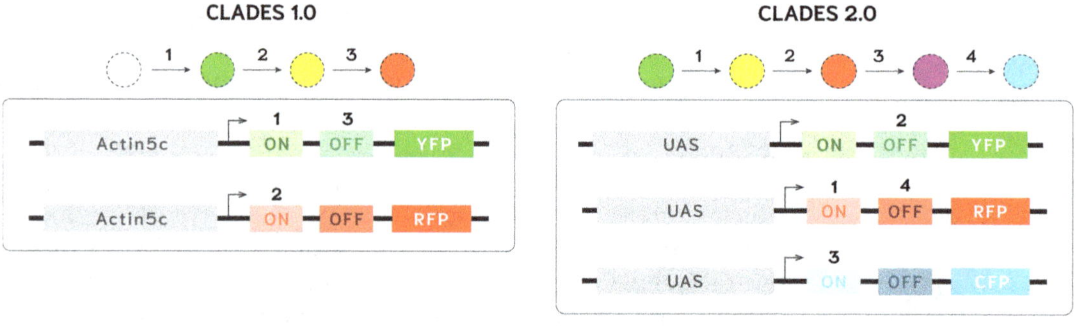

Fig. 4 CLADES versions. CLADES 1.0 features a dual-reporter system driven by the constitutive Actin5C promoter. The sequential reporter activation involves the following: YFP, YFP+ RFP, and RFP. In contrast, CLADES 2.0 utilizes a tri-reporter system governed by UAS, which enables reporter gene activation exclusively in the presence of the GAL4 transcription factor. Notably, CLADES 2.0 ensures immediate lineage marking due to the preactivation of the first reporter. The expression sequence in this version progresses as follows: YFP, YFP+RFP, RFP, RFP+CFP, and CFP

Moreover, the Actin5C promoter does not provide sufficiently robust labeling to resolve detailed axonal processes. While CLADES 1.0 offers utility for certain experimental paradigms, lineage tracing in neuronal lineages often requires stronger labeling to delineate intricate axonal processes. Additionally, CLADES 1.0 labels all cells, complicating the differentiation of distinct neural lineages.

To address these limitations, we introduced CLADES 2.0. This advanced version employs UAS sequences to regulate the reporter, facilitating integration with the Gal4 system. This modification not only enhances the visualization of fine axonal processes but also improves the selectivity of cellular targeting. Moreover, CLADES 2.0 incorporates an additional construct for cyan fluorescent protein (CFP) expression, tagged with the HA epitope, improving lineage tracing by expanding the available color palette. A significant upgrade in CLADES 2.0 is the preactivation of the ON cassette in the YFP construct, enabling immediate lineage labeling from the beginning of the lineage. The order of color progression in CLADES 2.0 is as follows: (1) green (YFP), (2) yellow (YFP +RFP), (3) red (RFP), (4) purple (RFP+CFP), and (5) blue (CFP).

5 Cascade Progression

In the majority of neuronal progenitor lineages within the *Drosophila melanogaster* brain, the neuroblast undergoes an asymmetric division, yielding another neuroblast and a GMC. This progenitor subsequently differentiates into two distinct neurons. Tracing these asymmetric cell lineages with CLADES requires a stringent regulation of cascade progression exclusively within the neuroblast. This precise control is imperative; should the cascade extend into the postmitotic neuron, each neuron would progress to the last color in the color sequence. By confining the cascade's progression to the neuroblast, CLADES ensures that the resultant progeny cells inherit a stable and distinctive mark, reflective of their birth order.

To confine cascade progression to the neuroblast, CLADES employs the *deadpan* promoter (DpnEE) to drive Cas9 expression exclusively in these cells. Consequently, as the neuroblast undergoes division and advances through the cascade, its progeny inherits the corresponding color marker, facilitating precise lineage tracing.

The continuous expression of Cas9 in the neuroblast using the DpnEE promoter provides no control over the cascade progression. The stochastic nature of cascade progression can exhaust the cascade prematurely in some cases and not completely progress to the final step in some other cases. Not knowing the exact time of progression for each of the steps in the cascade makes birth dating impossible. To overcome these limitations, we developed a

CLADES version in which Cas9 was controlled by a heat-shock promoter (Hsp). This approach ensures that color progression can only occur in a well-defined timing when the promoter is initiated by heat shock. In the first design of the Hsp-Cas9 construct, we detected Cas9 leaky expression even in the absence of heat shock. To correct this failure, we added a ribozyme to the 5′ untranslated region in the Cas9 gene. In noninduction conditions, this ribozyme catalyzes its own excision, which uncaps the Cas9 mRNA reducing translation and thus avoiding leaky expression. After induction, the Cas9 expression is strong enough to overcome the action of the ribozyme, therefore resulting in Cas9 expression. The Hsp is active in all cells that are induced by heat shock, resulting in ubiquitous Cas9 expression. To restrict Cas9 to proliferating cells (neural stem cells and intermediate progenitors), we fused Cas9 to the geminin domain which removes the protein when the cell is not in the G2 through M phases. With the Hsp-Ribozyme-Cas9 construct working as expected, we were able to deduce the exact timing of cascade progression, therefore birth-dating neurons with great precision.

6 CLADES for Lineage Tracing

CLADES has been specifically designed to address the challenges associated with determining the birth order of neurons originating from asymmetric (type I) lineages. Compared to established methods such as twin-spot MARCM, the current gold standard in the field, CLADES offers the ability to assemble an entire neuronal lineage, while requiring fewer brain specimens.

Prior to tracing the birth order of a neuronal population of interest, it is crucial to establish that these neurons originate from the same neural stem cell. Since *Drosophila* neurons do not migrate within the brain, spatial proximity of neuronal somata can serve as a good indicator of neurons belonging to the same cell lineage. To confirm whether two neurons arise from the same neural stem cell, one can induce mosaic clonal labeling using CaSSA or recombinase-based strategies, such as Twin-Spot MARCM or Flip-Out [5]. When two neurons are consistently co-labeled under conditions that favor sparse mosaic labeling, it is highly unlikely that they originated from separate progenitors.

When designing a lineage-tracing experiment using CLADES, a necessary consideration is the method of labeling the neuronal population of interest. CLADES 2.0 requires one or more overlapping GAL4 drivers that comprehensively label the entire neuronal population. If the study includes the analysis of a dense neuronal population, distinguishing individual cells can be challenging. To mitigate this, employing multiple GAL4 drivers that label overlapping subpopulations can expedite the identification of individual neurons. With each GAL4 driver labeling only a fraction of the entire population, labeling is sparse which facilitates clearer traceability of axonal projections. The overlapping neurons with expression of two or more GAL4 drivers work as references to relate birth order between the different neuronal subpopulations. In scenarios where overlapping GAL4 lines are unavailable, an alternative approach is to utilize clonal, mosaic labeling to reduce the number of labeled neurons. This can be achieved using lineage-specific genetic drivers to control a CaSSA switchable Cas9 variant, driven by the DpnEE promoter. This approach targets fewer lineages, thus amplifying the sparsity of the labeling, which is indispensable for distinguishing neurons based on their axonal projections or the expression of specific molecular markers.

Once it has been established that the neurons in question belong to the same lineage and appropriate GAL4 drivers are in place to label the entire target population, implementing CLADES becomes straightforward. Initially, flies harboring each of the GAL4 drivers must be crossed with a CLADES line that expresses the three CLADES 2.0 constructs alongside DpnEE-Cas9. Standard immunostaining protocols [13–15] are then employed to label the neurons with antibodies against V5, HA, and RFP. Following immunostaining, the brains are imaged using a confocal microscope, enabling the identification of the specific combination of reporters expressed by each neuron. Subsequently, each labeled cell is assigned a birth-dating score, which reflects the sequential order of the reporter combination that labels the cell (*see* **Note 5**). To minimize potential errors associated with residual GMC cascade progression (*see* **Note 6**), the analysis of an entire neuronal lineage typically requires a minimum of 50 brains. The birth order of neurons is determined by ranking each cell based on its average birth-dating score, providing a reliable measure of lineage sequence.

Utilizing DpnEE-Cas9 to drive the progression of the cascade enables the deduction of the relative birth order of neurons. However, this method does not specify the exact temporal window during which each neuron is born. To determine the precise timing of neuronal birth during development, we recommend conducting a subsequent experiment using hs-Cas9 instead of DpnEE-Cas9 (Fig. 5). This experiment requires the application of heat shocks at various developmental stages in synchronized animals, in accordance with established protocols ([13, 14], Fig. 6). The synchronization interval should be tailored based on the desired temporal resolution and the specific cell cycling length of the lineage under study. In our research, we implemented 2-h intervals for the harvesting of larvae to achieve optimal temporal precision.

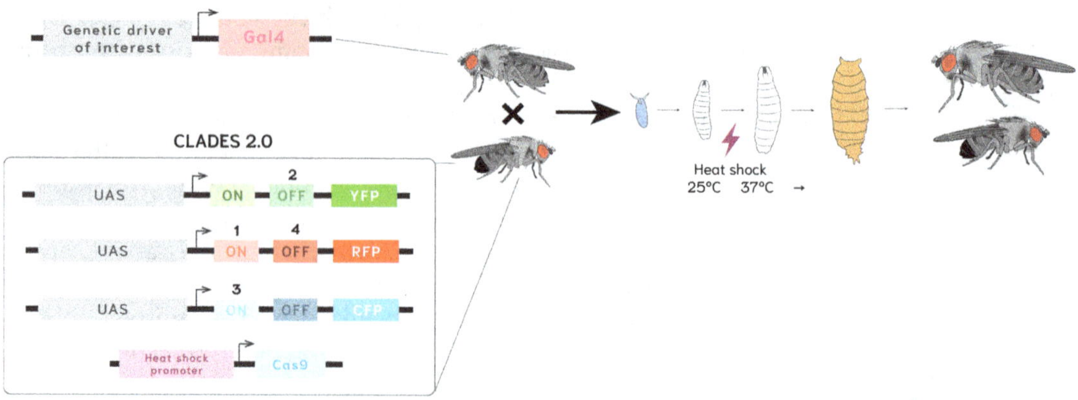

Fig. 5 Lineage tracing with CLADES in a single fly cross. The most efficient strategy for lineage tracing based on CLADES relies on a single cross of the following: (1) a GAL4 line with expression in the lineage of interest, and (2) a line bearing the three CLADES 2.0 constructs and a Cas9 driven by a heat-shock promoter. Then, the synchronized progeny carrying all transgenes are heat shocked at different stages, one or multiple times for each experiment

7 CLADES Beyond Lineage Tracing

From a broad perspective, CLADES functions as a recorder of biological events. By linking Cas9 with various genetic regulators, the progression of the cascade can serve as an indicator for other biological phenomena beyond cell cycles. The versatility of Cas9, which can be activated by various promoters, expands CLADES's utility to monitor diverse biological phenomena. Leveraging this versatility, we have developed CLADES to track generations of *Drosophila* flies by directing Cas9 expression with the *Bam* promoter [16]. This promoter exhibits transient activity within a narrowly defined time frame in the germ line, which excludes the germ stem cell. This precise activity of the *Bam* promoter leads to a controlled and limited progression of the cascade. If the *Bam* promoter was active in germ stem cells, which exist throughout an organism's lifespan, there would be the potential for these cells to progress to the last color in the cascade. The system ensures a restrained cascade progression, excluding Cas9 expression in the germ stem cells, offering a measure of fly generations. While the female cascade advances only a single step, males exhibit one or two steps of cascade progression per generation. By analyzing the distribution of color markers across the fly population, we can then infer the number of generations a given fly population has experienced. In addition to shedding light on population dynamics and generational changes in *Drosophila*, this approach serves as a crucial demonstration of CLADES's utility beyond lineage tracing.

Another example of the utility of CLADES beyond lineage tracing could be using Cas9 driven by a *cFos* promoter as a reporter of neuronal activity [17]; higher neuronal activity would result in

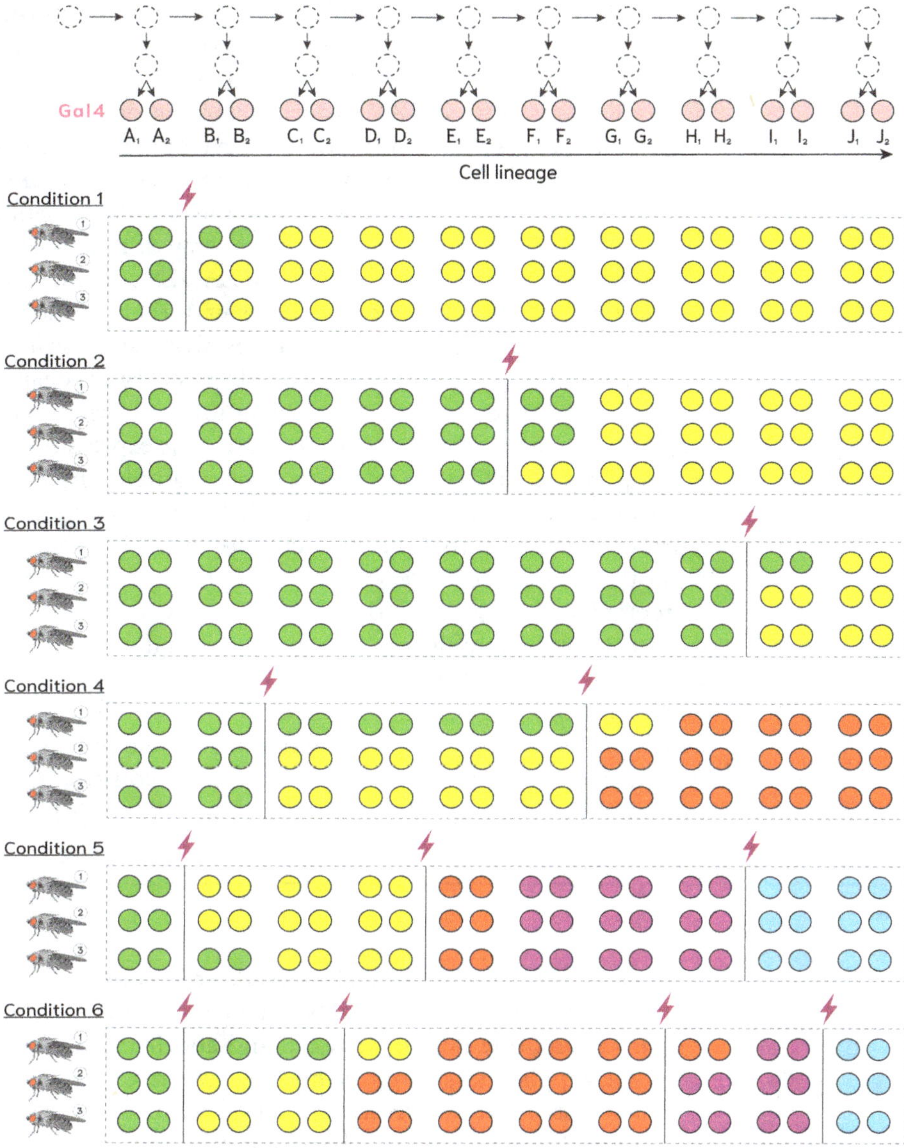

x1 (green)	18	12	8	7	7	6	3	3	1	0
x2 (yellow)	0	6	10	9	5	6	7	6	8	9
x3 (orange)	0	0	0	2	6	3	5	4	3	3
x4 (purple)	0	0	0	0	0	3	3	5	3	0
x5 (blue)	0	0	0	0	0	0	0	0	3	6
Birth-dating Score	18	24	28	31	35	39	44	47	53	57

Fig. 6 Reconstruction of a complete neuronal lineage Using CLADES. This figure illustrates the process for assembling a comprehensive neuronal lineage following the imaging of brain samples from the experiment detailed in Fig. 5. Each cell within the lineage of interest is systematically identified and recorded across multiple brain samples. A birth-date score is computed by multiplying the position of the color in the cascade (e.g., one for green, four for purple) by the frequency of that color's appearance in the sampled cells. The

increased Cas9 expression and thus more cascade progression. However, the main limitation of CLADES as a biological event reporter is that it requires dividing cells since that is where SSA predominantly occurs [18]. Future iterations of CLADES could potentially address this limitation by developing versions that do not rely on SSA. Alternatively, innovative designs might explore mechanisms to activate SSA in nondividing cells, expanding the utility of CLADES in capturing a wider array of biological events.

Beyond its use in lineage tracing, CLADES could also facilitate the on-demand generation of cell types with therapeutic potential. The development of specific cell types from stem cells to fully differentiated cells involves multiple cell fate decisions characterized by a regulated sequence of gene activation and inactivation. This sequential regulation ensures appropriate gene expression at precise developmental stages, preventing, for example, neurons from expressing stem cell genes or stem cells from initiating axogenesis. Current in vivo cell therapy protocols are limited in their ability to replicate these intricate developmental pathways, often resulting in generated cell types that differ significantly from those lost to disease, thereby diminishing their restorative efficacy. By employing CLADES to orchestrate the timely activation and inactivation of genes, it should be possible to develop in vivo differentiation protocols that closely emulate natural developmental processes, potentially enhancing the ability of these cells to contribute to lesion recovery.

8 Notes

1. It is crucial to understand that CLADES is optimally suited for lineages where the neural stem cell divides asymmetrically, producing one neural stem cell and another distinct cell type. When applied to symmetric lineages, where each division of the neural stem cell results in two neural stem cells, CLADES may not perform as intended. In such cases in which both daughter cells exhibit DpnEE activity, both resulting clones would continuously express Cas9, leading the cascade to progress to its final color marker in the entire resulting progeny.
2. After DNA cleavage events, there remains a small probability that sequences will be repaired via nonhomologous end joining (NHEJ) rather than SSA. In such instances, the cascade is likely to collapse. This collapse has a more negative impact during

Fig. 6 (continued) lineage order is then established by arranging the progeny according to these scores. Note that heat shock, used to trigger color change, does not invariably promote progression through the color cascade in all samples

steps involving reporter activation compared to those for reporter inactivation. For reporter activation, the random mutation introduced by NHEJ leads to reporter inactivation in two out of three events, when the mutation aligns the reporter out of frame. The reverse scenario occurs during reporter inactivation, where a random mutation is more likely to properly inactivate the reporter, aligning it out of frame in two out of three events. In any case, the cascade is likely to collapse at the last step, since the gRNA in the ON or OFF cassette will not be properly activated unless SSA occurs.

3. In CLADES constructs, the T2A peptide sequence facilitates the detachment of the reporter from the nonfunctional peptide encoded by the ON and OFF cassettes. Certain peptide localization signals, such as mCD8, require placement at the N-terminus of the entire ORF to function effectively. Therefore, when these signals are fused to the N-terminus of the reporter, they do not function as intended, as this position is not the N-terminus in the ORF context. As a result, we utilized the CAAX motif, which is effective at the C-terminus, to ensure proper localization of the reporter to the plasma membrane.
4. The development of CLADES constructs in novel animal models should begin with the characterization of three essential controls: (1) the standard CLADES construct, which ideally should show no reporter expression before activation; (2) a construct with a preactivated ON cassette, expected to exhibit robust reporter expression; and (3) another construct with a preactivated OFF cassette, which should demonstrate an absence of reporter expression. Although constructs 2 and 3 will not be utilized in the final experimental setup, these constructs are critical to validate and ensure the effective functioning of the system.
5. On extremely rare occasions, the cascade may erroneously lead to the expression of incorrect combinations of reporters. For instance, cells may be labeled with all three reporters or with two reporters that, according to the cascade design, should not coexist, such as the first and last reporters in the sequence. These anomalies are likely due to infrequent recombination events and should be excluded from the analysis to maintain accuracy.
6. The DpnEE promoter is highly effective in restricting Cas9 transcription to neuroblasts. However, Cas9 protein can persist post-cell division, leading to unintended residual progression of the cascade in the GMC. The core issue with these GMC events lies in the asynchronous progression: there are instances where GMC advances to a subsequent cascade step while the neuroblast remains at the prior step. This desynchronization

results in GMC-derived cells acquiring color markers indicative of later cascade stages, in contrast to neuroblast-derived progeny, which may retain markers from earlier stages. Such discrepancies can obscure the temporal lineage relationships, presenting a challenge in the unambiguous interpretation of developmental trajectories within the *Drosophila* neural lineage. Conducting repeated experiments to increase the sample size can help mitigate the impact of this asynchrony, as GMC cascade progression occurs randomly and much less frequently compared to neuroblasts.

Acknowledgments

Work produced with the support of a 2022 Leonardo Grant for Researchers and Cultural Creators, BBVA Foundation. J.G.M. was supported by the contract RyC2020-029831-I, funded by MICIU/AEI /10.13039/501100011033 and by the "European Union NextGenerationEU/PRTR." I.G.A. was supported by the grant PID2021-128285NA-I00, funded by MICIU/AEI/ 10.13039/501100011033 and by ERDF/UE. B.M.P. was supported by the PEJ-2021-AI/BMD-22918 contract, funded by the regional government of Madrid and the *European Social* Fund of the EU. All experimental work involved in the implementation of CLADES was performed at the laboratory of Prof. Tzumin Lee. We thank all members of his laboratory for their help and discussions.

References

1. Yu HH, Awasaki T, Schroeder MD et al (2013) Clonal development and organization of the adult Drosophila central brain. Curr Biol 23: 633–643
2. Urbach R, Schnabel R, Technau GM (2003) The pattern of neuroblast formation, mitotic domains and proneural gene expression during early brain development in Drosophila. Development 130:3589–3606
3. Yu HH, Chen CH, Shi L et al (2009) Twin-spot MARCM to reveal the developmental origin and identity of neurons. Nat Neurosci 12: 947–953
4. Garcia-Marques J, Espinosa-Medina I, Ku KY et al (2020) A programmable sequence of reporters for lineage analysis. Nat Neurosci 23:1618–1628
5. Nern A, Pfeiffer BD, Rubin GM (2015) Optimized tools for multicolor stochastic labeling reveal diverse stereotyped cell arrangements in the fly visual system. Proc Natl Acad Sci USA 112:E2967–E2976
6. Hampel S, Chung P, McKellar CE et al (2011) Drosophila Brainbow: a recombinase-based fluorescence labeling technique to subdivide neural expression patterns. Nat Methods 8: 253–259
7. Garcia-Marques J, Yang CP, Espinosa-Medina I et al (2019) Unlimited genetic switches for cell-type-specific manipulation. Neuron 104: 227–238
8. Gasiunas G, Barrangou R, Horvath P, Siksnys V (2012) Cas9-crRNA ribonucleoprotein complex mediates specific DNA cleavage for adaptive immunity in bacteria. Proc Natl Acad Sci USA 109:E2579–E2586
9. Jinek M, Chylinski K, Fonfara I et al (2012) A programmable dual-RNA-guided DNA endonuclease in adaptive bacterial immunity. Science 337:816–821
10. Bhargava R, Onyango DO, Stark JM (2016) Regulation of single-strand annealing and its role in genome maintenance. Trends Genet 32:566–575

11. Briner AE, Donohoue PD, Gomaa AA et al (2014) Guide RNA functional modules direct Cas9 activity and orthogonality. Mol Cell 56: 333–339
12. Hancock JF, Cadwallader K, Paterson H et al (1991) CAAX or a CAAL motif and a second signal are sufficient for plasma membrane targeting of ras proteins. EMBO J 10:4033–4039
13. Lai SL, Awasaki T, Ito K et al (2008) Clonal analysis of Drosophila antennal lobe neurons: diverse neuronal architectures in the lateral neuroblast lineage. Development 135:2883–2893
14. Lin S, Kao CF, Yu HH et al (2012) Lineage analysis of Drosophila lateral antennal lobe neurons reveals notch-dependent binary temporal fate decisions. PLoS Biol 10:e1001425
15. Awasaki T, Kao CF, Lee YJ et al (2014) Making Drosophila lineage-restricted drivers via patterned recombination in neuroblasts. Nat Neurosci 17:631–637
16. Chen D, McKearin DM (2003) A discrete transcriptional silencer in the *bam* gene determines asymmetric division of the *Drosophila* germline stem cell. Development 130:1159–1170
17. Krukoff TL (2008) c-fos expression as a marker of functional activity in the brain. In: Boulton AA, Baker GB, Bateson AN (eds) Cell neurobiology techniques, 1st edn. Humana Press, New Jersey
18. Al-Minawi AZ, Saleh-Gohari N, Helleday T (2007) The ERCC1/XPF endonuclease is required for efficient single-strand annealing and gene conversion in mammalian cells. Nucleic Acids Res 36:1–9

Index

Jorge García-Marqués and Tzumin Lee (eds.), *Lineage Tracing: Methods and Protocols*, Methods in Molecular Biology, vol. 2886, https://doi.org/10.1007/978-1-0716-4310-5,

D

E

F

G

H

I

www.ingramcontent.com/pod-product-compliance
Lightning Source LLC
Chambersburg PA
CBHW082228060225
21590CB00005B/83
9781071643099